Y0-AEU-263

METALLOPROTEINS

TOPICS IN MOLECULAR
AND STRUCTURAL BIOLOGY

Series Editors:
Watson Fuller, University of Keele
Stephen Neidle, King's College University of London

1 S. Neidle (ed.)
Topics in Nucleic Acid Structure, Part 1

2 S. Neidle (ed.)
Topics in Nucleic Acid Structure, Part 2

3 S. Neidle/M. J. Waring (eds)
Molecular Aspects of Anti-Cancer Drug Action

4 D. Chapman (ed.)
Biomembrane Structure and Function

5 D. W. L. Hukins (ed.)
Connective Tissue Matrix

6 P. M. Harrison (ed.)
Metalloproteins, Part 1

7 P. M. Harrison (ed.)
Metalloproteins, Part 2

Weinheim
Deerfield Beach, Florida
Basel

METALLOPROTEINS

Part 2: Metal Proteins with Non-redox Roles

edited by

PAULINE M. HARRISON

University of Sheffield

Weinheim
Deerfield Beach, Florida
Basel

Pauline M. Harrison
Department of Biochemistry
University of Sheffield
SHEFFIELD S10 2TN
UK

ISBN 3-527-26137-0 (Weinheim; Basel)
ISBN 0-89573-211-4 (Deerfield Beach, Florida)

Published in North America and Continental Europe
by Verlag Chemie GmbH, D-6940 Weinheim
First published 1985 outside North America and
Continental Europe by The Macmillan Press Ltd,
London and Basingstoke

Typeset by RDL Artset Ltd, Sutton, Surrey, England
Printed in Hong Kong

The Contributors

J. Brock,
Department of Bacteriology and Immunology,
Western Infirmary,
Glasgow G11 6NT,
UK

M. Brunori,
Department of Biochemistry,
II University of Rome,
Tor Vergata,
Rome,
Italy

G. Chin,
Department of Biochemistry and Molecular Biology,
Harvard University,
Divinity Avenue,
Cambridge, Mass. 02138,
USA

M. Coletta,
CNR Centre of Molecular Biology,
Institute of Chemistry and Biological Chemistry,
Faculty of Medicine,
I University of Rome,
La Sapienza–P. le A. Moro,
2–00185 Rome,
Italy

Michael Forgac,
Department of Biochemistry and Molecular Biology,
Harvard University,
Divinity Avenue,
Cambridge, Mass. 02138,
USA

B. Giardina,
Department of Biochemistry,
II University of Rome,
Tor Vergata,
Rome,
Italy

R. J. A. Grand,
Cancer Research Campaign Laboratories,
Department of Cancer Studies,
University of Birmingham,
The Medical School,
Birmingham B15 2TJ
UK

T. Hofmann,
Department of Biochemistry,
University of Toronto,
Toronto,
Ontario,
Canada M5S 1A8

P. E. Hunziker,
Institute of Biochemistry,
University of Zurich,
Winterthurerstrasse 190,
CH-8057 Zurich,
Switzerland

J. H. R. Kägi
Institute of Biochemistry,
University of Zurich,
Winterthurerstrasse 190,
CH-8057 Zurich,
Switzerland

Contents

Preface

In this second volume on metalloproteins we have brought together six classes of protein to illustrate a variety of important physiological processes in which metals are involved in association with protein. We exclude metalloproteins participating in oxidation-reduction reactions from consideration here since these were extensively discussed in Volume 6 of this series entitled *Metalloproteins, Part 1*.

The processes illustrated now can be divided, broadly speaking, into two types: those in which the metal centre is required for a specific biological activity, and those in which the metal is being transported or sequestered and the proteins concerned control the provision of the metal in the right place at the right time in the right amount. The three activities which are illustrated are *enzyme catalysis*, where the metal plays an essential role in enzyme activity; *triggering*, where metal binding to a protein (calcium to calmodulin or troponin C) induces structural changes in the protein which in turn modulate the activities of other proteins; and oxygen *transport*, where again the metals are at the active centres of the carriers, their action being modulated by the proteins to which they are bound. Under the second type of metalloprotein, involved in metal 'metabolism', we look at the passage of metals (sodium, potassium and calcium) across cell membranes, the transport of a metal (iron) in the blood plasma, and the intracellular sequestration of metals (mainly zinc and copper) including toxic metals (such as cadmium and mercury) by the protein metallothionein.

Several possible alternatives present themselves as examples of metalloenzymes. Metalloproteinases were selected for several reasons. In the first place a considerable amount of structural information is available for several members of this class. The three-dimensional structures determined by X-ray crystallography for carboxypeptidases A and B and for thermolysin have stimulated a large volume of research aimed at understanding the function of these enzymes in relation to their known structures. The carboxypeptidases are exopeptidases, and thermolysin is an endopeptidase, but both are zinc enzymes and there are similarities in their active-site ligands and enzymic mechanism which Dr Theo Hofmann nicely illuminates. Thermolysin also has four bound calcium ions, at least one of which may be partially responsible for the thermostability of this enzyme. Although these are the best characterised members of this class, many other metalloproteinases are reviewed in this chapter. Most of these are also zinc enzymes. They include collagenases, several amino peptidases, angiotensin converting enzyme, and the D,D-carboxypeptidase of *Streptomyces albus* which specifically catalyses the hydrolysis of D,D-dipeptides. The three-dimensional structure of this enzyme appears to be unique. Most of the metalloproteinases are extracellular enzymes, but some of the aminopeptidases are membrane proteins and the three-dimensional-structure determination of a cytosolic amino peptidase from bovine eye lens is in progress. As chapter 1 shows, research in metalloproteinases is still a very active area.

There is also much activity directed to determining the role of calcium as a second messenger, and the way that calmodulin, acting as a receptor for calcium, regulates a host of cellular processes as calcium concentrations vary between 10^{-7} and 10^{-5} M. It is fascinating that calmodulin, itself a highly conserved protein, seems to act as a regulatory subunit in several different enzyme complexes. Troponin C is the major target protein for calcium released from the sarcoplasmic reticulum and it regulates muscle contraction by modulating the interaction of actin with myosin. Two alternative theories of how this is achieved are critically assessed. Although the three-dimensional structures of calmodulin and troponin C are not yet available, they are known to be related to parvalbumin (the structure of which has been determined) and to the intestinal calcium binding proteins of uncertain function. Dr Grand concentrates on the two members of this class with well established trigger roles.

To perform functions within cells, metals must traverse cell membranes, and, if calcium is to perform its regulatory functions, its concentration within cells and cell compartments must be carefully controlled. Calcium transport through the eukaryote plasma membrane and through the sarcoplasmic reticulum is coupled to hydrolysis of ATP, and the (Ca^{2+})-ATPase of the sarcoplasmic reticulum accounts for much of the protein present in this membrane. The plasma membrane of all animal cells contains a (Na^+, K^+)-ATPase responsible for maintaining relatively low levels of sodium and high levels of cytoplasmic potassium, which is required for activity by several enzymes. Sodium gradients may also be used to generate action potentials or may be coupled to the uptake of sugars and amino acids. It is fascinating to find, as described by Drs Chin and Forgac, that the (Na^+, K^+)- and (Ca^{2+})-ATPases are both structurally and mechanistically similar, even though the former consist of two protein subunits and the latter of a single large subunit approximating in size to the sum of the two subunits of the (Na^+, K^+)-ATPases. Although the structural homologies between these proteins and their general membrane topologies have been deduced, their actual three-dimensional structures are unknown.

In the remaining three chapters the proteins of interest are associated with transition metals. Metallothioneins, the subject of chapter 4, are a group of small molecules forming a unique protein family. Of the 61 amino acids in mammalian metallothioneins, 20 are cysteine and aromatic amino acids are absent. These proteins are induced by zinc and copper and are responsible for the homeostasis of these metals, but they also bind several other metals (including cadmium, silver, mercury and bismuth). It is of considerable interest that cadmium, a toxic metal, also induces metallothionein biosynthesis, so that formation of the cadmium–metallothionein complex plays an important role in detoxification of the metal. The amino acid sequences of several metallothioneins, including isoproteins from the same species, are known, and exhibit clearly defined and conserved patterns of cysteine distribution, suggesting their sulphur atoms are metal ligands. There are as yet no three-dimensional structures of these proteins, but based on ^{113}Cd-NMR or EXAFS measurements, alternative models have been proposed for the geometries of the metal ions bound. The ^{113}Cd-NMR data have been interpreted by two metal clusters of four and three cadmiums with thiolate ligands. As the authors of chapter 4, Drs Hunziker and Kägi, point out, with the exception of the iron-storage protein ferritin (a much larger protein), metallothionein has the highest metal content of any protein known.

A purist might argue that the (Na^+, K^+)- and (Ca^{2+})-ATPases are not strictly metalloproteins in that their affinities for the metals they bind are fairly low compared with those of many metalloproteins and that they are concerned only in metal translocation. In contrast, transferrin (discussed by Dr Brock in chapter 5), the plasma transport protein which delivers iron to the bone marrow and other tissues for synthesis of haem and other iron-containing molecules, binds two ferric iron atoms with association constants of about 10^{24} M^{-1} at physiological pH. The two metal sites are structurally similar although not identical, and probably each has two tyrosines and one or two histidines as well as carbonate or bicarbonate and water or hydroxo ion as iron ligands. Curiously, a hypothesis that the two iron atoms were functionally different stimulated an enormous volume of work which has highlighted differences in chemical behaviour but failed to prove functional heterogeneity. Chemical differences have been shown by use of a variety of spectroscopic methods, coupled in some cases with substitution of iron by vanadyl, terbium or other metals, and, more simply, by studies of metal binding as a function of pH. The three-dimensional structure of rabbit transferrin under examination at 3.5 Å resolution is eagerly awaited. A protein similar to transferrin, ovotransferrin, is found in avian egg white, and another, lactoferrin, in milk and body secretions (the iron-binding protein of blood, sweat and tears). Its function, and indeed a secondary function of other members of the family, may be a bacteriostatic one: depriving bacteria of the iron they need for growth by virtue of the very high affinity of these proteins for this metal. This high affinity poses a problem for the release of iron and there has been much controversy over how iron is delivered. Transferrin is probably taken into cells by endocytosis but what happens after that is by no means certain.

The last chapter of the book is on oxygen carriers. Haemoglobin is the most studied of all metalloproteins, and its structure, with and without bound oxygen, is well characterised, as is that of the monomeric myoglobin. The Perutz model of oxygen binding and the role of protons as allosteric effectors (the Bohr effect) is well known. However, this does not prevent a vigorous discussion by the authors, Drs Brunori, Coletta and Giardina, of this model, some aspects of which are still open to question. The authors nicely show how comparative studies of haemoglobins, from species as diverse as man, trout and crocodile, contribute to an understanding of molecular function. Their chapter also contains a brief but interesting account of speculations relating to the possible origin of the haemoglobin fold from three functionally different parts, delineated by introns in the gene. Haemoglobins are not the only oxygen carriers discussed. Haemerythrins and haemocyanins have different metal centres – respectively two iron or two copper atoms per subunit – which provide the ability to combine reversibly with dioxygen, and a variety of quaternary structures, which in the haemocyanins are large and complex. These oxygen carriers are usually discussed separately in different reviews or different books, but here the authors attempt to provide a unified view of these proteins. Haemocyanins and haemoglobins, although totally different in structure, show several similarities in properties, including co-operativity in oxygen binding and dissociation and accompanied by structural changes, and modulations of binding by changes in pH. The chapter ends with a discussion of the evolutionary relationships between these molecules and the possibility, albeit a remote one, that these proteins, different as they are, may have evolved from a common ancestor. The Italians dedicate their chapter to the memory of Eraldo Antonini, who contributed much to this field, but who sadly is no longer with us.

The authorship of this volume is an international one, represented by Canada, Italy, Switzerland and the United States as well as the UK, and reflects the widespread involvement of scientists in this field of research. I thank the authors for the trouble they have taken to provide six fresh and stimulating reviews.

Sheffield, 1984 P. M. H.

1
Metalloproteinases

Theo Hofmann

1.1 INTRODUCTION

Proteolytic enzymes are divided into four major classes on the basis of the amino acids that make up the active site and by implication confer a common mechanism on each class. The metalloproteinases constitute one of these classes, the others being the serine proteinases, the cysteine proteinases and the aspartic proteinases (formerly acid proteinases). An apparent involvement of metal ions in the action of proteinases has been shown for a large number of enzymes on the basis of activation by metal ions or of inhibition by metal-chelating agents. However, unambiguous evidence for an essential role of metal ions has been presented for only a limited number of them. To qualify as a true metalloproteinase, the metal ion should be at the 'centre' of the active site and be involved in the catalytic event. The only two metalloproteinases for which detailed structural and mechanistic information is available are the endoproteinase thermolysin from *Bacillus thermoproteolyticus* Rokko and the exoproteinase carboxypeptidase A from bovine pancreas. In this review I shall concentrate on describing recent studies on the three-dimensional structures of these enzymes and their complexes with substrate analogues and other ligands, as well as studies on their mechanism of action. Other less well characterised enzymes will be mentioned more briefly at the end. Owing to limitations of space, no attempt will be made to present a comprehensive review. For earlier studies the reader is referred to a number of reviews on the individual enzymes mentioned below. Carboxypeptidase A and thermolysin along with other metalloenzymes are also the subjects of a recent review (Lipscomb, 1983). In addition a glossary and bibliography on mammalian metalloendoproteinases has been published (Barrett and McDonald, 1980), as have glossaries and bibliographies on carboxypeptidase A and other mammalian exopeptidases (McDonald and Barrett, 1984).

1.2 BOVINE CARBOXYPEPTIDASE A

For earlier studies of carboxypeptidase A the reader should consult two detailed reviews: Hartsuck and Lipscomb (1971) and Ludwig and Lipscomb (1973). Also, the preparation and many properties of the precursor complexes and the active enzyme have been summarised by Petra (1970).

Carboxypeptidase A is produced by the bovine pancreas as an inactive precursor in a complex consisting of two or three subunits (Brown *et al.*, 1961, 1963*a,b*). One of these (subunit I) is the immediate precursor of carboxypeptidase A. This precursor is activated by trypsin and by the trypsin-activated subunit II (a serine endopeptidase) through the removal of N-terminal peptides of varying lengths as indicated in figure 1.1 to give up to four forms of the active

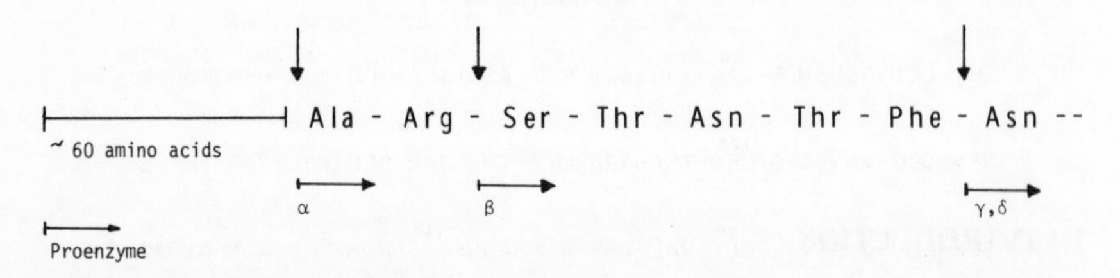

Figure 1.1 N-terminal sequence of bovine carboxypeptidase A. The arrows indicate the sites of activation cleavages; α, β, γ and δ denote the different forms of the active enzyme (according to Sampath-Kumar *et al.*, 1964).

enzyme, depending on the conditions used for activation.* However, the precise mechanism of activation is still unknown. This applies especially to the molecular event which generates the active enzyme from the apparently inactive precursor. In fact, Behnke and Vallee (1972) found that when zinc was substituted by cobalt in procarboxypeptidase A, the spectral properties were very similar to those of the active enzyme (Vallee and Williams, 1968), and the cobalt procarboxypeptidase showed as much activity with certain substrates as the native enzyme.

There are two allelic variants of the enzyme which differ in positions 179, 228 and 305 in the sequence shown in figure 1.2 (Bradshaw *et al.*, 1969). Most of the recent studies have been carried out with the A_α-form. Rather surprisingly, the only published sequence of an activation peptide of a carboxypeptidase is that the rat procarboxypeptidase A, which has been obtained from the nucleotide sequence of cloned cDNA of the preprocarboxypeptidase (Quinto *et al.*, 1982). The propeptide consists of 94 amino acids and shows 70 per cent identity with that of the bovine propeptide whose preliminary sequence has been determined by Neurath and co-workers (M. G. Hass, R. Wade, J. Gagnon, L. Erickson, H. Neurath and K. Walsh, unpublished results, quoted by Quinto *et al.*, 1982).

*The standard three-letter abbreviations for amino acid residues are used in table 1.1 and elsewhere; other abbreviations used are: Ac, acetyl; BMBP, (–)2-benzyl-3-*p*-methoxybenzoylpropionic acid; dansyl, 5-dimethylamino-naphthalene-1-sulphonyl; <Glu, pyroglutamyl; mansyl, (*N*-methylanilino)-2-naphthalene-1-sulphonyl; Z, benzyloxycarbonyl.

5 10 15

ALA-ARG-SER-THR-ASN-THR-PHE-ASN-TYR-ALA-THR-TYR-HIS-THR-LEU-

20 25 30

ASP-GLU-ILE-TYR-ASP-PHE-MET-ASP-LEU-LEU-VAL-ALA-GLN-HIS-PRO-

35 40 45

GLU-LEU-VAL-SER-LYS-LEU-GLN-ILE-GLY-ARG-SER-TYR-GLU-GLY-ARG-

50 55 60

PRO-ILE-TYR-VAL-LEU-LYS-PHE-SER-THR-GLY-GLY-SER-ASN-ARG-PRO-

65 70 75

ALA-ILE-TRP-ILE-ASP-LEU-GLY-ILE-HIS-SER-ARG-GLU-TRP-ILE-THR-

80 85 90

GLN-ALA-THR-GLY-VAL-TRP-PHE-ALA-LYS-LYS-PHE-THR-GLU-ASN-TYR-

95 100 105

GLY-GLN-ASN-PRO-SER-PHE-THR-ALA-ILE-LEU-ASP-SER-MET-ASP-ILE-

110 115 120

PHE-LEU-GLU-ILE-VAL-THR-ASN-PRO-ASN-GLY-PHE-ALA-PHE-THR-HIS-

125 130 135

SER-GLU-ASN-ARG-LEU-TRP-ARG-LYS-THR-ARG-SER-VAL-THR-SER-SER-

140 145 150

SER-LEU-CYS-VAL-GLY-VAL-ASP-ALA-ASN-ARG-ASN-TRP-ASP-ALA-GLY-

155 160 165

PHE-GLY-LYS-ALA-GLY-ALA-SER-SER-SER-PRO-CYS-SER-GLU-THR-TYR-

170 175 180

HIS-GLY-LYS-TYR-ALA-ASN-SER-GLU-VAL-GLU-VAL-LYS-SER-ILE/VAL-VAL-

185 190 195

ASP-PHE-VAL-LYS-ASN-HIS-GLY-ASN-PHE-LYS-ALA-PHE-LEU-SER-ILE-

200 205 210

HIS-SER-TYR-SER-GLN-LEU-LEU-LEU-TYR-PRO-TYR-GLY-TYR-THR-THR-

215 220 225

GLN-SER-ILE-PRO-ASP-LYS-THR-GLU-LEU-ASN-GLN-VAL-ALA-LYS-SER-

230 235 240

ALA-VAL-ALA/GLU-ALA-LEU-LYS-SER-LEU-TYR-GLY-THR-SER-TYR-LYS-TYR-

245 250 255

GLY-SER-ILE-ILE-THR-THR-ILE-TYR-GLN-ALA-SER-GLY-GLY-SER-ILE-

260 265 270

ASP-TRP-SER-TYR-ASN-GLN-GLY-ILE-LYS-TYR-SER-PHE-THR-PHE-GLU-

275 280 285

LEU-ARG-ASP-THR-GLY-ARG-TYR-GLY-PHE-LEU-LEU-PRO-ALA-SER-GLN-

290 295 300

ILE-ILE-PRO-THR-ALA-GLN-GLU-THR-TRP-LEU-GLY-VAL-LEU-THR-ILE-

305

MET-GLU-HIS-THR-VAL/LEU-ASN-ASN

Figure 1.2 Amino acid sequence of bovine carboxypeptidase A_α (with permission from Bradshaw *et al.*, 1969).

The sequence of the active rat carboxypeptidase appears to be 78 per cent identical with the bovine enzyme. The activation peptide of porcine procarboxypeptidase A has recently been isolated but not yet sequenced (Segundo *et al.*, 1982). It is not clear whether or not the complex of procarboxypeptidase with subunits II and III as isolated has any physiological significance (Koide and Yoshizawa, 1981), but it should be noted that porcine procarboxypeptidase occurs naturally in either a monomeric form or as a complex with a pro-serine proteinase (proteinase E), which is the equivalent to the bovine subunit II. Monomeric procarboxypeptidase has also been found in the spiny pacific dogfish (Lacko and Neurath, 1967).

Carboxypeptidase A is an exopeptidase that specifically cleaves carboxy-terminal amino acids from proteins or peptides. The negatively charged carboxylate ion is an absolute requirement. The enzyme will also cleave peptide analogues in which the C-terminal acid is an α-hydroxyacid, such as in benzoyl-glycyl-L-phenyllactate (esterase activity). There are some important kinetic differences between the peptidase and esterase activities which will be discussed in section 1.2.2.1.

Carboxypeptidase A has a molecular mass of 35 472 Da as deduced from the sequence (Bradshaw *et al.*, 1969) and contains one zinc atom per molecule.

1.2.1 Three-dimensional structure of carboxypeptidase A

1.2.1.1 Native protein

The three-dimensional structure of carboxypeptidase at a nominal resolution of 2.0 Å has been described in several publications (Reeke *et al.*, 1967; Lipscomb *et al.*, 1968, 1970; Hartsuck and Lipscomb, 1971; Ludwig and Lipscomb, 1973). A stereoview of the C_α positions is shown in figure 1.3. The molecule is nearly spherical with approximate dimensions 50 x 42 x 38 Å. Lipscomb *et al.* (1970) suggested that the molecule is made up of three structural domains, consisting

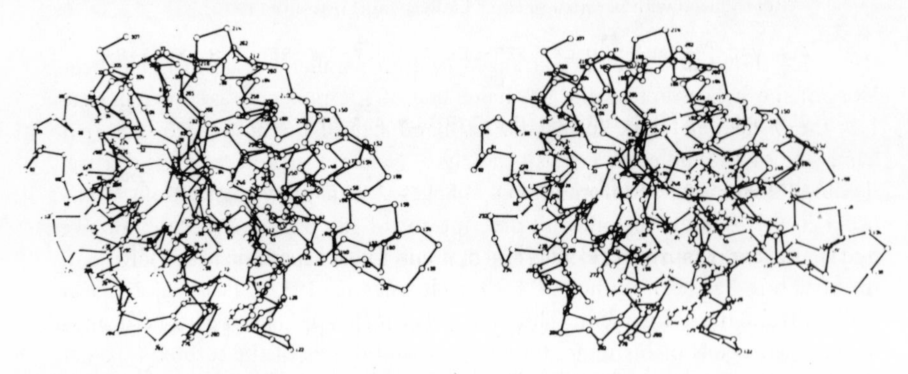

Figure 1.3 Stereoview (OR-TEP drawing) of the C^α positions of carboxypeptidase A (with permission from Lipscomb *et al.*, 1968).

of amino acids 1-127, 128-189 and 190-307, respectively. These domains do not correlate with the gene structure. The nucleotide sequence of the DNA of the rat procarboxypeptidase gene is interrupted by nine intervening sequences and the boundaries of the exons do not coincide with any of the structural or functional features of the enzyme (Quinto *et al.*, 1982). This is in contrast to immunoglobulins (Sakano *et al.*, 1979) or globins (Craik *et al.*, 1980), where the exons delineate the functional domains of the proteins.

The zinc atom which is liganded to the side chains of histidine-69, glutamic acid-72 and histidine-196 is located in the centre near the bottom of the molecule as viewed in figure 1.3. There is a well-defined pocket near the zinc, which has been identified as the substrate binding site; this binding site extends from the pocket along a groove which accommodates elongated peptide chains (Lipscomb *et al.*, 1968). There are a number of water molecules in this substrate binding pocket. About half of the 307 residues are involved in helical and β-sheet structures; the remainder are in structures which may be termed 'random' in that they do not correspond to any well-defined model (Ludwig and Lipscomb, 1973). Table 1.1 lists the residues that are involved in helices and in β-sheets;

Table 1.1 Secondary structure in carboxypeptidase[a]

Residue	Structure	Residue	Structure
14–28	Helix	173–187	Helix
32–36	β-sheet	190–196	β-sheet
49–53	β-sheet	200–204	β-sheet
60–66	β-sheet	215–231	Helix
72–80, 82–88	Helix	239–241	β-sheet
94–103	Helix	254–262	Helix
104–109	β-sheet	265–271	β-sheet
112–122	Helix	285–306	Helix
122–174	'Random'		

[a]Reproduced with permission from Ludwig and Lipscomb (1973).

figure 1.4 shows how they are arranged in both parallel and antiparallel sheets. Most of the helices are on the left-hand side of the molecule as shown in figure 1.3; the β-sheet network forms the left-hand side of the active site. The right-hand side of the molecule consists mostly of 'random' structure and is relatively flexible. Most of the conformational changes associated with substrate binding occur in this flexible region. Full descriptions of the overall structure are given by Hartsuck and Lipscomb (1971) and by Ludwig and Lipscomb (1973). Recently the structure has been refined at 1.75 Å (Rees *et al.*, 1981). Two major changes emerge from the refinement. The zinc ion, which was reported to be liganded to the δ nitrogens of histidines-69 and -196 and to one of the carboxyl oxygens of glutamic acid-72, as well as to a water molecule (Ludwig and Lipscomb, 1973) in a distorted tetrahedral configuration, now appears to be in a penta-

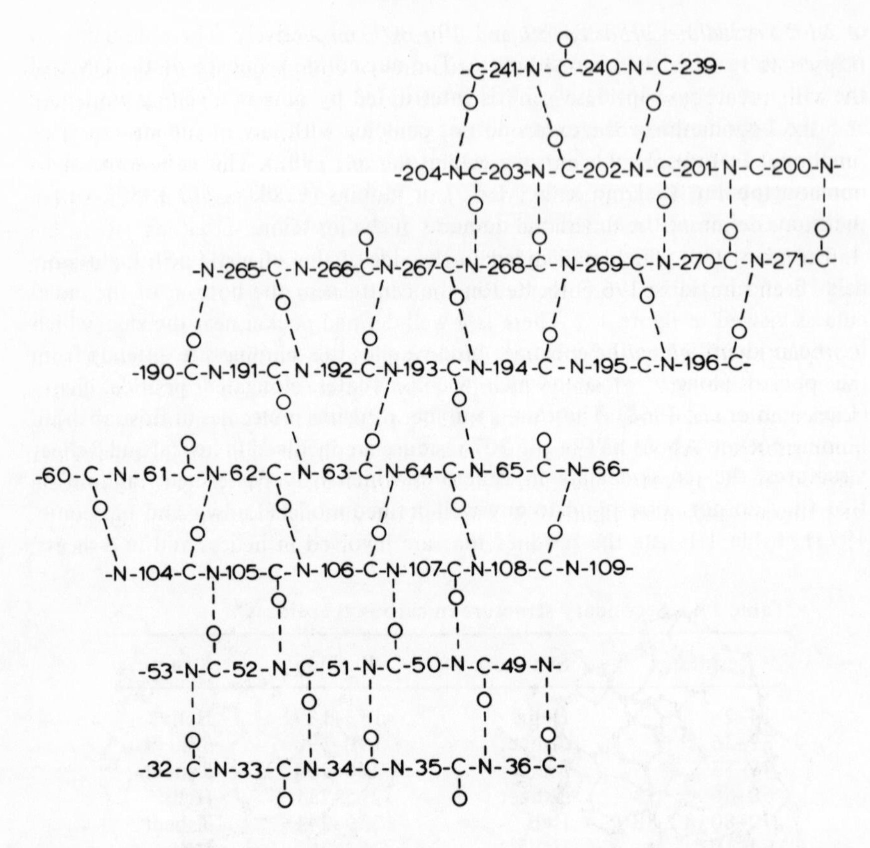

Figure 1.4 A diagram of the β-sheet found in carboxypeptidase A. The suggested hydrogen bonding scheme is shown by dashed lines. Arrows indicate the parallel or antiparallel arrangement of adjacent segments (with permission from Lipscomb *et al.*, 1968).

valent coordination. The fifth ligand is the second oxygen of glutamic acid-72 (Rees *et al.*, 1981). In the complex of the enzyme with the inhibitor Gly-Tyr, the water molecule is replaced and the zinc coordination number may vary from five to six, with the carbonyl oxygen and the amino nitrogen of Gly-Tyr being bound to zinc. The amino nitrogen can also bind to glutamic acid-270. The other interesting feature is the discovery of two *cis* peptide bonds which were not detected in the original structure. They are located between proline-205 and tyrosine-206, and between arginine-272 and aspartic acid-273. The previously located *cis* peptide bond is between serine-197 and tyrosine-198. None of these *cis* bonds involves the imino group of a proline residue. Although all three *cis* peptide bonds are located in the vicinity of the active site, there is no indication that they possess functional significance (Rees *et al.*, 1981).

1.2.1.2 Crystalline carboxypeptidase A-ligand complexes

Lipscomb's group has described the structures of complexes of carboxypeptidase A with glycyl-tyrosine, a very poor substrate (Lipscomb *et al.*, 1968); with 2-benzyl-3-*p*-methoxybenzoylpropionic acid, a substrate analogue of *N*-(*p*-methoxybenzoyl) phenylalanine in which the NH of the peptide bond has been replaced by a CH_2 group (Sugimoto and Kaiser, 1978; Rees *et al.*, 1980) and with a 39 amino acid residue peptide inhibitor from potatoes (Rees and Lipscomb, 1980). The detailed binding modes of these ligands with the enzyme have been summarised recently by Rees and Lipscomb (1981).

Complex with glycyl-L-tyrosine The complex of carboxypeptidase A with glycyl-tyrosine was originally described by Lipscomb *et al.* (1968) and after refinement to a nominal 2.0 Å by Rees *et al.* (1981). Glycyl-tyrosine is a very poor substrate whose turnover in the crystal is on the same time scale as that needed for X-ray data collection. The major interactions of the peptide with the enzyme are shown in figure 1.5. The carbonyl oxygen of the peptide bond is

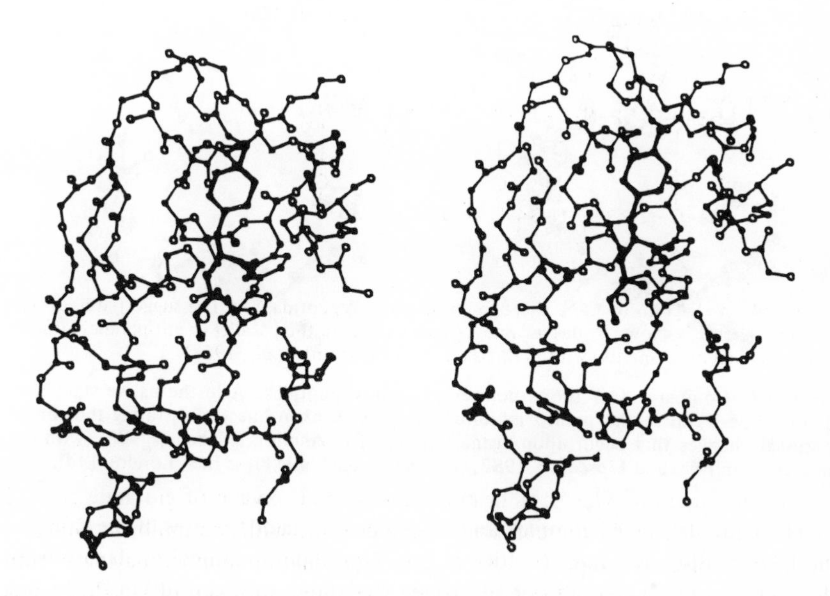

Figure 1.5 Active site region of the carboxypeptidase A–Gly–Tyr complex. The Gly–Tyr molecule is indicated by the darker bonds (with permission from Rees and Lipscomb, 1981).

liganded to the zinc, the C-terminal carboxylate group interacts with the positive charge on the side chain of arginine-145, and the side chain of the tyrosine is placed into the subsite S_1' (as defined by Schechter and Berger, 1967). No specific interactions seem to occur in this binding pocket, which is large enough to accommodate a tryptophan residue. The side chains of isoleucine-243, isoleucine-248, isoleucine-255 and asparagine-256 are in the vicinity of the ring (Ludwig and Lipscomb, 1973) and provide the predominantly hydrophobic environment indicated by the specificity of the enzyme (Ambler, 1972). The N-terminal amino group binds alternately in a salt linkage to glutamate-270 or to a coordination site on the zinc. These alternate binding modes involve a rotation of about 140° of the glycine C^{α}-carbonyl carbon bond. The binding of the peptide is associated with the displacement of several bound water molecules and several localised conformational changes, the most prominent of which is the movement of the ring of tyrosine-248 by rotation from an 'up' position to a 'down' position, as shown also in the complex of carboxypeptidase A with a 39 amino acid residue peptide inhibitor (see figure 1.6). Other significant changes involve arginine-145, isoleucine-247 and glutamate-270.

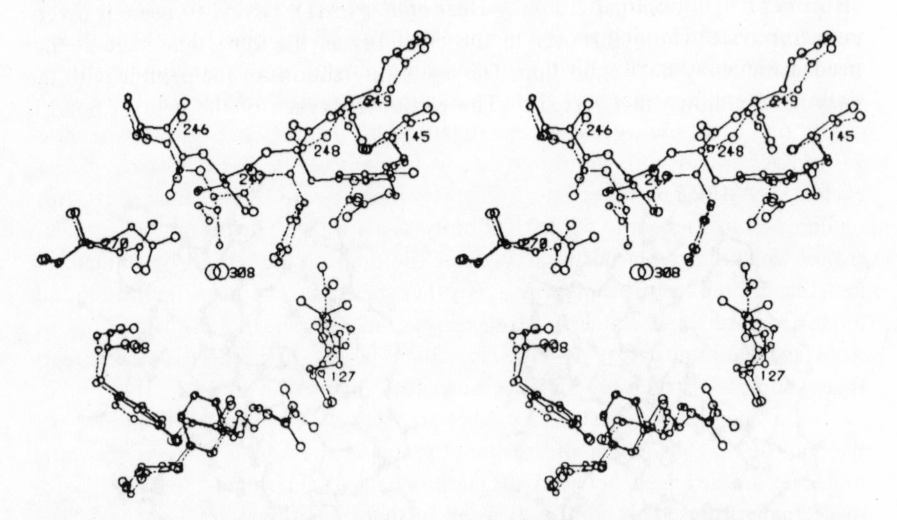

Figure 1.6 Comparison of the structures of carboxypeptidase A in the native state (continuous lines) and in the potato inhibitor complex (broken lines), illustrating the conformational changes that occur upon ligand binding. The zinc ion is labelled '308' (with permission from Rees and Lipscomb, 1982; copyright: Academic Press Inc., London Ltd).

The structure of the complex and the proposed major features of the catalytic mechanism discussed later (section 1.2.2.3) provide a rational explanation for the fact that Gly-Tyr is a poor substrate. The statistical distribution of the free amino group between the alternate positions on glutamate-270 and on the zinc interferes with the functioning of some of the mechanistic events. When the amino nitrogen binds to the zinc it displaces a water molecule from the zinc; this water molecule is then unable to take part in the catalytic step until it displaces the amino nitrogen. On the other hand, when the amino nitrogen interacts with glutamate-270, the latter is unable to act as either a nucleophile or a general base catalyst. Thus hydrolysis can only take place when the peptide is not bound to either of the two observed positions. These alternate nonproductive binding modes observed with Gly-Tyr do not occur with acylated dipeptides or with longer substrates (Rees *et al.*, 1981), because of the absence of the free amino group.

*Complex with 2-benzyl-3-*p*-methoxybenzoylpropionate* Sugimoto and Kaiser (1978, 1979) studied the carboxypeptidase-catalysed enolisation of BMBP, a substrate analogue in which the amide group of a substrate, *p*-methoxybenzoyl-phenylalanine, has been replaced by a CH_2 group. The enzyme catalyses the stereospecific hydrogen–deuterium exchange of one of the hydrogens in the CH_2 group and suggests the involvement of glutamate-270 in the exchange reaction. The X-ray analysis of the complex between BMBP and carboxypeptidase A (Rees *et al.*, 1980) shows that this compound – which is not cleaved by the enzyme – binds to the enzyme in a manner very similar to Gly-Tyr, with the carboxylate forming an anionic interaction with arginine-145, the carbonyl group forming a ligand with the zinc and the phenyl ring occupying subsite S_1', while the *p*-methoxybenzoyl group is located near tyrosine-198. The methylene group, one of whose hydrogens is exchanged by the enzyme, is near glutamate-270, in agreement with the proposed involvement of the latter in the hydrogen exchange. BMBP resembles ester substrates more closely than amide substrates, retaining as it does some of the flexibility of its ester analogue around the CH_2 group, which has replaced the oxygen of the ester or the NH of the amide. The fact that it binds in the same way as Gly-Tyr speaks against the different binding modes for ester and amide substrates suggested by Vallee *et al.* (1968) and by Auld and Holmquist (1974). However, the analysis of the BMBP–carboxypeptidase complex structure does not allow the distinction between the binding modes of peptide and ester substrates respectively proposed by Breslow and Wernick (1977), in which the carbonyl group of the scissile bond and a water molecule are arranged between the carboxylate of glutamate-270 and the zinc in alternate orders. This point is dicussed further in section 1.2.2.3.

Complex with carboxypeptidase inhibitor from potatoes The study of the structure of complexes of protein inhibitors of serine proteases with the active enzymes has helped greatly in describing the exact geometry of the active site of these enzymes in their transition states (Huber and Bode, 1978). Similarly, the recent description at 2.5 Å (Rees and Lipscomb, 1980, 1982) of the complex between carboxypeptidase A and the inhibitor from potatoes has been useful in confirming previous conclusions on the interactions between substrate and enzyme at the catalytic site and has provided information on the interactions in the subsites S_1, S_2 and S_3.

Protein inhibitors for carboxypeptidases have been found in potatoes (Rancour and Ryan, 1968) and in tomatoes (Hass and Ryan, 1980*a*). The inhibitors are small proteins consisting of 39 amino acids, and are believed to function as storage proteins and as a defence mechanism against proteolytic attack by invading microorganisms (Hass and Ryan, 1981). The amino acid sequence of the potato inhibitor used for the X-ray studies is as follows (Hass *et al.*, 1975).

<Glu–Gin–His–Ala–Asp–Pro–Ile–Cys–Asn–Lys–Pro–Cys–Lys–Thr–His–

Asp–Asp–Cys–Ser–Gly–Ala–Trp–Phe–Cys–Gin–Ala–Cys–Trp–Asn–Ser–

Ala–Arg–Thr–Cys–Gly–Pro–Tyr–Val–Gly

Like the inhibitors for serine proteases, the carboxypeptidase inhibitor forms a tight complex with the enzyme, with $K_I \cong 3 \times 10^{-9}$ M.

When the inhibitor associates with the enzyme, the C-terminal glycine residue is cleaved rapidly (Hass and Ryan, 1980*b*), but the liberated glycine residue remains trapped in the active site pocket of the crystalline complex (Rees and Lipscomb, 1980). The refined structure of the complex at 2.5 Å has recently been described (Rees and Lipscomb, 1982). The alpha carbon trace is shown in figure 1.7. The binding interactions near the catalytic site are in general agreement with those found for the small ligands. The S_1' subsite is occupied by the free glycine residue (it should be noted that because the glycine is cleaved off but not released, the structure of the inhibitor–enzyme complex represents the product complex). One of the oxygens of the carboxylate group of valine-38 is coordinated to the zinc, while the other is hydrogen bonded to tyrosine-248. (Numbers below 39 in this and the next paragraph refer to residues of the inhibitor, the others to residues of carboxypeptidase A (Rees and Lipscomb, 1982).) As in the complexes with small ligands, tyrosine-248 is in the 'down' position. Tyrosine-248 also receives a hydrogen bond from the amide proton of valine-38. Although the carboxylate of glutamate-270 points towards the carboxyl group

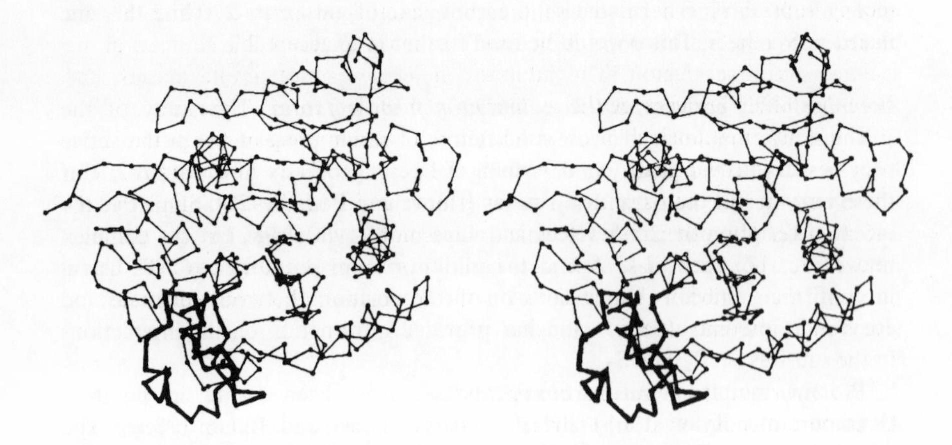

Figure 1.7 Alpha carbon trace of the carboxypeptidase A-potato inhibitor complex. The view is down the active site groove with the inhibitor molecule traced in darker lines (with perrmission from Rees and Lipscomb, 1982; copyright: Academic Press Inc., London Ltd).

of valine-38, the nearest oxygen of glutamate-270 is 4.5 Å from the carboxylate carbon of valine-38 and thus cannot form an anhydride bond in the complex. As in the other complexes, the zinc has five coordinate ligands; there is no high-occupancy water molecule bound to the zinc. The side chain of valine-38 in the S_1 subsite is in van der Waals' contact with the side chains of several residues: isoleucine-247, tyrosine-198 and -248, phenylalanine-279 and with the carbonyl oxygens of serine-197 and tyrosine-198. Tyrosine-37 interacts in the S_2 subsite through a major hydrogen bond between its carbonyl oxygen and the guanidinium group of arginine-71 while its side chain carbons are in contact with the side chains of threonine-164 and tyrosine-248. The interactions in subsite S_3 are not extensive. The C^{γ}- and C^{δ}-methylene carbons of the proline-36 side chain are in contact with the ring of tyrosine-198. Kinetic studies by Abramowitz *et al.* (1967) suggested that at least five subsites were involved in the binding of peptide substrates. However, the peptide chain on the inhibitor bends away from the active site cleft after proline-36. Hence the fifth binding site cannot be seen and the primary interactions of the inhibitor with carboxypeptidase involve only subsites S_1' and S_1–S_3. There are, however, secondary contacts between the two proteins, which are described in detail by Rees and Lipscomb (1982) and involve histidine-15, phenylalanine-22, and tryptophan-28 to serine-30 of the inhibitor on one hand and the residues from 246 to 248 of the enzyme on the other.

The structure of the potato inhibitor–carboxypeptidase complex shows why the inhibitor is not a substrate (after the initial cleavage of the C-terminal glycine). This point is discussed extensively by Rees and Lipscomb (1982). Their conclusions may be briefly summarised as follows. Although, in terms of the known specificity of carboxypeptidase, the bond between tyrosine-37 and valine-38 would be cleaved rapidly, in the inhibitor it is completely resistant. This implies that the inhibitor would bind nonproductively if valine-38 were to occupy subsite S_1'. The model of the complex suggests that if valine-38 were placed in S_1', the inhibitor would be forced to make unacceptable contacts in the secondary contact region. The inhibitor molecule cannot unfold because the three disulphide bridges (see the sequence given earlier) form a tight knot.

The conformational changes associated with the binding of the potato inhibitor to carboxypeptidase A are shown in figure 1.6. They involve mostly the movement of the tyrosine-248 ring and movements in the adjoining residues 246 to 249. The other residues which move more than about 1 Å are the arginines -71, -127 and -145 and the carboxyl group of glutamate-270. Upon binding, the inhibitor buries 677 Å^2 of the surface of carboxypeptidase A (Rees and Lipscomb, 1982).

1.2.1.3 Structure in crystal and in solution

An important point that arises from the study of the potato inhibitor–carboxypeptidase complex is that, apart from the active site region, the structures of the carboxypeptidase A molecule determined from the native crystals and the crystals of the complex are essentially the same (Rees and Lipscomb, 1981). The mean square deviations of the main chain atom coordinates are 0.26 Å and those for all atoms 0.36 Å. Since the native enzyme and the complex crystallise in different forms, the lattice interactions are different. It appears, therefore, that the effect of the crystal packing on the conformation is small and that there is probably little or no difference between the structures in the crystal and in solution, apart from the less restricted mobility of flexible side chains and the presumably greater internal motility in the less rigid parts of the molecule of the solution structure. Further support for the near identity of the active sites in the crystal and in solution comes from a comparison of the active sites of carboxypeptidases A and B (Schmid and Herriott, 1976), which have almost identical conformations except for the residues that determine the specificity in subsite S_1', yet have very different intermolecular contacts in the crystals.

The problem of the relevance of the structure deduced from X-ray analysis to the structure of the active form of an enzyme in solution is highlighted by carboxypeptidase A. Deductions of the mechanism based on the X-ray structures of the native enzyme and of ligand complexes have been questioned repeatedly (see, for example, Harrison *et al.*, 1975*b*; Spilburg *et al.*, 1977; Scheule *et al.*, 1980) on the basis of the different behaviour of arsaniloazotyrosine-248 carboxypeptidase A in solution and in crystals of different forms, as well as on the basis of different kinetic effects, such as substrate inhibition and activation (Spilburg *et al.*, 1977). A great deal of attention has been paid to this aspect since the time when Johansen and Vallee (1971) raised this issue. However, the differences have not been resolved and the controversy remains (Lipscomb, 1982). Quiocho *et al.* (1972) showed that their crystals of carboxypeptidase A_α which had been used for the X-ray analysis differed from those of A_γ in two respects. They had different habits and cell dimensions, although both were monoclinic, and differed in their activity, with the A_α having one-third and the A_γ form only one-threehundredth of the activity of the enzyme in solution with carbobenzoxyl-glycyl-L-phenylalanine as substrate. Thus Rees and Lipscomb (1981) ascribe the observed differences in spectral and kinetic behaviour to the different crystal forms used. However, Spilburg *et al.* (1977) found that their crystals of carboxypeptidase A_α and A_γ behaved kinetically identically towards a variety of substrates and did not show differences like those found by Quiocho *et al.* (1972). Unfortunately, no information on the unit cell dimensions of the crystals used by Spilburg *et al.* (1977) is available, but the habit of the crystals of A_α and A_γ is reported to be different. Some aspects of the apparent differences between crystals and solution are discussed further in section 1.2.2.2.

1.2.2 Enzymatic studies

The great number of kinetic studies on a variety of substrates and inhibitors carried out up to the end of the 1960s have been reviewed in detail (Petra, 1970; Hartsuck and Lipscomb, 1971; Ludwig and Lipscomb, 1973). In the present review I shall confine myself to a brief discussion of aspects of the enzymology of carboxypeptidase that bear on its mechanism of action and on the structure of the enzyme. I shall therefore reluctantly omit most of the numerous interesting studies on the effect of replacing the zinc with other metals and the effect of chemical modifications. These have also been reviewed (Petra, 1970; Hartsuck and Lipscomb, 1971; Ludwig and Lipscomb, 1973; Riordan, 1974; Auld, 1979).

1.2.2.1 Ester versus peptide substrates

The ability of carboxypeptidase A to cleave ester bonds as well as peptide bonds has been known for a long time (Snoke *et al.*, 1948). The basic structural requirements for ester substrates are the same as for peptide substrates, namely the presence of a metal ion, the L-configuration and a free carboxylate at the C-terminal residue. In general, k_{cat} values for analogous derivatives of β-phenyllactic acid and phenylalanine are within a factor of five of one another; the K_m values differ by up to two orders of magnitude. Many kinetic anomalies have been observed both with native and with modified enzymes. These have been discussed in great detail in several reviews (Petra, 1970; Hartsuck and Lipscomb, 1971; Ludwig and Lipscomb, 1973; Riordan, 1974; Auld, 1979). For the purpose of this review it will suffice to cite a few examples to illustrate the differences between peptidase and esterase activities. There are large differences in the pH rate profiles of both k_{cat} and k_{cat}/K_m (reviewed by Hartsuck and Lipscomb, 1981). The kinetic deuterium isotope effect is two for the esterase, but there is no effect for the peptidase activity. Substitution of the zinc by other metals has very different effects on the two activities. For a series of metal substitutes the values for k_{cat} of peptide hydrolysis [benzoyl](glycyl)$_2$-phenylalanine] range from 6000 min^{-1} for the cobalt enzyme to 43 min^{-1} for the cadmium enzyme, whereas for the corresponding ester substrate k_{cat} is essentially independent of the metal. The opposite holds true for K_m values. For the peptide substrate, K_m is almost completely independent of the metal, whereas for the ester it varies from 0.33 mM for the cobalt enzyme to about 8 mM for the cadmium enzyme (Auld and Holmquist, 1974). Using an elegant stopped-flow fluorescence technique, the same authors showed that peptides such as dansyl(glycyl)$_3$-phenylalanine bind to the metal-free carboxypeptidase (although they are not hydrolysed) but the ester analogue dansyl(glycyl)$_3$-phenyllactate does not bind. Acetylation of two tyrosine residues (probably tyrosines-198 and -248 according to Hartsuck and Lipscomb, 1971) decreases peptidase activity and increases esterase activity (Riordan and Vallee, 1963; Simpson *et al.*, 1963). And lastly, the inhibition of carboxypeptidase by phenylacetate and similar compounds is noncompetitive for peptides, but competitive for esters (Auld and Holmquist, 1974). These results led Vallee *et al.* (1968) to propose different binding sites and different catalytic mechanisms for the two types of substrate. The major difference in binding would be that the carboxylate of the C-terminal amino acid would bind to the zinc in the ester substrates instead of to arginine-145 as shown for the peptide substrates above.

Unfortunately, no support for this hypothesis has so far been obtained from X-ray crystallography. As discussed earlier, BMBP, which resembles an ester more than a peptide, binds like a peptide with its carboxyl group interacting with arginine-145. An explanation in structural terms of the differences between peptide and ester substrates must await further investigations.

1.2.2.2 Studies on carboxypeptidase in which tyrosine-248 has been diazotised
When carboxypeptidase A is coupled with diazotised *p*-arsanilic acid, a mono-substituted derivative can be obtained in which the ring of tyrosine-248 is specifically substituted in position 3 (Johansen and Vallee, 1971, 1973; Johansen *et al.*, 1972). This modification has no significant effect on the catalytic properties of the enzyme. Extensive studies with this arsanilazocarboxypeptidase have been carried out. Ultraviolet and visible absorption spectroscopy, circular dichroism studies (Johansen and Vallee, 1971, 1973) and resonance Raman spectroscopy (Scheule *et al.*, 1980) indicate that three pH-dependent different conformational states involving the chromophoric probe exist. The findings suggest that in one of these states an intramolecular complex between the zinc and the phenolic hydroxyl group of tyrosine-248 is formed. More recently a study of ^{15}N nuclear magnetic resonance, where the ^{15}N was incorporated into either one of the azo nitrogens, adds further evidence for the zinc-tyrosine interaction (Bachovchin *et al.*, 1982). It has been suggested on the basis of these findings that in the native (unmodified) enzyme the same interaction between the tyrosine hydroxyl and the zinc existed (Johansen and Vallee, 1975). However, the X-ray analysis of crystals of unmodified enzyme grown at pH values between 7.5 and 9.0 showed no evidence for such an interaction (Lipscomb, 1982). Lipscomb suggests that the tyrosine-zinc binding in arsanilazocarboxypeptidase is brought about by additional interactions of the arsanilazo group in the active site groove, and is not relevant to the native enzyme.

Arsanilazocarboxypeptidase A has also been used for an intensive investigation of the complex binding phenomena that the kinetic anomalies obtained with many dipeptide and small ester substrates suggest. The interactions with substrate analogues, inhibitors and other modifiers have been studied by circular dichroism (Alter and Vallee, 1978) and resonance Raman spectroscopy (Scheule *et al.*, 1981). The results suggest that there are up to four different inhibitor binding sites, some of which show strong interactions, while others interact only weakly or not at all. These binding sites may be located in the subsites S_2–S_4, as has been suggested by Lipscomb *et al.* (1968) and by Lipscomb (1980). Unfortunately, as pointed out earlier, X-ray analysis of carboxypeptidase–ligand complexes so far shows only one binding mode and no evidence for additional binding (Lipscomb, 1982). Further developments in this area are eagerly awaited. The exact localisation of additional binding sites for noncompetitive inhibitors and activators should go a long way towards enabling the understanding of the complex kinetics that are observed for small substrates.

Kinetic studies with arsanilazocarboxypeptidase A by stopped-flow and temperature jump methods (Harrison *et al.*, 1975*a,b*; Harrison and Vallee, 1978) have provided evidence for slow changes in conformation of enzyme–ligand complexes after their initial rapid formation.

1.2.2.3 Mechanism of action of carboxypeptidase A

The mechanism of action of carboxypeptidase A has been reviewed extensively in recent years (Breslow and Wernick, 1977; Lipscomb, 1980, 1982, 1983). In spite of the large volume of work and the availability of a high resolution structure at 1.75 Å resolution, some major aspects of the mechanism remain unclear. The most obvious one concerns the question as to whether catalysis proceeds via a general base mechanism or via a nucleophilic attack by a group on the enzyme on the scissile bond with the formation of a covalent acyl intermediate.

The X-ray analysis and chemical and kinetic studies, however, have clearly established that the sole direct participants in the catalytic event are the side chains of glutamate-270 and tyrosine-248, the zinc ion, a water molecule and of course the scissile peptide or ester bond of the substrate.

Role of the zinc ion As discussed in section 1.2.1, the zinc was originally believed to be coordinated to histidines-69 and -196, to glutamic acid-72 and to a water molecule (Lipscomb *et al.*, 1968) in a tetravalent coordination. However, the most recent refined structure shows that in the native enzyme it has a pentavalent coordination with the second oxygen of glutamate-72 acting as the fifth ligand whereas in a complex with a substrate analogue the coordination may vary from five to six (Rees *et al.*, 1981). The pentavalency is supported by two recent studies. Kuo and Makinen (1982) investigated the hydrolysis of the ester substrate *O*-(*trans-p*-chlorocinnamoyl)-L-β-phenyllactate by cobalt-carboxypeptidase A in ^{17}O-enriched water using electron paramagnetic resonance and cryoenzymological methods. They concluded that the acyl intermediate formed (see below) requires a pentacoordinate metal ion. Bertini *et al.* (1982) studied the binding of the inhibitor β-phenylpropionate to cobalt carboxypeptidase A by proton nuclear magnetic resonance and concluded that their results support the hypothesis of a five-coordinate metal ion, but suggest that in the free enzyme the five ligands are two histidines, one glutamic acid and two water molecules, one of which is replaced by the inhibitor. The exact role of the metal ion in the catalytic event is still unclear. In peptide substrates, it acts as ligand to the carbonyl group of the scissile bond (Breslow and Wernick, 1977; Lipscomb, 1980) and polarises it to increase its susceptibility to nucleophilic attack. This hypothesis receives support from the experiments of Mock *et al.* (1981) who compared the actions of zinc and cadmium carboxypeptidase A on a substrate with a thioamide linkage, benzoylglycyl-(thioglycyl)-phenylalanine. Whereas the thioamide substrate is only poorly cleaved by the zinc enzyme, the cadmium enzyme hydrolyses it some 30 times faster. The authors suggest that the cadmium–sulphur synergism supports a mechanism in which the metal ion activates the scissile bond through the carbonyl or thiocarbonyl coordination.

Another suggested function is that a zinc-bound water or hydroxyl ion could act as a nucleophile (Lipscomb, 1980) in the formation of the transition state or in the deacylation of an anhydride intermediate (Makinen *et al.*, 1979; Rees *et al.*, 1981).

Acyl intermediates Direct evidence for the involvement of an acyl intermediate has been obtained for two ester substrates, *O*-(*trans-p*-chlorocinnamoyl-L-β-phenyllactate (Makinen *et al.*, 1976) and *O*-3-(2,2,5,5,-tetramethyl-1-oxo-pyrollinyl)propen-2-oyl-β-phenyllactate (Suh and Kaiser, 1976*b*; Koch *et al.*, 1979). Cryoenzymological studies showed that at subzero temperatures the deacylation step becomes rate limiting and allows the kinetic and spectral demonstration of an accumulated intermediate at –60°C. At room temperature the evidence indicates that the acylation step is rate limiting (Makinen *et al.*, 1976). The covalent intermediate has been isolated recently by gel filtration at –60°C (Makinen *et al.*, 1982).

Are ester and peptide substrates hydrolysed by the same mechanism? Whereas the evidence for an acyl intermediate obtained with the rather complex substrates discussed in the previous paragraph is quite strong, the acyl intermediate has not been demonstrated for other ester substrates. Evidence against a covalent intermediate in ester and peptide hydrolysis comes from the observation by Breslow and Wernick (1976) that carboxypeptidase catalyses the exchange of ^{18}O from the carboxyl group of benzoylglycine only in the presence of phenylalanine, but not in the presence of phenyllactic acid. This exchange is due to resynthesis of a substrate and thus eliminates an acyl intermediate in peptide bond hydrolysis (Breslow and Wernick, 1976, 1977). Furthermore, during the hydrolysis of Gly-Gly-Leu in ^{18}O-enriched water, no ^{18}O is incorporated into glutamate-270, the only possible acceptor for an acyl residue (Nau and Riordan, 1975); this again eliminates a covalent intermediate in peptide bond hydrolysis. The question as to whether all ester hydrolyses catalysed by carboxypeptidase A proceed via an acyl intermediate, or whether the substrates studied in the laboratories of Kaiser and Makinen form a special case, must await further studies. However, in a recent study Galdes *et al.* (1983) provided evidence against the involvement of acyl intermediates in the hydrolysis of ester and peptide substrates. These authors used low-temperature stopped-flow techniques with fluorescent peptides and depsipeptides to demonstrate the existence of two intermediates during both peptide and ester hydrolysis. None of these was apparently a covalent acyl enzyme compound. This suggests that the demonstration of covalent intermediates with substrates containing a *trans*-cinnamoyl group (Makinen *et al.*, 1979, 1982) is not relevant to the general mechanism of carboxypeptidase hydrolysis, but may represent a unique case which is due to the properties of the *trans*-cinnamoyl group. In actual fact, in an even more recent paper, Hoffman *et al.* (1983) show by multichannel resonance Raman spectroscopy that, in apparent contradiction to Makinen *et al.* (1979, 1982), there is no evidence for the *accumulation* of anhydride intermediates with *O*-(*trans-p*-(dimethylamino) cinnamoyl)-L-β-phenyllactate as substrate, even at very low temperatures in cryosolvents, although the *transient* formation of acyl intermediates is not ruled out. Breslow and Wernick (1977) have proposed a simple model which would not only explain the difference between the two types of substrate in terms of the intermediates but could also begin to explain the many kinetic differences and anomalies observed. This model arranges the substrate carbonyl group and a water molecule in different orders between glutamate-270 and the zinc ion (figure 1.8), allowing a general base mechanism mediated by glutamate-270 for peptide hydrolysis and a covalent mechanism with glutamate-270 as the

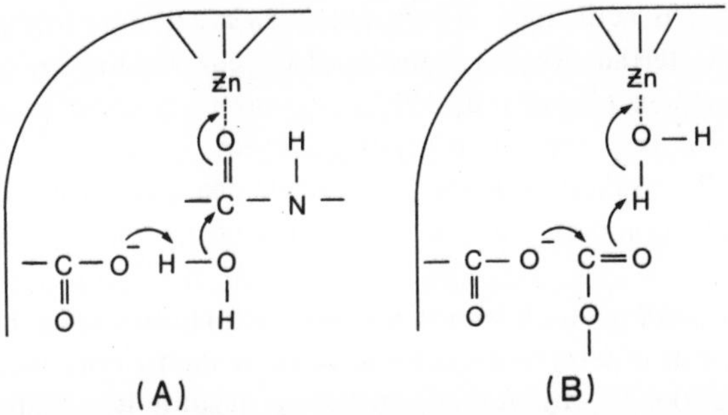

Figure 1.8 Alternate binding modes of the scissile bond of peptide (A) and ester (B) substrates, respectively, in the active site of carboxypeptidase A (according to Breslow and Wernick, 1977).

nucleophile for ester hydrolysis. The differences in binding and in kinetic parameters between ester and peptide substrates do not necessitate the different sites for catalysis proposed by Vallee *et al.* (1968). Cleland (1977) suggests that differences in rate-controlling steps could adequately explain the kinetic differences. This author proposes a single mechanism in which the rate-limiting step for peptide bond cleavage is a rotation about the peptide bond into a precatalytic complex that resembles the transition state for nucleophilic attack. This bond rotation would be more facile for ester substrates so that a subsequent step, such as a nucleophilic attack, would become rate limiting. The study of the kinetic parameters of the hydrolysis of two pairs of peptide and ester substrates, benzoylglycyl-glycyl-phenylalanine and benzoylglycyl-thioglycyl-phenylalanine, and benzoylglycyl-glycyl-phenyllactate and benzoylglycyl-thioglycyl-phenyllactate, respectively, lends support to this hypothesis (Campbell and Nashed, 1982). The kinetic parameters for the ester pair are very similar, but for the peptide pair there is a large difference in k_{cat}, although not in K_m. The reason for the difference in k_{cat} can be ascribed to the greater energy required for the rotation of the C–N bond in thioamides than in amides.

The role of glutamate-270 From its position in the three-dimensional structure, glutamate-270 is best suited to act either as general base in peptide hydrolysis, or as the nucleophile in attacking the scissile bond of an ester. Additionally, Breslow and Wernick (1977) suggested that glutamate-270 could donate a proton to the tyrosinate-248 after the latter donates a proton to the NH leaving group during peptide hydrolysis.

The role of tyrosine-248 From the three-dimensional structure of carboxypeptidase–ligand complexes (Rees and Lipscomb, 1981; Lipscomb, 1982) two major functions can be ascribed to tyrosine-248: it acts as proton donor to the leaving group and it is involved in hydrogen bond formation with the NH group of the penultimate peptide bond and the newly formed carboxylate of the product, as well as with the C-terminal carboxylate group of the substrate while the latter is transferred along the subsites of the enzyme before forming the final productive complex (Lipscomb, 1982). However, tyrosine is not required as proton donor during ester hydrolysis (Simpson *et al.*, 1963; Quiocho and Lipscomb, 1971, Suh and Kaiser, 1976*a*; Urdea and Legg, 1979). In ester hydrolysis the proton presumably comes directly from a water molecule.

Transition state analogues One of the major factors responsible for the catalytic rate enhancements of enzymes is considered to be their ability to bind 'transition states' more strongly – by several orders of magnitude – than the substrates (see, for example, the review by Wolfenden, 1976). This is based on the hypothesis, supported by chymotrypsin and other enzymes, that the structure of the active site is complementary to the transition state rather than the native substrate. Transition state analogues have been helpful in providing structures of the transition state in serine proteinases (see, for example, Steitz and Shulman, 1982) and other enzyme systems. So far, no three-dimensional structures of transition state analogues of carboxypeptidase have been determined and it is difficult to assess whether or not carboxypeptidase favours binding of the transition state (Lipscomb, 1982). An inhibitor that resembles a tetrahedral transition state has been described by Jacobsen and Bartlett (1981). It has the following structure:

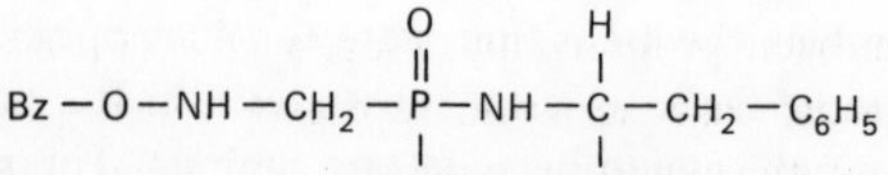

$$\mathrm{Bz-O-NH-CH_2-\overset{\overset{\displaystyle O}{\|}}{\underset{\underset{\displaystyle O^-}{|}}{P}}-NH-\overset{\overset{\displaystyle H}{|}}{\underset{\underset{\displaystyle COO^-}{|}}{C}}-CH_2-C_6H_5}$$

This phosphoramidate inhibits the enzyme with $K_I = 9 \times 10^{-8}$ M (pH 7.5) and $K_I = 6 \times 10^{-9}$ M (pH 6). The carboxylate presumably binds to arginine-145 and the $P-O^-$ to the zinc.

1.2.2.4 Use of carboxypeptidases A and B in sequencing

Carboxypeptidases A and B were the first exopeptidases to be used successfully in the sequencing of peptides obtained from enzymatic digests. General procedures for their application are reviewed by Ambler (1972) and Allen (1981) as well as in several monographs on sequencing. A semiquantitative estimate of the rate of release of different amino acids has been made by Ambler (1982). He found that the most rapidly released amino acids by carboxypeptidase A were the hydrophobic amino acids tryptophan, tyrosine, phenylalanine, leucine, isoleucine and methionine, as well as threonine, glutamine, histidine, valine and homoserine. Asparagine, serine and lysine were slowly released, glycine, aspartic acid and glutamic acid very slowly and proline, arginine and hydroxyproline not at all. However, the rate of release is also influenced by pH, temperature, ionic strength and the nature of the penultimate amino acid. The primary specificity deduced from sequence studies on small peptides accords well with the presence of the large hydrophobic binding pocket in subsite S_1' discussed in section 1.2.1.

Carboxypeptidase B cleaves predominantly the two basic amino acids, lysine and arginine, but will also remove some nonbasic ones. This is not due to contamination with carboxypeptidase A, but appears to be an inherent property of the enzyme (Wintersberger *et al.*, 1962).

Carboxypeptidases A and B are usually treated with diisopropyl fluorophosphonate before use in order to eliminate endopeptidase activity, although with small peptides low levels of contamination are not likely to lead to ambiguities in interpretation. However, when carboxypeptidases are used for the determination of C-terminal amino acids and partial sequences on intact proteins, the removal of every trace of endopeptidase activity is of the utmost importance. Even then the most careful analysis can lead to erroneous results. An excellent example is the pig intestinal calcium-binding protein which has been studied in our laboratory. A careful analysis of the C-terminal sequence of the intact protein – which consists of 80 amino acids – with carboxypeptidase A and penicillocarboxypeptidases S_1 and S_2 (Hofmann, 1976) in ten experiments appeared to give the following sequence: . . .Ala-Ile-Val(Phe, Ser)Leu-Lys-Gln (Dorrington *et al.*, 1974). When the complete sequence was determined, however, the C-terminal sequence was found to be . . .Lys-Ile-Ser-Glu-Lys-Gln (Hofmann *et al.*, 1979). The reasons for the erroneous results with the intact protein are not clear.

A promising extension of the use of carboxypeptidases A and B into the subnanomolar level has been proposed by Tsugita *et al.* (1982). They show that small samples of partial carboxypeptidase digests can be analysed directly and without separation of the digestion products by 'direct field desorption mass spectrometry'. The method is simple and rapid. No interference from background due to the enzyme has been observed.

1.3 CARBOXYPEPTIDASE B

Carboxypeptidase B is very similar to carboxypeptidase A in many of its properties, but differs most significantly in its preference for the cleavage of arginine and lysine residues from the C-terminal of peptides, although it can also cleave hydrophobic residues. The enzyme is secreted as an inactive precursor by the pancreas. The proenzymes from the cow and from the spiny dogfish are monomeric (Cox *et al.*, 1962; Wintersberger *et al.*, 1962). They are rapidly converted to the active enzymes by trypsin (Folk and Gladner, 1958).

The early studies on carboxypeptidase B from several species have been reviewed by Folk (1971). Since then the complete amino acid sequence of the active bovine enzyme has been determined (Titani *et al.*, 1975). This sequence is 49 per cent identical with that of bovine carboxypeptidase A shown in figure 1.2. Most notably, the amino acid residues that have been identified in the latter enzyme as being involved in various aspects of the active site occupy analogous positions in the amino acid sequence of carboxypeptidase B and in the three-dimensional structure (Schmid and Herriott, 1976; see below).

The three-dimensional structure of bovine carboxypeptidase B has been determined under the assumption that the tertiary structures of carboxypeptidases A and B would be very similar because of the similarity in their sequences and in their general enzymic properties. Tollin (1966) has suggested that the rotation function of Rossmann and Blow (1962) could be used to find the relationship between similar molecules in different crystal lattices. Schmid and Herriott (1976) used the known structure of carboxypeptidase A as a beginning model and after refinement obtained a structure at 2.8 Å resolution. This structure confirms the overall similarity of the two carboxypeptidases. The differences can all be accounted for by the known differences in amino acid sequence. The active sites appear identical except for the replacement in the hydrophobic primary binding site S_1' of isoleucine-255 in carboxypeptidase A by an aspartic acid residue in carboxypeptidase B. This is presumably the residue that determines the specificity of the enzyme for lysine and arginine side chains. The retention of the remainder of the hydrophobic nature of S_1' explains why carboxypeptidase B has an intrinsic activity that has a specificity similar to that of carboxypeptidase A (Sokolovsky and Zisapel, 1974; Wintersberger *et al.*, 1962). As in the latter enzyme, histidines-69 and -196 and glutamate-72 are liganded to the zinc. Arginine-145, whose side chain binds the C-terminal carboxyl group, and glutamate-270, which is involved in peptide bond cleavage, occupy positions in carboxypeptidase B compatible with their proposed function. Tyrosine-248, whose side chain is very flexible in carboxypeptidase A, is disordered in the carboxypeptidase B crystals. The basic arrangements of the internal helices and the β-pleated sheet are almost identical in the two enzymes. An interesting difference, however, involves tyrosine-198. This residue, which is located in subsite S_1 (Rees and Lipscomb, 1981) in carboxypeptidase A and which has been thought responsible for some of its kinetic anomalies, is rotated by 180° in the B enzyme and deeply buried. This would be compatible with the suggestion that tyrosine-198 is responsible – at least partly – for such effects as excess substrate inhibition in carboxypeptidase A, an effect which is not observed with comparable substrates in the B enzyme (Wolff *et al.*, 1962).

It may be noted that the presence of glutamate-270 in the active site of carboxypeptidase B has been predicted from the affinity labelling experiments with *N*-bromoacetyl-*N*-methyl-L-phenylalanine, which labels both carboxypeptidases A and B by reacting specifically with glutamate-270 (Hass and Neurath, 1971; Hass *et al.*, 1972) and with α-*N*-bromoacetyl-D-arginine, which specifically inactivates carboxypeptidase B at glutamate-270 (Plummer, 1972).

1.4 OTHER CARBOXYPEPTIDASES

A considerable number of enzymes from various sources with carboxypeptidase or dipeptidyl-carboxypeptidase activity have been described. Many of these belong to the superfamily of the pancreatic carboxypeptidases and include enzymes not only from mammalian sources, but also from the Pacific dogfish, African lungfish and invertebrates such as shrimp and crayfish (Hass *et al.*, 1980). Their general properties can be expected to be similar to those of the enzymes discussed in the preceding two sections.

There are, however, enzymes that cleave C-terminal amino acids or dipeptides that are not related to the pancreatic enzymes, but which are also metalloproteinases. Three of these will now be discussed (sections 1.4.1, 1.4.2 and 1.4.3).

In addition, mention should be made of a series of carboxypeptidases from fungi, higher plants and mammalian lysosomes (cathepsin A) (see Hayashi, 1976; Hofmann, 1976; Zuber, 1976). These enzymes are inhibited by diisopropylfluorophosphonate, which reacts with an active site serine residue. They also require a histidine, function presumably by a mechanism of action that is similar to that of the trypsin family of serine proteinases, and are appropriately called serine carboxypeptidases.

1.4.1 Dipeptidyl carboxypeptidase

Dipeptidyl carboxypeptidase is an enzyme that is also known as angiotensin converting enzyme, kininase II, peptidase, P, carboxycathepsin or peptidyl dipeptide hydrolase. It is found in many mammalian tissues, including human lung and kidney (Erdos and Yang, 1967; Yang and Erdos, 1967). It cleaves dipeptides from the C-terminus of peptide chains and functions physiologically in converting angiotensin I (Asp–Arg–Val–Tyr–Ile–His–Pro–Phe–His–Leu) into angiotensin II by removing the C-terminal His–Leu. It also inactivates bradykinin (Skeggs *et al.*, 1954; Yang and Erdos, 1967). It thus plays an important role in the control of blood pressure. The enzyme is being investigated widely, because inhibitors against it are useful clinically as drugs for treating certain kinds of hypertensive patients (Erdos, 1979).

Properties Many properties of angiotensin converting enzyme (or converting enzyme), which is the commonly used name for the most extensively studied dipeptidyl carboxypeptidase, have been reviewed recently (Ondetti and Cushman, 1982). It is a single chain glycoprotein of molecular mass 155 000 Da

(Lanzillo and Franburg, 1977; Stewart *et al.*, 1981*b*). It is a metalloproteinase and contains one zinc per molecule (Das and Soffer, 1975). Whereas the zinc is tightly bound at pH $\geqslant$ 7.5, at lower pH there is a slow spontaneous release of the metal and added zinc is required for enzyme studies (Bunning and Riordan, 1983). One of the earliest properties of the enzyme to be described was the apparently absolute requirement for chloride ions (Skeggs *et al.*, 1954), which has been confirmed in a recent detailed study of the effect of monovalent anions on the converting enzyme (Bunning and Riordan, 1983). The hydrolysis of bradykinin, however, while not absolutely requiring chloride, is strongly activated by it (Dorer *et al.*, 1974). This anion effect is largest for halides, is less for other monovalent anions and is not due to ionic strength (Bunning and Riordan, 1983).

Erdos (1979) reported that converting enzyme cleaves dipeptides from the C-terminus of tripeptides and longer chains, as long as the α-N on the P_1 residue is substituted. There is no specificity for P_1' or P_2', except that the enzyme will not cleave peptides in which P_1' is proline, or peptides in which P_2' is aspartic or glutamic acid. The proline restriction explains why the hydrolysis of angiotensin I is confined to the release of a single dipeptide. The requirement for a free terminal carboxyl group on P_2' is absolute.

The purification and many general properties of the converting enzyme have been reviewed by Stewart *et al.* (1981*a*). In a recent study, detailed kinetic and specificity properties of the enzyme acting on a series of furylacryloyl-tripeptides have been described (Bunning *et al.*, 1983). Furylacryloyl–Phe–Gly–Gly is a convenient substrate for routine assays of converting enzyme.

The role of converting enzyme in the regulation of blood pressure led to an early search for inhibitors as potential drugs for the control of hypertension. The first of these to be used clinically was a nonapeptide, <Glu–Trp–Pro–Arg–Pro–Gln–Ile–Pro–Pro (Ondetti *et al.*, 1971). Subsequently, the inhibitor 3-mercapto-2-methyl-D-propanoyl-L-proline was developed (Cushman *et al.*, 1977) and marketed as 'Captopril'. Patchett *et al.* (1980) have recently described another series of angiotensin converting enzyme inhibitors with potential as antihypertensive drugs. Most of these are derivatives of the following compound:

$$\begin{array}{l} R_1-\mathrm{CH}-\mathrm{Ala}-\mathrm{Pro} \\ \quad\;\;\; | \\ \quad\;\mathrm{COOH} \end{array}$$

and are considered to be transition state analogues. Their inhibition constants are in the range of 10^{-6} to 10^{-9} M.

The most widely studied enzyme is the converting enzyme from human lung tissue. Only limited structural information is at present available (Ng and Vane, 1967). Recently, some molecular and catalytic properties of a dipeptidyl carboxypeptidase from rabbit testicles have been reported (El-Dorry *et al.*, 1982). This enzyme differs from human lung converting enzyme in size (100 000 Da) and in its N- and C-terminal amino acids. The two enzymes are immunologically related and catalytically similar.

Bunning *et al.* (1978) suggested that the mechanism of action of converting enzyme was similar to that of carboxypeptidase A. Inactivation by acetyl imidazole and tetranitromethane suggested a role for tyrosine, reaction with butanedione a role for arginine, and the partial loss of activity after treatment with a carbodiimide reagent implicated an aspartic or glutamic acid residue.

1.4.2 Arginine carboxypeptidase

Arginine carboxypeptidase, also called kininase I or carboxypeptidase N, is physiologically related to angiotensin converting enzyme in that it, too, inactivates bradykinin. It is, however, specific for cleaving lysine and arginine rather than dipeptides from the C-terminal position of peptide chains (Belew *et al.*, 1980). It was originally discovered as an enzyme that inactivated kallidin as well as bradykinin (Erdos and Sloane, 1962). It is produced in the liver and can be isolated from plasma (Plummer and Hurwitz, 1978).

Only a limited amount of information is available at present on its molecular and enzymatic properties. These have been reviewed recently by Plummer and Erdos (1981). Plasma arginine carboxypeptidase, a zinc enzyme, has a molecular mass of around 270 000 Da and consists of two pairs of subunits with molecular masses of around 90 000 Da and 55 000 Da each (Plummer and Hurwitz, 1978). In addition to its kininase activity it has also been suggested that it acts on other physiologically active peptides such as the anaphylatoxins from the complement system C3a and C5a (Bokisch and Mueller-Eberhard, 1970) and C4a (Gorski *et al.*, 1979).

An enzyme with arginine carboxypeptidase activity has also been isolated from human urine (Figueiredo and Marquezini, 1978). Its molecular mass is 210 000 Da. It has a cobalt rather than a zinc ion at the active site. This enzyme cleaves arginine from bradykinin, but not from benzoylglycyl-arginine. It does not act on angiotensin I. Its relationship to the liver-plasma enzyme is not clear.

1.4.3 Muramoyl pentapeptide carboxypeptidases

A considerable number of different enzymes are involved in the system that crosslinks peptidoglycans in bacterial cell walls. In the context of this review, only those enzymes that show activity on peptide bonds will be mentioned. They are known as muramoyl pentapeptide carboxypeptidases, D–Ala–D–Ala carboxypeptidases or DD-carboxypeptidases, and are unique in their ability to cleave D-alanine from the C-terminus of UDP-N-acetylmuramoyl-L–Ala–γ–D–Glu–X–D–Ala–D–Ala where X is either an ϵ-N substituted lysine or LL-meso-diaminopimelic acid (Ghuysen *et al.*, 1980). These enzymes occur in a number of different forms and are either membrane bound or secreted. Their involvement in cell wall synthesis and their interaction with penicillin and related antibiotics have been reviewed extensively (Ghuysen *et al.*, 1979, 1980,

1981). At present two major classes can be distinguished on the basis of their enzymatic action and their mechanism. The first class includes the extracellular DD-carboxypeptidases from *Streptomyces* (Frere *et al.*, 1976), which catalyse both hydrolysis and the transfer of muramoyl pentapeptides by transpeptidation. It also includes the membrane-bound DD-carboxypeptidases of several bacilli (Georgopapadakou *et al.*, 1977; Yocum *et al.*, 1979). This class of enzymes functions via a covalent mechanism involving an acyl intermediate linked to an active site serine residue (Rasmussen and Strominger, 1978). The other class which is at present represented by only one enzyme, the extracellular DD-carboxypeptidase from *Streptomyces albus G* (Ghuysen *et al.*, 1970), acts as a DD-carboxypeptidase and has endopeptidase activity, but no transpeptidase activity.

Streptomyces albus G DD-carboxypeptidase is a zinc metalloprotease of 22 076 Da that binds zinc with a very high affinity ($K_A \sim 2 \times 10^{14}\ M^{-1}$). The zinc is essential for the enzyme activity and for binding benzylpenicillin (Dideberg *et al.*, 1980*b*).

Amino acid sequence The complete amino acid sequence of muramoyl pentapeptide carboxypeptidase has recently been published and is shown in figure 1.9 (Joris *et al.*, 1983). It consists of 212 amino acid residues and has three disulphide bridges between Cys-3 and Cys-80, Cys-93 and Cys-141, and Cys-169 and Cys-210. The N-terminal dipeptide sequence Asn–Gly appears to be partly cyclised. It is not clear, however, whether this is an artifact of the purification procedure of the enzyme or whether cyclisation occurs physiologically. The most interesting feature of the sequence is that it shows no homology with any

```
 1                          10                                                           25
Asn-Gly-Cys-Tyr-Thr-Trp-Ser-Gly-Thr-Leu-Ser-Glu-Gly-Ser-Ser-Gly-Glu-Ala-Val-Arg-Gln-Leu-Gln-Ile-Arg-

 26                         35                                                           50
Val-Ala-Gly-Tyr-Pro-Gly-Thr-Gly-Ala-Gln-Leu-Ala-Ile-Asp-Gly-Gln-Phe-Gly-Pro-Ala-Thr-Lys-Ala-Ala-Val-

 51                         60                                                           75
Gln-Arg-Phe-Gln-Ser-Ala-Tyr-Gly-Leu-Ala-Ala-Asp-Gly-Ile-Ala-Gly-Pro-Thr-Phe-Asn-Lys-Ile-Tyr-Gln-Leu-

 76                         85                                                          100
Gln-Asp-Asp-Asp-Cys-Thr-Pro-Val-Asn-Phe-Thr-Tyr-Ala-Glu-Leu-Asn-Arg-Cys-Asn-Ser-Asp-Trp-Ser-Gly-Gly-

101                        110                                                          125
Lys-Val-Ser-Ala-Ala-Thr-Ala-Arg-Ala-Asn-Ala-Leu-Val-Thr-Met-Trp-Lys-Leu-Gln-Ala-Met-Arg-His-Ala-Met-

126                        135                                                          150
Gly-Asp-Lys-Pro-Ile-Thr-Val-Asn-Gly-Gly-Phe-Arg-Ser-Val-Thr-Cys-Asn-Ser-Asn-Val-Gly-Gly-Ala-Ser-Asn-

151                        160                                                          175
Ser-Arg-His-Met-Tyr-Gly-His-Ala-Ala-Asp-Leu-Gly-Ala-Gly-Ser-Gln-Gly-Phe-Cys-Ala-Leu-Ala-Gln-Ala-Ala-

176                        185                                                          200
Arg-Asn-His-Gly-Phe-Thr-Glu-Ile-Leu-Gly-Pro-Gly-Tyr-Pro-Gly-His-Asn-Asp-His-Thr-His-Val-Ala-Gly-Gly-

201                                        212
Asp-Gly-Arg-Phe-Trp-Ser-Ala-Pro-Ser-Cys-Gly-Ile
```

Figure 1.9 Amino acid sequence of zinc-containing muramoyl pentapeptide carboxypeptidase according to Joris *et al.* (1983).

mechanistically or functionally related enzymes, such as the serine DD-carboxypeptidases from *Bacillus subtilis* and *B. stearothermophilus* (partial sequences: Waxman *et al.*, 1980); carboxypeptidase A (Bradshaw *et al.*, 1969: see figure 1.2); thermolysin (Titani *et al.*, 1972: see figure 1.10); carbonic anhydrase (Andersson *et al.*, 1972) and alcohol dehydrogenase (Jornvall, 1970); or the zinc-requiring β-lactamase II of *Bacillus cereus* (Ambler, quoted by Joris *et al.*, 1983) and the serine-dependent β-lactamases (Ambler, 1980; Jaurin and Grundstrom, 1981).

Three-dimensional structure The three-dimensional structure of *S. albus G* DD-carboxypeptidase has been determined at a nominal 2.5 Å resolution (Dideberg *et al.*, 1980*a*, 1982). The enzyme, whose dimensions are 48 x 34 x 28 Å, consists of two globular domains. The N-terminal domain of 76 residues contains three α-helices (43 per cent) and is connected to the C-terminal domain of 136 residues through the residues Leu-75 to Cys-80. The latter domain also contains three α-helices as well as five short β-strands which form the lining of one side of the active site cavity. The cavity resembles the open cleft of the active site of thermolysin more than the rather closed structure of carboxypeptidase A. The open cleft probably explains the ability of the enzyme to act as an endopeptidase (see below) as well as an exopeptidase since it would allow binding of extended chains on both sides of the scissile bond.

The overall structure appears to differ from that of any other protein (Joris *et al.*, 1983). This strengthens the conclusion drawn from the sequence comparison that this enzyme is not related to any other known enzyme through evolution.

The zinc atom is located in the active site cavity and is liganded to the side chains of histidines-153, -194 and -196. Arginine-136 is in a position where it could bind the C-terminal carboxyl group of a substrate and determine its specificity as a carboxypeptidase.

Enzymatic properties Studies on the carboxypeptidase activity of *S. albus G* DD-carboxypeptidase (Leyh-Bouille *et al.*, 1970) established that the enzyme is absolutely specific for D-amino acids in the C-terminal position with D-alanine showing the highest specificity constants. However, C-terminal glycine, D-lysine, D-leucine and D-glutamic acid were also readily cleaved. The penultimate position is more restricted: D-alanine is required, although glycine-containing substrates are hydrolysed slowly. No activity was found with other D-amino acids or with L-alanine. The dipeptide Ac–D–Ala–D–Ala is not cleaved and is in fact an inhibitor (Nieto *et al.*, 1973). A minimal length of three amino acids is required. The tripeptide α-*N*,ε-*N*-diacetyl-L–Lys–D–Ala–D–Ala is routinely used as the standard substrate for enzyme assays.

As an endopeptidase the enzyme acts on the bond between a D-alanine residue and an amino group of diaminopimelic acid in cell wall polymers (Ghuysen *et al.*, 1980).

Of all the DD-carboxypeptidases the zinc-containing enzyme is the least sensitive to penicillin. It is inhibited weakly by the antibiotic (Ghuysen *et al.*, 1979) and it also has weak penicillinase (β-lactamase) activity.

Mechanism of action Although insufficient information is at present available for the formulation of a definite mechanism, the three-dimensional structure suggests that the mechanism resembles those of carboxypeptidase A and thermolysin (Dideberg *et al.*, 1982). This tentative conclusion is drawn because histidine-190, which could function as a proton donor, and aspartic acids-159 and -192, one of which could function as a general base or as a nucleophile, are located in the active site near the zinc. However, a more refined structure of the native protein and of ligand complexes is required before more definite proposals can be formulated.

1.4.4 Zinc-containing β-lactamase

β-Lactamases or penicillinases have been extensively studied because they are responsible for the resistance of many microorganisms to penicillin and related antibiotics. Most of these enzymes act by a mechanism that involves a serine residue at the active site and thus fall outside this review. One enzyme, β-lactamase II, however, which has been isolated from *Bacillus cereus* (Sabath and Abraham, 1966; Davies *et al.*, 1974), is different. Although not a proteolytic enzyme, it nevertheless deserves mention here since it acts on the imino bond of the lactam ring of penicillin and similar compounds, and may therefore have a mechanism that is related to those of metalloproteinases.

β-Lactamase II has a molecular mass of 22 000 Da (Davies *et al.*, 1974) and binds two zinc atoms, one of which is required for the hydrolysis of benzylpenicillin (Davies and Abraham, 1974). The zinc can be replaced by cobalt, cadmium or manganese to give an enzyme with lower activity. The active site zinc atom is liganded to the single cysteine residue as shown by spectral studies of cobalt and cadmium enzymes (Davies and Abraham, 1984). Proton nuclear magnetic resonance experiments implicated three histidine residues as additional ligands of the zinc in the higher affinity site and one histidine as a ligand for the second zinc (Baldwin *et al.*, 1978). The presence of a thiol ligand shows a real difference in the ligand pattern from that of the proteolytic enzymes and from carbonic anhydrase. However, the zinc in liver alcohol dehydrogenase is also liganded to the side chains of cysteine residues.

1.5 THERMOLYSIN

Thermolysin is an extracellular neutral proteinase which has been isolated from the culture medium of the thermophilic microorganism *Bacillus thermoproteolyticus* (Endo, 1962). Many similar enzymes have been demonstrated in and isolated from *Bacillus, Pseudomonas, Streptomyces* and *Aspergillus* species, among others (Matsubara and Feder, 1971). The neutral proteinase from a mesophilic strain of *B. subtilis* (see section 1.5.3) has an amino acid sequence that is very similar to that of thermolysin and shows that the enzymes are homologous (Pangburn *et al.*, 1976). Thermolysin has a molecular mass of 34 000 Da and contains one zinc and four calcium ions. It is an endopeptidase whose specificity is rather broad, but primarily directed towards peptide bonds in which the nitrogen is contributed by an amino acid residue with a hydrophobic side chain (Matsubara and Feder, 1971). Thermolysin and the related metalloproteinases resemble pancreatic carboxypeptidases in their molecular size and the presence of an essential zinc at the catalytic site. However, there is no evidence for any sequence homology between the neutral proteinases and the mammalian carboxypeptidases (Titani *et al.*, 1972; Pangburn *et al.*, 1976). The four bound calcium ions (Latt *et al.*, 1969) were originally thought to be responsible for the heat stability of the enzyme, but many of the homologous bacterial metalloproteinases (e.g. one of the enzymes from *B. subtilis*: Pangburn *et al.*, 1976) also contain bound calcium, but show much lower denaturation temperatures.

Thermolysin is a representative of a group of metalloproteinases of evolutionary origin distinct from that of other well-characterised classes of proteolytic enzymes.

1.5.1 The three-dimensional structure of thermolysin

The primary and tertiary structures were described simultaneously by the laboratories of Neurath (Titani *et al.*, 1972) and Matthews (Matthews *et al.*, 1972*a,b*) respectively. The amino acid sequence is shown in figure 1.10. In view of the high thermal stability, the most notable feature is the absence of any cystine bridges. There is no other unique feature of the sequence that could be related to the thermal stability of the enzyme (Titani *et al.*, 1972).

1.5.1.1 Native thermolysin

The overall three-dimensional structure is shown in figure 1.11 (Matthews *et al.*, 1972*a,b*; Colman *et al.*, 1972). An improved detailed structure has been described by Matthews *et al.* (1974). The slightly aspherical molecule shows a distinct cleft across the centre which accommodates the zinc, and a pocket that represents the hydrophobic primary binding site (see below), which determines the specificity. The upper part of the bilobal structure is made up entirely of the N-terminal half, which is connected to the C-terminal of the molecule by an internal helix (residues 137–150) that runs through the centre of the molecule and is connected via a loop to a similar helix (residues 160–180) in the C-terminal half. There are five additional stretches of helix from 8 to 15 residues long throughout the molecule, giving the protein a 34 per cent helical content. The bulk of defined β-structure is located in the N-terminal half of the molecule and involves about 70 residues in both parallel and antiparallel arrangements (Colman *et al.*, 1972). One glutamic acid and three aspartic acid residues are

1 5 10 15 20
Ile-Thr-Gly-Thr-Ser-Thr-Val-Gly-Val-Gly-Arg-Gly-Val-Leu-Gly-Asp-Gln-Lys-Asn-Ile-

25 30 35 40
Asn-Thr-Thr-Tyr-Ser-Thr-Tyr-Tyr-Tyr-Leu-Gln-Asp-Asn-Thr-Arg-Gly-Asp-Gly-Ile-Phe

45 50 55 60
Thr-Tyr-Asp-Ala-Lys-Tyr-Arg-Thr-Thr-Leu-Pro-Gly-Ser-Leu-Trp-Ala-Asp-Ala-Asp-Asn-

65 70 75 80
Gln-Phe-Phe-Ala-Ser-Tyr-Asp-Ala-Pro-Ala-Val-Asp-Ala-His-Tyr-Tyr-Ala-Gly-Val-Thr-

85 90 95 100
Tyr-Asp-Tyr-Tyr-Lys-Asn-Val-His-Asn-Arg-Leu-Ser-Tyr-Asp-Gly-Asn-Asn-Ala-Ala-Ile-

105 110 115 120
Arg-Ser-Ser-Val-His-Tyr-Ser-Gln-Gly-Tyr-Asn-Asn-Ala-Phe-Trp-Asn-Gly-Ser-Glu-Met-

125 130 135 140
Val-Tyr-Gly-Asp-Gly-Asp-Gly-Gln-Thr-Phe-Ile-Pro-Leu-Ser-Gly-Gly-Ile-Asp-Val-Val-

145 150 155 160
Ala-His-Glu-Leu-Thr-His-Ala-Val-Thr-Asp-Tyr-Thr-Ala-Gly-Leu-Ile-Tyr-Gln-Asn-Glu-

165 170 175 180
Ser-Gly-Ala-Ile-Asn-Glu-Ala-Ile-Ser-Asp-Ile-Phe-Gly-Thr-Leu-Val-Glu-Phe-Tyr-Ala-

185 190 195 200
Asn-Lys-Asn-Pro-Asp-Trp-Glu-Ile-Gly-Glu-Asp-Val-Tyr-Thr-Pro-Gly-Ile-Ser-Gly-Asp-

205 210 215 220
Ser-Leu-Arg-Ser-Met-Ser-Asp-Pro-Ala-Lys-Tyr-Gly-Asp-Pro-Asp-His-Tyr-Ser-Lys-Arg-

225 230 235 240
Tyr-Thr-Gly-Thr-Gln-Asp-Asn-Gly-Gly-Val-His-Ile-Asn-Ser-Gly-Ile-Ile-Asn-Lys-Ala-

245 250 255 260
Ala-Tyr-Leu-Ile-Ser-Gln-Gly-Gly-Thr-His-Tyr-Gly-Val-Ser-Val-Val-Gly-Ile-Gly-Arg-

265 270 275 280
Asp-Lys-Leu-Gly-Lys-Ile-Phe-Tyr-Arg-Ala-Leu-Thr-Gln-Tyr-Leu-Thr-Pro-Thr-Ser-Asn-

285 290 295 300
Phe-Ser-Gln-Leu-Arg-Ala-Ala-Ala-Val-Gln-Ser-Ala-Thr-Asp-Leu-Tyr-Gly-Ser-Thr-Ser-

305 310 315
Glx-Glx-Val-Ala-Ser-Val-Lys-Gln-Ala-Phe-Asp-Ala-Val-Gly-Val-Lys

Figure 1.10 Amino acid sequence of thermolysin. (Reprinted by permission from *Nature, Lond.* **238**, 35. Copyright © 1972, Macmillan Journals Ltd.)

buried and take part in ionic interactions with lysine and arginine residues and with calcium ions. There are about 24 β-bends (or hairpin bends) in thermolysin and one so-called γ-turn (Matthews, 1972) whose possible existence in proteins had been predicted by Nemethy and Printz (1972). The β-turn is defined as a hairpin bend in which the direction of the polypeptide chain changes by 180° and which contains *at least* four α-carbon atoms. The γ-turn is a conformation in which the peptide chain also turns around 180° but which contains *only three* α-carbon atoms. The γ-turn in thermolysin has been fully documented by Matthews (1972). The overall arrangement of the different discernible structural

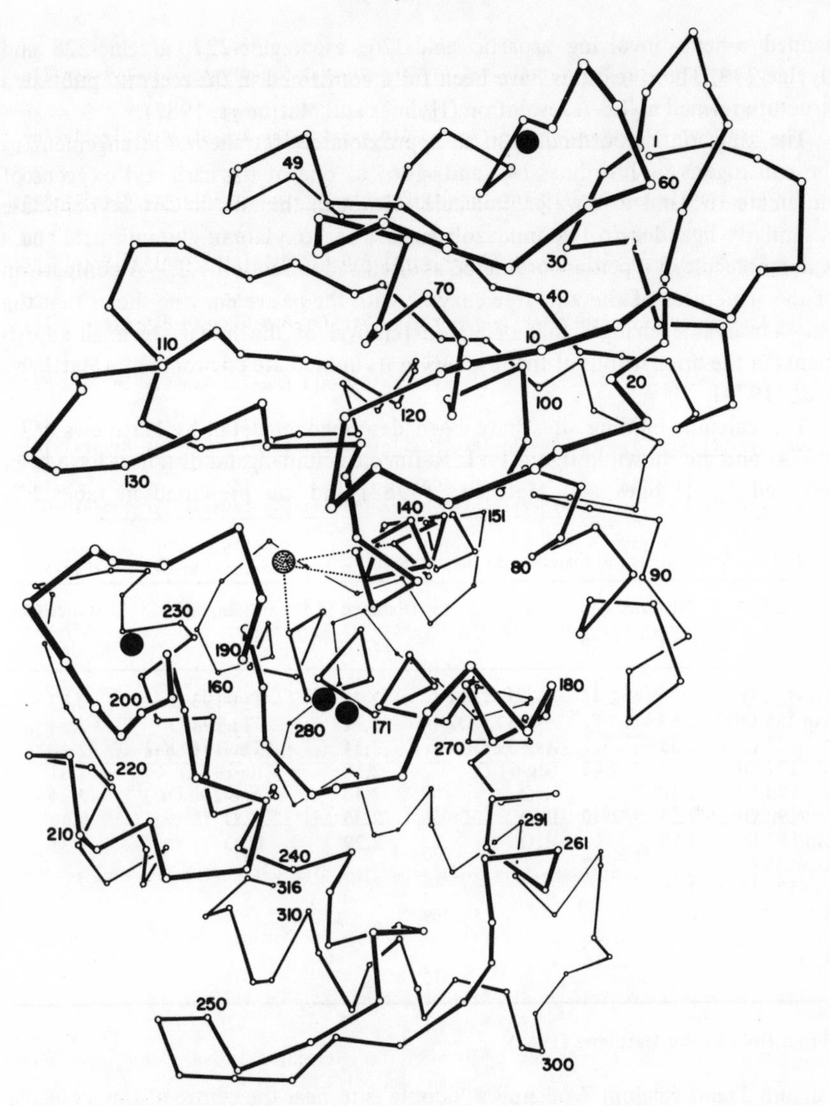

Figure 1.11 Projection drawing of the backbone conformation of thermolysin. The open circles correspond to the α-carbon positions obtained from electron density maps at 2.3 Å resolution. The zinc ion is stippled; its three protein ligands are shown as dotted lines. The four calcium ions are shown as black filled circles. Sites 1 and 2 are in the centre near the zinc, site 3 is in the N-terminal and site 4 in the C-terminal part of the molecule. (Reprinted by permission from *Nature, Lond.* **238**, 37. Copyright © 1972, Macmillan Journals Ltd.)

features is quite similar to those encountered in other proteins, such as lysozyme, chymotrypsin, ribonuclease S and carboxypeptidase A (Matthews *et al.*, 1974) and does not appear to offer any clues to the high thermal stability of thermolysin. There is one *cis*-peptide bond at proline-51 and a single turn of a left-handed α-helix involving aspartic acid-226, asparagine-227, glycine-228 and glycine-229. These features have been fully confirmed in the recently published structure refined at 1.6 Å resolution (Holmes and Matthews, 1982).

The zinc ion is coordinated in an approximately tetrahedral arrangement to the ε-nitrogens of histidines-142 and -146, to one of the carboxyl oxygens of glutamate-166 and to a water molecule. Although the zinc in carboxypeptidase is similarly liganded to two imidazole rings, a carboxylate of glutamic acid and a water molecule, it is pentacoordinated as discussed in section 1.2.1. A comparison of the structures of the zinc-free enzyme with the native enzyme shows that the only observable changes that occur on removal of the metal are small adjustments in the orientations of the residues in its immediate environment (Matthews *et al.*, 1974).

The calcium binding sites have been described in detail by Matthews *et al.* (1974) and are shown in figure 1.11. Refined calcium–ligand distances have been reported by Holmes and Matthews (1982) and are presented in table 1.2.

Table 1.2 Calcium–ligand distances in thermolysin[a]

Ligand	Distance (Å)		Ligand	Distance (Å)	Ligand	Distance (Å)
	Ca-1	Ca-2		Ca-3		Ca-4
Asp-185 OD1		2.43	Asp-57 OD1	2.23	Tyr-193 O	2.43
Asp-185 OD2	2.52		Asp-57 OD2	2.71	Thr-194 O	2.54
Glu-177 OE1	2.53		Asp-59 OD1	2.34	Thr-194 OG1	2.48
Glu-177 OE2		2.42	Glu-61 O	2.18	Ile-197 O	2.31
Glu-190 OE1	2.40		H_2O	2.40	Asp-200 OD1	2.24
Glu-190 OE2	2.51	2.40	H_2O	2.36	H_2O	2.42
Glu-187 O	2.29		H_2O	2.29	H_2O	2.35
Asp-183 O		2.37				
Asp-138 OD1	2.46					
H_2O	2.62					
H_2O		2.46				
H_2O		2.16				

[a]From Holmes and Matthews (1982).

Calcium 1 and calcium 2 occupy a 'double' site near the centre of the molecule as drawn in figure 1.11, in the N-terminal portion, and calcium 4 is to the 'left' of the zinc binding site. The coordination number for calcium 2 is six, that for the other three calcium ions is seven. The affinity of the enzyme for all four calcium ions is relatively high, with all four sites being saturated at calcium ion concentrations above 10^{-4} M (Voordouw and Roche, 1974). Two of the calcium ions dissociate between 10^{-4} and 10^{-5} M; the other two have higher affinity. Voordouw and Roche (1974, 1975) have attempted to measure the individual binding constants for the four calcium ions. They used a zinc-free enzyme in order to eliminate autolysis and concluded that at 25°C, pH 9.0, the binding isotherms best fit a model in which there is complete cooperativity between two calcium ions, with an affinity constant $K_{1,2} = 2.8 \times 10^9\ M^{-1}$. The other two calciums bind independently with $K_3 = K_4 = 3.2 \times 10^6\ M^{-1}$. (Owing to lack of sufficient data it has only been assumed that $K_3 = K_4$.) At 6°C the affinity of thermolysin for calcium decreases and the data can be fitted equally well to a model in which all four sites are independent or one in which sites 1 and 2 are fully cooperative. Weaver *et al.* (1976) studied semiquantitatively the relative affinities of the calcium sites by soaking crystals at pH 7.2 in varying concentrations of EDTA in the presence of the inhibitor phosphoramidon. They examined difference electron density maps at various stages of calcium removal and concluded that calcium 2 dissociates first, followed by calcium 4 and then calcium 3. No dissociation of calcium 1 was observed. This leads to a different order of the affinities from that obtained by Voordouw and Roche (1974, 1975). However, the differences may well be due to the different experimental conditions (apoenzyme versus zinc enzyme, pH 9.0 versus pH 7.2, and solution versus crystals). Further details on calcium binding as well as binding of other ions are given in a review by Roche and Voordouw (1978) where they point out that much more work is required before the binding properties of thermolysin for calcium and other ions will be understood.

1.5.1.2 Crystalline thermolysin–ligand complexes

Binding of dipeptide inhibitors Kester and Matthews (1977*a*) have described the binding of three dipeptides, β-phenylpropionyl-phenylalanine, alanyl-phenylalanine and phenylalanyl-phenylalanyl amide, as well as of phenylalanine and carbobenzoxyphenylalanine, in crystals of thermolysin at 2.3 Å resolution. The details of the binding of β-phenylpropionyl-phenylalanine are as follows. The phenyl ring of the phenylalanine residue binds in the hydrophobic pocket which has been described earlier (Colman *et al.*, 1972) and is believed to represent the subsite S_1', the determinant site of the primary specificity for amino acids with hydrophobic side chains. This pocket is lined by the side chains and α-carbons of the following hydrophobic amino acids, phenylalanine-130, leucine-133, valine-139, isoleucine-188, glycine-189, valine-192 and leucine-202. The carbonyl oxygen of the peptide bond of the inhibitor binds to the zinc and displaces a water molecule. There is a water molecule close to the carbonyl carbon; it is hydrogen bonded to glutamate-143 and to the peptide bond nitrogen of tryptophan-115. The nitrogen of the peptide bond of β-phenylpropionyl-phenylalanine is hydrogen bonded to the carbonyl oxygen of alanine-113. The free carboxyl group forms a salt link with arginine-203 and a hydrogen bond with the N^{δ} of asparagine-112. Although no detailed analysis was made of the binding of alanyl-phenylalanine, phenylalanyl-phenylalanine amide and free phenylalanine, the difference electron density maps showed that these compounds bound in an exactly analogous manner with the phenyl ring of the C-terminal residue being located in the S_1' binding pocket. In striking contrast is the binding of carbobenzoxyphenylalanine, which binds in a 'reverse' mode with the phenyl ring of the carbobenzoxy group being located in the S_1' binding site and the free carboxyl group binding to the zinc. This is in spite of the fact that β-phenyl-propionyl-phenylalanine and carbobenzoxyphenylalanine differ only in the replacement of the α-CH_2 group of the former by an oxygen. The binding of carbobenzoxyphenylalanine represents nonproductive binding (Kester and Matthews, 1977*a*).

Binding of phosphoramidon A proposal for binding of an extended substrate chain (shown in figure 1.12) was made on the basis of the X-ray analysis of a complex with phosphoramidon (Weaver *et al.*, 1977). Phosphoramidon (*N*-(α-L-rhamnopyranosyl-oxyhydroxyphosphinyl)-L-leucyl-L-tryptophan) was isolated from the culture medium of *Streptomyces tanashiensis* (Suda *et al.*, 1973) and found to be a powerful inhibitor of thermolysin ($K_I = 2.8 \times 10^{-8}$ M) with little

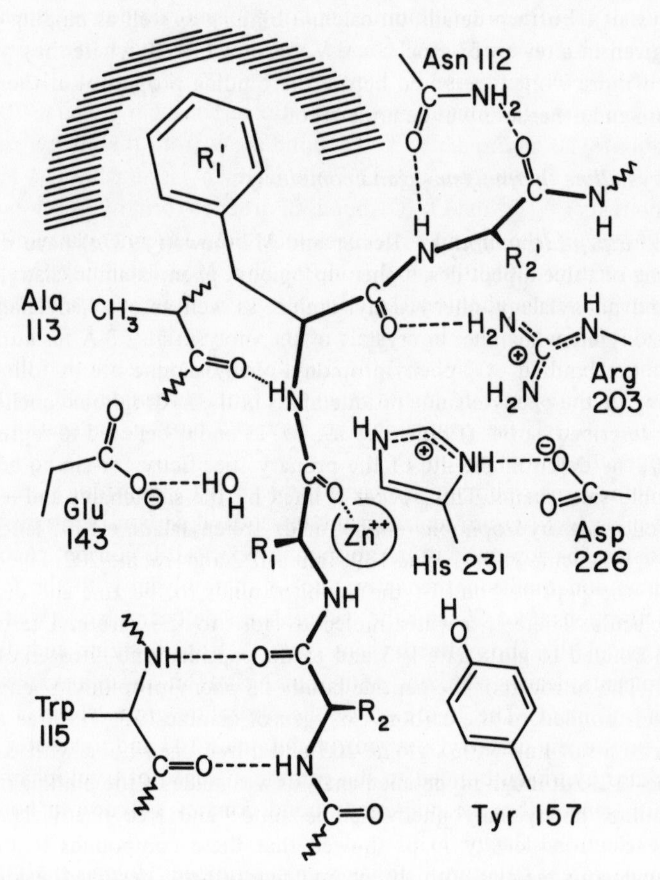

Figure 1.12 Schematic drawing illustrating the binding of an extended substrate to thermolysin, inferred from the binding of inhibitors (with permission from Kester and Matthews, 1977*a*; copyright 1977, American Chemical Society).

inhibitory effect on trypsin, chymotrypsin, papain and pepsin (Komiyama *et al.*, 1975). The observed binding of phosphoramidon closely resembles that of the dipeptide inhibitors discussed above. The results are interpreted by Weaver *et al.* (1977) to show that the active site groove of thermolysin contains at least four subsites, S_2, S_1, S_1' and S_2'. These are shown schematically in figure 1.12, where S_2 is shown to interact through two backbone hydrogen bonds of tryptophan-115 with the backbone of the substrate at residue P_2; the side chain of residue P_1 interacts with phenylalanine-114 in S_1; S_1' is the specificity pocket described previously; and S_2' is adjacent to the side chains of leucine-202 and phenylalanine-130. As yet, no complex of thermolysin with a long peptide inhibitor or other long substrate analogue has been analysed by X-ray crystallography. Therefore no information is available on the additional binding sites postulated by Morihara and Tsuzuki (1970) who found that up to six residues of a polypeptide chain interact with the enzyme.

Binding of the dipeptide inhibitors and of phosphoramidon is associated with several localised conformational changes restricted to the active site region. In contrast to carboxypeptidase A, they involve shifts of at most a few tenths of an Ångstrom unit. The major changes are experienced by leucine-202, which rotates about 120° around its $C^{\alpha}-C^{\beta}$ bond away from the phenyl ring of the inhibitor in the binding pocket. The side chain of asparagine-112 appears to rotate about its $C^{\alpha}-C^{\beta}$ and $C^{\beta}-C^{\gamma}$ bonds in order to form hydrogen bonds with the backbone of the ligand. The backbone between phenylalanine-114 and asparagine-116 shifts slightly away from the zinc. Model building does not show any evidence for an additional subsite S_3' for a long substrate (Kester and Matthews, 1977*a*).

The analyses of the native and the ligand structures suggest that three groups on the enzyme are involved in the catalytic event, glutamate-143, histidine-231 and the zinc ion. These groups will be discussed later in connection with the proposed mechanism of action.

The inhibitor phosphoramidon can be considered a transition state analogue, with the phosphinyl group replacing the tetrahedral carbon of a transition state intermediate. The low inhibition constant supports this concept. Accordingly, phosphoramidon binds in the active site of thermolysin with the leucyl side chain in the pocket of S_1' and the tryptophan side chain in S_2' (Weaver *et al.*, 1977). One of the phosphite oxygens is 2.0 Å from the zinc and displaces a water molecule, leaving the zinc with four ligands in an approximately tetrahedral arrangement. The hydroxyl of the phosphite group forms a hydrogen bond to one of the carboxyl oxygens of glutamate-143 and to a water molecule which is also hydrogen bonded to the peptide nitrogen of tryptophan-115. The amide nitrogen of the phosphoramide bond donates a hydrogen bond to the peptide carbonyl of alanine-113.

The most recent structure that has been determined (Holmes *et al.*, 1983) is a covalent complex between thermolysin and the active-site directed inhibitor $ClCH_2CO-D,L-(N-OH)Leu-OCH_3$, which alkylates glutamate-143 (Rasnick and Powers, 1978). The binding of this inhibitor is similar to that described earlier for other ligands, with two exceptions. The interaction of the hydroxamate group with the zinc is quite distorted compared with that seen with L-leucine hydroxamate (Holmes and Matthews, 1981) where the hydroxamate and the carbonyl functions are both liganded to the zinc in a pentacoordination. In the covalent complex the distances between the two oxygens and the zinc are greater (2.5 and 2.8 Å) and the complex is distorted. The second difference is a conformational change in which the peptide chain between alanine-113 and asparagine-116 has been moved away from the inhibitor by up to 0.3 Å. This is the section of the peptide chain which is in a β-conformation and is believed to contribute to the interaction of the protein with extended substrate chains (Kester and Matthews, 1977*a*). The cause for this change is purely steric, in that without the shift the carbonyl group of alanine-113 would be too close to the methylene carbon of the chloroacetyl moiety of the inhibitor. The implications of this for the mechanism of action will be discussed below.

1.5.2 Enzymatic studies

1.5.2.1 Action and application of thermolysin

Specificity As has been mentioned, the primary specificity of thermolysin is determined by the residue in the P_1' position. The predominant cleavages that have been observed during studies of synthetic peptides as well as during the use of the enzyme in sequence determinations involve isoleucine, valine, leucine and phenylalanine residues (see review by Heinrikson, 1977). However, the specificity is rather broad and cleavages have been observed on the N-terminal side of most amino acid residues except lysine, arginine and cysteine. In the sequence determination of penicillopepsin, several high-yield peptides that had been cleaved at tyrosine, alanine and threonine were particularly noticeable (Rao and Hofmann, 1976). The indole ring of tryptophan appears to be slightly too bulky to fit into the binding site S_1'; cleavages at this residue are rare and up to 1977 only one had been observed (Bradshaw, 1969; Heinrikson, 1977). Although thermolysin shows preference for aromatic and other hydrophobic residues in P_1 (Morgan and Fruton, 1978), it will cleave on the carboxyl side of almost any other residue including proline (Sterner and Heinrikson, 1975).

Thermolysin is a true endopeptidase since it does not cleave peptides with either free amino or free carboxyl groups (Matsubara, 1966).

The secondary specificity as determined by the kinetic effects of amino acid residues in positions P_2 and P_3, and P_2' to P_4' has been studied by Matsubara (1966), Morihara and Oka (1968) and Morihara and Tsuzuki (1966, 1970). The specificity requirements in these sites are rather broad. Whereas there is a strict requirement for the L-configuration in P_1 and P_1', D-amino acids are tolerated in the other subsites. Interestingly, peptides with proline in the P_2' position are completely resistant (Blumberg and Vallee, 1975; Ambler and Medway, 1968).

A study of tripeptides blocked with a furyl-acryloyl group (Blumberg and Vallee, 1975) shows that hydrophobic residues are also preferred in the P_2' position. The relatively small importance of secondary binding has been demonstrated in the studies of Morgan and Fruton (1978), who studied kinetic parameters of a series of substrates of the type A–Phe–Leu–Ala where A was Z–Gly, Z–Gly–Gly, mansyl–Gly or mansyl–Gly–Gly. In contrast to similar studies with pepsin (Fruton, 1976) and papain (Mattis and Fruton, 1976), there was no evidence that secondary binding of the substrates extended on the N-terminal side made a significant contribution to either k_{cat} or K_m. This is in contrast to the suggestion by Morihara and Tsuzuki (1970) that the subsite specificity of thermolysin is extensive and allows up to six residues of a substrate to bind.

Action on ester substrates A comparison of peptide and ester substrate pairs by Holmquist and Vallee (1976) shows that thermolysin as well as the related neutral metalloproteinases from *Bacillus subtilis* and *Aeromonas proteolytica* are effective esterases. The specificity constants k_{cat}/K_m of the esters, however, are between three- and eight-fold lower than those of the peptides because of lower k_{cat} or higher K_m values depending on the substrate. The side chain specificity requirements, the pH rate profiles, the inhibitors and the kinetic deuterium isotope effects are the same for both types of substrate and suggest that, in contrast to carboxypeptidase A, there is no difference in the mechanism of action between the two types of substrate (Holmquist and Vallee, 1976).

Peptide bond synthesis Thermolysin also catalyses the synthesis of peptides from a variety of benzyloxycarbonyl amino acids and amino acid methyl esters or amides (Isowa *et al.*, 1977; Isowa and Ichikawa, 1979). In accordance with the known specificity for the P_1' position, the best amino acids, esters or amides for the synthesis are those of isoleucine, valine, leucine, phenylalanine and methionine. For the amino acids in the P_1 position, hydrophobic amino acids are suitable. High substrate and enzyme concentrations (0.1 M to > 1 M and 10 μM, respectively) are needed and the equilibrium of the reaction is shifted towards synthesis by the insolubility of the products (Oka and Morihara, 1980).

Use of thermolysin in protein structure studies Thermolysin has been very useful in the cleavage of protein chains for complete sequence determinations. A detailed discussion of the strategy for the use of thermolysin is given by Heinrikson (1977). Thermolysin, because of its rather broad specificity, is more useful for the secondary cleavage of peptides isolated from other digests than as a tool for the hydrolysis of long peptide chains. However, thermolysin has been used in a limited way to cleave selectively peptide bonds in some native proteins which are resistant towards other proteinases (Heinrikson, 1977). An interesting study involved the haemoglobin-haptoglobin complex (D'Udine and Bernini, 1974). A comparison of the susceptibility of haemoglobin and haptoglobin chains alone and in the complex showed that the β-chain of haptoglobin and the α-chain of haemoglobin are in close association in the complex where they are resistant to thermolytic attack.

1.5.2.2 Thermolysin inhibitors

In recent years there has been considerable interest in inhibitors for metalloproteinases, particularly zinc proteinases, because of the demonstration that the angiotensin converting enzyme was a zinc enzyme (Das and Soffer, 1975). Angiotensin converting enzyme (see section 1.4.1) is a dipeptidyl carboxypeptidase that converts the hexapeptide angiotensin I into the vasoconstricting octapeptide angiotensin II and also inactivates the hypertensive peptide bradykinin. The design of inhibitors with high affinity and specificity is expected to lead to ever more useful pharmacological agents for the control of hypertension. Such inhibitors also inhibit collagenase, which has been implicated in arthritis (Harris and Krane, 1974), and other metalloproteinases. The principle of the design of specific metalloproteinase inhibitors is the combination of a moiety with a function that will bind strongly to the zinc and a peptide chain that will satisfy the specificity requirement of the enzymes. Series of compounds have been synthesised in which the zinc ligand functions are phosphoramides, hydroxamates, carboxylates or mercaptan functions (see, for example, Komiyama *et al.*, 1975; Cushman *et al.*, 1977; Holmquist, 1977; Nishino and Powers, 1978, 1979; Kam *et al.*, 1979; Patchett *et al.*, 1980; Jacobsen and Bartlett, 1981; Maycock *et al.*, 1981). A discussion of these and related papers is considered to be beyond the scope of this review.

Active site directed irreversible inhibitors Rasnick and Powers (1978) have synthesised two active site directed irreversible inhibitors ('affinity labels') for thermolysin, $ClCH_2CO$–D,L–(N–OH)Leu-OCH_3(I) and 2-(*N*-bromoacetyl-*N*-hydroxamino)-4-methylpentanonitrile. These compounds react in a 1 : 1 stoichiometric manner with thermolysin. They contain an alkylating function which reacts covalently with glutamate-143 and a hydroxamate group that binds to the zinc. They also contain hydrophobic amino acyl residues or analogues that satisfy the primary specificity site. The binding mode as determined by X-ray analysis has been described earlier. Inhibitor I also specifically inhibits the neutral metalloproteinases A and B from *Bacillus subtilis*, but reacts only very slowly with carboxypeptidase A and not at all with chymotrypsin and subtilisin. Conversely, the active site directed inhibitor of carboxypeptidases A and B (Hass and Neurath, 1971; Hass *et al.*, 1972), *N*-bromoacetyl-L-*N*-methylphenylalanine, does not inhibit thermolysin. The effectiveness of the compounds with a hydroxyamino function is presumably due to its coordination with the zinc atom (Rasnick and Powers, 1978), as has been confirmed by X-ray analysis (Holmes *et al.*, 1983).

1.5.2.3 Chemical modification of thermolysin

The study of pH dependences of thermolysin-catalysed reactions and of chemical modification with ethoxyformic anhydride (Pangburn and Walsh, 1975) suggested that a histidine residue was involved in the catalysis which has been identified as histidine-231 on the basis of the tertiary structure. Blumberg *et al.* (1973, 1974) had previously found that the inhibition by ethoxyformic anhydride could be reversed by hydroxylamine, and that if the reaction with ethoxyformic anhydride was carried out in the presence of a reversible inhibitor, thermolysin was converted into an enzyme whose activity was increased by more than an order of magnitude towards both peptide and ester substrates. This 'superactivation' could also be reversed by treatment with hydroxylamine. Similar activity increases were observed when thermolysin and the neutral metalloproteinases from *Bacillus subtilis, Bacillus megaterium* and *Aeromonas proteolytica* were modified with *N*-hydroxysuccinimide esters of acyl amino acid derivatives (Blumberg and Vallee, 1975). Tyrosine-110 was identified by Blumberg (1979) as the site of modification with *N*-acetyl-*p*-(2,4-dinitroanilino)-L-phenylalanine *N*-hydroxysuccinimide ester, one of the 'superactivating' agents. The cause for the 'superactivation' is not clear, but Blumberg (1979) suggests that acylation of tyrosine-110 affects the interactions between the substrates and the peptide chain from asparagine-112 to tryptophan-115 and that the acyl moieties attached to the tyrosine hydroxyl have the same effect as a substrate phenyl ring in the P_1 position of a substrate (it should be stressed that only substrates with glycine, alanine and other small side chains in P_1 show the activation phenomenon).

1.5.2.4 Mechanism of action of thermolysin

As in carboxypeptidase A, four functional components appear to be involved in the mechanism of action of thermolysin, the zinc ion, glutamate-143, histidine-231 and a water molecule. Unlike in carboxypeptidase, however, all reactions catalysed can unambiguously be described by a single mechanism.

The three-dimensional structures of native thermolysin and of its inhibitor complexes show that the zinc ion is in a tetravalent coordination in contrast to carboxypeptidase A. Based on these structures and on kinetic and chemical modification studies, Pangburn and Walsh (1975) and Kester and Matthews (1977*a*) proposed a mechanism which can be summarised as follows. The substrate binds into the specificity site(s) with the carbonyl group of the scissile bond displacing a water molecule from the zinc ion which, by becoming liganded to it, polarises the peptide bond. Glutamate-143 then promotes a nucleophilic attack by a water molecule upon the scissile bond. Histidine-231 simultaneously or subsequently donates a proton to the amide nitrogen to form a tetrahedral intermediate which is stabilised by hydrogen bonds to glutamate-143 and histidine-231 and by interaction with the zinc. The last step is the breaking of the C–N bond to yield the products. This mechanism is very similar to that proposed by Lipscomb *et al.* (1968) for carboxypeptidase A where glutamate-270 is the equivalent of glutamate-143 and tyrosine-248 plays the role suggested for histidine-231 in thermolysin. Kester and Matthews (1977*a*) also discuss an alternative mechanism in which glutamate-143 acts as the nucleophile rather than as a general base forming a tetrahedral intermediate on the glutamate carboxyl. The transfer of a proton from histidine-231 would then lead to a covalent acyl intermediate which would be hydrolysed to products by a water molecule. This is similar to the alternative mechanism proposed for carboxypeptidase A discussed in section 1.2.2.3, which is thought to be a probable mechanism for ester catalysis by carboxypeptidase A. Kester and Matthews (1977*a*), however, tend to rule out a covalent mechanism on the basis of steric restrictions which are discussed in full in their paper. A refinement and modification of this mechanism has recently been proposed by Holmes and Matthews (1981). The X-ray analysis of complexes of thermolysin with inhibitors carrying a hydroxamic acid function shows that in these complexes the zinc is pentacoordinated, with the two nonprotein ligands coming from the hydroxamate group. This suggests the possibility that the catalytic events also involve a pentacoordinate rather than a tetracoordinate zinc. A mechanism proposed on this basis is shown in figure 1.13. The incoming substrate would not displace the water molecule from the

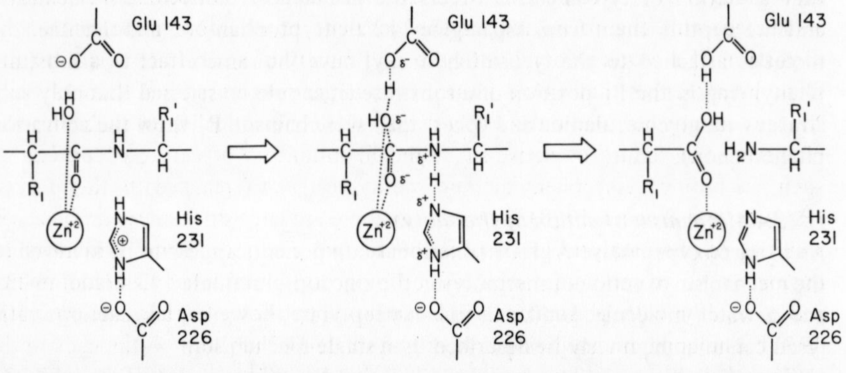

Figure 1.13 Schematic illustration of the mechanism of action of thermolysin (with permission from Holmes and Matthews, 1981; copyright 1981, American Chemical Society).

zinc, which would accept the carbonyl group of the scissile peptide bond as an additional ligand. The water molecule would probably move into line with glutamate-143 as indicated by model-building studies. This arrangement would lead to a mechanism which would involve both a general base catalysis and a direct participation of the zinc. Zinc would thus play two roles: it would polarise the peptide bond and also help in the correct alignment of the carbonyl bond and the attacking nucleophile. The remaining steps would be the same as those suggested previously. The pentacoordinate mechanism is supported by experiments in which the zinc has been replaced by other metals. Whereas the substitution of the zinc by cobalt (Holmquist and Vallee, 1974; Bigbee and Dahlquist, 1974) leads to an enzyme that is at least as active as the unmodified one, substitution by copper, cadmium or mercury gives inactive enzymes (Holmquist and Vallee, 1974). Zinc and cobalt can change their coordination state more readily than other metals (Vallee and Williams, 1968) and more often assume a trigonal bipyramidal coordination.

The possibility of a pentacoordinate zinc in zinc-containing metalloenzymes has been suggested by Rossmann and Argos (1978) on the basis of a structural comparison of carbonic anhydrase, alcohol dehydrogenase, carboxypeptidase A and thermolysin. Model studies by Woolley (1975) indicated that a zinc-bound water molecule could form a good nucleophile. On the basis of studies of the pH and temperature dependences of kinetic parameters of good substrates for thermolysin, Kunugi *et al.* (1982) suggested, independently of Holmes and Matthews (1981), that the zinc would be pentacoordinated in the enzyme-substrate complex and in the transition state.

Holmes *et al.* (1983) provide the strongest evidence against the involvement of a covalent intermediate in thermolysin catalysis in their recent X-ray analysis of thermolysin covalently labelled with the inhibitor $ClCH_2CO$–D,L–(N–OH)-Leu-OCH_3 discussed earlier. This inhibitor alkylates glutamate-143 and thus at first sight could be considered an analogue of a covalent intermediate. However, as Holmes *et al.* (1983) show, there are considerable conformational changes associated with the formation of the covalent bond. More importantly, the formation of a covalent bond with the acyl moiety of a substrate would require steric interferences that would induce even larger conformational changes. Such changes cannot be ruled out *a priori* if they were energetically not unfavourable. However, the failure of active-site directed inhibitors of carboxypeptidase A, such as bromoacetyl phenylalanine methyl ester, to react with thermolysin (Rasnick and Powers, 1978) suggests that there are large differences in the accessibilities of the catalytic glutamates in the two enzymes. These inhibitors are analogous to compounds such as bromoacetyl-methylphenylalanine, which reacts readily with glutamate-270 in carboxypeptidase A (Hass and Neurath, 1971). The apparent inaccessibility of glutamate-143 in thermolysin leads to the inference that the non-covalent binding of the inhibitor is insufficient for the induction of a conformational change needed to render the glutamate accessible to the alkylating function.

In a recent paper Hangauer *et al.* (1984) modify and extend the proposed mechanism for the action of thermolysin (Kester and Matthews, 1977*a*; Holmes and Matthews, 1981). With the use of extensive interactive computer graphics on putative enzyme–substrate and enzyme–inhibitor complexes, Hangauer *et al.* (1984) construct models that support the major features of the mechanism proposed earlier and provide many additional insights.

1.5.2.5 Thermostability of thermolysin

One of the interesting features of thermolysin is its thermostability. Upon heating to 80°C for 1 h it retains half its activity (Endo, 1962). Pangburn *et al.* (1976) showed that thermolysin had a half-life of 15 min at 84°C whereas the homologous neutral metalloproteinase from *Bacillus subtilis* had a half-life of 15 min at only 59°C. Calcium ions were implicated in the thermal stability because in the presence of EDTA thermolysin is unstable below 40°C (Endo, 1962; Feder *et al.*, 1971; Fontana *et al.*, 1976). Vourdouw and Roche (1975) investigated the temperature dependence of inactivation as a function of calcium ion concentration and found that the native conformation is protected from inactivation by only a single calcium ion and that this calcium ion was remote from the active site. They suggested that it might be the calcium in site 3 (see figure 1.11). This accords well with the observation of Pangburn *et al.* (1976), who found that the much less thermostable neutral metalloproteinase from the mesophilic *Bacillus subtilis* also bound calcium, but only two atoms per molecule. A comparison of the partial sequences available indicated that the two calcium ions in the double site near the zinc were most probably those that were also present in the *B. subtilis* enzyme, but that site 4 in the C-terminal part of the molecule did not exist because aspartic acid-200 was replaced by a proline. The catalytic zinc does not contribute to the thermal stability, which is the same for the holo- and the apoenzymes (Holmquist and Vallee, 1974).

The question of which factors are the determinants of the thermostability has been discussed most recently by Holmes and Matthews (1982). As these authors point out, thermolysin is one of the few structures available at present that can contribute towards our understanding of the mechanisms of thermostability. From the discussion in the preceding paragraph it is clear that calcium is only one factor. The single calcium is not sufficient to explain the difference between thermolysin and the homologous proteinase from *B. subtilis*. Rather, as Matthews *et al.* (1974) and Weaver *et al.* (1976) suggested earlier, the thermostability arises from a combination of small differences in hydrophobic character, hydrogen bonding, ionic interactions and metal binding. It is interesting in this context to mention the recent studies of the temperature-sensitive mutants of phage lysozyme (Gruetter *et al.*, 1979; Schellman *et al.*, 1981). These studies show the extreme sensitivity of the thermal stability to replacement of only one amino acid in the protein, even when no changes in charge and only small changes in the bulk of the side chain are involved in the substitution. Holmes and Matthews (1982) conclude that hydrophobic interactions make the major contribution to the stability of a protein and that there is no requirement for special structural features in thermostable proteins like thermolysin.

1.5.3 Thermolysin-related neutral proteinase from *Bacillus subtilis*

The neutral proteinase from *Bacillus subtilis* is also a zinc metalloproteinase. It is homologous with thermolysin, as shown by a comparison of partial sequences (Pangburn *et al.*, 1976). Of the 171 amino acids that have been determined for the *B. subtilis* enzyme, 49 per cent are identical with the corresponding amino acids of thermolysin. The enzymes are also similar in other respects. The primary specificity of both enzymes is directed towards hydrophobic residues in the P_1'

position (Morihara, 1974), their action towards peptide and ester substrates is very similar (Holmquist and Vallee, 1976), and they are inhibited by the same reversible and irreversible inhibitors (Rasnick and Powers, 1978; Nishino and Powers, 1979). Isowa *et al.* (1978) also showed that *B. subtilis* neutral proteinase, like thermolysin, synthesises peptide bonds under suitable conditions. The main difference between them is the thermostability, which has been mentioned previously.

1.6 AMINOPEPTIDASES

An enzyme from pig intestinal mucosa that could cleave peptides with leucine in the N-terminal position was first described by Linderstrom-Lang (1929). The enzyme was called 'leucine aminopeptidase', a name that is still commonly used at present, although it was shown early on that the enzyme has rather broad specificity (Smith and Hill, 1960). Since then, aminopeptidase activity has been demonstrated in most tissues and cells from mammalian to bacterial sources and many of the enzymes have been isolated and at least partially characterised. A comprehensive review of aminopeptidases was published by Delange and Smith (1971). Since then there has been an ever increasing interest in these enzymes as shown by the voluminous literature, and a comprehensive review is far beyond the scope of this chapter. I shall confine myself to a discussion of recent work which is leading to an understanding of the complex structures of these enzymes.

The aminopeptidases are the only proteolytic enzymes which occur mostly as large oligomers with molecular masses in the hundred thousand daltons range and which are made up of as many as twelve subunits. For a long time, progress on their molecular characterisation was slow because of their size, because of the instability of many of them, and because they are membrane proteins (e.g. microsomal aminopeptidase). Furthermore, many are glycoproteins and occur in multiple forms (Himmelhoch and Peterson, 1968). With very few exceptions (e.g. dipeptidyl aminopeptidase) they are metalloproteinases or at least metal-activated proteinases (aminopeptidase P; Yaron and Berger, 1970). Little is known about their physiological function. A partial list of some of the enzymes that have been best characterised in molecular terms and as regards their metal requirements is given in table 1.3.

1.6.1 Bovine eye lens leucine aminopeptidase

The aminopeptidase originally obtained from the bovine eye lens has probably been the most extensively studied (Carpenter and Vahl, 1973; Lasch *et al.*, 1973). It is one of the classical 'leucine aminopeptidases' and occurs in all tissues and body fluids of vertebrates. Its purification and many of its properties have been reviewed by Hanson and Frohne (1976). Although it is still commonly called leucine aminopeptidase, a more appropriate name is cytosolic amino-

Table 1.3 Aminopeptidases[a]

Enzyme name	Molecular mass (Da)	Number and size of subunits	Metal; type and number per subunit	Reference
Aminopeptidase, cytosolic	310 146	6; 51 640 Da	Zn^{2+}; 2	Hanson and Frohne, 1976
Aminopeptidase, microsomal	280 000[b]	2[b], 140 000 Da	Zn^{2+}; 2	Sjostrom *et al.*, 1978
Alanine aminopeptidase, human liver	235 000	2; 118 000 Da	Zn^{2+}; 2	Starnes and Behal, 1974
Thermophilic aminopeptidase I[c]	395 000	12; (~30 000 Da)	Co^{2+}, Zn^{2+}; 2	Roncari *et al.*, 1976
Aeromonas proteolytica aminopeptidase	29 500	1	Zn^{2+}; 2	Prescott and Wilkes, 1976
Clostridium histolyticum aminopeptidase	340 000	6; 58 000	Mn^{2+}, Co^{2}; ?[d]	Kessler and Yaron, 1976
Aminopeptidase P from *E. coli*	230 000	?	Mn^{2+}[e]	Yaron and Berger, 1970

[a] Only reasonably well characterised enzymes have been included.
[b] May vary for purified enzymes depending on purification method.
[c] From *Bacillus stearothermophilus*.
[d] Zinc ions inhibit this enzyme.
[e] Metal needed for activity has relatively low affinity, but excess metal ions inhibit.

peptidase, a name suggested by the Nomenclature Committee of the International Union of Biochemistry (Bielka *et al.*, 1978) to indicate its widespread occurrence and to distinguish it from the microsomal aminopeptidase from which it differs in specificity and immunological behaviour (Hanson and Frohne, 1976).

Primary structure Cytosolic aminopeptidase from the eye lens is composed of six identical subunits. The recently determined amino acid sequence (figure 1.14, from Cuypers *et al.*, 1982*a*) shows that the subunit is a chain of 478 amino acids with a molecular mass of 51 691 Da. Determination of the sequence confirmed the absence of the disulphide bridges suggested by Melbye and Carpenter (1971) and showed that there are seven free thiol groups per subunit. Their reactivity has been determined by Cuypers *et al.* (1982*b*); in the native enzyme, which contains two zinc atoms per subunit, and in an enzyme that contains one zinc and one manganese ion, only cysteine-344 reacts readily with iodoacetic acid; in an enzyme that contains two cobalt ions, only cysteine-412 reacts. In an enzyme form with one zinc and one magnesium, two cysteine residues react, while in the presence of mercaptoethanol, the apoenzyme has up to five cysteines that react, albeit at different rates. All seven cysteine residues react with iodo-

Figure 1.14 Amino acid sequence of bovine eye lens aminopeptidase according to Cuypers *et al.* (1982*a*).

acetic acid in 8 M urea and in the absence of reducing agents. The authors conclude that cysteine-344 is not essential for activity, and that one or more of the cysteines may be involved as metal ligands in the metal binding sites, although there is as yet no conclusive evidence for this.

Six stretches of 25 amino acids each have been compared with the sequences in the data base of the *Atlas of Protein Sequence and Structure* (Dayhoff, 1976). No evidence for any homology with existing sequences was found and Cuypers *et al.* (1982*a*) conclude that cytosolic aminopeptidase represents a new superfamily of proteins.

Three-dimensional structure The X-ray analysis of lens cytosolic aminopeptidase is in progress. Suitable crystals of the enzyme have been described by Jurnak *et al.* (1977). They are hexagonal, space group $P6_322$ with unit cell dimension of a = 132 Å and b = 122 Å. The asymmetric unit consists of one chemical subunit of molecular mass 51 000 Da.

Electron microscopic studies indicate that the subunits are bilobal, with a major and a minor lobe (Taylor *et al.*, 1979). The subunits are located at the vertices of a triangular prism.

Aminopeptidase binds two zinc ions per subunit (Carpenter and Vahl, 1973). One of them is essential for activity and is located in the 'structural' site. The zinc at the other site, called the 'activation' site, can readily be exchanged with other ions, such as magnesium or manganese, with a significant enhancement of the activity. Both zinc ions can be replaced by cobalt to yield an active cobalt enzyme. A recent nuclear magnetic resonance study with a zinc-manganese enzyme shows that the second metal binding site is located close to the carbonyl oxygen of a competitive inhibitor bound to the enzyme and hence may be close to a carbonyl oxygen of a substrate in the active site (Taylor *et al.*, 1982). It is not clear, however, what the role of this second metal ion is. Fittkau *et al.* (1979), using electron spin resonance, found that the spin-labelled inhibitor

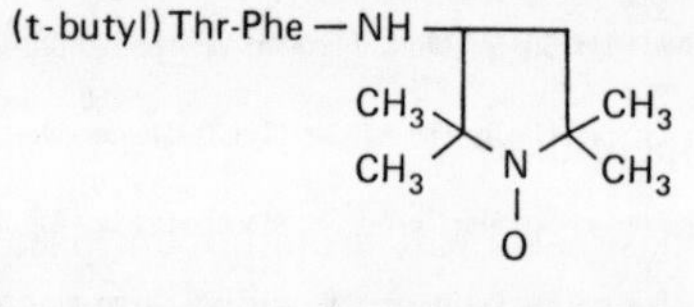

(t-butyl) Thr-Phe — NH
CH_3 CH_3
CH_3 N CH_3
O

was bound in two different sites on each subunit. The second inhibitor site is presumably not associated with the second metal binding site in view of the close proximity of the latter to the structural site.

Enzymatic properties These have been reviewed by Delange and Smith (1971) and Hanson and Frohne (1976). An extensive bibliography is also contained in a book by McDonald and Barrett (1984).

The enzyme has an absolute requirement for a free α-amino group (in the unprotonated form) and for the L-form of the N-terminal amino acid (Smith and

Hill, 1960). Peptides with D-amino acids or proline in the penultimate position are hydrolysed hardly at all (Smith and Spackman, 1955). The primary specificity shows a preference for amino acids with large hydrophobic side chains, and, although all residues including charged residues or proline are cleaved, the latter are poor substrates (Smith and Hill, 1960).

Aminopeptidase also has esterase activity and hydrolyses esters of leucine and tryptophan especially readily (Kleine and Hanson, 1961).

Cytosolic aminopeptidase is inhibited by metal chelating agents and also by some chloromethylketones of amino acids and peptides which carry a free amino group in accord with the specificity requirements (Fittkau *et al.*, 1974). It is also inhibited by the naturally occurring inhibitor bestatin, 3-amino-2-hydroxy-4-phenylbutanoyl-L-leucine ($K_I = 2 \times 10^{-8}$ M; Suda *et al.*, 1976).

Mechanism of action The mechanism of action of cytosolic aminopeptidase is as yet only poorly understood. Thus, there is evidence both for and against the involvement of covalent intermediates. Hanson and Lasch (1967) found that the enzyme catalysed transpeptidation reactions. This was made use of by Antonov *et al.* (1979), who showed that if the transpeptidation of leucine amide to give Leu–Leu amide were carried out in $H_2{}^{18}O$, about 50 per cent of the oxygen in the CO group of the Leu–Leu peptide bond originated from the ^{18}O of water. They concluded that the transpeptidation proceeded by a general base mechanism and did not involve a covalent intermediate. Similarly, preliminary studies with cytosolic kidney aminopeptidase at low temperature in methanol-water solvents provided no evidence for a covalent intermediate (Lin and Van Wart, 1982). On the other hand, Andersson *et al.* (1982) showed that L-leucinal was a very tightly binding inhibitor and could be considered a transition state analogue. The tight binding ($K_I = 6 \times 10^{-8}$ M, compared with the $K_m = 7.7 \times 10^{-4}$ M for the substrate L-leucine *p*-nitroanilide) suggested that the inhibitor was bound covalently in a complex that was analogous to unstable intermediates during substrate hydrolysis.

1.6.2 Microsomal aminopeptidase

Microsomal aminopeptidase, also called aminopeptidase M or N, is a membrane-bound enzyme and, like the cytosolic aminopeptidase, occurs widely in vertebrate tissues. The enzymes from the brush border of the small intestine (Maroux *et al.*, 1973) and from the brush border of the kidney (Pfleiderer and Celliers, 1963) have been purified and characterised. Short reviews by Delange and Smith (1971) and Pfleiderer (1970) have been published. Microsomal aminopeptidase clearly differs in its specificity and its immunological behaviour, as well as in molecular size (table 1.3) from the cytosolic enzyme.

The molecular mass of microsomal aminopeptidase as isolated is usually 280 000 Da (Auricchio and Bruni, 1964) although when isolated from the kidney with Triton-X it can have a mass of 10^6 Da. The molecular mass of the monomeric subunit is 140 000 Da (Wacker *et al.*, 1976). It is a glycoprotein with about 20 per cent carbohydrate and contains two zinc atoms (Wacker *et al.*, 1971). Although the enzymes from both intestinal mucosa and kidney have often been isolated in forms that have a subunit structure consisting of three different chains of different length, recent work has shown that these are artifacts due to autolysis or trypsin-like cleavage (Sjostrom *et al.*, 1978; Benajiba and Maroux, 1981). Microsomal aminopeptidase is attached to the brush border membrane as a dimer through the hydrophobic N-terminal tail (Feracci *et al.*, 1982).

1.6.3 Alanine aminopeptidase from human liver

The alanine aminopeptidase from human liver has been purified by Starnes and Behal (1974). It is a glycoprotein with 17 per cent carbohydrate and occurs as a dimer of two identical subunits of molecular mass 118 000 Da. It contains one zinc ion per subunit and is activated by a low affinity cobalt ion. This enzyme differs from cytosolic aminopeptidase in that it is more active on alanine-containing substrates than on leucine-containing substrates (Little *et al.*, 1976). It occurs in blood in the form of several isoenzymes which differ presumably in their carbohydrate composition. Starnes *et al.* (1982) recently suggested that its physiological function may be the formation of bradykinin from Met-Lys-bradykinin by the stepwise release of methionine and lysine. Further degradation is prevented by the presence of the proline residue in the penultimate position (from the N-terminal) of bradykinin (Arg–Pro–Pro–Gly–Phe–Ser–Pro–Phe–Arg).

1.6.4 Thermophilic aminopeptidase

Bacillus stearothermophilus and related strains produce three types of aminopeptidases, one of which has been characterised (Roncari *et al.*, 1976). Thermophilic aminopeptidase I, as it has been designated, is a complex oligomer (M_r = 400 000) that consists of 12 subunits, which are of two types, α and β. They differ significantly in electrophoretic behaviour and in their specificities, but they are homologous. The first 30 amino acids of the N-terminal sequences of the α and β forms are 67 per cent identical or chemically similar (Stoll *et al.*, 1973). The two types of subunit can combine to give dodecamer hybrids containing different ratios of the α and β subunits (e.g. $\alpha_{10}\beta_2$, $\alpha_8\beta_4$ and $\alpha_6\beta_6$).

Thermophilic aminopeptidase is a metalloenzyme which contains cobalt after isolation, possibly because one of the purification steps involves a heat treatment in the presence of cobalt. It is not known which metal is at the active site under physiological conditions. The metal ions, of which there appear to be two per subunit (Roncari *et al.*, 1976), are essential for activity. The cobalt enzyme is considerably more active than the zinc enzyme.

1.6.5 *Aeromonas proteolytica* aminopeptidase

Aeromonas aminopeptidase has been isolated as a heat-stable enzyme from the marine bacterium *Aeromonas proteolytica* (Prescott and Wilkes, 1966). It is a monomeric metalloenzyme (29 500 Da) with two bound zinc ions per chain and is the only single chain, low molecular weight aminopeptidase characterised that has so far been described. Two half-cystine residues are present as a disulphide bridge. There is no free thiol group.

Like all the other aminopeptidases mentioned, *Aeromonas* aminopeptidase has a broad specificity but does not cleave at acidic amino acid residues (Wilkes *et al.*, 1973). Some of its properties have been reviewed by Prescott and Wilkes (1976).

1.6.6 *Clostridium histolyticum* aminopeptidase

The extracellular aminopeptidase from *Clostridium histolyticum* is another aminopeptidase which deserves a brief comment because it, too, illustrates the wide variety in the types of aminopeptidase that occur. This enzyme is oligomeric (34 000 Da) and consists of six subunits. Its purification and properties have been reviewed by Kessler and Yaron (1976). It is a metalloproteinase which is most active in the cobalt or manganese form. What distinguishes it from the other aminopeptidases is that zinc not only fails to activate the apoenzyme but in fact inhibits the active manganese enzyme.

1.7 Collagenases

Enzymes with the ability to cleave intact collagen molecules have been demonstrated in many tissues and cells from both eucaryotic and procaryotic organisms. A great deal of attention has been paid to the collagenases because of their importance in development (Grillo and Gross, 1967) and their suspected involvement in certain disease states, and because of other possible physiological functions. Many aspects of collagenases have been reviewed by Harper (1980). At the present time, however, our knowledge of their molecular properties is rather scant. One reason is that collagenases are not normally present in significant concentrations in mammalian tissues. Even when present they are difficult to detect. They are, however, found in the media from cell cultures of connective tissue and also in tissues that undergo rapid resorption, such as the involuting uterus (Cawston and Murphy, 1981). Most, if not all, *mammalian* collagenases appear to be metalloproteinases (Seltzer *et al.*, 1977), but the collagenolytic enzymes from microorganisms and from invertebrate sources seem to belong to any one of three major classes of proteinases: the serine proteinases, thiol proteinases and metalloproteinases (Keil, 1979).

1.7.1 Mammalian collagenases

Mammalian collagenases have recently been reviewed by Cawston and Murphy (1981). They are zinc enzymes and usually also require calcium for activation. They are most easily purified from the culture media of a variety of connective tissue cells, such as fibroblasts and chondrocytes (Werb and Burleigh, 1974; Cawston and Tyler, 1979).

Collagenases are usually present in a latent form and require activation. The nature of the latency of collagenases is not yet understood. Although a number of methods for activation have been described (short treatment with trypsin or trypsin-related enzymes; addition of mercurials or chaotropic agents, or prolonged incubation at 37°C), the mechanism of activation is unknown. Latent collagenases could be proenzymes that require activation by a proteolytic cleavage, or they could be present in complexes with inhibitors that can be removed by proteolysis or by adding chemical agents. As yet no latent form of collagenase has been adequately characterised, although some, like the one from human skin fibroblasts (Stricklin *et al.*, 1977), have been obtained in pure form. An extensive discussion of the latency problem is found in Murphy and Sellers (1980).

A wide range of molecular sizes for mammalian collagenase have been reported, but the most recent estimates vary from 33 000 to 60 000 Da (Cawston and Murphy, 1981). The lower molecular size enzyme may originate from a 60 000 Da protein (Sakamoto *et al.*, 1978; Cawston and Tyler, 1979).

Mammalian collagenases have a very restricted specificity. They cleave the collagen triple helix in the area of the 'weak' helix which is located at a point about three-quarters of the length of the helix from the N-terminal residue. The susceptible bond is Gly–Leu or Gly–Ile (Harper, 1980; Cawston and Murphy, 1981). Contrary to earlier reports that collagenases do not act on non-collagen proteins, recent studies have shown that proteins like casein undergo limited cleavage, which is, however, not easily detected because the cleavage products are insoluble in trichloracetic acid (Murphy and Sellers, 1980). Type I collagen is most rapidly cleaved, followed by type III and type II.

1.7.2 Bacterial collagenases

Because of its ready availability, much of the early work on collagenases was carried out with an enzyme from *Clostridium histolyticum.* One of the collagenolytic enzymes from this organism is a zinc-metalloproteinase (Seifter and Harper, 1971). Another zinc enzyme has been isolated from *Achromobacter iophagus* (Welton and Woods, 1975). Both enzymes, along with collagenolytic serine proteinases from a fungus and an insect, are being studied at present in the laboratory of Keil (1979).

Most of the earlier studies with *C. histolyticum* were made with commercially available or similar preparations which, largely because of autolysis, are very heterogeneous (Keil, 1980). A homogeneous preparation of high specific activity has now been obtained (Emod *et al.*, 1981). Keil (1979, 1980) suggests that the multiple forms of *C. histolyticum* collagenase (see, for example, Grant and Alburn, 1959) were mostly active autolysis products, although clostripain, a collagenolytic thiol proteinase, is obviously a distinct and different enzyme (Emod and Keil, 1977). *Clostridium histolyticum* collagenase has a molecular mass of 68 000 Da and has so far resisted dissociation into subunits, in contrast to *A. iophagus* collagenase whose molecular mass is 70 000 Da but which clearly dissociates into two identical subunits of 35 000 Da (Keil-Dlouha and Keil, 1978). However, Keil (1980) suggests that both enzymes may belong to the thermolysin superfamily and speculates that the *C. histolyticum* enzyme may eventually be found to consist of two subunits.

The collagenase from *A. iophagus* has also been obtained in large amounts in homogeneous form (Lecroisey *et al.*, 1975; Keil-Dlouha, 1976). It, too, is a zinc metalloproteinase. The reasons for Keil's suggestion that this enzyme belongs to the thermolysin superfamily are the following. *Achromobacter iophagus* collagenase is a zinc metalloproteinase; the molecular size of its subunit is about the same as that of thermolysin (35 000 Da); it is inhibited by phosphoramidon; and chemical modification studies implicate one tyrosine, two carboxyl groups and two histidines in the activity (Keil, 1979, 1980). However, at this time, conclusive evidence is lacking. Also, it is rather puzzling to find that the enzyme contains only one zinc per two subunits (Keil-Dlouha, 1976).

Earlier reports indicated that the degradation of collagen by microbial enzymes differed in its mechanism from that by the mammalian enzymes. Thus, Stark and Kuhn (1968) found that *C. histolyticum* collagenase degrades collagen fairly extensively from the N- and C-terminals. However, studies with homogeneous collagenase from *A. iophagus* and with purified collagenase from *C. histolyticum* show a much more limited cleavage. The *A. iophagus* enzyme attacks bonds in the same 'weak' region of the collagen helix as the mammalian collagenases, namely about three-quarters down the collagen helix from the N-terminal (Lecroisey and Keil, 1979). The enzyme from *C. histolyticum* also shows much more limited initial cleavage, mostly in the central region of the helical collagen molecule (Lecroisey and Keil, 1979), but these authors suggest that if pure collagenase had been used, the cleavage sites might have been even more restricted and similar to those of the *A. iophagus* enzyme. Nevertheless, the peptide bond specificities of the two enzymes are different, as shown by their action on synthetic peptides and on the peptide chain of β-casein (Gilles and Keil, 1976). β-Casein, which is 209 amino acids long (Ribadeau-Dumas *et al.*, 1972), is cleaved in two places by the *C. histolyticum* collagenase (at a . . . Ser$\overset{\downarrow}{-}$Leu–Val–Tyr. . . and a. . .Ser$\overset{\downarrow}{-}$Leu–Thr–Leu. . . bond) and in four places by the *A. iophagus* enzyme (at $X\overset{\downarrow}{-}$Gly–Pro . . . , three bonds, and at $X\overset{\downarrow}{-}$Ala-Pro. . . , one bond). These experiments also show that the microbial enzymes, like the mammalian ones, cleave a limited number of bonds in non-collagen proteins, although this activity is not detected when standard assays with protein substrates are performed.

1.8 CALCIUM-ACTIVATED PROTEINASES FROM MAMMALIAN TISSUES

In recent years a great deal of interest has been shown in the calcium-activated proteinases whose presence has been demonstrated in a large number of different tissues and cells (reviewed by Waxman, 1981, and Murachi, 1983). Although these enzymes are not metalloproteinases in the strict sense, but belong to the thiol-proteinase family (Murachi *et al.*, 1981), they nevertheless deserve a brief mention in this chapter because they appear to have an absolute requirement for calcium ions for their activity. There are two groups of these enzymes; one is half-activated by calcium concentrations of approximately 1 mM, whereas the other, more recently discovered, group (Mellgren, 1980) requires only about 40 μM calcium for half-activation. No metal ion other than calcium can activate the enzymes, but in the presence of calcium at levels below those needed for activity, the addition of other metal ions generates activity (Suzuki and Tsuji, 1982). The role of the metal ion in the catalysis is not clear.

The question of the molecular structure is also still open. Some methods of purification yield heterodimers for both forms of the calcium-activated proteinases consisting of two subunits, 80 000 Da and 30 000 Da (e.g. Dayton *et al.*, 1976) whereas other preparations yield only an 80 000 Da protein (Ishiura *et al.*, 1978). The 'low' and the 'high' calcium-requiring proteinases appear to be closely related. Suzuki *et al.* (1981) and Mellgren *et al.* (1982) have shown that the high calcium-requiring form can be converted to the low calcium-requiring form by limited proteolysis. On the other hand, Wheelock (1982) found that whereas the 30 000 Da subunits are probably identical for the two forms, the 80 000 Da subunits are substantially different as shown by peptide mapping and immunological techniques. A similar conclusion regarding the large subunit of low calcium-requiring proteinase from pig heart muscle was reached very recently by Otsuka and Tanaka (1983), who show that the large subunit cannot arise from the large subunit of the high calcium-requiring proteinase.

Various names have been suggested for the calcium-activated mammalian proteinases, such as calcium-activated factor (CAF), calcium-activated neutral protease (mCANP and μCANP), kinase-activating factor (KAF) and calpain I and II (Murachi *et al.*, 1981).

Waxman and Krebs (1978) showed that the calcium-activated proteinase in all tissues is accompanied by a high-molecular mass inhibitor (270 000 Da). The name 'calpastatin' has been suggested for this protein (Murachi *et al.*, 1981).

1.9 COMPARISON OF MECHANISM OF ACTION OF PROTEINASES

1.9.1 Comparison of carboxypeptidase A and thermolysin

In the discussions on carboxypeptidase A in section 1.2 and on thermolysin in section 1.5 of this chapter, some of the similarities and differences between the two enzymes have already been mentioned. In this section, I wish to give a brief summary of the major points. Most of the information will be taken from the paper by Kester and Matthews (1977*b*), who describe a detailed structural comparison of the two enzymes.

A comparison of the amino acid sequences (Bradshaw *et al.*, 1969; Titani *et al.*, 1975) and the overall three-dimensional structures shows that carboxypeptidase A and thermolysin are very different and therefore by implication have different evolutionary origins. When, however, the active sites and especially the zinc binding sites are compared, some striking similarities emerge. In each case the zinc ion is bound to the enzyme through two histidine residues and one glutamic acid residue. These residues, however, are in different positions relative to each other and the zinc is pentacoordinate in native carboxypeptidase A and tetracoordinate in native thermolysin, with the single non-protein ligand being provided by a water molecule (figure 1.15). In each enzyme there is a glutamic acid residue that acts as a general base (or in the case of ester hydrolysis by carboxypeptidase A as a nucleophile). They occupy similar positions relative to the zinc, but differ substantially in the proposed distance from the scissile bond with glutamate-270 of carboxypeptidase A being considerably closer than glutamate-145 of thermolysin.

In each enzyme there is also an arginine residue in the active site which in carboxypeptidase A acts as one of the determinants of its specificity as an exopeptidase (arginine-145) whereas arginine-203 in thermolysin is not in a position where it would allow the binding of either a free carboxyl or an extended peptide chain. The function of arginine-203 in thermolysin is not known. The major side chain specificity of both enzymes is determined by hydrophobic pockets in subsite S_1' which are similar, except that the one in carboxypeptidase is slightly larger.

The putative donors of a proton to the leaving group of the substrate, tyrosine-248 in carboxypeptidase and histidine-231 in thermolysin, are in quite different locations (figure 1.15) with the hydroxyl group of tyrosine and the ϵ-nitrogen of histidine being 4.3 Å apart in the superimposed structures. It is not clear how this difference affects their function in the mechanism of action. In any case, the role of tyrosine-248 and histidine-231 as proton donors is not firmly established and the possibility that the proton originates from a water molecule has not been ruled out for either enzyme.

Other differences which have already been discussed include the mechanism of ester hydrolysis, the differences in the magnitude of the conformational changes that are observed on binding of substrate analogues, and the binding of extended substrate chains.

The resemblance of the active sites and the differences in the primary and tertiary structures indicate strongly that the similarities arise from a convergent evolution which has its counterpart in the well-documented case of the serine proteinases. The trypsin-related and the subtilisin-related superfamilies, respectively, also show extensive similarities in the active sites but no evidence for evolutionary homology (see, for example, Robertus *et al.*, 1972).

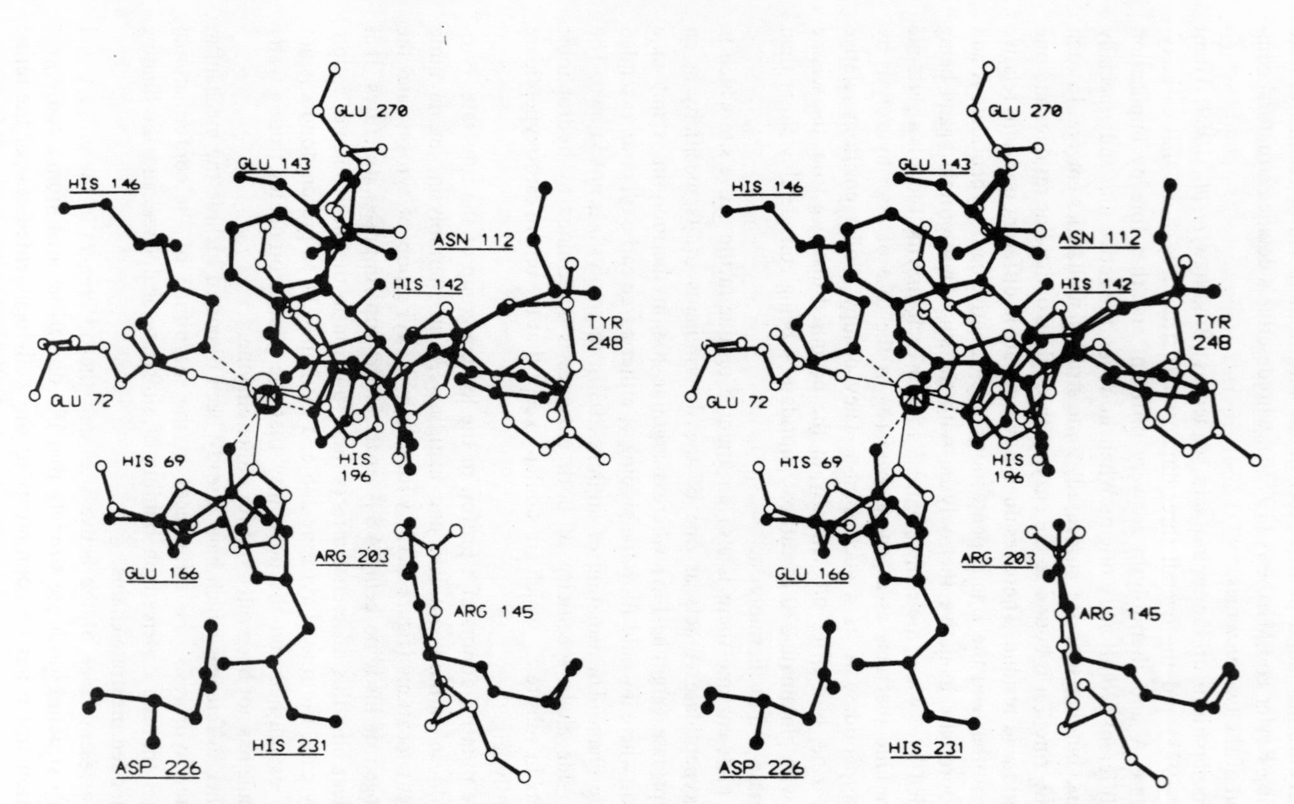

Figure 1.15 Superposition of the active site of thermolysin (filled circles) on the active site of carboxypeptidase A (open circles) with a bound dipeptide. Residue names of thermolysin are underlined (with permission from Kester and Matthews, 1977*b*).

A comparison of the geometries of the zinc at the active site of the zinc proteinases has also been made with carbonic anhydrase and liver alcohol dehydrogenase (Argos *et al.*, 1978). In the dehydrogenases the zinc is also coordinated by three groups on the protein, with a water molecule providing a fourth ligand. In all four enzymes there is a group (glutamic acid in the proteinases) that abstracts a proton from water during catalysis and occupies a very similar spatial orientation to the zinc atom.

1.9.2 Comparison of metalloproteinases with serine, thiol and aspartic proteinases

The basic mechanism by which peptide bonds are cleaved by proteolytic enzymes is the same. It includes an electrophilic component which increases the polarisation of the –C=O bond of the scissile bond, some form of a nucleophile, which forms a tetrahedral intermediate with the carbonyl carbon atom, and a proton donor, which puts a proton on the nitrogen of the leaving group (James, 1980). However, the amino acid residues and other components that provide these three functions differ widely between the different enzymes.

In this chapter I wish to highlight briefly the present state of our understanding of the mechanism of action of the four major classes of proteinases. It is very obvious by now that most of our insights into enzyme mechanisms come from the details provided by three-dimensional structures determined by X-ray crystallography and most recently by neutron crystallography (for the monoisopropyl-phosphate ester of trypsin: Kossiakoff and Spencer, 1980). The serine proteinases form an excellent example.

The *serine proteinases* are by far the best understood of all classes of enzymes. This is especially true for the serine proteinases of the trypsin superfamily. The subtilisin superfamily is not so well understood, and much less is known about the serine carboxypeptidases because there are no three-dimensional structures available as yet. The depth of our knowledge on the trypsin superfamily comes from the detailed structures of several native mammalian and bacterial enzymes, and of their complexes with molecules representing analogues of the various intermediate stages of the catalytic pathway. An excellent summary based on X-ray crystallographic and nuclear magnetic resonance studies has been published recently (Steitz and Shulman, 1982).

Briefly, the reaction pathway for substrate hydrolysis proceeds via a Michaelis complex, a tetrahedral intermediate, an acyl enzyme, another tetrahedral intermediate and an enzyme–product complex. Good analogues of the Michaelis complex are some of the trypsin–trypsin-inhibitor complexes (Huber and Bode, 1978). Analogues of the first tetrahedral intermediates are the diisopropylphosphoryl complexes (Sigler *et al.*, 1968) and the boronyl complexes (Matthews *et al.*, 1975). Indoleacryloyl chymotrypsin, an intermediate that is stable at pH 4, provides the only example of an acyl intermediate complex for which a high resolution X-ray analysis is available, although a putative acyl intermediate of elastase has been examined at 3.5 Å resolution at −55°C (Alber *et al.*, 1976). Finally, N-formyl tryptophan is a virtual substrate and provides an example of a product complex. These structures show that the conformation of the active site of the enzyme is complementary to a tetrahedral intermediate adduct of the carbonyl carbon to the oxygen of serine-195, which is the transition state. They also show that no significant conformational changes occur in the enzyme upon binding of the substrate, and in combination with nuclear magnetic resonance studies they show that little, if any, distortion occurs in the substrate molecules.

The catalytic groups of the serine proteinases consist of a triad of aspartic acid–histidine–serine, where the serine residue acts as the nucleophile; the histidine residue receives the proton from the serine and subsequently donates it to the leaving group; and the aspartic acid functions to position the histidine. The aspartic acid also provides the correct electrostatic environment, in concert with the histidine, for the proton transfer from serine via histidine to the leaving group (James, 1980; Steitz and Shulman, 1982). The electrophilic component is provided by the so-called 'oxyanion hole' (Henderson, 1970) where the backbone –NH– groups of glycine-193 and serine-195 provide two hydrogen bonds. The oxyanion hole is particularly suited to accommodate the tetrahedral intermediate. A water molecule hydrogen-bonded to histidine-57 acts as the nucleophile in the deacylation of the acyl enzyme.

Most of the evidence for the mechanism has been obtained from studies with enzymes of the trypsin superfamily, but it is reasonable to assume that the mechanism for the subtilisin superfamily is essentially the same. The structure of subtilisin (Kraut, 1977) shows that there is an aspartic acid–histidine–serine triad and an 'oxyanion hole' with two hydrogen bond donors. The major difference is that subtilisin does not have a specificity pocket that is as well defined as that for chymotrypsin and trypsin.

The mechanism of action of the *thiol proteinases* of the papain superfamily is similar to that of the serine proteinases in that it also proceeds via a well-defined acyl intermediate except that cysteine-25 is the residue that is acylated (Stockwell and Smith, 1957; Hinkle and Kirsch, 1971). The catalytic triad of the serine proteinases has an analogue in the triad asparagine–histidine–cysteine in papain (Arenth *et al.*, 1975). In this case the strong negative charge of the aspartic acid is not needed to assist in the proton transfer because of the acidic nature of the thiol group. The structure of papain inhibited with the chloromethyl ketone of *N*-benzyloxycarbonyl-L-phenylalanine-L-alanine (Drenth *et al.*, 1976) shows that there is an equivalent of the oxyanion hole, i.e. a binding site for the carbonyl group of the scissile peptide bond where the NH groups of the backbone of cysteine-25 and glutamine-19 provide one hydrogen bond each to the carbonyl group. Unfortunately, some details of the catalytic events are not yet known. In particular, there is no good evidence for the role of histidine-159 in the acylation step, although on the deacylation step it presumably acts as a general base catalyst. A review of the mechanism, the structure of the active site, and substrate and inhibitor binding is given by Fersht (1977).

A great deal of information is available on *aspartic proteinases* (Fruton, 1970,

1976). Possible mechanisms for pepsin have been discussed extensively by Knowles (1970) and by Clement (1973). Three-dimensional structures of four enzymes: pig pepsin (Andreeva *et al.*, 1978), penicillopepsin (Hsu *et al.*, 1977), *Rhizopus* pepsin (Jenkins *et al.*, 1977; Subramanian *et al.*, 1977*b*) and *Endothia* pepsin (Subramanian *et al.*, 1977*a, b*) are available. The structure of penicillopepsin has been refined to 1.8 Å resolution (James and Sielecki, 1983). In spite of this there is no definitive evidence as to whether the catalytic event involves covalent intermediates or proceeds via a general base mechanism. Indirect evidence for covalent intermediates comes from transpeptidation reactions in which either amino moieties (Neumann *et al.*, 1959; Wang and Hofmann, 1976) or acyl moieties (Takahashi and Hofmann, 1975) are transferred. The enzyme-bound intermediates (acyl or amino enzymes) cannot exchange their donor groups (amino acids) with free enzymes and the corresponding free amino acid in solution. Also, free amino acids cannot become involved in the transpeptidation reaction at a detectable level (M. Blum and T. Hofmann, unpublished observations). On the other hand, all attempts to find evidence for covalent intermediates – other than that obtained from the transpeptidation reactions – have failed. Furthermore, Antonov *et al.* (1981) observed ^{18}O exchange reactions during transpeptidation that are not readily explainable in terms of covalent intermediates. A mechanism involving general base catalysis has been proposed by James *et al.* (1977, 1981). It is based mostly on structural considerations and the fact that pepstatin, a high-affinity inhibitor of aspartic proteinases ($K_I \sim 10^{-11}$ M), can be considered as a transition state analogue for a noncovalent tetrahedral intermediate (Marciniszyn *et al.*, 1976). The only established fact is that the catalysis involves two aspartic acid residues and that substantial conformational changes occur when a pepstatin-homologue binds to the enzyme (James *et al.*, 1982). At present there is no evidence that amino acids other than the two aspartic acids take part directly in the catalytic event. In the noncovalent mechanism proposed by James *et al.* (1977, 1981) and modified by James and Sielecki (1983), the electrophilic component is provided by the proton that is shared by the two catalytic aspartic acids and the nucleophile is a water molecule that is activated by one of the aspartic acids. Originally, tyrosine-75 was proposed as the proton donor (James *et al.*, 1977, 1981). However, this appears to be ruled out now that the structure of a pepstatin-homologue complex has been determined. It turns out that tyrosine-75 is not located in a position where it could act as a proton donor because of the large movement the whole 'tyrosine-75 flap' undergoes upon binding of the inhibitor (James *et al.*, 1982).

This brief comparison shows that our understanding of the mechanism of action of metalloproteinases falls somewhere in between that of the serine and thiol proteinases and that of the aspartic proteinases. In the metalloproteinases the electrophilic component is provided by the zinc ion and the nucleophilic component by a glutamic acid directly (i.e. in the esterase activity in carboxypeptidase A) or by a water molecule activated by a glutamic acid. There is a good likelihood that the proton donor is a tyrosine in carboxypeptidase and a histidine in thermolysin, although conclusive evidence for this is still lacking.

1.10 CONCLUSION

From the discussion in this chapter on the various groups of metalloproteinases it is clear that these enzymes belong to several different superfamilies on the basis of their evolutionary relationships. A number of enzymes that appear to involve metals and that have not been mentioned are listed in the *Handbook on Enzyme Nomenclature* (Bielka *et al.*, 1978). Insufficient information is available about the molecular properties of these enzymes to warrant their inclusion. Some of them, like the dipeptidases from mammalian kidney and from *Escherichia coli*, have been well characterised in terms of their enzymatic action, but their molecular structure is relatively poorly understood. Reviews on dipeptidases have been published by Campbell (1970), Delange and Smith (1971) and Patterson (1976).

Proteinases from the venoms of many different species of snake have been isolated. They belong either to the serine proteinases or to the metalloproteinases. The latter are zinc enzymes of molecular mass in the range of 25 000 Da (Nikai *et al.*, 1982). They have been used extensively in research related to blood coagulation (Tu, 1977), and have even been used to a limited extent for clinical purposes. Some properties of the metalloproteinases from *Agkistrodon species* (Iwanaga *et al.*, 1976; Prescott and Wagner, 1976) have been reviewed.

Lastly, recent studies have demonstrated the widespread occurrence of a large metalloproteinase (300 000 Da) in mammalian tissues (Beynon *et al.*, 1981; Kirschner and Goldberg, 1981, 1983). This is an endoproteinase that is distinctly different from previously characterised mammalian proteinases.

Doubtless as investigations of the various metalloproteinases continue, we shall learn more about the differences in their molecular properties and about the mechanism of action of the different classes of metalloproteinases, and find out whether or not the metal ions in the different families contribute the same or a similar function to all the enzymes.

REFERENCES

Abramowitz, N., Schechter, I. and Berger, A. (1967) *Biochem. Biophys. Res. Commun.* **29**, 862.

Alber, T., Petsko, G. A. and Tsernoglou, D. (1976) *Nature, Lond.* **263**, 297.

Allen, G. (1981) in *Sequencing of Proteins and Peptides.* Vol. 9 of *Laboratory Techniques in Biochemistry and Molecular Biology* (Work, T. S. and Burdon, R. H., eds). Elsevier Biomedical Press, Amsterdam, p. 8.

Alter, G. M. and Vallee, B. L. (1970) *Biochemistry* **17**, 2212.

Ambler, R. P. (1972) *Meth. Enzymol.* **25**, 143, 262.

Ambler, R. P. (1980) *Phil. Trans. Roy. Soc. Lond. B* **289**, 321.

Ambler, R. P. and Medway, R. J. (1968) *Biochem. J.* **108**, 893.

Andersson, B., Nyman, P. O. and Strid, L. (1972) *Biochim. Biophys. Acta* **48**, 270.

Andersson, L., Isley, T. C. and Wolfenden, R. (1982) *Biochemistry* **21**, 4177.

Andreeva, N. S., Federov, A. A., Gushchina, A. E., Riskulov, R. R., Shutskever, N. E. and Safro, M. G. (1978) *Mol. Biol. (Moscow)* **12**, 922.

Antonov, V. K., Yavashev, L. P., Volkova, L. I., Sadowskaya, V. L. and Ginodman, L. M. (1979) *Bioorg. Khim.* **5**, 1427.

Antonov, V. K., Ginodman, L. M., Rumsch, L. D., Kapitannikov, T. N., Barshevskaya, T. N., Yavashev, L. P., Gurova, A. G. and Volkova, L. I. (1981) *Eur. J. Biochem.* **117**, 195.
Arenth, A., Swen, H. M., Hoogenstraan, W. and Sluyterman, L. A. A. E. (1975) *Kon. Ned. Akad. Wet. Ser. C* **78**, 104.
Argos, P., Garavito, R. M., Eventoff, W., Rossmann, M. G. and Branden, C. I. (1978) *J. Mol. Biol.* **126**, 141.
Auld, D. S. (1979) *Adv. Chem. Ser.* **172**, 112.
Auld, D. S. and Holmquist, B. (1974) *Biochemistry* **13**, 4355.
Auricchio, F. and Bruni, C. B. (1964) *Biochem. Z.* **340**, 321.
Bachovchin, W. W., Kanamori, K., Vallee, B. L. and Roberts, J. D. (1982) *Biochemistry* **21**, 2885.
Baldwin, G. S., Galdes, A., Hill, H. A. O., Smith, B. E., Waley, S. G. and Abraham, E. P. (1978) *Biochem. J.* **175**, 441.
Barrett, A. J. and McDonald, J. K. (1980) *Mammalian Proteases, a Glossary and Bibliography. Vol. 1, Endopeptidases.* Academic Press, London.
Behnke, W. D. and Vallee, B. L. (1972) *Proc. Natl Acad. Sci. USA* **69**, 2442.
Belew, M., Gerdin, B., Lindeberg, G., Porath, J., Saldeen, T. and Wallin, R. (1980) *Biochim. Biophys. Acta* **621**, 169.
Benajiba, A. and Maroux, S. (1981) *Biochem. J.* **197**, 573.
Bertini, I., Canti, G. and Luchinat, C. (1982) *J. Am. Chem. Soc.* **104**, 4943.
Beynon, R. J., Shannon, J. D. and Bond, J. S. (1981) *Biochem. J.* **199**, 591.
Bielka, H., Horecker, B. L., Jacoby, W. B., Karlson, P., Keil, B., Liebecq, C., Lindtberg, B. and Webb, E. C. (1978) *Enzyme Nomenclature 1978*, Academic Press, New York, p. 300.
Bigbee, W. L. and Dahlquist, F. W. (1974) *Biochemistry* **13**, 3542.
Blumberg, S. (1979) *Biochemistry* **18**, 2815.
Blumberg, S. and Vallee, B. L. (1975) *Biochemistry* **14**, 2410.
Blumberg, S., Holmquist, B. and Vallee, B. L. (1973) *Biochem. Biophys. Res. Commun.* **51**, 987.
Blumberg, S., Holmquist, B. and Vallee, B. L. (1974) *Israel J. Chem.* **12**, 643.
Bokisch, V. A. and Mueller-Eberhard, H. J. (1970) *J. Clin. Invest.* **49**, 2427.
Bradshaw, R. A. (1969) *Biochemistry* **8**, 3871.
Bradshaw, R. A., Ericsson, L. H., Walsh, K. A. and Neurath, H. (1969) *Proc. Natl Acad. Sci. USA* **63**, 1389.
Breslow, R. and Wernick, D. L. (1976) *J. Am. Chem. Soc.* **98**, 259.
Breslow, R. and Wernick, D. L. (1977) *Proc. Natl Acad. Sci. USA* **74**, 1303.
Brown, J. R., Cox, D. J., Greenshields, R. N., Walsh, K. A., Yamasaki, M. and Neurath, H. (1961) *Proc. Natl Acad. Sci. USA* **47**, 1554.
Brown, J. R., Greenshields, R. N., Yamasaki, M. and Neurath, H. (1963*a*) *Biochemistry* **2**, 867.
Brown, J. R., Yamasaki, M. and Neurath, H. (1963*b*) *Biochemistry* **2**, 877.
Bunning, P. and Riordan, J. F. (1983) *Biochemistry* **22**, 110.
Bunning, P., Holmquist, B. and Riordan, J. F. (1978) *Biochem. Biophys. Res. Commun.* **83**, 1442.
Bunning, P., Holmquist, B. and Riordan, J. F. (1983) *Biochemistry* **22**, 103.
Campbell, B. J. (1970) *Meth. Enzymol.* **19**, 722.
Campbell, P. and Nashed, N. T. (1982) *J. Am. Chem. Soc.* **104**, 5221.
Carpenter, F. H. and Vahl, J. M. (1973) *J. Biol. Chem.* **248**, 294.
Cawston, T. E. and Murphy, G. (1981) *Meth. Enzymol.* **80**, 711.
Cawston, T. E. and Tyler, J. A. (1979) *Biochem. J.* **183**, 647.
Cleland, W. W. (1977) *Adv. Enzymol.* **45**, 273.
Clement, G. E. (1973) *Prog. Biorg. Chem.* **2**, 178.
Colman, P. N., Jansonius, J. N. and Matthews, B. W. (1972) *J. Mol. Biol.* **70**, 701.
Cox, D. J., Wintersberger, E. and Neurath, H. (1962) *Biochemistry* **1**, 1078.
Craik, C. S., Buchman, S. R. and Beychock, S. (1980) *Proc. Natl Acad. Sci. USA* **77**, 1384.
Cushman, D. W., Cheung, H. S., Sabo, B. F. and Ondetti, M. A. (1977) *Biochemistry* **16**, 5484.
Cuypers, H. T., van Loon-Klaassen, L. A. H., Vree Egberts, W. T. M., de Jong, W. W. and Bloemendal, H. (1982*a*) *J. Biol. Chem.* **257**, 7077.

Cuypers, H. T., van Loon-Klaassen, L. A. H., Vree Egberts, W. T. M., de Jong, W. W. and Bloemendal, H. (1982*b*) *J. Biol. Chem.* **257**, 7086.
Das, M. and Soffer, R. L. (1975) *J. Biol. Chem.* **250**, 6762.
Davies, R. B. and Abraham, E. P. (1974) *Biochem. J.* **143**, 129.
Davies, R. B., Abraham, E. P. and Melling, J. (1974) *Biochem. J.* **143**, 115.
Dayhoff, M. O. (1976) *Atlas of Protein Sequence and Structure.* Vol. 5, National Biomedical Research Foundation, Washington, D.C., p. 89.
Dayton, W., Goll, D., Zeece, M. and Robson, W. (1976) *Biochemistry* **15**, 2150.
Delange, R. J. and Smith, E. L. (1971) in *The Enzymes* (Boyer, P. D., ed.) 3rd Edn, Vol. 3. Academic Press, New York, p. 81.
Dideberg, O., Charlier, P., Dupont, L., Vermere, M., Frere, J. M. and Ghuysen, J. M. (1980*a*) *FEBS Lett.* **117**, 212.
Dideberg, O., Joris, B., Frere, J. M., Ghuysen, J. M., Weber, G., Robaye, R., Delbrouck, J. M. and Roelandts, I. (1980*b*) *FEBS Lett.* **117**, 215.
Dideberg, O., Charlier, P., Dive, G., Joris, B., Frere, J. M. and Ghuysen, J. M. (1982) *Nature, Lond.* **299**, 469.
Dorer, F. E., Kahn, J. R., Lentz, K. E., Levine, M. and Skeggs, L. T. (1974) *Circul. Res.* **34**, 824.
Dorrington, K. J., Hui, A., Hofmann, T., Hitchman, A. J. W. and Harrison, J. E. (1974) *J. Biol. Chem.* **249**, 199.
Drenth, J., Kalk, K. H. and Swen, H. M. (1976) *Biochemistry* **15**, 3731.
D'Udine, B. and Bernini, L. F. (1974) *Protides Biol. Proc. Coll.* **22**, 603.
El-Dorry, H. A., Bull, H. G., Iwata, L., Thornberry, N. A., Cordes, E. H. and Soffer, B. L. (1982) *J. Biol. Chem.* **257**, 14128.
Emod, I. and Keil, B. (1977) *FEBS Lett.* **77**, 51.
Emod, I., Tong, N. T. and Keil, B. (1981) *Biochim. Biophys. Acta* **659**, 283.
Endo, S. (1962) *J. Ferment. Techn.* **40**, 346.
Erdos, E. G. (1979) *Handbook Exp. Pharmacol.* **25**, suppl. 428.
Erdos, E. G. and Sloane, E. M. (1962) *Biochem. Pharmacol.* **11**, 585.
Erdos, E. G. and Yang, H. Y. T. (1967) *Life Sci.* **6**, 569.
Feder, J., Garrett, L. R. and Wildi, B. W. (1971) *Biochemistry* **10**, 4552.
Feracci, H., Maroux, S., Bonicel, J. and Desnuelle, P. (1982) *Biochim. Biophys. Acta* **684**, 133.
Fersht, A. (1977) *Enzyme Structure and Mechanism.* W. H. Freeman & Co., San Francisco, p. 312.
Figueiredo, A. F. S. and Marquezini, A. J. (1978) *Adv. Exp. Med. Biol.* **120B**, 589.
Fittkau, S., Forster, U., Pascual, S. and Schunck, W. H. (1974) *Eur. J. Biochem.* **44**, 523.
Fittkau, S., Kammerer, G. and Damerau, W. (1979) *Ophthal. Res.* **11**, 377.
Folk, J. E. (1971) in *The Enzymes* (Boyer, P. D., ed.) 3rd Edn, Vol. 3. Academic Press, New York, p. 57.
Folk, J. E. and Gladner, J. A. (1958) *J. Biol. Chem.* **231**, 379.
Fontana, A., Boccu, E. and Veronese, F. M. (1976) in *Enzymes and Proteins from Thermophilic Microorganisms* (Zuber, H., ed.) *Experientia Suppl.* **26**, Birkhäuser, Basel, p. 55.
Frere, J. M., Duez, C., Ghuysen, J. M. and Vandekerchove, J. (1976) *FEBS Lett.* **70**, 257.
Fruton, J. S. (1970) *Adv. Enzymol.* **33**, 401.
Fruton, J. S. (1976) *Adv. Enzymol.* **44**, 1.
Galdes, A., Auld, D. S. and Vallee, B. L. (1983) *Biochemistry* **22**, 1888.
Georgopapadakou, N., Hammarstrom, S. and Strominger, J. L. (1977) *Proc. Natl. Acad. Sci. USA* **74**, 1009.
Ghuysen, J. M., Leyh-Bouille, M., Bonaly, R., Nieto, M., Perkins, H. R., Schleifer, K. H. and Kandler, O. (1970) *Biochemistry* **9**, 2955.
Ghuysen, J. M., Frere, J. M., Leyh-Bouille, M., Coyette, J., Dusart, J. and Nguyen-Disteche, M. (1979) *Ann. Rev. Biochem.* **48**, 73.
Ghuysen, J. M., Frere, J. M., Leyh-Bouille, M., Perkins, H. R. and Nieto, M. (1980) *Phil. Trans. Roy. Soc. Lond. B* **289**, 285.
Ghuysen, J. M., Frere, J. M., Leyh-Bouille, M., Dideberg, O., Lamotte-Brasseur, J., Perkins, H. R. and De Coen, J. L. (1981) in *Topics in Molecular Pharmacology* (Burgen, A. S. V. and Roberts, G. C. K., eds). Elsevier-North Holland, Amsterdam, p. 64.
Gilles, A. M. and Keil, B. (1976) *FEBS Lett.* **65**, 369.

Gorski, J. D., Hugli, T. A. and Mueller-Eberhard, H. J. (1979) *Proc. Natl Acad. Sci. USA* **76**, 5299.
Grant, N. H. and Alburn, H. E. (1959) *Arch. Biochem. Biophys.* **82**, 245.
Grillo, H. C. and Gross, J. (1967) *Dev. Biol.* **15**, 300.
Gruetter, M. G., Hawkes, R. B. and Matthews, B. W. (1979) *Nature, Lond.* **277**, 667.
Hangauer, D. G., Monzingo, A. F. and Matthews, B. W. (1984) *Biochemistry*, in press.
Hanson, H. and Frohne, M. (1976) *Meth. Enzymol.* **45**, 504.
Hanson, H. and Lasch, J. (1965) *Hoppe-Seyler's Z. Physiol. Chem.* **348**, 1525.
Harper, E. (1980) *Ann. Rev. Biochem.* **49**, 1063.
Harris, E. D. and Krane, S. M. (1974) *New Engl. J. Med.* **291**, 557, 605, 652.
Harrison, L. W. and Vallee, B. L. (1978) *Biochemistry* **17**, 4359.
Harrison, L. W., Auld, D. S. and Vallee, B. L. (1975*a*) *Proc. Natl Acad. Sci. USA* **72**, 3930.
Harrison, L.W., Auld, D. S. and Vallee, B. L. (1975*b*) *Proc. Natl Acad. Sci. USA* **72**, 4356.
Hartsuck, J. A. and Lipscomb, W. N. (1971) in *The Enzymes* (Boyer, P. D., ed.) 3rd Edn, Vol. 3. Academic Press, New York, p. 1.
Hass, G. M. and Neurath, H. (1971) *Biochemistry* **10**, 3535, 3541.
Hass, G. M. and Ryan, C. A. (1980*a*) *Phytochemistry* **19**, 1329.
Hass, G. M. and Ryan, C. A. (1980*b*) *Biochem. Biophys. Res. Commun.* **97**, 1481.
Hass, G. M. and Ryan, C. A. (1981) *Meth. Enzymol.* **80**, 778.
Hass, G. M., Govier, M. A., Grahn, D. T. and Neurath, H. (1972) *Biochemistry* **11**, 3787.
Hass, G. M., Nau, H., Biemann, K., Grahn, D. T., Ericsson, L. H. and Neurath, H. (1975) *Biochemistry* **14**, 1334.
Hass, G. M., Ager, S. P. and Makus, P. J. (1980) *Fed. Proc., Fed. Am. Soc. Exp. Biol.* **39**, 1689.
Hayashi, R. (1976) *Meth. Enzymol.* **45**, 568.
Heinrikson, R. L. (1977) *Meth. Enzymol.* **47**, 175.
Henderson, R. (1970) *J. Mol. Biol.* **54**, 341.
Himmelhoch, S. R. and Peterson, E. A. (1968) *Biochemistry* **7**, 2085.
Hinkle, P. M. and Kirsch, J. F. (1971) *Biochemistry* **10**, 2717.
Hoffman, S. J., Chu, S. S. T., Lee, H. H., Kaiser, E. T. and Carey, P. R. (1983) *J. Am. Chem. Soc.* **105**, 6971.
Hofmann, T. (1976) *Meth. Enzymol.* **45**, 587.
Hofmann, T., Kawakami, M., Hitchman, A. J. W., Harrison, J. E. and Dorrington, K. J. (1979) *Can. J. Biochem.* **57**, 737.
Holmes, M. A. and Matthews, B. W. (1981) *Biochemistry* **20**, 6912.
Holmes, M. A. and Matthews, B. W. (1982) *J. Mol. Biol.* **160**, 623.
Holmes, M. A., Tronrud, D. E. and Matthews, B. W. (1983) *Biochemistry* **22**, 236.
Holmquist, B. (1977) *Biochemistry* **16**, 4591.
Holmquist, B. and Vallee, B. L. (1974) *J. Biol. Chem.* **249**, 4601.
Holmquist, B. and Vallee, B. L. (1976) *Biochemistry* **15**, 101.
Hsu, I. N., Delbaere, L. T. J., James, M. N. G. and Hofmann, T. (1977) *Nature, Lond.* **266**, 140.
Huber, R. and Bode, W. (1978) *Acc. Chem. Res.* **11**, 114.
Ishiura, S., Morofushi, H., Suzuki, K. and Imahori, K. (1978) *J. Biochem.* **84**, 225.
Isowa, Y. and Ichikawa, T. (1979) *Bull. Chem. Soc. Japan* **60**, 796.
Isowa, Y., Ohmori, M., Ichikawa, T., Kuritia, H., Sato, M. and Mori, K. (1977) *Bull. Chem. Soc. Japan* **50**, 2762.
Isowa, Y., Ichikawa, T. and Ohmori, M. (1978) *Bull. Chem. Soc. Japan* **51**, 271.
Iwanaga, S., Oshima, G. and Suzuki, T. (1976) *Meth. Enzymol.* **45**, 459.
Jacobsen, N. E. and Bartlett, P. A. (1981) *J. Am. Chem. Soc.* **103**, 654.
James, M. N. G. (1980) *Can. J. Biochem.* **58**, 251.
James, M. N. G. and Sielecki, A. R. (1983) *J. Mol. Biol.* **163**, 299.
James, M. N. G., Delbaere, L. T. J. and Hsu, I. N. (1977) *Nature, Lond.* **267**, 808.
James, M. N. G., Hsu, I. N., Hofmann, T. and Sielecki, A. (1981) in *Structural Studies on Molecules of Biological Interest* (Dodson, G., Glusker, J. P. and Sayre, D., eds). Clarendon Press, Oxford, p. 350.
James, M. N. G., Sielecki, A. R., Salituro, F., Rich, D. H. and Hofmann, T. (1982) *Proc. Natl Acad. Sci. USA* **79**, 6137.
Jaurin, B. and Grundstrom, M. T. (1981) *Proc. Natl Acad. Sci. USA* **78**, 4897.

Jenkins, J. A., Tickle, I., Sewell, T., Ungaretti, L., Wollmer, A. and Blundell, T. (1977) in *Acid Proteases, Structure, Function and Biology* (Tang, J., ed.). Plenum Press, New York, p. 43.
Johansen, J. T. and Vallee, B. L. (1971) *Proc. Natl Acad. Sci. USA* **68**, 2532.
Johansen, J. T. and Vallee, B. L. (1973) *Proc. Natl Acad. Sci. USA* **70**, 2006.
Johansen, J. T. and Vallee, B. L. (1975) *Biochemistry* **14**, 649.
Johansen, J. T., Livingston, D. M. and Vallee, B. L. (1972) *Biochemistry* **11**, 2584.
Joris, B., van Beeumen, J., Casagrande, F., Gerday, C., Frere, J. M. and Ghuysen, J. M. (1983) *Eur. J. Biochem.* **130**, 53.
Jornvall, H. (1970) *Eur. J. Biochem.* **16**, 25.
Jurnak, J., Rich, A., van Loon-Klassen, L. A. H., Bloemendal, H., Taylor, A. and Carpenter, F. H. (1977) *J. Mol. Biol.* **112**, 149.
Kam, C. M., Nishino, N. and Powers, J. C. (1979) *Biochemistry* **18**, 3032.
Keil, B. (1979) *Mol. Cell. Biochem.* **23**, 87.
Keil, B. (1980) in *Enzyme Regulation and Mechanism of Action.* FEBS Symposia Vol. 60. (Mildner, P. and Ries, B., eds). Pergamon, Oxford, p. 351.
Keil-Dlouha, V. (1976) *Biochim. Biophys. Acta* **429**, 239.
Keil-Dlouha, V. and Keil, B. (1978) *Biochim. Biophys. Acta* **522**, 218.
Kessler, E. and Yaron, A. (1976) *Meth. Enzymol.* **45**, 544.
Kester, W. R. and Matthews, B. W. (1977*a*) *Biochemistry* **16**, 2506.
Kester, W. R. and Matthews, B. W. (1977*b*) *J. Biol. Chem.* **252**, 7704.
Kirschner, R. J. and Goldberg, A. L. (1981) *Meth. Enzymol.* **80**, 702.
Kirschner, R. J. and Goldberg, A. L. (1983) *J. Biol. Chem.* **258**, 967.
Kleine, R. and Hanson, H. (1961) *Hoppe-Seyler's Z. Physiol. Chem.* **326**, 106.
Knowles, J. R. (1970) *Phil. Trans. Roy. Soc. Lond. B* **257**, 135.
Koch, T. R., Kuo, L. C., Douglas, E. G., Jaffer, S. and Makinen, M. W. (1979) *J. Biol. Chem.* **254**, 12310.
Koide, A. and Yoshizawa, M. (1981) *Biochem. Biophys. Res. Commun.* **100**, 1091.
Komiyama, T., Suda, H., Aoyagi, T., Takeuchi, T., Umezawa, H., Fujimoto, K. and Umezawa, S. (1975) *Arch. Biochem. Biophys.* **171**, 727.
Kossiakoff, A. A. and Spencer, S. A. (1980) *Nature, Lond.* **288**, 414.
Kraut, J. (1977) *Ann. Rev. Biochem.* **46**, 331.
Kunugi, S., Hirohara, H. and Ise, N. (1982) *Eur. J. Biochem.* **124**, 157.
Kuo, L. C. and Makinen, M. W. (1982) *J. Biol. Chem.* **257**, 24.
Lacko, A. G. and Neurath, H. (1967) *Biochem. Biophys. Res. Commun.* **26**, 272.
Lanzillo, J. J. and Franburg, B. L. (1977) *Biochemistry* **16**, 5491.
Lasch, J., Kudernatsch, N. and Hanson, H. (1973) *Eur. J. Biochem.* **34**, 53.
Latt, S. A., Holmquist, B. and Vallee, B. L. (1969) *Biochem. Biophys. Res. Commun.* **37**, 333.
Lecroisey, A. and Keil, B. (1979) *Biochem. J.* **179**, 53.
Lecroisey, A., Keil-Dlouha, V., Woods, D. R., Perrin, D. and Keil, B. (1975) *FEBS Lett.* **59**, 167.
Leyh-Bouille, M., Ghuysen, J. M., Bonayl, R., Nieto, M., Perkins, H. R., Schleifer, K. H. and Kandler, O. (1970) *Biochemistry* **9**, 2961.
Lin, S. H., and Van Wart, H. E. (1982) *Biochemistry* **21**, 5528.
Linderstrom-Lang, K. (1929) *Z. Physiol. Chem.* **182**, 151.
Lipscomb, W. N. (1980) *Proc. Natl Acad. Sci. USA* **77**, 3875.
Lipscomb, W. N. (1982) *Acc. Chem. Res.* **15**, 238.
Lipscomb, W. N. (1983) *Ann. Rev. Biochem.* **52**, 17.
Lipscomb, W. N., Hartsuck, J. A., Reeke, G. N., Quiocho, F. A., Bethge, P. A., Ludwig, M. L., Steitz, T. A., Muirhead, H. and Coppola, J. C. (1968) *Brookhaven Symp. Biol.* **21**, 24.
Lipscomb, W. N., Reeke, G. N., Hartsuck, J. A., Quiocho, F. A. and Bethge, P. A. (1970) *Phil. Trans. Roy. Soc. Lond. B* **257**, 177.
Little, G. H., Starnes, W. L. and Behal, F. J. (1976) *Meth. Enzymol.* **45**, 495.
Ludwig, M. L. and Lipscomb, W. N. (1973) in *Inorganic Biochemistry* (Eichhorn, G. L., ed.) Vol. 2. Elsevier, Amsterdam, p. 438.
McDonald, J. K. and Barrett, A. J. (1984) *Mammalian Proteases, a Glossary and Bibliography. Vol. 2, Exopeptidases.* Academic Press, London.

Makinen, M. W., Yamamura, K. and Kaiser, E. T. (1976) *Proc. Natl Acad. Sci. USA* **73**, 3882.
Makinen, M. W., Kuo, L. C., Dymowski, J. J. and Shaffer, S. (1979) *J. Biol. Chem.* **254**, 356.
Makinen, M. W., Fukuyama, J. M. and Kuo, L. C. (1982) *J. Am. Chem. Soc.* **104**, 2667.
Marciniszyn, J., Hartsuck, J. A. and Tang, J. (1976) *J. Biol. Chem.* **251**, 7088.
Maroux, S., Louvard, D. and Baratti, J. (1973) *Biochim. Biophys. Acta* **321**, 282.
Matsubara, A. (1966) *Biochem. Biophys. Res. Commun.* **24**, 427.
Matsubara, A. and Feder, J. (1971) in *The Enzymes* (Boyer, P. D., ed.) 3rd Edn, Vol. 3. Academic Press, New York, p. 721.
Matthews, B. W. (1972) *Macromolecules* **5**, 818.
Matthews, B. W., Jansonius, J. N., Colman, P. M., Schoenborn, B. P. and Dupourque, D. (1972*a*) *Nature New Biol. Lond.* **238**, 37.
Matthews, B. W., Colman, P. M., Jansonius, J. N., Titani, K., Walsh, K. A. and Neurath, H. (1972*b*) *Nature New Biol. Lond.* **238**, 41.
Matthews, B. W., Weaver, L. H. and Kester, W. R. (1974) *J. Biol. Chem.* **249**, 8030.
Matthews, D. A., Alden, R. A., Birktoft, J. J., Freer, S. T. and Kraut, J. (1975) *J. Biol. Chem.* **250**, 7120.
Mattis, J. A. and Fruton, J. S. (1976) *Biochemistry* **15**, 2191.
Maycock, A. L., De Sousa, D. L., Payne, L. G., ten Broeke, J., Wu, M. T. and Patchett, A. A. (1981) *Biochem. Biophys. Res. Commun.* **102**, 963.
Melbye, S. W. and Carpenter, F. H. (1971) *J. Biol. Chem.* **246**, 2459.
Mellgren, R. L. (1980) *FEBS Lett.* **109**, 129.
Mellgren, R. L., Repetti, A., Muck, T. C. and Easly, J. (1982) *J. Biol. Chem.* **257**, 7203.
Mock, W. L., Chen, J. T. and Tsang, J. W. (1981) *Biochem. Biophys. Res. Commun.* **102**, 389.
Morgan, G. and Fruton, J. S. (1978) *Biochemistry* **17**, 3562.
Morihara, K. (1974) *Adv. Enzymol.* **41**, 179.
Morihara, K. and Oka, T. (1968) *Biochem. Biophys. Res. Commun.* **30**, 625.
Morihara, K. and Tsuzuki, H. (1966) *Biochim. Biophys. Acta* **118**, 215.
Morihara, K. and Tsuzuki, H. (1970) *Eur. J. Biochem.* **15**, 374.
Murachi, T. (1983) *Trends in Biochem. Sci.* **8**, 167.
Murachi, T., Tanaka, K., Hatanaka, M. and Murakami, T. (1981) *Adv. Enz. Regul.* **19**, 407.
Murphy, G. and Sellers, A. (1980) in *Collagenase in Normal and Pathological Connective Tissues* (Woolley, D. E. and Evanson, J. M., eds). Wiley, New York, p. 65.
Nau, H. and Riordan, J. F. (1975) *Biochemistry* **14**, 5285.
Nemethy, G. and Printz, M. P. (1972) *Macromolecules* **5**, 755.
Neumann, H., Levin, Y., Berger, A. and Katchalski, E. (1959) *Biochem. J.* **73**, 33.
Ng, K. K. F. and Vane, J. R. (1967) *Nature, Lond.* **216**, 762.
Nieto, M., Perkins, H. R., Leyh-Bouille, M., Frere, J. M. and Ghuysen, J. M. (1973) *Biochem. J.* **131**, 163.
Nikai, T., Ishizaki, H., Tu, A. Y. and Sugihara, H. (1982) *Comp. Biochem. Physiol.* **72C**, 103.
Nishino, N. and Powers, J. C. (1978) *Biochemistry* **17**, 2846.
Nishino, N. and Powers, J. C. (1979) *Biochemistry* **18**, 4340.
Oka, T. and Morihara, K. (1980) *J. Biochem. (Tokyo)* **88**, 807.
Ondetti, M. A. and Cushman, D. W. (1982) *Ann. Rev. Biochem.* **51**, 283.
Ondetti, M. A., Williams, N. J., Sabo, E. F., Pluscec, J., Weaver, E. R. and Kocy, O. (1971) *Biochemistry* **10**, 4033.
Otsuka, Y. and Tanaka, H. (1983) *Biochem. Biophys. Res. Commun.* **111**, 700.
Pangburn, M. K. and Walsh, K. A. (1975) *Biochemistry* **14**, 4050.
Pangburn, M. K., Levy, P. L., Walsh, K. A. and Neurath, H. (1976) in *Enzymes and Proteins from Thermophilic Microorganisms* (Zuber, H., ed.) *Experientia Suppl.* **26**, 19, Birkhäuser, Basel.
Patchett, A. A., Harris, E., Tristram, E. W., Wyvratt, M. J., Wu, M. T., Taub, D., Peterson, E. R., Ikeler, T. J., ten Broeke, J., Payne, L. G., Ondeyka, D. L., Thorsett, E. D., Greenlee, W. J., Lohr, N. S., Hoffscommer, R. D., Joshua, H., Ruyle, W. V., Rothrock, J. W., Aster, S. D., Maycock, A. L., Robinson, F. M., Hirschmann, R., Sweet, C. S., Ulm, E. H., Gross, D. M., Vassil, T. C. and Stone, C. A. (1980) *Nature, Lond.* **288**, 280.

Patterson, E. K. (1976) *Meth. Enzymol.* **45**, 377, 386.
Petra, P. H. (1970) *Meth. Enzymol.* **19**, 460.
Petra, P. H., Bradshaw, R. A., Walsh, K. A. and Neurath, H. (1969) *Biochemistry* **8**, 2762.
Pfleiderer, G. (1970) *Meth. Enzymol.* **19**, 514.
Pfleiderer, G. and Celliers, P. G. (1963) *Biochem. Z.* **339**, 186.
Plummer, T. H. (1972) *J. Biol. Chem.* **247**, 7864.
Plummer, T. H. and Erdos, E. G. (1981) *Meth. Enzymol.* **80**, 442.
Plummer, T. H. and Hurwitz, M. Y. (1978) *J. Biol. Chem.* **253**, 3907.
Prescott, J. M. and Wagner, F. W. (1976) *Meth. Enzymol.* **45**, 397.
Prescott, J. M. and Wilkes, S. H. (1966) *Arch. Biochem. Biophys.* **117**, 328.
Prescott, J. M. and Wilkes, S. H. (1976) *Meth. Enzymol.* **45**, 530.
Quinto, C., Quiroga, M., Swain, W. F., Nikovits, W. C., Standring, D. N., Pictet, R. L., Valenzuela, P. and Rutter, W. J. (1982) *Proc. Natl Acad. Sci. USA* **79**, 31.
Quiocho, F. A. and Lipscomb, W. N. (1971) *Adv. Prot. Chem.* **25**, 1.
Quiocho, F. A., Murray, C. H. and Lipscomb, W. N. (1972) *Proc. Natl Acad. Sci. USA* **69**, 2850.
Rancour, J. M. and Ryan, C. A. (1968) *Arch. Biochem. Biophys.* **125**, 380.
Rao, L. and Hofmann, T. (1976) *Can. J. Biochem.* **54**, 885.
Rasmussen, J. R. and Strominger, J. L. (1978) *Proc. Natl Acad. Sci. USA* **75**, 84.
Rasnick, D. and Powers, J. C. (1978) *Biochemistry* **17**, 4363.
Reeke, E. N., Hartsuck, J. A., Ludwig, H. L., Quiocho, F. A., Steitz, T. A. and Lipscomb, W. N. (1967) *Proc. Natl. Acad. Sci. USA* **58**, 2220.
Rees, D. C. and Lipscomb, W. N. (1980) *Proc. Natl Acad. Sci. USA* **77**, 4633.
Rees, D. C. and Lipscomb, W. N. (1981) *Proc. Natl Acad. Sci. USA* **78**, 5455.
Rees, D. C. and Lipscomb, W. N. (1982) *J. Mol. Biol.* **160**, 475.
Rees, D. C., Honzatko, R. B. and Lipscomb, W. N. (1980) *Proc. Natl Acad. Sci. USA* **77**, 3288.
Rees, D. C., Lewis, M., Honzatko, R. B., Lipscomb, W. N. and Hardman, K. D. (1981). *Proc. Natl Acad. Sci. USA* **78**, 3408.
Ribadeau-Dumas, B., Brignon, G., Grosclaude, F. and Mercier, J. C. (1972) *Eur. J. Biochem.* **25**, 505.
Riordan, J. F. (1974) *Adv. Chem. Ser.* **136**, 227.
Riordan, J. F. and Vallee, B. L. (1963) *Biochemistry* **2**, 1460.
Robertus, J. D., Alden, R. A., Birktoft, J. J., Kraut, J., Powers, J. C. and Wilcox, P. E. (1972) *Biochemistry* **11**, 2439.
Roche, R. S. and Voordouw, G. (1978) *C.R.C. Crit. Rev. Biochem.* **5**, 1.
Roncari, G., Stoll, E. and Zuber, H. (1976) *Meth. Enzymol.* **45**, 522.
Rossmann, M. G. and Argos, P. (1978) *Mol. Cell. Biochem.* **21**, 161.
Rossmann, M. G. and Blow, D. M. (1962) *Acta Crystallogr.* **15**, 24.
Sabath, L. D. and Abraham, E. P. (1966) *Biochem. J.* **98**, 11c.
Sakamoto, S., Sakamoto, M., Goldhaber, P. and Glimcher, M. (1978) *Arch. Biochem. Biophys.* **188**, 438.
Sakano, H., Rogers, J. H., Huppi, K., Brack, C., Traunecker, A., Maki, R., Wall, R. and Tonegawa, S. (1979) *Nature, Lond.* **277**, 627.
Sampath-Kumar, K. S. V., Clegg, J. B. and Walsh, K. A. (1964) *Biochemistry* **3**, 1728.
Schechter, I. and Berger, A. (1967) *Biochem. Biophys. Res. Commun.* **27**, 157.
Schellman, J. A., Lindorfer, M., Hawkes, R. B. and Gruetter, M. (1981) *Biopolymers* **20**, 1979.
Scheule, R. K., Van Wart, H. E., Vallee, B. L. and Scheraga, H. A. (1980) *Biochemistry* **19**, 759.
Scheule, R. K., Han, S. L., Van Wart, H. E., Vallee, B. L. and Scheraga, H. A. (1981) *Biochemistry* **20**, 1778.
Schmid, M. F. and Herriott, J. R. (1976) *J. Mol. Biol.* **103**, 175.
Segundo, S., Martinez, M. C., Vilanova, M., Cuchillo, C. M. and Aviles, F. X. (1982) *Biochim. Biophys. Acta* **207**, 74.
Seifter, S. and Harper, E. (1971) in *The Enzymes* (Boyer, P. D., ed.) 3rd Edn, Vol. 3. Academic Press, New York, p. 662.
Seltzer, J. L., Jeffrey, J. J. and Eisen, A. Z. (1977) *Biochim. Biophys. Acta* **485**, 179.
Sigler, P. B., Blow, D. M., Matthews, B. W. and Henderson, R. (1968) *J. Mol. Biol.* **35**, 143.

Simpson, R. T., Riordan, J. F. and Vallee, B. L. (1963) *Biochemistry* **2**, 616.
Sjostrom, H., Noren, O., Jeppesen, L., Stann, M., Svenson, B. and Christiansen, L. (1978). *Eur. J. Biochem.* **88**, 503.
Skeggs, L. T., Marsh, W. H., Kahn, J. R. and Shuway, N. P. (1954) *J. Exp. Med.* **99**, 275.
Smith, E. L. and Hill, R. L. (1960) in *The Enzymes* (Boyer, P. D., Hardy, H. and Myrback, K., eds) 2nd Edn, Vol. 4, Part A. Academic Press, New York, p. 37.
Smith, E. L. and Spackman, D. H. (1955) *J. Biol. Chem.* **212**, 271.
Snoke, J. E., Schwert, G. W. and Neurath, H. (1948) *J. Biol. Chem.* **175**, 7.
Sokolovsky, M. and Zisapel, N. (1974) *Isr. J. Chem.* **12**, 631.
Spilburg, C. A., Bethune, J. L. and Vallee, B. L. (1977) *Biochemistry* **16**, 1142.
Stark, M. and Kuhn, K. (1968) *Eur. J. Biochem.* **6**, 542.
Starnes, W. L. and Behal, F. J. (1974) *Biochemistry* **13**, 3221.
Starnes, W. L., Szechinski, J. and Behal, F. J. (1982) *Eur. J. Biochem.* **124**, 363.
Steitz, T. A. and Shulman, R. G. (1982) *Ann. Rev. Biophys. Bioeng.* **11**, 419.
Sterner, R. and Heinrikson, R. L. (1975) *Arch. Biochem. Biophys.* **168**, 693.
Stewart, T. A., Weare, J. A. and Erdos, E. G. (1981*a*) *Meth. Enzymol.* **80**, 450.
Stewart, T. A., Weare, J. A. and Erdos, E. G. (1981*b*) *Peptides* **2**, 145.
Stockell, A. and Smith, E. L. (1957) *J. Biol. Chem.* **227**, 1.
Stoll, E., Ericsson, L. H. and Zuber, H. (1973) *Proc. Natl Acad. Sci. USA* **70**, 3781.
Stricklin, G. P., Bauer, E. A., Jeffrey, J. J. and Eisen, A. Z. (1977) *Biochemistry* **16**, 1607.
Subramanian, E., Liu, M., Swan, I. D. A. and Davies, D. R. (1977*a*) in *Acid Proteases, Structure, Function and Biology* (Tang, J., ed.). Plenum, New York, p. 33.
Subramanian, E., Swan, I. D. A., Liu, M., Davies, D. R., Jenkins, J. A., Tickle, I. J. and Blundell, T. (1977*b*) *Proc. Natl Acad. Sci. USA* **74**, 556.
Suda, H., Aoyagi, T., Takeuchi, T. and Umezawa, H. (1973) *J. Antibiotics* **26**, 621.
Suda, H., Aoyagi, T., Takeuchi, T. and Umezawa, H. (1976) *Arch. Biochem. Biophys.* **177**, 196.
Sugimoto, T. and Kaiser, E. T. (1978) *J. Am. Chem. Soc.* **100**, 7750.
Sugimoto, T. and Kaiser, E. T. (1979) *J. Am. Chem. Soc.* **101**, 3946.
Suh, J. and Kaiser, E. T. (1976*a*) *J. Am. Chem. Soc.* **98**, 1940.
Suh, J. and Kaiser, E. T. (1976*b*) *J. Org. Chem.* **43**, 3311.
Suzuki, K. and Tsuji, S. (1982) *FEBS Lett.* **140**, 16.
Suzuki, K., Tsuji, S., Kobota, S., Kimura, Y. and Imahori, K. (1981) *J. Biochem. (Tokyo)* **90**, 275.
Takahashi, M. and Hofmann, T. (1975) *Biochem. J.* **147**, 549.
Taylor, A., Carpenter, F. H.and Wlodawar, A. (1979) *J. Ultrastr. Res.* **68**, 92.
Taylor, A., Sawan, S. and James, T. L. (1982) *J. Biol. Chem.* **257**, 11571.
Titani, K., Hermodson, M. A., Ericsson, L. H., Walsh, K. A. and Neurath, H. (1972) *Nature New Biol., Lond.* **238**, 35.
Titani, K., Ericsson, L. H., Walsh, K. A. and Neurath, H. (1975) *Proc. Natl Acad. Sci. USA* **72**, 1666.
Tollin, P. (1966) *Acta Crystallogr.* **21**, A165.
Tsugita, A., Van den Brock, R. and Przybylski, M. (1982) *FEBS Lett.* **137**, 19.
Tu, A. T. (1977) *Venoms, Chemistry and Molecular Biology.* Wiley, New York, p. 329.
Urdea, M. S. and Legg, J. I. (1979) *J. Biol. Chem.* **254**, 11868.
Vallee, B. L. and Williams, R. J. P. (1968) *Proc. Natl Acad. Sci. USA* **59**, 498.
Vallee, B. L., Riordan, J. F., Bethune, J. L., Coombs, T. L., Auld, D. S. and Sokolovsky, M. (1968). *Biochemistry* **7**, 3547.
Voordouw, G. and Roche, R. S. (1974) *Biochemistry* **13**, 5017.
Voordouw, G. and Roche, R. S. (1975) *Biochemistry* **14**, 4667.
Wacker, H., Lehky, P., Fischer, E. H. and Stein, E. A. (1971) *Helv. Chim. Acta* **54**, 473.
Wacker, H., Lehky, P., Vanderhaege, F. and Stein, E. A. (1976) *Biochim. Biophys. Acta* **429**, 546.
Wang, T. T. and Hofmann, T. (1976). *Biochem. J.* **153**, 691.
Waxman, L. (1981) *Meth. Enzymol.* **80**, 664.
Waxman, L. and Krebs, E. G. (1978) *J. Biol. Chem.* **253**, 5888.
Waxman, D. J., Yocum, R. R. and Strominger, J. L. (1980) *Phil. Trans. Roy. Soc. B* **289**, 257.
Weaver, L. H., Kester, W. R., Ten Eyck, L. F. and Matthews, B. W. W. (1976) in *Enzymes*

and Proteins from Thermophilic Microorganisms (Zuber, H., ed.) *Experientia Suppl.* **26**, 31, Birkhäuser, Basel.
Weaver, L. H., Kester, W. R. and Matthews, B. W. W. (1977) *J. Mol. Biol.* **114**, 119.
Welton, R. C. and Woods, D. R. (1975) *Biochim. Biophys. Acta* **384**, 228.
Werb, Z. and Burleigh, M. C. (1974) *Biochem. J.* **137**, 373.
Wilkes, S. H., Bayliss, M. E. and Prescott, J. M. (1973) *Eur. J. Biochem.* **34**, 459.
Wintersberger, E., Cox, D. J. and Neurath, H. (1962) *Biochemistry* **1**, 1069.
Wheelock, M. J. (1982) *J. Biol. Chem.* **257**, 12471.
Wolfenden, R. (1976) *Ann. Rev. Biophys. Bioeng.* **5**, 271.
Wolff, E. C., Schirmer, E. W. and Folk, J. E. (1962) *J. Biol. Chem.* **237**, 3094.
Woolley, P. R. (1975) *Nature, Lond.* **258**, 677.
Yang, H. Y. T. and Erdos, E. G. (1967) *Nature, Lond.* **215**, 1402.
Yaron, A. and Berger, A. (1970) *Meth. Enzymol.* **19**, 521.
Yocum, R. R., Waxman, D. J., Rasmussen, J. R. and Strominger, J. L. (1979) *Proc. Natl Acad. Sci. USA* **76**, 2730.
Zuber, H. (1976) *Meth. Enzymol.* **45**, 561.

2

A comparative study of the occurrence, structure and function of troponin C and calmodulin

Roger J. A. Grand

2.1 INTRODUCTION

The importance of calcium in the regulation of a very wide range of eukaryotic cellular enzymes and processes has become evident over the past two decades. The actions of the metal ion are mediated through a limited number of related specific calcium binding proteins which act as receptors within the cell. One of these proteins, calmodulin, has been the subject of intensive study in recent years, largely as a result of its ubiquitous distribution and the fact that it is involved in the control of the majority of the calcium-regulated processes. However, in spite of the efforts of a large number of researchers, its mode of action is not well understood at the molecular level. On the other hand, the mechanism by which a second calcium binding protein, troponin C, is able to 'turn off' or 'turn on' striated muscle in response to changes in calcium ion concentration is very well characterised.

In this chapter I have limited detailed discussion to certain aspects of these two members of the family of calcium binding proteins (Kretsinger, 1980) and in particular I have attempted to demonstrate how their primary, secondary and tertiary structures enable them to perform their varied roles within the cell. This has meant that large areas of our knowledge of calmodulin have been dealt with only briefly or ignored altogether; readers in need of a more general discussion of the properties of this protein are referred to the many excellent reviews available (Wolff and Brostrom, 1979; Cheung, 1980*a*; Klee *et al.*, 1980; Means and Dedman, 1980; Klee and Vanaman, 1982; Means *et al.*, 1982) as well as compilations containing articles on various aspects of calmodulin biochemistry (Cheung, 1980*b*; Siegel *et al.*, 1980; Watterson and Vincenzi, 1980; Carafoli, 1981*a*).

2.1.1 The family of calcium binding proteins

The intracellular calcium binding proteins are related by their common evolutionary origin and by similarities in amino acid sequence and three-dimensional structure. The group includes, as well as calmodulin and troponin C, the phosphorylatable and alkali light chains of myosin (see Grand, 1982, for review), the parvalbumins (see Pechere, 1977, and Wnuk *et al.*, 1982, for reviews), the intestinal calcium binding proteins (Wasserman *et al.*, 1978; Wasserman, 1980) and S-100 protein from nervous tissue (Isobe and Okuyama, 1978; Klevitt *et al.*, 1981*a*). The relationship, in terms of amino acid sequence, between certain members of the family is somewhat remote, but they probably all adopt conformations which have been termed by Kretsinger (1980) 'EF hands' (discussed in detail in section 2.4.3.1). This structure seems to have been retained in these proteins even when the ability to bind divalent metal ions has been lost – for example, by the alkali light chains of myosin.

Goodman (1980) has constructed a genealogical tree for the calcium binding proteins using the maximum parsimony algorithm and this is shown in figure 2.1. As can be seen, the parvalbumins form a closely related subgroup – as do the myosin light chains, troponin Cs and calmodulins. It is interesting to note that the myosin phosphorylatable (regulatory) light chains are more closely related to the troponin Cs whereas the myosin alkali light chains are more closely related to the calmodulins. It is now generally recognised that the calcium binding proteins originated in a primitive single domain calcium binding polypeptide which, as a result of a double gene duplication, has given rise to the four domain proteins listed in figure 2.1, in which domain one is most closely related to domain three and domain two is most similar to domain four (Kretsinger, 1972; Collins *et al.*, 1973; Collins, 1974, 1976; Weeds and McLachlan, 1974; Barker *et al.*, 1977; Watterson *et al.*, 1980*a*) (shown diagrammatically in figure 2.2). However, a number of modifications to the structures have occurred during evolution to account for the diversity in molecular weight and calcium binding properties which may now be distinguished, for example the loss of domains three and four from the intestinal calcium binding proteins and domain one from the parvalbumins.

2.1.2 General consideration of the implications of metal ion binding

Kretsinger (1977) has suggested a number of postulates which provide a useful background to any consideration of calcium binding proteins. Those points which are particularly relevant to this discussion may be summarised as follows:

(1) The concentration of free calcium ions within the cytosol of eukaryotic cells is maintained at 10^{-7} to 10^{-8} M.

(2) The sole function of calcium in the cytosol is to transmit information (to act as a second messenger).

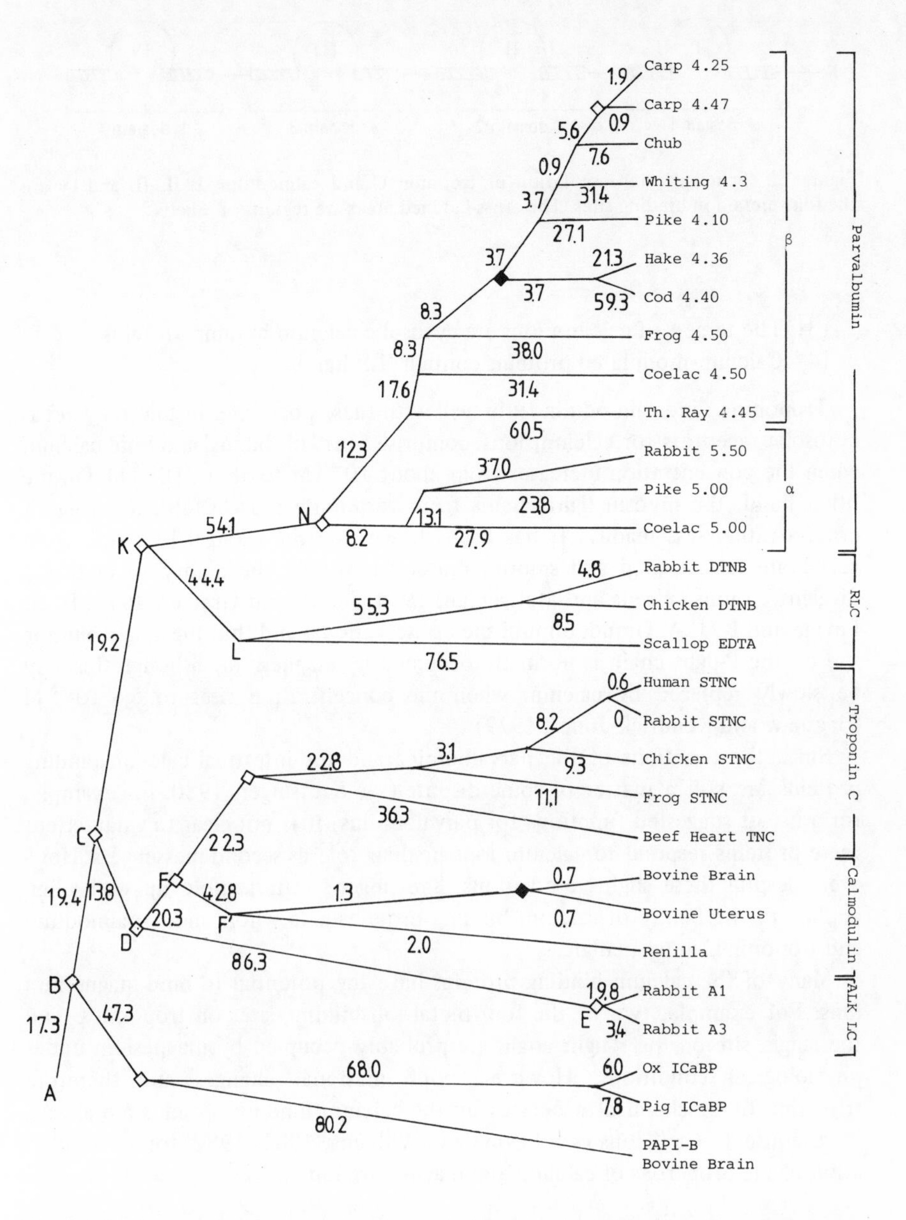

Figure 2.1 Genealogical tree of the calcium binding protein family. Numbers on the links are the nucleotide replacements per 100 codons. The values are corrected for superimposed mutations and were obtained by putting together the results of the most parsimonious tree constructed for the separate domains as well as from any N-terminal chain positions coming before the first domain. Symbol for obvious gene duplication ◇. ◆ is used whenever there is only circumstantial evidence for gene duplication; ALK LC, myosin alkali light chains; RLC, myosin regulatory light chains. (Reproduced with permission from Goodman (1980). Copyright 1980, Elsevier Science Publishing Co., Inc.)

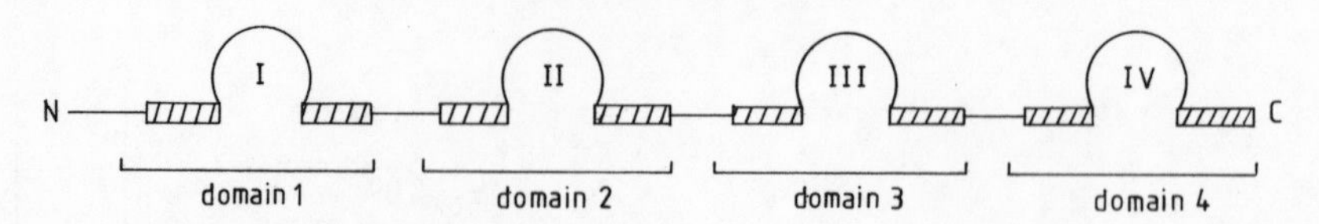

Figure 2.2 Schematic representation of troponin C and calmodulin. I, II, III and IV are the four metal ion binding sites. The cross-hatched areas are regions of α-helix.

(3) The targets of calcium ions are cytosolic calcium binding proteins.
(4) Calcium-modulated proteins contain 'EF hands'.

Troponin C and calmodulin fit in well with these postulates in that they act as cytosolic receptors for calcium ions, comprise four 'EF hands' and bind calcium when the concentration increases from about 10^{-7} M to about 10^{-5} M. On the other hand, the myosin light chains from vertebrate muscle fulfil Kretsinger's criteria rather less readily. It has been shown that the alkali light chains from vertebrate fast striated and smooth muscles have lost the ability to bind any divalent cations (Weeds and McLachlan, 1974; Alexis and Gratzer, 1978; B. A. Levine and R. J. A. Grand, unpublished observations) and that the single binding site on the P-light chain is normally occupied by magnesium, although this may be slowly replaced by calcium when the concentration rises to 5×10^{-6} M (Bagshaw and Kendrick-Jones, 1979).

Since the functions of the parvalbumins and the intestinal calcium binding proteins are still a matter of some dispute (see Kretsinger, 1980, for example, for a list of suggested functions for parvalbumins) it is not clear to what extent these proteins respond to calcium ions in their role as second messengers. However, despite these slight reservations, Kretsinger's criteria hold up well when applied to the family of calcium binding proteins as a whole and to calmodulin and troponin C in particular.

Many of the calcium binding proteins have the potential to bind magnesium ions. For example, two of the four metal ion binding sites on troponin C and the single site on the P-light chain are probably occupied by magnesium under physiological conditions. However, conformational changes and therefore triggering of biochemical processes by the calcium binding proteins are always in response to Ca^{2+} ions (see Levine and Williams, 1981, 1982, for a consideration of the properties of calcium and magnesium ions).

2.2 PURIFICATION PROCEDURES

Most of the methods widely used for the preparation of calmodulin and troponin C make use of two properties common to both proteins: first, the preponderance of acidic amino acids, which means that they bind very much more strongly to cation ion exchange media than do most other proteins and, secondly, the fact that neither protein appears to be deleteriously affected by exposure to extreme denaturing conditions (for example, high temperatures, very high concentrations of urea or organic solvents).

2.2.1 Troponin C (TNC)

Generally, preparative procedures for troponin C rely on the isolation of the troponin complex from striated muscle followed by separation of the calcium binding protein under denaturing conditions. In the original method of Ebashi *et al.* (1968, 1971) troponin was extracted from mammalian fast striated muscle in the presence of LiCl following removal of myosin. An alternative approach has been to denature and insolubilise myosin with organic solvents and then to extract the regulatory proteins in the presence of 1 M KCl (Hartshorne and Mueller, 1969; Greaser and Gergely, 1971). In both of these procedures tropomyosin was removed by iso-electric precipitation at pH 4.6 (Bailey, 1948). Following ammonium sulphate fractionation, both procedures give a product in which the major proteins are the three troponin components. The troponin can be further purified by gel filtration or ion exchange chromatography under non-denaturing conditions (van Eerd and Kawasaki, 1973). The troponin components are normally fractionated by chromatography on DEAE cellulose (Perry and Cole, 1974) or DEAE-Sephadex (Greaser and Gergely, 1971) run in the presence of 6 M urea. Troponin C binds most strongly under these conditions and is eluted at a salt concentration of about 0.3 M. Further purification is not normally required as SDS-polyacrylamide gel electrophoresis indicates few, if any, impurities.

A number of alternatives and adaptations of these basic procedures have been suggested – for example, most of the troponin I and T may be removed from troponin C by dialysis of whole troponin against 1 M KCl/1 M HCl (Hartshorne and Mueller, 1968). However, ion-exchange chromatography is still required as a final purification step (Wilkinson, 1974). Cox *et al.* (1981*a*) have described a rather different method which, unlike those mentioned above, does not subject the protein to denaturing conditions. There is general agreement that troponin C is not affected in any obvious way by repeated denaturation and renaturation, but it is quite possible that very subtle changes could occur in the secondary and tertiary structures of the protein as appears to happen with calmodulin (Klevitt *et al.*, 1981*b*). If this were the case, a procedure that uses only mild conditions is of obvious importance.

Because of the high concentrations of troponin C in fast striated muscle, yields from all of these procedures, although rather low, are quite acceptable. However, when similar methods have been applied to slow and cardiac muscle, they are very much less efficient. Tsukui and Ebashi (1973) have used an adaptation of the original skeletal muscle troponin preparation to isolate troponin I (TNI), troponin T (TNT) and TNC from cardiac muscle. Yields of troponin C tend to be low from beef hearts and particularly poor from the hearts of small mammals, for example rat and rabbit. Faced with this problem a number of workers have introduced methods which do not rely on the purification of whole troponin. Head *et al.* (1977) have used an affinity chromatographic method in which troponin I is immobilised on Sepharose 4B and then a whole muscle homogenate in 9 M urea is added to the column. After washing to remove non-specifically bound protein, TNC is eluted by removal of Ca^{2+} ions with EGTA. Wilkinson (1980) found that good yields of troponin C could be obtained from rabbit slow muscle if no attempt was made to separate tropomyosin from the troponin until the final DEAE cellulose fractionation. However, when applied to cardiac muscle, yields were little better than those obtained with the original procedure of Tsukui and Ebashi (1973). Troponin C may be purified from cardiac muscle by an adaptation of the calmodulin preparative procedure of Grand *et al.* (1979). Washing of the muscle with low ionic strength buffers prior to solubilisation in 9 M urea removes much of the calmodulin, which would otherwise co-purify with troponin C (Dhoot *et al.*, 1979; Grand *et al.*, 1979).

2.2.2 Calmodulin

As calmodulin has been studied by numerous different research groups, it is not surprising that the preparative procedures available are many and various. Most of them give quite acceptable results, in terms of yield and purity, when applied to brain. However, with the 'tougher' tissues, smooth muscle for example, some of the methods have proved rather less effective. Watterson *et al.* (1976*a*), Klee (1977) and Wolff *et al.* (1977), as well as a number of other workers, have described procedures for the purification of brain calmodulin which avoid the use of denaturing conditions. In these a low ionic strength extract of the tissue obtained in the presence of EGTA is subjected to ion-exchange and gel filtration chromatography. The procedures may be simplified by the use of a heat denaturation step which removes much of the contaminating protein without seriously harming the calmodulin (Teo *et al.*, 1973; Lin *et al.*, 1974; Dedman *et al.*, 1977*a*). Considerably higher yields have been obtained using a modification of the procedure of Grand *et al.* (1979). In this method bovine brain acetone powder is solubilised in 9 M urea. After ethanol fractionation the calmodulin is purified by chromatography on DEAE cellulose and Sephadex G100. Yields of 300–400 mg kg^{-1} from bovine brain have been obtained routinely. The method seems to work well for large- and small-scale preparations from all vertebrate tissues, with the exception of striated muscle where the major product is troponin C.

Because of the similarity of calmodulin to troponin C it is possible to purify the former protein from non-muscle sources using the troponin I Sepharose affinity method described above (Head *et al.*, 1977; Grand *et al.*, 1979). However, yields are poor and a much more effective affinity chromatographic method has been described by Charbonneau and Cormier (1979) and Jamieson and Vanaman (1979). This takes advantage of the highly specific calcium-dependent interaction between calmodulin and the psychoactive phenothiazine drugs. In this procedure a low ionic strength extract is passed through a column of either fluphenazine (Charbonneau and Cormier, 1979) or 2-chloro-10-(3-aminopropyl)phenothiazine (Jamieson and Vanaman, 1979) immobilised on a Sepharose 4B matrix. Non-specifically bound proteins are removed with 0.5 M NaCl. Calmodulin may be eluted in high yield with buffers containing 5 mM EGTA. The advantages of such a rapid and efficient method are obvious, particularly as the protein is not exposed to denaturing conditions, and it has been used to prepare calmodulin from various lower organisms (Jamieson *et al*., 1979), plants (Charbonneau and Cormier, 1979; Watterson *et al*., 1980*b*) and fungi (Cormier *et al*., 1980) where conventional techniques had proved ineffective. (Detailed descriptions of various preparative methods have recently been published in *Methods in Enzymology*, 1983.)

Different samples of mammalian brain calmodulin prepared by any of the methods described above or by many of the other available procedures appear to be indistinguishable judged by the two most widely adopted criteria. They all give similar patterns on polyacrylamide gels run in the presence of SDS and they all activate bovine brain cAMP-dependent phosphodiesterase to a similar extent. However, Klevitt *et al.* (1981*b*) have shown that the proton magnetic resonance (PMR) spectra of these different calmodulin preparations vary considerably. Although it is not possible to judge the full significance of these differences at present, they imply that calmodulin may be subtly affected by prolonged exposure to denaturing conditions and those methods of preparation which do not employ a heat denaturation step may be preferable. Exposure to solutions containing 8 M urea (Grand *et al*., 1979) does not seem to cause changes in the PMR spectrum of calmodulin if care is taken to keep the time to a minimum (D. Dalgarno, B. A. Levine, R. J. A. Grand, unpublished results).

2.3 OCCURRENCE AND DISTRIBUTION OF TROPONIN C AND CALMODULIN

The occurrence of troponin C is restricted to those muscles in which contraction is regulated via the thin filament. Its presence has been confirmed in mammalian fast striated, slow striated and cardiac muscles, as well as in muscles from birds (Wilkinson, 1976), reptiles, amphibians and fish (Demaille *et al*., 1974). In vertebrate smooth muscle and non-muscle contractile systems, troponin C is replaced as the target for calcium ions by calmodulin.

Detailed information on the presence of troponin C in invertebrate muscles is limited and somewhat confusing mainly because there still remains some confusion over which species have thick filament (myosin)-linked regulation (Kendrick-Jones *et al*., 1970), which have thin filament (actin)-linked regulation and which have dual control. Lehman and Szent-Gyorgyi (1975) examined a large number of organisms and concluded that thin filament-linked regulation occurs in the fast muscles of the decapods and mysidacea whereas thick filament-linked control occurs in molluscs, echinoderms, brachiopods, echiuroids and nemertine worms. Dual regulation was found in insects, annelids, crustaceans, cirripedes, stomatopods, amphipods and isopods. However, later investigations have shown that a number of animals which were originally classified as having either only thick or thin filament-linked regulation have, in fact, dual regulation. For example, myosin-linked regulation was detected in decapod fast striated muscles when the ATPase assays were performed at high ionic strength (Lehman, 1977). Conversely, troponin I- and troponin C-like proteins have been isolated and characterised from the scallop *Aequipecten irradians* (Goldberg and Lehman, 1978; Lehman *et al*., 1980) and have also been shown to be present in the squid, *Ommastrephes sloani pacificus* (Konno, 1978; Tsuchiya *et al*., 1978). Since striated muscles in both molluscs and decapods are, in fact, dual regulated, it is possible that such systems operate, perhaps universally, in invertebrate muscles. If this were the case, it is likely that troponin C would be present in the muscles of all invertebrates.

The troponin-like proteins which have been isolated from invertebrate muscles have properties similar, although not identical, to those of their vertebrate counterparts. The ratio of actin : tropomyosin : troponin C, for example, in scallop and *Limulus* thin filaments is 7 : 1 : 1 (Lehman *et al*., 1976, 1980) and a hybrid troponin containing *Limulus* troponin C and rabbit troponin I and troponin T confers calcium sensitivity on rabbit actomyosin (Lehman, 1975). Troponin C from different invertebrates has a similar molecular weight (about 18 000) to the rabbit muscle proteins, but in the lower invertebrates (for example, annelids and sipunculids) appears to be rather less acidic (Lehman and Ferrell, 1980). The calcium binding properties of scallop, lobster and horseshoe crab troponin Cs also appear to be rather different from the rabbit protein, in that the former three proteins bind only about one atom of calcium per protein molecule compared with four atoms for the rabbit fast muscle troponin C (Regenstein and Szent-Györgyi, 1975; Lehman *et al*., 1976, 1980).

Nachmias and Asch (1974) and Kato and Tonomura (1975) have reported that actomyosin from the plasmodia of *Physarum* is regulated through a troponin-tropomyosin-like system. However, no protein corresponding to troponin C appears to be present (Kato and Tonomura, 1975). A number of reports have been published in which it has been claimed that troponin C has been isolated from various vertebrate smooth and non-muscle sources. In retrospect it seems probable that the proteins were in fact calmodulin. There appears to be little doubt that troponin components are absent from smooth muscle where regulation occurs through the myosin light chain kinase/calmodulin system (see Small and Sobiesek, 1980; Hartshorne and Siemankowski, 1981; Grand, 1982, for reviews).

Whereas troponin C has a relatively limited distribution (some or all animal striated muscles), calmodulin appears to be ubiquitous in eukaryotes. Originally discovered as the activator of cyclic nucleotide phosphodiesterase in brain (Cheung, 1970; Kakuichi and Yamazaki, 1970) it has gradually become apparent that calmodulin is present in all mammalian tissues and organs (Smoake *et al*.,

1974; Grand *et al*., 1979; Klee and Vanaman, 1982). The protein is most abundant in brain and testes (approximately 0.5 per cent of the protein) (Smoake *et al*., 1974; Watterson *et al*., 1976*a*; Grand and Perry, 1979) but it accounts for at least 0.2 per cent of the protein in a number of other mammalian tissues (Grand and Perry, 1979). Calmodulin has been purified from other vertebrates (Childers and Siegel, 1975; Cartaud *et al*., 1980; Watterson *et al*., 1980*c*) and has been shown to be present in particularly high concentrations (5 per cent) in the electroplax of the electric eel (Munjaal *et al*., 1980).

An initial survey of a wide range of invertebrate species by Waisman *et al.* (1975) indicated the presence of calmodulin in members of the following phyla: Cnidaria (see anemone), Mollusca (clam and snail), Annelida (earthworm), Arthropoda (blue crab and mealworm), Echinoderma (starfish) and Porifera (sponge). These observations have been confirmed by subsequent reports in which calmodulins have been purified and characterised from annelids (earthworm, *Lumbricus terrestris*: Waisman *et al*., 1978), molluscs (octopus: Giuditta *et al*., 1977; Seamon and Moore, 1980; and scallop: Yazawa *et al*., 1980), nematodes (*Caenorhabditis elegans*: Jamieson *et al*., 1980*a*), insects (locust, *Schistocerca gregaria*: Cox *et al*., 1981*b*), echinoderms (sea urchin, *Arbacia punctulata*: Head *et al*., 1979) and coelenterates (sea pansy, *Renilla reniformis*: Jones *et al*., 1979; Jamieson *et al*., 1980*b*; and sea anemone, *Metridium senile*: Yazawa *et al*., 1980). The invertebrate calmodulins are very similar although not identical to the mammalian protein. All will bind 4 mol of calcium with high affinity (possibly with the exception of earthworm calmodulin, which is reported to bind only 2 mol of calcium (Waisman *et al*., 1978)) and will activate cyclic nucleotide phosphodiesterase or myosin light-chain kinase. Amino acid analysis and sequence data (see section 2.4.1) indicate minor differences between the primary structures of mammalian and sea invertebrate calmodulins. Although the amino acid substitutions are all conservative, the replacement of the tyrosine residue 98 by phenylalanine in the lower organisms produces marked effects on the spectral properties of the proteins (Seamon and Moore, 1980; Kilhoffer *et al*., 1980*a*, 1981).

The presence of calmodulin has also been confirmed in a number of protozoans: *Amoeba* and *Euglena* (Kuznicki *et al*., 1979), *Paramecium* (Maihle *et al*., 1981) and *Tetrahymena* (Jamieson *et al*., 1979; Suzuki *et al*., 1979; Yazawa *et al*., 1981). This latter protein has been shown to be similar to mammalian calmodulin although a number of substitutions in the amino acid sequence have been observed (Yazawa *et al*., 1981). *Tetrahymena* calmodulin binds calcium with high affinity (Shimizu *et al*., 1982) and activates brain phosphodiesterase although there is not full agreement on whether it does so as effectively as the mammalian protein (Jamieson *et al*., 1979; Kakiuchi *et al*., 1981*a*). The ability of *Tetrahymena* calmodulin to activate *Tetrahymena* guanylate cyclase is not shared with calmodulins from mammals, molluscs or coelenterates (Kakiuchi *et al*., 1981*a*) and this is particularly interesting in view of the highly conserved nature of the protein.

Initial observations by Anderson and Cormier (1978) suggested that NAD kinase in plants was activated by the presence of calmodulin. Subsequent investigation has confirmed that the calcium binding protein is as widespread in plants as in animals. Calmodulins from spinach leaves (Watterson *et al.*, 1980*b*) and peanut seeds (Anderson *et al.*, 1980) are similar to mammalian calmodulin in their activation of phosphodiesterase and their ability to bind troponin I and the phenothiazine drugs (Anderson *et al.*, 1980; Watterson *et al.*, 1980*b*; reviewed by Cormier *et al.*, 1980, 1981). However, there appear to be certain differences between the animal and plant proteins in molecular weight (peanut calmodulin has a slightly lower molecular weight than bovine calmodulin), amino acid analysis and primary structure (see section 2.4.1). The most notable difference in the amino acid analyses is the presence of cysteine in peanut calmodulin. However, it is unlikely that this is of special significance as the proteins behave so similarly under assay conditions. Calmodulin prepared from barley shoots (Grand *et al.*, 1980*a*) appears to differ somewhat from other plant proteins so far studied in that it is considerably less effective than bovine calmodulin in the activation of phosphodiesterase and lacks trimethyllysine. At present no explanation can be offered for these observations.

Although considerable attention has been focused on its ability to activate NAD kinase (Cormier *et al.*, 1980, 1981; Jarrett *et al.*, 1980), it is probable that calmodulin has the same diversity of function in plants as it has in animals; for example, Dieter and Marmé (1980, 1981) have shown the importance of calmodulin in the regulation of a microsomal ATPase and in microsomal Ca^{2+} uptake in *Zea* and *Cucurbita*, respectively.

There seems to be less general agreement on the properties of fungal calmodulins than is the case for the plant and animal proteins. Cormier *et al.* (1980) found that all three types of calmodulin would activate phosphodiesterase to an equal extent and that the amino acid compositions of the proteins varied very little. However, Grand *et al.* (1980*a*), Ortega Perez *et al.* (1981) and Cox *et al.* (1982) noted marked differences in the properties of fungal calmodulins in that they were considerably less effective in the activation of phosphodiesterase and they contained no trimethyllysine residues. Similar properties have been observed for the calmodulin from *Dictyostelium discoideum* (Bazari and Clarke, 1981). It is possible that these groups could have isolated calmodulin-like proteins which have some properties in common with true calmodulins, or possibly the highly conserved nature of calmodulin is less marked in the basidiomycetes, ascomycetes or the myxomycetes.

Recent observations indicate the presence of calmodulin in yeast (Hubbard *et al.*, 1982) although a number of previous studies had concluded that the protein was absent (Clarke *et al.*, 1980; Grand *et al.*, 1980*a*). Similarly, it had been assumed that calmodulin was not present in prokaryotes (Clarke *et al.*, 1980; Grand *et al.*, 1980*a*; Yates and Vanaman, 1981), but Iwasa *et al.*, (1981*a*,*b*) have partially purified a factor from *Escherichia coli* which will activate mammalian phosphodiesterase, myosin light chain kinase and Ca^{2+}–Mg^{2+} ATPase in the presence of calcium. It will, however, be necessary to purify this activator to homogeneity before it can be classified as a prokaryotic calmodulin. Wolff *et al.* (1980) have also noted that mammalian calmodulin will activate prokaryotic adenylate cyclase. Despite these observations, the presence of calmodulin in prokaryotes is not unequivocally established.

2.4 STRUCTURAL ASPECTS OF CALMODULIN AND TROPONIN C

2.4.1 Amino acid sequence

All of those proteins which have been included in the general classification of 'intracellular calcium binding proteins' (Kretsinger, 1980) have similar amino acid compositions and this is particularly evident when considering calmodulins and troponin Cs. Both of these sets of proteins have a preponderance of aspartyl and glutamyl residues (slightly less than one-third of the total) giving a pI of approximately 4. Tryptophan is entirely absent and most of the calmodulins and troponin Cs have a single histidine and two tyrosine residues (one in invertebrates) (figure 2.3); they also have a relatively large number of phenylalanine residues (8–10), which with the tyrosines give a characteristic ultraviolet spectrum with peaks at 254, 259, 265, 269 and 278 nm (Watterson *et al.*, 1976*a*; Richman and Klee, 1979). (Note that the numbering of all amino acid sequences discussed in this section is taken from figure 2.3.) The most obvious difference between the two sets of proteins, in terms of amino acid composition, is the presence of trimethyllysine (residue 128 in figure 2.3) in all calmodulins, although it is not clear what significance attaches to its presence since Van Eldik *et al.* (1980) have shown that trimethylation has little effect on the physical properties of the protein. (The suggestion that certain invertebrate (Molla *et al.*, 1981), plant (Grand *et al.*, 1980*a*) and fungal (Grand *et al.*, 1980*a*; Bazari and Clarke, 1981) calmodulins do not contain trimethyllysine seems to be erroneous in view of conflicting evidence – compare, for example, Seamon and Moore, 1980; Cormier *et al.*, 1980, 1981).) Cysteine is absent from animal calmodulins whereas it is present in troponin Cs, but it has recently been confirmed by Charbonneau and Vanaman (unpublished observation quoted in Klee and Vanaman, 1982) that position 26 in plant calmodulins is occupied by cysteine (Anderson *et al.*, 1980; Cormier *et al.*, 1981).

Virtually complete amino acid sequences have been determined for calmodulins from nine diverse sources. They exhibit a remarkable degree of homology, differing only in a few residues, and confirm that calmodulin is one of the most conservative proteins so far studied. The primary structure of mammalian calmodulin is given in figure 2.3 (sequence 1). This probably represents the sequence occurring *in vivo* and is based on evidence obtained from studies of a number of different proteins. It is likely that all of these calmodulins are, in fact, identical and observed differences are due to sequencing errors or modifications occurring during preparation of the proteins or peptides. Bovine brain calmodulin (Watterson *et*

```
                                      10                  20                            30                40
                                                            n         n  n              n  X     Y     Z  G     I -X
1:                                   Ac-A D  Q  L  T  E  E  Q  I  A  E  F  K  E  A  F  S  L  F  D  K  D  G  D  G  T  I  T  T  K
2:                                   Ac-A D  Q  L  T  E  E  Q  I  A  E  F  K  E  A  F  S  L  F  D  K  D  G  D  G  T  I  T  T  K
3:                                    [-A D  Q  L  T  E  E  Q  I  A  E  F  K  E  A  F  S  L  F  D  K  D  G  D  G  T  I  T  T  K
4:

5:                 Ac-D T  Q  Q  A  E  A  R  S  Y  L  S  E  E  M  I  A  E  F  K  A  A  F  D  M  F  D  A  D  G  G  G  D  I  S  V  K
6:   [-(A, S, M, T) D  Q  Q  A  E  A  R  A  F  L  S  E  E  M  I  A  E  F  K  A  A  F  D  M  F  D  A  D  G  G  G  D  I  S  T  K
7:   Ac-A Q  P  T  D  Q  Q  M  D  A  R  S  F  L  S  E  E  M  I  A  E  F  K  A  A  F  D  M  F  D  T  D  G  G  G  D  I  S  T  K
8:         Ac-M D  D  I  Y  K  A  A  V  E  Q  L  T  E  E  Q  K  N  E  F  K  A  A  F  D  I  F  V_L G  A  E  D  G  C  I  S  T  K

                           50                            60                      70                            80
    -Z n        n  n        n                          n           n  n        n  X     Y     Z  G     I -X     -Z n
1:  E  L  G  T  V  M  R  S  L  G  Q  N  P  T  E  A  E  L  Q  D  M  I  N  E  V  D  A  D  G  N  G  T  I  D  F  P  E  F  L  T
2:  E  L  G  T  V  M  R  S  L  G  Q  N  P  T  E  A  E  L  Q  D  M  I  N  E  V  D  A  D  G  N  G  T  I  D  F  P  E  F  L  T
3:  E  L  G  T  V  M  R  S  L  G  Q  N  P  T  E  A  E  L  Q  D  M  I  N  E  V  D  A  D  G  D* G  T  I  D  F  P  E  F  L  S*
4:

5:  E  L  G  T  V  M  R  M  L  G  Q  T  P  T  K  E  E  L  D  A  I  I  E  E  V  D  E  D  G  S  G  T  I  D  F  E  E  F  L  V
6:  E  L  G  T  V  M  R  M  L  G  Q  N  P  T  K  E  E  L  D  A  I  I  E  E  V  D  E  D  G  S  G  T  I  D  F  E  E  F  L  V
7:  E  L  G  T  V  M  R  M  L  G  Q  T  P  T  K  E  E  L  D  A  I  I  E  E  V  D  E  D  G  S  G  T  I  D  F  E  E  F  L  V
8:  E  L  G  K  V  M  R  M  L  G  Q  N  P  T  A  E  E  L  Q  E  M  I  D  E  V  D  E  D  G  S  G  T  V  D  F  D  E  F  L  V
```

Figure 2.3 Amino acid sequences for: (1) mammalian calmodulin (see text) (the amino acid sequence of calmodulin from the eel (*Electrophorus electricus*) differs from the mammalian protein in having K instead of R at position 84 and D instead of N at position 142 (Lagacé *et al.*, 1983)); (2) calmodulin from *Renilla reniformis* (Jamieson *et al.*, 1980*b*), *Metridium senile* (Takagi *et al.*, 1980) and *Patinopecten* (Toda *et al.*, 1981) (residue 156 is K in *Renilla* and *Metridium* proteins but T in *Patinopecten* calmodulin); (3) calmodulin from *Tetrahymena pyriformis* (Yazawa *et al.*, 1981); (4) partial sequence of spinach calmodulin (Iverson *et al.*, 1981); (5) troponin C from rabbit fast striated muscle (Collins *et al.*, 1973); (6) troponin C from chicken fast striated muscle (Wilkinson, 1976); (7) troponin C from frog striated muscle (van Eerd *et al.*, 1978); (8) troponin C from bovine cardiac muscle (van Eerd and Takahashi, 1976) and rabbit slow muscle (Wilkinson, 1980) and probably bovine slow striated muscle and rabbit cardiac muscle (Wilkinson, 1980) (in the rabbit protein residue 116 is D instead of E). Residues which are identical in the eight sequences are surrounded by a solid line, residues identical in the troponin Cs are surrounded by a dotted line.

```
                               90                            100                           110                           120
    n  n        n                                      n        n  n        n  X     Y     Z  G     I  -X       -Z n
1:  M  M  A  R  K  M  K  D  T  D  -  -  -  S  E  E  E  I  R  E  A  F  R  V  F  D  K  D  G  N  G  Y  I  S  A  A  E  L  R  H
2:  M  M  A  R  K  M  K  D  T  D  -  -  -  S  E  E  E  I  R  E  A  F  R  V  F  D  K  D  G  D  G  F  I  S  A  A  E  L  R  H
3:  L  M  A  R  K  M  K  D  T  D  -  -  -  S  E  E  E  L  I  E  A  F  K  V  F  D  R  D  G  D  G  L  I  T  A  A  E  L  R  H
    *        *                                         *  *           *           *        *     *     *
4:                                                                                                                       H

5:  M  M  V  R  Q  M  K  E  D  A  K  G  K  S  E  E  E  L  A  E  C  F  R  I  F  D  R  N  A  D  G  Y  I  D  A  E  E  L  A  E
6:  M  M  V  R  Q  M  K  E  D  A  K  G  K  S  E  E  E  L  A  D  C  F  R  I  F  D  K  N  A  D  G  F  I  D  I  E  E  L  G  E
7:  M  M  V  R  Q  M  K  Q  D  A  E  G  K  S  E  E  E  L  A  E  C  F  R  I  F  D  K  N  A  D  G  Y  I  D  S  E  E  L  G  E
8:  M  M  V  R  C  M  K  D  D  S  K  G  K  S  E  E  E  L  S  D  L  F  R  M  F  D  K  N  A  D  G  Y  I  D  L  E  E  L  K  I

                               130                           140                           150                           160
    n  n        n                          n        n  n        n  X     Y     Z  G     I  -X       -Z n        n  n        n
1:  V  M  T  N  L  G  E †K  L  T  D  E  E  V  D  E  M  I  R  E  A  N  I  D  G  D  G  Q  V  N  Y  E  E  F  V  Q  M  M  T  A  K
2:  V  M  T  N  L  G  E †K  L  T  D  E  E  V  D  E  M  I  R  E  A  D  I  D  G  D  G  Q  V  N  Y  E  E  F  V  K  M  M  T  S  K
3:  V  M  T  N  L  G  E †K  L  T  D  E  E  V  D  E  M  I  R  E  A  D  I  D  G  D  G  H  I  N  Y  E  E  F  V  R  M  M  -  A  K
4:  V  M  T  N  L  G  E †K  L  T  D  E  E  V  D  E  M  I  R  E  A  D  V  D  G  D  G  Q  I  N  Y  E  E  F  V  K  V  M  M  A  K
                                                                   *  *              *  *                    *        *  *

5:  I  F  R  A  S  G  E  H  V  T  D  E  E  I  E  S  L  M  K  D  G  D  K  N  N  D  G  R  T  D  F  D  E  F  L  K  M  M  E  G  V  Q
6:  I  L  R  A  T  G  E  H  V  T  E  E  D  I  E  D  L  M  K  D  S  D  K  N  N  D  G  R  I  D  F  D  E  F  L  K  M  M  E  G  V  Q
7:  I  L  R  S  S  G  E  S  I  T  D  E  E  I  E  E  L  M  K  D  G  D  K  N  N  D  G  K  I  D  F  D  E  F  L  K  M  M  E  G  V  Q
8:  M  L  Q  A  T  G  E  T  I  T  E  D  D  I  E  E  L  M  K  D  G  D  K  N  N  D  G  R  I  D  Y  D  E  F  L  E  F  M  K  G  V  E
```

Figure 2.3 (continued) * indicates positions where variations occur in the calmodulin sequences; [represents an unknown blocking group; – indicates where a deletion has occurred; † this lysine is trimethylated in all calmodulins; 'n' are those hydrophobic residues forming the inner aspects of helices E and F (Kretsinger, 1980); X, Y, Z, -X and -Z indicate residues potentially capable of binding calcium or magnesium ions. I and G in the column headings indicate those positions normally occupied by isoleucine and glycine residues in the calcium binding proteins (see text). Numbering of residues is from the N-terminus of chicken and frog troponin C.

al., 1980*a*), bovine uterus calmodulin (Grand and Perry, 1978), rabbit skeletal muscle calmodulin (Nairn *et al.*, 1984), rabbit skeletal muscle phosphorylase kinase δ subunit (Grand *et al.*, 1981) and human brain calmodulin (Sasagawa *et al.*, 1982) differ only in a few amide assignments. These are mainly aspartic acid residues and it is likely that deamidation occurred at some stage during the sequence determination (discussed by Grand *et al.*, 1981).

Rat testis calmodulin has also been sequenced but differs from the other mammalian proteins at a number of positions. Apart from a considerable number of amide differences, Dedman *et al.* (1978*a*) also found the sequence around residue 70 to be N–D–G–A whereas in all other calmodulins this has been determined as A–D–G–N. It is probable that the former represents a sequencing error since these residues form part of calcium binding site II which is highly conserved throughout all the calmodulins and troponin Cs.

Although the weight of evidence favours the view that all mammalian calmodulins are identical, there appear to be slight differences in the proteins present in the lower vertebrates. Based on data obtained from the nucleic acid sequence of the calmodulin gene isolated from *Electrophorus electricus*, Lagacé *et al.* (1983) have shown that there are two amino acid differences between calmodulins from the eel and cow. In the former, positions 84 and 142 in the sequence are lysine and aspartic acid residues, respectively, whereas in the latter protein they are arginine and asparagine, respectively. The primary structures of invertebrate calmodulins also differ from their mammalian counterparts. Jamieson *et al.* (1980*b*) have shown that calmodulin from the marine coelenterate *Renilla reniformis* (sea pansy) contains a number of replacements when compared with bovine brain calmodulin (sequence 2, figure 2.3). The sea pansy protein has a phenylalanine residue at position 112, a lysine at position 156 and a serine at position 160 in place of tyrosine, glutamine and alanine, respectively (figure 2.3). It is probable that none of these replacements affects the three-dimensional structure or calcium binding properties of the protein to any appreciable extent since they are all of a reasonably conservative nature. Furthermore, the V → F replacement occurs in chicken striated muscle troponin C and the Q → K replacement in all of the skeletal muscle troponin Cs without any apparent effect on the physico-chemical properties of the proteins. In a number of positions in the invertebrate calmodulin, aspartic acid is present rather than asparagine (in the mammalian proteins) (figure 2.3). Although this may represent the situation *in vivo*, it is possible that deamidation could have occurred during protein preparation. The amide groups on asparagine residues in the amino acid sequence –N–G– are often lost under relatively mild conditions (discussed by Grand *et al.*, 1981) and therefore it is possible that the mammalian and invertebrate calmodulins are rather more similar than has been claimed.

The primary structures of calmodulins from scallop (*Patinopecten*) (Toda *et al.*, 1981) and the sea anemone (*Metridium senile*) (Takagi *et al.*, 1980) have also been determined, although many of the peptides in the latter protein have only been ordered on the basis of homology. *Metridium* calmodulin has similar substitutions to the *Renilla* calmodulin at positions 112, 156 and 160. Unfortunately, many of the amides in this sequence have not been assigned; therefore it is not possible to confirm the data of Jamieson *et al.* (1980*b*) on this point (it has been assumed in figure 2.3 that all amides in calmodulins from the sea anemone are identical to those in the scallop (Toda *et al.*, 1981) and sea pansy (Klee and Vanaman, 1982). Scallop calmodulin has a threonine at position 156, compared with glutamine in mammals, lysine in the other sea invertebrates and spinach (Iverson *et al.*, 1981), and arginine in *Tetrahymena* (Yazawa *et al.*, 1981). Again this substitution will have little effect on the structure of the protein.

Calmodulin from the protozoan *Tetrahymena pyriformis* (sequence 3, figure 2.3) has rather more amino acid substitutions, when compared with the mammalian protein, than the invertebrates (Yazawa *et al.*, 1981), but most are of a conservative nature (for example K → R, 107, and S → T, 114). However, since most of these substitutions occur around calcium binding sites three and four, it is possible that they could account for the differences in calcium binding properties observed by Yazawa *et al.* (1981). On the whole, however, primary structure differences between mammalian and protozoan calmodulins are very slight, yet it has been reported (Kakiuchi *et al.*, 1981*a*) that only the latter protein is capable of serving as the calcium binding subunit for *Tetrahymena* guanylate cyclase. The partial amino acid sequence for spinach leaf calmodulin (Iverson *et al.*, 1981) is also presented in figure 2.3 (sequence 4).

Of all the different calcium binding proteins which are homologous to calmodulin, the group which is most closely related, both in terms of primary structure and calcium binding properties, is the troponin Cs (Goodman, 1980), with about half the residues in mammalian calmodulin and rabbit fast striated muscle troponin C being identical. Complete amino acid sequences have been determined for TNC from rabbit fast striated muscle (Collins *et al.*, 1973, 1977), rabbit slow striated and cardiac muscle (Wilkinson, 1980), bovine cardiac muscle (van Eerd and Takahashi, 1976), human fast striated muscle (Romero-Herrera *et al.*, 1976), frog striated muscle (van Eerd *et al.*, 1978) and chicken fast striated muscle (Wilkinson, 1976) and these are shown in figure 2.3. On the basis of observations by Wilkinson (1980) and Romero-Herrera *et al.* (1976) it has been concluded that slow striated muscle and cardiac muscle troponin Cs are the product of a single gene for any particular mammalian species. Wilkinson (1980) showed that the sequence of rabbit slow muscle troponin C differed from bovine cardiac troponin C only in a D for E replacement at residue 116 (protein 8, figure 2.3). This probably represents a species difference rather than a tissue difference. The author argued that if the cardiac and slow muscle TNCs were the products of different genes, they would have diverged about 10^9 years ago and, given a mutation rate for troponin C of about 1.6 PAM units/10^8 years (Barker *et al.*, 1977), the two sequences would vary by about 30 per cent. As this is obviously not the case, it is likely that in most, if not all, mammals the slow and cardiac troponin Cs are identical. Similarly, Romero-Herrera *et al.*

(1976) found that a minor species present in their preparation of human skeletal muscle TNC was identical to bovine cardiac TNC over at least half its length. This probably represents the slow muscle form of human TNC since the proteins were prepared from muscles containing a mixed population of slow and fast fibres.

Recent evidence obtained from a study of mRNAs encoding quail striated muscle proteins (Hastings and Emmerson, 1982) has confirmed the presence of DNA coding for a protein virtually identical to rabbit slow/cardiac troponin C in striated muscle. The authors concluded that it represented the slow muscle form of quail TNC.

In spite of the similarities evident in figure 2.3, the amino acid sequences of troponin Cs are much less highly conserved than calmodulins. Troponin Cs from rabbit fast and slow muscles, bovine cardiac muscle and frog and chicken fast muscle only have 50 per cent of their residues in common, compared with calmodulin which seems to be invariant in vertebrates. The similarities between the two groups of calcium binding proteins are very marked and about 42 per cent of the mammalian calmodulin residues are common to all the troponin Cs (figure 2.3). Similarities in the sequences are most pronounced in the calcium binding sites II, III and IV (cardiac TNC has lost the ability to bind divalent cations at site I – see section 2.4.2) and at the positions 'n' in figure 2.3, which are hydrophobic residues representing the inner aspects of the α-helices E and F (Kretsinger, 1980; see section 2.4.2). Where there are variations between the proteins, these tend to be conservative replacements which are unlikely to affect the three-dimensional shape or charge distribution except at the N-terminus where about 12 residues are present, which obviously have mutated at a rather higher rate than the rest of the proteins. It seems probable that this part of troponin C is of little importance in the functioning of the calcium binding proteins; indeed calmodulin is ten residues shorter than chicken and frog muscle troponin Cs, having lost this area totally. In the troponin Cs the amino acid sequences between calcium binding sites I and II (residues 50-60) and sites II and III (residues 88-98) are as highly conserved as the helices and divalent cation binding sites and have been implicated in the binding of troponin I to troponin C (see section 2.5.1).

It is possible that the presence of amino acids which are common to all troponin Cs but are not similar to those present in the calmodulins (i.e. non-conservative replacements) is of functional significance – for example, the D and T for E and A replacements at residues 89 and 116, respectively. Similarly, the deletion of three residues between binding sites II and III may facilitate the interaction of calmodulin with the calmodulin-dependent enzymes.

2.4.2 Nucleic acid sequences

Although considerable information is available on the amino acid sequences of different calmodulins, relatively little is known about the mRNAs coding for the protein or the organisation of the calmodulin gene. However, recent work by Lagacé *et al.* (1983) has started to provide solutions to these problems. These authors have used a partial calmodulin cDNA, originally isolated from the eel *Electrophorus electricus* by Munjaal *et al.* (1980), to prepare a full-length calmodulin cDNA, which has been cloned and sequenced. The clone was found to be composed of a 5′ nontranslated region (26 nucleotides), a region coding for calmodulin (the amino acid sequence is presented in figure 2.3) and a 3′ non-translated region (408 nucleotides). Three cytoplasmic mRNAs of varying lengths were also isolated, as well as a 5.5 kb nuclear RNA which probably represents the primary transcript. Lagacé *et al.* (1983) have concluded that the three cytoplasmic RNAs arise from the common precursor by differential polyadenylation. Because of the considerable length of the nuclear RNA it is probable that a number of intervening sequences (introns) are present in the calmodulin gene from the eel.

Lagacé *et al.* (1983) have also sequenced genomic clones of two chicken calmodulin genes. One of these was found to contain introns whereas the other did not. Preliminary data suggested that only the intron-containing gene was expressed in all chicken tissues.

cDNA clones coding for parts of two troponin Cs have also been sequenced. Hastings and Emmerson (1982) concluded that quail slow striated muscle troponin C differed from the rabbit protein in two conservative replacements between residues 93 and 117 (figure 2.3). Similarly, Garfinkel *et al.* (1982) concluded, on the basis of sequencing of cDNA, that rat and rabbit fast striated muscle troponin Cs were identical over much of their sequence (residues 75-119, figure 2.3).

In the near future the application of molecular biology to the study of calcium binding proteins should answer a number of crucial questions concerning calmodulin and troponin C genes. Specifically, it should provide information on the number of genes coding for each protein in any given genome: for example, whether the situation described by Lagacé *et al.* (1983) for chicken calmodulin genes is the general rule, or whether multiple gene copies (differing perhaps in the number and extent of introns and in the third bases present in some of the triplets) code for identical proteins. It should also allow the calmodulin amino acid sequences to be established unequivocally.

2.4.3 Three-dimensional structure and metal ion binding properties of troponin C and calmodulin

'EF-hands' in troponin C and calmodulin

In a crystallographic study of the carp calcium binding protein (CBP, parvalbumin), Kretsinger (1972) and Kretsinger and Nockolds (1973) noted marked similarities in the three-dimensional structure of two segments of the protein (residues 39-69 and 78-108) such that they were virtually superimposable. Each of these segments comprised a short section of α-helix (E), a calcium binding loop, and another short section of α-helix (F). A third segment of the protein which does not bind calcium ions does, however, contain two α-helices. Kretsinger concluded that the protein had arisen through gene triplication although the internal homologies in the amino acid sequence were very weak.

A comparison of the primary structures of carp CBP and rabbit fast striated muscle troponin C allowed Kretsinger and Barry (1975) to postulate a three-dimensional structure, based on the original observations of the carp CBP, for TNC, which has since been applied to the whole calcium binding protein family. They concluded that troponin C consisted of two pairs of 'EF-hands', each pair as found in carp CBP. The term 'EF-hand' was coined by Kretsinger *et al.* (1974) (see also Kretsinger and Barry, 1975; Kretsinger, 1977) to refer to the disposition of the polypeptide chains around the calcium binding site of parvalbumin. This region is composed of two turns of α-helix (E), a calcium binding loop in which the six calcium coordinating ligands in the primary structure must be occupied by appropriate amino acids, and a further two turns of α-helix (F) arranged at right angles to the first. Most 'EF-hands' found in the calcium binding proteins which retain the ability to bind divalent cations have a number of properties in common (Kretsinger, 1980). They comprise 31 residues and certain positions in the helical parts of the sequence (3, 6, 7, 10, 23, 26, 27 and 30) are occupied by hydrophobic residues – designated 'n' in figure 2.3; the calcium binding ligands are assigned to the vertices of an octahedron and are designated X, Y, Z, -X and -Z (figure 2.3). The sixth ligand, -Y, occurring between -X and -Z, is not shown because the oxygen atom responsible for metal ion binding is part of the polypeptide chain backbone and is not donated by a side chain. The other calcium coordinating residues generally have oxygen-containing side chains in those sites which bind metal ions *in vivo*, although some of these may also be contributed by backbone oxygens – for example in ICaBP calcium binding site 1 (Szebenyi *et al.*, 1981). In the 'EF-hands', positions 15 and 17 are normally occupied by glycine and isoleucine residues, respectively. Such a structure is consistent with the original observations of Kretsinger and Nockolds (1973) who found that most of the hydrophilic amino acids were on the surface of carp CBP whereas most of the hydrophobic amino acids were directed inwards to form a hydrophobic core.

The predictions of the three-dimensional structure for troponin C and calmodulin have been made solely on the basis of amino acid sequence data and although some preliminary results are available (Cook *et al.*, 1980; Kretsinger *et al.*, 1980; Strasburg *et al.*, 1980), there is no high-resolution crystallographic data on the structure of any of the calcium binding proteins with the exception of the carp CBP and the vitamin D-dependent calcium binding protein from bovine intestine (Szebenyi *et al.*, 1981). Studies of this latter protein have confirmed that the 'EF-hand' structure occurs in proteins other than the parvalbumins. In particular the ICaBP III-IV domain closely resembles the parvalbumin 'EF-hand'; however, the calcium binding loop of the I-II domain has been modified to some extent with two amino acid insertions (an alanine at position 15 and an asparagine at position 21) such that the first four calcium ligands are main-chain carbonyl groups.

The amino acid sequences of the four 'EF-hands' in the calmodulins meet the criteria suggested by Kretsinger for normal metal ion binding and for formation of the two areas of α-helix on either side of the calcium binding loop. The five side-chain calcium ligands are occupied by amino acids which contain oxygen atoms in their side groups. Similarly, in rabbit fast skeletal muscle troponin C the four sites contain sequences which will bind divalent cations. In the cardiac and slow muscle proteins, however, it is probable that the ability of the first loop to bind calcium or magnesium has been lost as a result of an amino acid insertion (valine or leucine) at residue 31 (figure 2.3) and the replacement of two of the calcium binding residues (positions X and Y) with inappropriate amino acids (valine or leucine and alanine). The suggestion, based on these sequence data, that cardiac and slow muscle troponin Cs only bind three metal ions has been confirmed by direct measurements (see below).

van Eerd and Takahashi (1976) noted that the relative mutation rate of the first 33 residues of cardiac troponin C (encompassing site I) was approximately twice that of the rest of the molecule, and they concluded that, once the ability to bind metal ions had been lost by this area of the protein, restrictions on the rate of mutation would be relaxed. However, since there appears to be no difference between rabbit and beef cardiac troponin C in this region (Wilkinson, 1980), it seems likely that although metal ion binding ability has been lost, site I is still important for the contribution which it makes to the three-dimensional structure of the protein.

2.4.3.2 Calcium and magnesium binding

Direct measurements of metal ion binding have confirmed predictions based on amino acid sequence data that calmodulin and fast skeletal muscle troponin C bind four metal ions each. Potter and Gergely (1975) originally reported that troponin C, when present in the troponin complex, has two sites that can bind calcium with high affinity ($K_a = 5 \times 10^8$ M^{-1}) and two sites that bind with lower affinity ($K_a = 5 \times 10^6$ M^{-1}). The high-affinity sites are also capable of binding magnesium ions ($K_a = 5 \times 10^4$ M^{-1}) competitively and these were generally referred to as the high-affinity Ca^{2+}-Mg^{2+} sites (table 2.1), whereas the low-affinity sites are calcium specific at physiological Mg^{2+} concentrations (1-2 mM). Studies using spin-labelled or dansylaziridine-labelled troponin C (Potter *et al.*, 1976, 1977), proteolytic fragments (Leavis *et al.*, 1977, 1978) and chemically modified troponin C (Sin *et al.*, 1978) have allowed the high-affinity sites to be assigned to metal ion binding sites III and IV (figures 2.2 and 2.3) and the low-affinity calcium-specific sites to be assigned to sites I and II.

On the basis of amino acid sequence data it is apparent that site I of cardiac troponin C does not bind metal ions and therefore by analogy with skeletal muscle troponin C it has been assumed that the cardiac protein possesses only one Ca^{2+}-specific site (II) and two Ca^{2+}-Mg^{2+} high-affinity sites (III and IV).

Table 2.1 Calcium and magnesium binding properties of skeletal and cardiac muscle troponin Cs and calmodulin

Protein or peptide	Metal binding side	Cation	K_a (M^{-1})	k_{off} (s^{-1})^a	References (for K_a)
Skeletal troponin C	I, II	Ca^{2+}	$2\text{–}4 \times 10^5$	300	Potter and Gergely (1975);
	III, IV	Ca^{2+}	2×10^7	3–4	Hincke *et al.* (1978);
	III, IV	Ca^{2+}I/PMg^{2+}	2×10^6	–	Johnson and Potter (1978);
	III, IV	Mg^{2+}	5×10^3	8	Potter *et al.* (1980);
					Potter and Johnson (1982)
Skeletal troponin	I, II	Ca^{2+}	5×10^6	23	Potter and Gergely (1975);
	III, IV	Ca^{2+}	5×10^8	0.6	Potter *et al.* (1980);
	III, IV	Ca^{2+}I/PMg^{2+}	5×10^6	–	Potter and Johnson (1982)
	III, IV	Mg^{2+}	$4\text{–}5 \times 10^4$	2.0	
Skeletal troponin C peptides					
Tr1	I, II	Ca^{2+}	$>10^5$	$10^2\text{–}10^3$	Leavis *et al.* (1978, 1980)
Tr2	III, IV	Ca^{2+}	5×10^7	<80	
Tr2	III, IV	Ca^{2+}I/PMg^{2+}	5×10^6	–	
Th1	I, II	Ca^{2+}	$>10^5$	–	
	III	Ca^{2+}	2×10^5	–	
CB8	II	Ca^{2+}	8×10^3	$>10^4$	
CB9	III	Ca^{2+}	2.2×10^5	<300	
Th2	IV	Ca^{2+}	2.5×10^4	$10^3\text{–}10^4$	
Cardiac troponin C	II	Ca^{2+}	2×10^4 to 2×10^5	300	Potter *et al.* (1977);
	III, IV	Ca^{2+}	$10^7\text{–}2 \times 10^7$	–	Hincke *et al.* (1978);
	III, IV	Ca^{2+}I/PMg^{2+}	3.6×10^6	–	Leavis and Kraft (1978);
	III, IV	Mg^{2+}	1.2×10^3		Holroyde *et al.* (1980)
Cardiac troponin	II	Ca^{2+}	2×10^6	20	Holroyde *et al.* (1980);
	III, IV	Ca^{2+}	3×10^8	0.3	Potter *et al.* (1980);
	III, IV	Ca^{2+}I/PMg^{2+}	2×10^7	–	Potter and Johnson (1982)
	III, IV	Mg^{2+}	3×10^3	3.3	

Calmodulin	High affinity (1)[b]	Ca^{2+}	3.3×10^5	–	Teo and Wang (1973)
	Low affinity (2)	Ca^{2+}	8.3×10^4	–	Teo and Wang (1973)
	Four sites	Ca^{2+}	2.5×10^5 to 5.6×10^4	–	Lin *et al.* (1974)
	High affinity (2)	Ca^{2+}	9×10^5	–	Watterson *et al.* (1976*a*)
	Low affinity (2)	Ca^{2+}	8.6×10^4	–	Watterson *et al.* (1976*a*)
	Equivalent (4)	Ca^{2+}	4.2×10^5	–	Dedman *et al.* (1977*a*); Potter *et al.* (1977)
	High affinity (3)	Ca^{2+}	2×10^7	–	Wolff *et al.* (1977)
	Low affinity (1)	Ca^{2+}	10^6	–	Wolff *et al.* (1977)
	High affinity (3)	Mg^{2+}	2×10^5	–	Wolff *et al.* (1977)
	Low affinity (3)	Mg^{2+}	1.4×10^4	–	Wolff *et al.* (1977)
	Equivalent (3)	$Ca^{2+}I/PMg^{2+}$	3×10^6	–	Wolff *et al.* (1977)
	Four nonequivalent sites	Ca^{2+}	8.6×10^5	–	Crouch and Klee (1980)
		Ca^{2+}	3×10^5	–	Crouch and Klee (1980)
		Ca^{2+}	1.2×10^5	–	Crouch and Klee (1980)
		Ca^{2+}	0.45×10^5	–	Crouch and Klee (1980)
		$Ca^{2+}I/PMg^{2+}$	2.1×10^5	–	Crouch and Klee (1980)
		$Ca^{2+}I/PMg^{2+}$	1.9×10^5	–	Crouch and Klee (1980)
		$Ca^{2+}I/PMg^{2+}$	0.4×10^5	–	Crouch and Klee (1980)
		$Ca^{2+}I/PMg^{2+}$	0.26×10^5	–	Crouch and Klee (1980)
	High affinity (3)	Ca^{2+}	1.7×10^5	–	Cox *et al.* (1981*c*)
	Low affinity (1)	Ca^{2+}	5×10^3	–	Cox *et al.* (1981*c*)
	High affinity (2)	Ca^{2+}	2.5×10^5	–	Kohse and Heilmeyer (1981)
	Low affinity (2)	Ca^{2+}	2.5×10^4	–	Kohse and Heilmeyer (1981)
	High affinity (2)	Mg^{2+}	1.3×10^3	–	Kohse and Heilmeyer (1981)
	Four nonequivalent sites	Ca^{2+}	9×10^7	–	Haiech *et al.* (1981)
		Ca^{2+}	6×10^7	–	Haiech *et al.* (1981)
		Ca^{2+}	1.7×10^7	–	Haiech *et al.* (1981)
		Ca^{2+}	0.67×10^7	–	Haiech *et al.* (1981)
		Mg^{2+}	1.5×10^3	–	Haiech *et al.* (1981)
		Mg^{2+}	8.7×10^3	–	Haiech *et al.* (1981)
		Mg^{2+}	10.6×10^3	–	Haiech *et al.* (1981)
		Mg^{2+}	3.7×10^3	–	Haiech *et al.* (1981)

[a] Values for k_{off} for the dissociation of calcium from the troponin components are taken from Potter *et al.* (1980) and Potter and Johnson (1982), and for the peptides from Leavis *et al.* (1980). Values for k_{off} for calmodulin are probably about 250 s^{-1} (Potter *et al.*, 1980).

[b] 'High affinity' and 'low affinity' refer to the type of binding sites; in most cases the authors have made no attempt to relate these to the actual binding sites on the protein. Numbers in parentheses refer to the number of sites.

Indeed these are essentially the results that have been obtained by Potter *et al.* (1977, 1980*a*), Leavis and Kraft (1978) and Holroyde *et al.* (1980) (table 2.1). It should be noted that the affinity of the Ca^{2+}-specific sites and the Ca^{2+}-Mg^{2+} sites in skeletal and cardiac muscle troponin and the troponin C–troponin I complex is approximately ten times greater than in isolated troponin C. (The binding of metal ions to troponin C has been considered fully by Potter and Johnson, 1982.)

Although the amino acid sequence of calmodulin is very similar in the calcium binding sites to that of troponin C (figure 2.3), it is possible that the metal ion binding properties of the two sets of proteins are appreciably different. However, at present there appears to be some disagreement on what these properties might be in the case of calmodulin.

It has been suggested (Dedman *et al.*, 1977*a*; Potter *et al.*, 1977, 1980*a*) that all four sites on calmodulin are equivalent and are specific for calcium, even at relatively high magnesium concentrations (3 mM). However, Wolff *et al.* (1977) and Wolff and Brostrom (1979) reported that only three sites are equivalent and the fourth unique site probably binds Mg^{2+} ions under physiological conditions. Another alternative scheme has been proposed by Kohse and Heilmeyer (1981) in which it is suggested that calmodulin binds divalent cations in the same way as troponin C, i.e. with two Ca^{2+}-Mg^{2+} sites and two Ca^{2+}-specific sites. Proton magnetic resonance studies (Seamon, 1980) have tended to support the view that there are two pairs of metal ion binding sites. By comparison with results obtained for troponin C the high-affinity sites were assigned to calcium binding sites III and IV (Seamon, 1980). However, Kilhoffer *et al.* (1980*a*,*b*) and Wang *et al.* (1982) on the basis of a study of terbium binding have concluded that, in contrast to troponin C, sites I and II are the high-affinity sites and that the order of metal ion binding is: sites I and II simultaneously, then III and finally site IV. Whether extrapolation can be made from these studies directly to calcium binding is not clear although the authors consider that binding of the two metals to calmodulin occurs in a similar manner. On the other hand Andersson *et al.* (1983) have suggested that the terbium fluorescence data may be an artifact produced by 'unexpected orientation of the tyrosines with respect to the calcium binding sites', and have assigned the high-affinity sites to sites III and IV. Since some of these latter studies were performed on proteolytic fragments of the protein (residues 78–148), they may provide a more reliable representation of the reality of the situation.

Recent evidence has suggested positive co-operativity between the metal ion binding sites (Crouch and Klee, 1980). However, Haiech *et al.* (1981) consider that the binding of K^+ is responsible for this apparent co-operativity. These authors found the binding constants for calcium for all four sites to be different. On this basis they suggested that binding of Ca^{2+} occurred sequentially, II → I → III → IV or II → III → I → IV. In the presence of physiological Na^+ or K^+ concentrations it is likely that magnesium is not bound by calmodulin, even in the absence of calcium (Kilhoffer *et al.*, 1981). The resolution of these contradictory findings will have to await rather more detailed and careful study. A number of criticisms have been aimed at some of the reports cited above, casting some doubt on the validity of the results. For example, there is now considerable evidence that EGTA binds strongly to the calcium binding proteins and it is probable, therefore, that studies using Ca^{2+}/EGTA buffers (Dedman *et al.*, 1977*a*; Potter *et al.*, 1977) may produce anomalous results.

There is some evidence to suggest that both troponin C and calmodulin can bind metal ions at sites other than those generally recognised as the divalent cation binding sites (designated in figure 2.3). Potter and Gergely (1975) found in their original study that two Mg^{2+}-specific sites could be detected both in troponin C and whole troponin. Recently it has been reported that two Mg^{2+}-specific sites also exist on calmodulin (Kohse and Heilmeyer, 1981). The position and significance of these will have to await further investigation.

It is probable that under resting conditions and during a single transient stimulation of a muscle fibre the high-affinity sites on troponin C will remain saturated with Mg^{2+}; after repeated pulses of Ca^{2+} a level of 35 per cent saturation of Ca^{2+} on these sites may be achieved (Potter *et al.*, 1980*a*). However, under all circumstances the off rate (even in the absence of Mg^{2+}) is too slow to allow them to play a regulatory role in muscle contraction. In contrast the on and off rates of the Ca^{2+}-specific sites are sufficiently fast for the bound calcium to follow the rises and falls in calcium concentration closely (Potter *et al.*, 1980*a*).

At present it is still far from clear which metal ion binding sites on calmodulin need to be occupied for interaction with the catalytic subunits of the different calmodulin-dependent enzymes; whether additional calcium ions can be bound after formation of the Ca^{2+}-calmodulin–enzyme complex, and whether mechanisms which have been worked out for one system are generally applicable. A consideration of the changes in three-dimensional structure which calmodulin undergoes on binding calcium ions may suggest some answers to these problems, but at present our knowledge is too limited to provide unequivocal answers. It does appear, however, that activation of phosphodiesterase requires calmodulin with three or four bound calciums (Crouch and Klee, 1980; Cox *et al.*, 1981*c*) whereas activation of myosin light chain kinase may only require one or two calcium atoms to be bound (Crouch *et al.*, 1981; Johnson, J. D. *et al.*, 1981).

2.4.3.3 Structure of troponin C and calmodulin in solution

Considerable information on the structure of the calcium binding proteins has been provided by following the changes that occur during the addition of calcium and magnesium using proton magnetic resonance. In the absence of divalent cations (apoprotein), skeletal muscle troponin C contains ordered areas, possibly similar to those in the native protein, but exists in a relatively extended conformation (Levine *et al.*, 1977; Seamon *et al.*, 1977). If calcium ions are titrated into a solution of the protein, a number of changes in tertiary structure can be observed. Initially these may be envisaged as an increase in the stabilising restraints imposed on the protein; Levine *et al.* (1977) and Seamon *et al.* (1977) have suggested that this coincides with the addition of Ca^{2+} ions to regions of preformed structure corresponding to binding sites III and IV (the high-affinity sites).

The major changes in the troponin C tertiary structure occur during the addition of approximately 0.7 to 1.8 mol of calcium per mole of protein – the polypeptide backbone is rearranged and the structure of the protein is generally 'tightened'. This stage presumably coincides with the filling of the remaining high-affinity sites, which obviously are of crucial importance in stabilising the tertiary structure of troponin C (rather than playing a regulatory role) (Levine *et al.*, 1977). Addition of the final 2 mol of calcium leads to more local changes in structure. In particular, glutamic and aspartic acid residues are affected as well as a cluster of hydrophobic side chains which are spatially close to one another (although not necessarily close in the amino acid sequences). There is also a further 'tightening' and stabilisation of the structure. These observations are generally consistent with earlier reports which noted a reduction in α-helix content from about 50 per cent to 35 per cent on removal of calcium from the high-affinity sites of troponin C (Murray and Kay, 1972; Mani *et al.*, 1974).

Conformation changes to troponin C during the initial stages of titration with Mg^{2+} are similar , although not identical, to the changes occurring during the addition of Ca^{2+} (Levine *et al.*, 1978; Andersson *et al.*, 1981). These differences in effect of the two cations may be seen in the extent of backbone folding and in the interaction between groups of hydrophobic residues. One interesting observation, based on the PMR study of troponin C in the presence of calcium and magnesium, was that when increasing concentrations of Ca^{2+} ions were added, in the presence of saturating Mg^{2+} concentrations, changes did not occur in discrete stages but were blurred together. Also, Ca^{2+} began to displace Mg^{2+} ions from the high-affinity sites before the low-affinity sites were filled. Levine *et al.* (1978) could detect no magnesium binding to the low-affinity sites – even at high concentrations – confirming the original proposal that these are calcium specific.

Further PMR studies using proteolytic fragments of rabbit skeletal muscle troponin C have allowed the changes occurring in the two halves of the molecule on metal ion binding to be more closely defined (Evans *et al.*, 1980*a*; Leavis *et al.*, 1980). In the absence of divalent cations a peptide covering the C-terminal half of troponin C (residues 89–159) has relatively little tertiary structure. The addition of Ca^{2+} or Mg^{2+} results in changes in both secondary and tertiary structure, producing a very stable conformation similar to that observed in troponin C (Leavis *et al.*, 1980). Of more interest are the conformational changes occurring in the peptide encompassing the two low-affinity calcium binding sites (residues 9–84 and 1–120). When Ca^{2+} is bound at sites I and II, a destabilisation occurs such that the hydrophobic amino acids (F 19, 23, 26) in the helical region of site I move away from hydrophobic residues (F 72 and 75) in the helical region of site II; there appears to be little change, however, in the actual α-helical content of the peptide. In the intact protein, this movement of the tertiary fold is constrained somewhat by the rest of the C-terminal half of the molecule such that weak contact is still maintained between the two hydrophobic areas. Leavis *et al.* (1980) have concluded 'this long range interaction between the two halves of the troponin C molecule influences the Ca (II) triggered alteration in tertiary structure which initiates the cascade of structural events correlated with actomyosin ATPase activity'. It is not clear whether the alteration in the hydrophobic areas of troponin C (Levine *et al.*, 1977, 1978; Johnson *et al.*, 1978; Leavis *et al.*, 1980) on calcium binding is analogous to the calcium-induced exposure of a hydrophobic area on the surface of calmodulin (La Porte *et al.*, 1980; and see section 2.5.2.1), but it is tempting to propose that, despite considerable differences between troponin C and calmodulin, both proteins undergo related conformational changes as a result of metal ion binding.

Other physico-chemical techniques besides PMR have been used to study structural changes in troponin C, although these have been rather less informative. For example, tyrosine fluorescence and circular dichroism measurements performed on the whole molecule and the C-terminal half of the protein have confirmed that sites III and IV are the high-affinity Ca^{2+}-Mg^{2+} sites and that binding of metal ions results in a large increase in helical content and tyrosine fluorescence of the protein (Johnson and Potter, 1978; Leavis *et al.*, 1978). Much less marked spectral changes could be associated with binding of Ca^{2+} to the low-affinity sites, either in troponin C (Johnson and Potter, 1978) or in a peptide covering only sites I and II. This result is consistent with the PMR data, which suggests subtle rather than gross changes associated with Ca^{2+} binding to the regulatory sites. Separate peptides containing sites III and IV bind Ca^{2+} relatively weakly (and Mg^{2+} not at all) and it has, therefore, been concluded that interactions between various parts of troponin C are crucial in determining its properties rather than simply amino acid sequences.

Although studies on the conformation changes of calmodulin in the presence of increasing concentrations of divalent cations have been more numerous than those of troponin C, there appears to be rather less detailed information on the molecular events that occur. Using tyrosine fluorescence and circular dichroism measurements it has been shown that the α-helical content of calmodulin increases by about 10 per cent on binding calcium (Dedman *et al.*, 1977*a*; Klee, 1977; Potter *et al.*, 1977; Wolff *et al.*, 1977) and magnesium (Wolff *et al.*, 1977). Estimates of the helical content of the apoprotein vary from about 30 per cent (Klee, 1977; Wolff *et al.*, 1977) to about 45 per cent (Dedman *et al.*, 1977*a*). The two tyrosine residues in calmodulin are present in different environments – tyrosine-99 is accessible to solvent and is therefore probably on the surface of the protein whereas tyrosine-138 is deeply buried in the hydrophobic core (Klee, 1977; Richman, 1978; Richman and Klee, 1978, 1979; Seamon, 1980). Use has been made of this difference in availability, with tyrosine-99 being used as a sensor in a number of investigations of calmodulin structure. As a result of spectral and PMR studies it has been concluded that the major conformational changes occur in calmodulin after binding two atoms of calcium (Klee, 1977; Crouch and Klee, 1980; Seamon, 1980).

Proton magnetic resonance studies on calmodulin (Seamon, 1980; Krebs, 1981) have confirmed that structural changes occur that are analogous to those that have been observed in skeletal muscle troponin C (Levine *et al.*, 1977; Seamon *et al.*, 1977), with 80 per cent of the changes occurring after the binding of two calciums. Seamon (1980) has found that magnesium binding to calmodulin causes resonance shifts similar to those observed on calcium binding but that large conformational changes occur only in the presence of calcium.

An alternative approach has been to examine the increased resistance of calmodulin to proteolysis in the presence of divalent cations. In the presence of calcium, calmodulin is not digested by thrombin (Wall *et al.*, 1981), but when calcium is removed, the peptide bond between residues 106 and 107 is rapidly cleaved suggesting that the amino acid sequence encompassing binding site III is exposed in the apoprotein. The addition of magnesium ions only marginally reduces the degradation rate, confirming that the conformation of calmodulin in the presence of Mg^{2+} is not the same as in the presence of Ca^{2+}. In contrast, troponin C is unaffected by thrombin in the presence of either Ca^{2+} or Mg^{2+} (Wall *et al.*, 1981).

The calcium-bound form of calmodulin is also much more resistant to tryptic digestion than is the apoprotein (Drabikowski *et al.*, 1977; Walsh *et al.*, 1977) although the peptide bond between residues 77 and 78 is cleaved at relatively high concentrations of trypsin. The resulting N-terminal peptide (1-77) is then rapidly digested, suggesting that this half of calmodulin is not highly structured in the absence of the remainder of the molecule.

Peptides corresponding to both the N- and C-terminal halves of calmodulin bind Ca^{2+} and undergo conformational changes (Drabikowski *et al.*, 1977; Walsh *et al.*, 1977; Brzeska *et al.*, 1980; Wall *et al.*, 1981) which are similar as determined by circular dichroism measurement and mobility changes on polyacrylamide gels (Brzeska *et al.*, 1980; Drabikowski *et al.*, 1980). It seems probable that peptides encompassing single binding sites also bind Ca^{2+} but do not show any pronounced conformational changes (Brzeska *et al.*, 1980; R. J. A. Grand, C. M. Wall and S. V. Perry, unpublished observations). Digestion of calmodulin with trypsin in the absence of calcium proceeds rapidly, but if the amount of proteolytic enzyme is reduced, it has been found that bonds other than 77-78 are split (106-107 and 90-91), suggesting that considerable conformational change occurs in the area around sites II and III when Ca^{2+} is bound.

2.4.3.4 Binding of other cations

Both troponin C and calmodulin have the ability to bind metals other than calcium and magnesium, for example manganese (Wolff *et al.*, 1977), cobalt and zinc (Teo and Wang, 1973), terbium (Kilhoffer *et al.*, 1980*b*) and various other lanthanides (Wang *et al.*, 1982). However, the only cations that are present in the cell in sufficient concentrations such that binding might occur *in vivo* are sodium and potassium. Using ^{23}Na nuclear magnetic resonance, Delville *et al.* (1980) have shown that Na^+ binds to troponin C and troponin C peptides (residues 9-84 and 89-159) in a specific and tight fashion. Sodium ions compete with calcium for the low-affinity sites, but the binding of Na^+ to the high-affinity sites, although it results in significant conformational changes, is non-competitive with calcium. In the presence of relatively high calcium concentrations, sodium appears to be expelled from the high-affinity sites totally but continues to bind to weak secondary sites elsewhere on the molecule. At present it is not clear what effect the presence of Mg^{2+} would exert on the sodium binding. Potassium ions bind competitively to the four sites on calmodulin (Haiech *et al.*, 1981), but are displaced by calcium ions.

2.5 PROTEIN-PROTEIN INTERACTIONS INVOLVING TROPONIN C AND CALMODULIN

It is in a consideration of the protein–protein interactions that calmodulin and troponin C can undergo that the differences between the two molecules become most striking. Troponin C only binds to the other components of the troponin complex (troponin I and troponin T) and although it will interact with a limited number of other proteins *in vitro*, it is probable that this is a function of its similarity to calmodulin (or its strongly anionic character) rather than a true representation of what occurs in the cell. Calmodulin, on the other hand, by virtue of its role as a ubiquitous intracellular calcium binding subunit, must bind to a large number of other proteins – specifically those enzymes which it regulates and also those proteins of unknown function which have been termed the calmodulin-binding proteins (CaM-BP). Our knowledge of what occurs during troponin C–troponin I interactions is considerably more detailed than is the case for any of the calmodulin–protein interactions and therefore emphasis will be placed on this aspect in the hope that it will help to demonstrate how the calcium binding proteins exert their influence as a result of metal ion binding.

2.5.1 Troponin C

As described in section 2.3, troponin C acts as the major calcium binding subunit in striated muscle. When the muscle is stimulated by a nervous impulse, calcium ions are released from the sarcoplasmic reticulum; their binding to the low-affinity calcium-specific binding sites on troponin C results in muscle contraction. Troponin C has been shown to interact directly with two other components of the thin filaments – troponin I and troponin T. Interaction with troponin I is not calcium dependent in the absence of denaturing agents (Head and Perry, 1974) although, as will be described later, different parts of the molecules appear to be in contact depending upon the calcium concentration (table 2.2).

Table 2.2 Protein–protein interactions of the troponin complex in striated muscle

Component	Site of interaction	Calcium sensitive	Urea stable	Number of interaction sites
Troponin C	Troponin I	+	+	2
	Troponin T	+	–	probably 1
Troponin I	Troponin C	+	+	2
	Troponin T	–	–	probably 1
	Actin	–	–	1
Troponin T	Troponin C	+	–	probably 1
	Troponin I	–	–	probably 1
	Tropomyosin	–	–	2

At least two interaction sites have been identified on both troponin C and troponin I. Large troponin C peptides which contain residues 89–100 (for example in rabbit skeletal muscle troponin C the CNBr fragment residues 84–135, the thrombic fragment residues 1–120 and residues 1–100 and the tryptic fragment residues 1–100 and residues 89–159) will all form complexes with troponin I in the presence of calcium (Leavis *et al.*, 1978, 1980; Weeks and Perry, 1978; Wall *et al.*, 1981). With the exception of peptide 84–135 this complex formation is also resistant to high urea concentrations. Therefore it is probable that the area between binding sites II and III represents a calcium-dependent binding site for troponin I. It is interesting to note that the proteolytic susceptibility of this region is considerably changed by Ca^{2+} binding to the protein and it seems likely that conformational changes at the low-affinity sites could affect troponin I binding in this region. Confirmatory evidence for this area being important in troponin C–troponin I interaction is provided by PMR data (Evans *et al.*, 1980*a*) and enzymic studies in which it was found that troponin C peptides covering residues 89–100 were able to relieve inhibition of actomyosin by troponin I (Weeks and Perry, 1978; Grabarek *et al.*, 1981). Evidence for the second troponin I binding site on troponin C is rather less strong, but it appears to be in the area between calcium binding sites I and II (around residues 48–54). Peptides covering various parts of the N-terminal half of troponin C (the CNBr fragment 46–78 and the tryptic fragment 9–84) bind to troponin I in the presence of calcium (Leavis *et al.*, 1978; Grabarek *et al.*, 1981). Proton magnetic resonance studies (Evans *et al.*, 1980*b*) have shown that interaction of the CNBr fragment 46–78 results in a marked reduction in segmental mobility of the side chains of threonines 49 and 51 and glutamic acid residues (53, 84). Lysine-52 also becomes more reactive in the absence of troponin I (Hitchcock, 1981).

It has recently been suggested (Grabarek *et al.*, 1981) that a third troponin I binding site is present towards the C-terminus of troponin C (residues 127–138), which is Ca^{2+} insensitive. Apart from the difficulty of rationalising this result with the fact that only two troponin C binding sites have been identified on troponin I, there is considerable evidence to suggest that peptides from troponin C C-terminal to residue 100 do not interact with troponin I (Leavis *et al.*, 1978; Wall *et al.*, 1981; R. J. A. Grand, C. M. Wall and S. V. Perry, unpublished results). Grabarek *et al.* (1981) have implied that this additional site is necessary to account for troponin I-troponin C complexes formed in the absence of Ca^{2+}. However, there is evidence that the troponin I CNBr peptide residues 96–116 are not dissociated from troponin C totally when Ca^{2+} is removed (Dalgarno *et al.*, 1982; B. A. Levine and R. J. A. Grand, unpublished results). It is possible, therefore, that the interaction between troponin I and residues 90–100 is calcium insensitive in the whole molecule.

The two troponin C binding sites on rabbit skeletal muscle troponin I have been closely defined (Cole and Perry, 1975; Syska *et al.*, 1976; Talbot and Hodges, 1979; Dalgarno *et al.*, 1982; Grand *et al.*, 1982; Moir *et al.*, 1983). One is close to the N-terminus and comprises the amino acids from about 6 to 14 (interaction site A) (Grand *et al.*, 1982) and the other encompasses approximately residues 97–105 (interaction site B) (see figure 2.4). Dalgarno *et al.* (1982) have shown that the N-terminal peptide of troponin I (site A) probably interacts with that area of troponin C between Ca^{2+} binding sites I and II whereas site B on troponin I interacts with that area of troponin C between calcium binding sites II and III. Site B on troponin I is located very close to the actin binding site (approximately residues 108–115) (Syska *et al.*, 1976; Grand *et al.*, 1982), and therefore it is probable that conformational changes in troponin C could exert a strong influence on the binding of troponin I to actin. The sequence of troponin I covering the actin binding site and the second troponin C binding site is very highly conserved in the four different types of troponin I for which sequence data are available (Wilkinson and Grand, 1978) with only a few very conservative substitutions. This supports the idea that the proteins bound at these sites are very similar in the different muscle types. Indeed there are very few differences in actins from fast, slow and cardiac muscles, or in the different troponin C sequences over residues 90–100 (figure 2.3). However, it might also be supposed that the amino acid sequences of the other troponin Is equivalent to site A on the rabbit skeletal muscle protein would all be very similar as there is little variation in troponin Cs from the different muscle types in the area around residues 50–60. It is, of course, possible that the relatively limited number of conserved residues in the troponin I site A sequences constitute the troponin C binding site, but confirmation of this will have to await further evidence. In summary it appears that in the absence of calcium ions troponin I is complexed to actin and to troponin C (residues 88–100) through site B (residues 97–115, figure 2.4). The binding of Ca^{2+} at the low-affinity sites I and II results in an additional interaction between troponin C (residues 50–60) and troponin I site A (residues 6–14). This results in a conformational change transmitted through either or both troponin C and troponin I, which

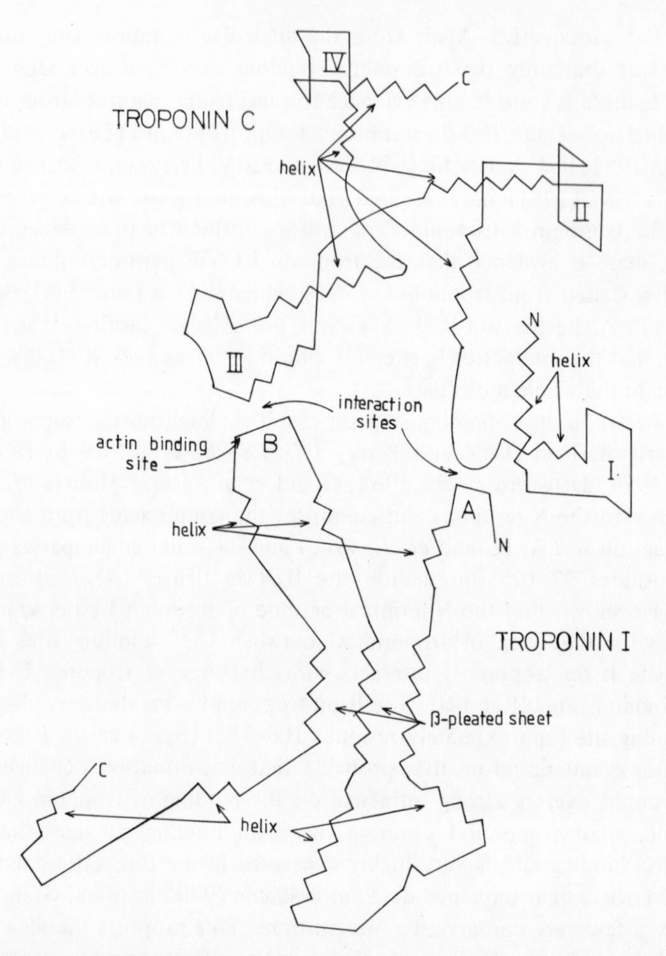

Figure 2.4 Schematic representation of the interaction of rabbit skeletal muscle troponin C and troponin I. The structure of troponin C is taken from Dalgarno *et al.* (1982) and is based on the predictions of Kretsinger and Barry (1975). The structure of troponin I is that predicted by Wilkinson and Grand (1975) using the method of Chou and Fasman (1974). I, II, III, IV are the metal ion binding sites on troponin C; A and B are the troponin C binding sites on troponin I (see text).

causes either a dissociation of troponin I from actin or, at least, a major conformational change in troponin I.

Troponin C and troponin I both contain a high proportion of charged amino acids and it is therefore reasonable to suppose that interactions between the two proteins are based on electrostatic interaction. However, it appears that although basic side chains on troponin I and acidic side chains on troponin C are important for the location of one protein against another, there is a considerable hydrophobic element involved in the formation and maintenance of the troponin I–troponin C complex (B. A. Levine, personal communication).

Evidence in favour of a troponin C-troponin T interaction is not as clear cut as that presented above for troponin C-troponin I complex formation. Initial observations by Ebashi *et al.* (1972), Margossian and Cohen (1973), van Eerd and Kawasaki (1973), Mani *et al.* (1974), Perry and Cole (1974) and Jackson *et al.* (1975) established that troponin T could bind to troponin C in a calcium-dependent manner. Troponin C peptides spanning residues 1-120 and 1-100 will solubilise troponin T (Leavis *et al.*, 1980; Grabarek *et al.*, 1981), suggesting that the troponin T binding site is located in the N-terminal two-thirds of the calcium binding protein; further investigation, possibly using PMR, should allow the site to be defined more closely. Troponin I and troponin C both bind to the C-terminal peptide (residues 159-259) produced by limited chymotryptic cleavage of rabbit skeletal muscle troponin T (Pearlstone and Smillie, 1978; Katayama, 1979; Ohtsuki *et al.*, 1981). Recently, Pearlstone and Smillie (1982) have shown that the presence of troponin C confers calcium sensitivity on the interaction between troponin T (or the peptide T_2, 159-259) and tropomyosin, such that in the presence of Ca^{2+} the strength of interaction between the troponin T peptide T_2 and troponin C is increased whereas that between T_2 and tropomyosin is much reduced. In the absence of calcium the T_2-troponin C interaction is much reduced and the T_2-tropomyosin interaction increased. Pearlstone and Smillie (1982) have concluded that 'an important element in the molecular mechanism by which Ca^{2+} exerts its control on thin filament protein assembly is the troponin C-troponin T interaction'. Observations that the interaction between cardiac troponin C and T is not calcium sensitive and is relatively weak and nonspecific have led Potter and Johnson (1982) to suggest that troponin T-troponin C interactions do not occur *in vivo*. However, the weight of evidence must favour the view of Pearlstone and Smillie (1982) that such interactions are very important in the regulation of the response of striated muscle to stimulation. (Interactions occurring within the troponin complex are summarised in table 2.2.)

It has recently been shown (Cohen *et al.*, 1979) that troponin C also interacts with phosphorylase kinase *in vitro* and the site of this interaction has been assigned to the β-subunit (Picton *et al.*, 1983). It has been suggested that such an interaction could occur *in vivo* as a way of synchronising the onset of muscle contraction and glycogenolysis, but at present the evidence in favour of this proposal is equivocal (see section 2.5.2.3).

2.5.2 Calmodulin

Our knowledge of interactions involving calmodulin is considerably less precise than for troponin C, but some information is available on a few of the protein-protein interactions in which it is involved. It is rather ironic that the interaction system about which most is known, at the molecular level, is one which does not occur *in vivo*. Before the ubiquity of calmodulin was realised it was noted that brain calmodulin could substitute for troponin C in skeletal muscle regulated actomyosin (Amphlett *et al*., 1976; Dedman *et al*., 1977*b*). In retrospect it is obvious that this system is able to function because calmodulin can interact with troponin I so as to modify the troponin I-actin interaction. Since that time a number of workers have investigated the troponin I-calmodulin system in the hope that it will illuminate other protein-protein interactions involving calmodulin (for example, see Keller *et al*., 1982; Kincaid *et al*., 1982). Although calmodulin can form a stable complex with troponin I, there are a number of differences from the troponin C–troponin I complex. The interaction is calcium dependent in the absence of denaturing agents whereas the troponin I–troponin C complex is not (Grand *et al*., 1979). Also the troponin I–calmodulin complex is dissociated at urea concentrations above 7 M even in the presence of Ca^{2+} (Grand *et al*., 1979; R. J. A. Grand, unpublished results). Using a similar approach to that adopted for troponin C it has been possible to locate the troponin I binding site on calmodulin by the use of isolated peptides spanning various parts of the calcium binding protein. A thrombic peptide (residues 1-106) (Grand *et al*., 1980*b*; Wall *et al*., 1981), various tryptic peptides (Drabikowski, 1980; Kuznicki *et al*., 1981), and a CNBr peptide (residues 77–124) (Vanaman and Perry, 1978; Perry *et al*., 1979) will interact with troponin I. It is therefore probable that the region on calmodulin homologous to the troponin I binding site on troponin C, between calcium binding sites II and III, is responsible for binding troponin I. However, it has been found that calmodulin peptide 1-90 fails to bind troponin I. It is possible, therefore, that the protein binding site extends rather closer to the C terminus on calmodulin than it does in troponin C (Grand *et al*., 1980*b*). It appears that there is only one binding site for troponin I on calmodulin and this is located between calcium binding sites II and III. Similarly there is only one calmodulin binding site on troponin I (Perry *et al*., 1979; Moir *et al*., 1983) and that is probably the TNC binding site B (i.e. around residues 95–103). It is interesting to note that the EDTA myosin light chain from scallop adductor muscle does not interact with troponin I (R. J. A. Grand, unpublished results), even though it has a very similar primary structure (Kendrick-Jones and Jakes, 1976) to calmodulin in the area of the proposed binding site. This casts some doubt on the suggestion (Vanaman, 1980; Klee and Vanaman, 1982) that the sequence S–E–E–E is the determining factor in TNI binding, although it is not an unequivocal result as the third glutamic acid residue is replaced by threonine in the light chain. However, it is quite possible that the rest of the protein is important in producing a favourable tertiary structure which will favour troponin I binding.

A number of other small basic proteins bind to calmodulin, some in a calcium-dependent manner. Myelin basic protein forms a 1 : 1 molar complex with calmodulin on polyacrylamide gels which is dissociated when calcium is removed (Grand and Perry, 1980; Itano *et al*., 1980). Histone H2B also binds to calmodulin in the presence of calcium, but a number of other histones, protamine and polylysine interact in a calcium-independent manner. It is probable that none of these interactions occurs *in vivo*, although it is quite likely that the binding sites on calmodulin, at least for myelin basic protein and H2B, are identical to that for TNI. (It should be noted, however, that there is no amino acid sequence homology in these three basic proteins.)

Although it is reasonable to assume that the site occupied by troponin I on calmodulin is in the area between the second and third calcium binding sites, it is not clear to what extent this represents the area of interaction with the calmodulin-regulated enzymes. Our knowledge of calmodulin-enzyme interactions is limited to three systems: cyclic nucleotide-dependent phosphodiesterase, myosin light chain kinase and phosphorylase kinase, and these will be discussed briefly in turn.

2.5.2.1 Cyclic nucleotide dependent phosphodiesterase (PDE)

The catalytic subunit of PDE has been purified to homogeneity from bovine brain (Klee *et al*., 1979*a*; Sharma *et al*., 1980; Tucker *et al*., 1981) and bovine heart (La Porte *et al*., 1979) and has a chain weight of about 60 000. It appears that in the cell the enzyme exists as a 120 000 molecular weight dimer in the absence of calcium, and in the presence of calcium it complexes with 2 mol of calmodulin to form a tetramer (α_2CaM_2) (table 2.3).

The following scheme is the simplest for the activation of phosphodiesterase consistent with the experimental observations:

$$n\mathrm{Ca}^{2+} + \text{calmodulin} = \mathrm{Ca}_n^{2+}\text{–calmodulin}$$
$$2\mathrm{Ca}_n^{2+}\text{–calmodulin} + \mathrm{PDE}_2 \text{ (less active)} = \mathrm{Ca}_n^{2+}\text{–calmodulin}_2\,\mathrm{PDE}_2 \text{ (more active)}$$

where n is either three or four (Crouch and Klee, 1980; Cox *et al*., 1981*c*).

Limited proteolysis of the PDE enzymic subunit has been known for many years to activate the enzyme irreversibly (Cheung, 1971) such that the activity of the degraded PDE in the absence of calmodulin and calcium is similar to that of the Ca^{2+}-calmodulin-PDE holoenzyme. Recently it has been shown that this activation is due to the removal of a terminal polypeptide of 15 000–22 000 molecular weight (Klee, 1980; Tucker *et al*., 1981). Presumably this peptide contains the calmodulin binding site on the enzymic subunit and its loss exposes the substrate binding site so that a fully active PDE is produced. In the normal situation the conformational change in the subunit to expose the substrate binding site is affected by calmodulin binding (Klee, 1980).

Attempts to locate the site of interaction of the PDE enzymic subunit on calmodulin have not been very successful. A calmodulin peptide comprising residues 1–106 activates PDE in the presence of calcium, but much less efficiently than undegraded calmodulin (Walsh *et al*., 1977; Kuznicki *et al*., 1981). The major criticism of experiments of this kind is that there is no way of proving unequivocally that very small amounts of whole calmodulin are not present as contaminants.

Calmodulin-activated PDE may be inhibited by troponin I (Grand *et al*., 1980*b*; La Porte *et al*., 1980), myelin basic protein and histone H2B (Grand

and Perry, 1980; Itano *et al*., 1980), 50 per cent inhibition occurring at molar ratios of troponin I : calmodulin of approximately 2 : 1. These results suggest that the troponin I binding site and the PDE binding site on calmodulin might be identical, but it is quite possible that conformational changes in calmodulin on binding to troponin I may make other sites on calmodulin no longer available to the enzyme.

La Porte *et al*. (1980) have noted that hydrophobic areas of calmodulin are exposed or formed on binding of calcium ions and that these hydrophobic domains are involved in both troponin I and PDE binding. Binding of a number of amphiphilic ligands, including the phenothiazine drugs (see below) and fluorescent dyes, also occurs at these sites such that calmodulin-troponin I and calmodulin–PDE complex formation is inhibited. The authors argue that this indicates a common hydrophobic binding site for proteins (or at least for troponin I and PDE) and dyes. However, since a number of molecules of fluorescent dye may be bound to calmodulin at any one time, it seems reasonable to assume that more than one such hydrophobic domain could become exposed on Ca^{2+} binding. Although trifluoperazine inhibits the binding of calmodulin to a troponin I-Sepharose column (La Porte *et al*., 1980), the analogous drug chlorpromazine does not affect troponin I-calmodulin interactions on polyacrylamide gels (R. J. A. Grand, unpublished results); therefore, although it is possible, it has certainly not been proven unequivocally that troponin I and PDE interact at a similar site on calmodulin.

2.5.2.2 Myosin light chain kinase (MLCK)

Different forms of MLCK exist in different muscle types (see Adelstein and Klee, 1980; Stull, 1980; Grand, 1982, for reviews), but all are totally dependent on calmodulin and calcium for activity. The catalytic subunit in striated muscle is a single polypeptide of molecular weight about 80 000 (Pires and Perry, 1977), whereas in smooth muscle and non-muscle systems it has a molecular weight in excess of 100 000 (Yerna *et al*., 1979; Adelstein and Klee, 1980, 1981) (table 2.3).

A similar scheme for the activation of skeletal muscle MLCK by calmodulin has been proposed to that for PDE, except that the active holenzyme is a dimer (αCaM) rather than a tetramer. In this case the number of calcium ions required for full activation is not clear. Blumenthal and Stull (1980) consider that all four sites on calmodulin must be occupied whereas Johnson, J. D. *et al*. (1981) and Crouch *et al*. (1981) found that activation could occur after binding to only one site. In the case of both PDE and MLCK and a number of other calmodulin-dependent enzymes, interaction between calmodulin and the catalytic subunit only occurs after calcium binding.

Relatively little is known about what happens to MLCK on calmodulin binding except that a conformational change occurs so that the binding sites for ATP and myosin light chain become available. This change can be monitored by studying changes in tryptophan fluorescence which appear to occur in two stages, one very rapid (60-70 s^{-1}) or one more slowly (5-7 s^{-1}) (Johnson, J. D. *et al.*, 1981).

As is the case with PDE, limited proteolysis of MLCK results in a calcium- and calmodulin-independent enzyme of high activity (Bremel and Shaw, 1978; Hathaway and Adelstein, 1979, Tanaka *et al.*, 1980). It is probable that removal of part of the subunit polypeptide chain exposes the substrate binding sites which are only normally available after the conformational changes induced by calmodulin binding. It is possible that such a mechanism occurs in all those calmodulin-activated enzymes which interact only with calmodulin-Ca^{2+}.

Various calmodulin peptides have been tested as activators of MLCK but most have little effect. The thrombic peptide spanning residues 1-106 will activate MLCK, but only when present at very high concentrations (100-fold molar excess); the somewhat smaller tryptic peptide (1-90) is even less effective at enzyme activation (Grand *et al.*, 1980*b*; Nairn *et al.*, 1980).

It is not clear what significance attaches to the activation of either PDE or MLCK by very high concentrations of calmodulin peptides except that there is a considerably greater requirement for calmodulin integrity for enzymic activation than is the case for troponin I binding. One significant difference between PDE and MLCK is that calmodulin activation of the latter is unaffected by the small basic calmodulin binding proteins such as troponin I, myelin basic protein and histone H2B (Grand *et al.*, 1980*b*; Nairn *et al.*, 1980) even when present in 100-fold molar excess over calmodulin. Such a result probably indicates different binding sites on calmodulin for MLCK and PDE.

2.5.2.3 Phosphorylase kinase (PK)

The structure of phosphorylase kinase is considerably different from that of MLCK and PDE in that it comprises four subunits and the smallest active species is the hexadecamer $(\alpha\beta\gamma\delta)_4$. The δ subunit has a molecular weight of 17 000 and has been identified as calmodulin (Cohen *et al.*, 1978; Shenolikar *et al.*, 1979; Grand *et al.*, 1981). The α, β and γ subunits have molecular weights of 145 000, 128 000 and 45 000, respectively. In the absence of calcium the δ subunit is not dissociated from the other subunits, indicating much tighter binding than is the case for other calmodulin-dependent enzymes.

Although the enzyme isolated in the absence of calcium comprises stoichiometric amounts of the four components, its activity can be considerably increased by the addition of more calmodulin (DePaoli-Roach *et al.*, 1979; Shenolikar *et al.*, 1979), troponin C or whole troponin (Cohen *et al.*, 1979). This activation by calmodulin (termed the δ' subunit) can be reversed with troponin I or trifluoperazine, suggesting that the integration of this subunit into the enzyme is not total. Cohen (1980*a*), in a detailed study of the system, has shown that 3-4 mol of calcium must be bound to the δ' subunit before activation can occur and similarly 3-4 mol of calcium must be bound to the δ subunit for activation of phosphorylase kinase *b* (the dephosphorylated form of PK). However, only 1-2 mol of calcium were bound by the δ subunit of phosphorylase kinase *a* for full activity. Cohen (1980*a*) has proposed that calmodulin (in the form of the δ subunit) is the calcium-dependent regulator of phosphorylase kinase *a* whereas troponin C is the physiological regulator of phosphorylase kinase *b*. By this means the regulation of glycogenolysis and muscle contraction may be integrated.

The δ subunit of phosphorylase kinase seems to be bound to the γ subunit *in vivo* (Picton *et al*., 1980) whereas the δ' subunit is bound to either the α or β subunits but only in the presence of calcium (Picton *et al*., 1980). Similarly, troponin C binds to the β subunit (Picton *et al*., 1983). The site of interaction on calmodulin (in its role as the δ' subunit) appears to lie in the same area as the troponin I binding site (around residues 80–100) (Kuznicki *et al*., 1981).

Phosphorylase kinase is unique among the calmodulin-dependent enzymes so far studied in that the calcium binding subunit remains part of the enzyme at all times. Although this situation probably exists for other calmodulin requiring enzymes, it has not yet been demonstrated unequivocally. If the δ' (calmodulin) subunit does play a regulatory role *in vivo* (although Cohen's (1980*a*) results tend to indicate that it does not), it behaves in much the same way as the regulatory subunit of PDE and MLCK, i.e. weakly bound only in the presence of calcium.

2.5.2.4 Calmodulin binding proteins (CaM-BP)

The generic term 'calmodulin binding proteins' has been applied to those proteins that have been shown to form complexes with calmodulin and yet have no detectable enzymic function. It has been known for some time that various proteins form complexes with calmodulin *in vitro* but do not interact *in vivo*. For example, as discussed above, troponin I binds to calmodulin in the presence of calcium (Amphlett *et al*., 1976; Dedman *et al*., 1977*b*; Grand *et al*., 1979). Similarly, histones and myelin basic protein have many properties in common with the higher molecular weight CaM-BPs, for example the ability to inhibit calmodulin-activated PDE (Grand and Perry, 1979; Itano *et al*., 1980).

There is no suggestion that any of these small, highly basic proteins are major calmodulin binding proteins *in vivo* since it may be simply demonstrated that they are not found near large calmodulin pools within the cell (Grand and Perry, 1980). However, the recognition that such interactions may occur should provide some warning against too readily classifying any protein as a Cam-BP without evidence that it and calmodulin interact within the cell.

One of the observations that has led researchers to expect appreciable amounts of CaM-BPs is the very high concentration of calmodulin relative to the calmodulin-regulated enzymes in almost all tissues. On teleological grounds, therefore, proteins capable of binding to calmodulin are predicted to occur at relatively high concentrations, since if whole tissue homogenates are electrophoresed under denaturing conditions, but in the absence of SDS, little free calmodulin can be observed unless calcium is removed (Grand *et al*., 1979). However, it is possible that disruption of the organelles within the cell creates an artifactual interaction between calmodulin and some other protein (e.g. histones from the nucleus).

Despite all of these reservations it is quite feasible that certain proteins may be present that perform a regulatory role through their interactions with calmodulin and these might legitimately qualify as 'calmodulin-binding proteins'.

The possibility that some, if not all, CaM-BPs are calmodulin-requiring enzymes of indeterminate function is suggested by recent investigations into the nature of calcineurin (also known as CaM-BP_{80}, inhibitory protein, modulator binding protein). This protein was first identified as a heat-labile inhibitor of PDE (Wang and Desai, 1977). Further study showed that it comprised two equimolar subunits, A with a molecular weight of 60 000, and B with a molecular weight of approximately 15 000 (Klee and Krinks, 1978; Wallace *et al.*, 1978; Sharma *et al.*, 1979). The larger subunit (A) of calcineurin (a name coined by Klee *et al.*, 1979*b*) is the calmodulin binding component; calcineurin B is a calcium binding protein quite distinct from calmodulin. Like calmodulin, however, it is capable of binding four calcium atoms per protein molecule with high affinity ($K_d < 10^{-6}$ M).

Because of the high affinity of calcineurin for calmodulin, it is capable of inhibiting many calmodulin-activated enzymes *in vitro*. Apart from PDE, calcineurin reduces the activities of adenylate cyclase (Wallace *et al.*, 1978) and phosphorylase kinase (Cohen *et al.*, 1979) to a basal level, counteracting any activating effect of added calmodulin.

Recently, however, it has been recognised that calcineurin and calmodulin-dependent protein phosphatase 2B are identical (Stewart *et al.*, 1982). This observation has been confirmed by Yang *et al.* (1982), who showed the two proteins to be immunologically indistinguishable. Therefore it is unlikely that calcineurin can function as part of a widespread calmodulin–binding protein regulatory complex, but is simply another calmodulin-dependent enzyme present at particularly high concentrations in the brain (Wallace *et al.*, 1980).

A number of other proteins have been suggested as calmodulin binding proteins – although the physiological significance of this binding has not been established. For example, brain proteins of molecular weight 70 000 (Sharma *et al.*, 1978), 60 000 and 40 000 (Watterson and Vanaman, 1976) and 50 000 (Maekawa and Abe, 1980) can form calcium-dependent complexes with calmodulin. Probably of more significance are the demonstrations that certain CaM-BPs are common to a number of tissues.

A calmodulin binding protein of molecular weight about 120 000 has been identified in a number of rabbit tissues and organs (Grand and Perry, 1980), in avian smooth muscle (Glenney and Weber, 1980) and in a number of rat tissues (Palfrey *et al.*, 1982). The function of this protein remains unknown but there is now considerable evidence to suggest that, in some cell types at least, an appreciable proportion of the calmodulin is complexed to a high molecular weight component of the cytoskeleton. Sobue *et al.* (1981*a*) showed that spectrin was a major CaM-BP of erythrocyte ghosts. Calmodulin binding proteins of similar size (240 000) have also been identified in brain microsomal fractions (Kakiuchi *et al.*, 1981*b*), isolated brush borders and microvilli of intestinal epithelial cells (Glenney *et al.*, 1982*a*) and a number of rat tissues (Palfrey *et al.*, 1982). Recently, Glenney *et al.*, (1982*b*) have suggested that spectrin (chicken erythrocyte), fodrin (rabbit and chicken brain) and the intestinal brush border protein TW260/240 all contained a common 240 000 molecular weight subunit which binds to calmodulin. The proteins are distinguished by possession of a second unique subunit. It is feasible, therefore, that a component of the cytoskeleton present in most cell types binds calmodulin and possibly acts as a regulatory protein for non-muscle actin organisation. From immunofluorescence studies using a calmodulin-specific antibody it has been shown that most of the cellular calmodulin is located in stress fibres (microfilaments) of the cell (Dedman *et al.*, 1978*b*). It has been suggested (Dedman *et al.*, 1978*b*; Means *et al.*, 1982) that these localisation studies show calmodulin bound to myosin light chain kinase. However, it is probable that the concentration of this enzyme in non-muscle tissues (and smooth muscle) is insufficient to account for all of the calmodulin which appears to be present on the filaments. In view of the results of Glenney *et al.* (1982*a,b*) it is possible that a controlling element of microfilaments is the calmodulin–240 000 subunit complex.

A number of tissue-specific CaM-BPs have been identified with the ^{125}I-labelled calmodulin–gel overlay technique used in the studies described previously. (This technique does not show any calmodulin–histone interaction.) Glenney and Weber (1980) and Palfrey *et al.* (1982) showed the major CaM-BPs of brain to be polypeptides of molecular weight approximately 50 000 and 60 000. These results have been confirmed using a slightly different method in which proteins are electrophoretically transferred from SDS gels on to cellulose nitrate filters which are then incubated with ^{125}I-calmodulin (R. J. A. Grand and M. K. Darby, unpublished results). The major CaM-BP of skeletal muscle is a 83 000 molecular weight protein (Palfrey *et al.*, 1982) which could be myosin light chain kinase. The idea that one or more major CaM-BPs are present in different tissues and organisms is a highly attractive one and it was originally envisaged that they would play a regulatory role in the cell's response to changes in calcium levels. The analogy has been drawn between the troponin C–troponin I complex of skeletal muscle and calmodulin–calmodulin binding protein of non-muscle systems. Despite this it now seems likely that cells do not require additional components for the regulation of the large number of calmodulin-dependent enzymes. In time it is expected that the relatively low molecular weight (<100 000) CaM-BPs will be shown to be calmodulin-dependent enzymes. The 240 000 polypeptide does, however, appear to be a possible CaM-BP with a specific role in the regulation of microfilaments.

2.5.2.5 Interaction of calmodulin with pharmacological agents

Apart from binding metal ions and other proteins, calmodulin has been shown to interact with a number of small hydrophobic compounds. These agents, many of which have pharmacological uses, have been valuable in the study of calmodulin in two ways: first, as relatively specific inhibitors of calmodulin-activated processes and, secondly, in the information they provide about the structure of calmodulin itself.

Levin and Weiss (1977, 1978, 1979) demonstrated that a number of antipsychotic drugs bound to calmodulin in the presence of calcium so as to inhibit the calmodulin activation of PDE. Since then, major interest has centred on the phenothiazines and in particular trifluoperazine (TFP) and chlorpromazine. Trifluoperazine binds to calmodulin at two distinct sets of sites: there are two high-affinity calcium-dependent binding sites per protein molecule ($K_d = 1\ \mu M$) (Levin and Weiss, 1978; Klevitt *et al.*, 1981*b*) and approximately 24 low-affinity calcium-independent sites ($K_d = 5$ mM).

A second set of calmodulin-binding drugs are the naphthalenesulfonamides (W compounds) prepared by Hidaka and his colleagues (Hidaka *et al.*, 1978, 1979), which are rather less hydrophobic than the phenothiazines but are equally efficient inhibitors of calmodulin-activated enzymes ($K_d = 10^{-6}$ to 10^{-7} M). Both sets of compounds have been shown to reduce the level of activity of PDE (Levin and Weiss, 1977), Ca^{2+}-Mg^{2+} ATPase from erythrocytes (Kobayashi *et al.*, 1979) and MLCK (Hidaka *et al.*, 1979). As with the phenothiazines there are two classes of binding site on calmodulin for W7: three calcium-dependent high-affinity sites ($K_{W7} = 11\ \mu M$) and nine low-affinity sites ($K = 200\ \mu M$) (Hidaka *et al.*, 1980). Chlorpromazine appears to compete with W7 for similar binding sites on the protein but other drugs, for example prenylamine, interact at different sites.

The physiological significance of these calmodulin–drug interactions is unresolved, but it seems likely that the antipsychotic action of the phenothiazines, for example, is not attributable to their calmodulin-binding properties (Norman *et al.*, 1979; Roufogalis, 1981). However, they have been useful in the elucidation of calmodulin-dependent processes such as the crucial role played by MLCK in the regulation of smooth muscle contraction (Hidaka *et al.*, 1979).

Tanaka and Hidaka (1980) have suggested that calcium binding to calmodulin causes a conformational change in the protein allowing a hydrophobic area on the surface to become available. In that study W7 suppressed fluorescence resulting from the interaction of 2-*p*-toluidinyl-naphthalene-6-sulfonate (a hydrophobic probe for proteins) with this region of calmodulin. It was suggested that the hydrophobic area is important in protein–protein interaction. Similarly, La Porte *et al.* (1980) reported hydrophobic domains on calmodulin which could serve as binding sites for troponin I and PDE.

Klevitt *et al.* (1981*b*) have confirmed that two high-affinity drug binding sites are present on calmodulin and have suggested, on the basis of PMR data, that one lies between calcium binding sites II and III whereas the other lies at the C-terminus of the protein. This has been partially confirmed by Head *et al.* (1982), who found that the calmodulin CNBr fragment covering residues

77–124 bound to a phenothiazine affinity column. It is quite possible that digestion of calmodulin with CNBr could have resulted in cleavage of the protein destroying the C terminal site (e.g. at methionines 144 and 145). These studies and a number of those on calmodulin–enzyme and calmodulin–troponin I interactions all implicate the area between calcium binding sites II and III as being important in drug and protein binding. However, it should be noted that troponin I-calmodulin interactions on polyacrylamide gels are not dissociated by chlorpramazine at concentrations below 1 mM (R. J. A. Grand, unpublished results). This result suggests that the phenothiazines may bind close to, but not at, the troponin I interaction site.

It appears that both hydrophobic and electrostatic forces are involved in the interaction of the phenothiazines with calmodulin (Prozialeck and Weiss, 1982). The hydrophobic nature of the phenothiazine ring is the major determinant of the efficiency of the drug in the inhibition of calmodulin-activated enzymes but an interaction also occurs between the drug amino group and an acidic residue on calmodulin (Prozialeck and Weiss, 1982).

A number of other pharmacological agents affect the calmodulin activation of enzymes. Among the most important of these are calcium agonists such as nimodipine and nicardipine (Epstein *et al.*, 1982) and No. 233 (N^2-dansyl-L-arginine-4-t-butylpiperidine amide) (Hidaka *et al.*, 1980), all of which inhibit calmodulin-activated phosphodiesterase. One particularly useful result of the investigation of drug–calmodulin interaction has been the development of affinity media in which phenothiazine derivatives have been coupled to Sepharose (Charbonneau and Cormier, 1979; Jamieson and Vanaman, 1979). Use of such columns has allowed the purification of calmodulin from various organisms using a one-step procedure (see section 2.2). Recent observations (Moore and Dedman, 1982) have indicated, however, that the phenothiazine affinity columns are not absolutely specific for calmodulins and that other proteins bind to them in a calcium-dependent manner.

2.6 FUNCTION OF TROPONIN C AND CALMODULIN

The functions of troponin C and calmodulin are similar in that both act as triggers for various intracellular processes. In both cases the binding of calcium to the protein results in a conformational change which is transmitted to one or more other proteins, eventually switching on or modulating a biochemical process. The major difference between the two proteins lies in the systems which they regulate. Troponin C is present only in striated muscle and is the binding subunit of only one enzyme system, the actomyosin Mg^{2+} ATPase, whereas calmodulins are present in the tissues and organs of all eukaryotes and control a very large number of biochemical processes. Limited space precludes a full consideration of the enzymes for which troponin C and calmodulin act as calcium-binding subunits, but a brief outline follows.

2.6.1 Troponin C

In resting vertebrate muscle the calcium concentration is probably less than 10^{-7} M. However, as a result of nervous stimulation, calcium ions are released from the sarcoplasmic reticulum increasing the effective concentration to about 10^{-5} M. The major target protein for these calcium ions is troponin C (calmodulin is also present although at a lower level than in any other tissue), which is part of the troponin complex (see, for example, Perry, 1979; Adelstein and Eisenberg, 1980; Potter and Johnson, 1982, for reviews). Troponin comprises three subunits present in equimolar amounts: troponin C; troponin I, the inhibitory component (Wilkinson and Grand, 1978); and troponin T, the tropomyosin-binding subunit (Jackson *et al.*, 1975; Pearlstone *et al.*, 1976). Protein-protein interactions within troponin are somewhat complex, but are summarised in table 2.2, and those involving troponin C are discussed fully in section 2.5.1.

The globular component of the troponin complex (probably made up mainly of troponin C and troponin I (Flicker *et al.*, 1982)) is localised at a periodicity of 38.5 nm (Huxley and Brown, 1967) along the I filament, giving a molar ratio of troponin : actin : tropomyosin of 1 : 7 : 1. Since each actin monomer is potentially able to interact with a myosin head, any theory of the regulation of muscle contraction must explain how troponin can control the interaction of all the actins with myosin. The most widely accepted model for this has been the 'steric blocking hypothesis' (Haselgrove, 1972; Huxley, 1972; Parry and Squire, 1973) which is, to a considerable extent, a 'structural model' based on X-ray diffraction and electron microscopy data. This theory explains the regulation of contraction by a translocation of tropomyosin across the surface of the actin molecules. In the 'off state', tropomyosin is positioned on the thin filaments close to the site on actin at which interaction with the myosin head occurs; but in the 'on state' (i.e. in the presence of calcium), tropomyosin is displaced from the site of actin-myosin interaction. It has therefore been suggested that, in the absence of calcium, tropomyosin physically blocks the interaction of actin and myosin. One way in which troponin might regulate this process has been summarised by Squire (1981*a*). In the absence of calcium, troponin I interacts strongly with actin locking tropomyosin in the blocking position through the TNC-TNT-TM series of interactions. When calcium is bound to troponin C, however, the I/C interaction is strengthened at the expense of I/actin. This also alters the conformation of troponin T and tropomyosin, the latter moving towards the groove of the actin helix allowing actin-myosin interaction. In spite of the necessity of fundamental reinterpretations of the data over the past 3 years (Seymour and O'Brien, 1980; Squire, 1981*b*; Taylor and Amos, 1981), this theory continues to attract widespread support.

Although a limited number of reports, over the last decade, have presented data that were not readily explicable on the basis of the 'steric blocking hypothesis', it is only recent evidence derived from enzyme kinetics and protein chemical studies which has seriously questioned the validity of the theory (Eaton, 1976; Chalovitch *et al.*, 1981; Johnson, P. *et al.*, 1981; Chalovitch and

Eisenberg, 1982). On the basis of this and PMR studies (B. A. Levine, R. J. A. Grand, G. Henry, A. J. G. Moir, S. V. Perry, I. P. Trayer, unpublished data) an alternative scheme may be proposed. In the absence of calcium, conformational constraints on the whole thin filament, caused by tight binding of troponin I to one or perhaps two actin monomers, may prevent actin–myosin interaction or ATP hydrolysis by actomyosin. The binding of calcium to troponin C at the low-affinity calcium-specific sites causes a conformational change so that a second interaction between that protein and troponin I can take place. The binding of troponin C to troponin I at a site very close to the actin binding site probably causes the dissociation, or at least modification, of the troponin I-actin complex. The change in actin conformation must then be transmitted directly from actin to actin along the I filament, allowing interactions with the myosin heads to take place or the ATP hydrolysis reaction to go to completion (Chalovitch *et al.*, 1981). In this view the function of tropomyosin is largely to hold the actin filaments in the correct conformation – the observed change in its position in the actin groove is then a result of actin movement.

Evidence is not sufficiently strong to allow the dismissal of either of these explanations. However, both agree on assigning to troponin C the triggering role for striated muscle contraction. Thus the conformational changes caused by calcium binding to the calcium-specific low-affinity sites (I and II) causes a conformational change which is transmitted, through protein–protein interaction, to the other troponin components.

2.6.2 Calmodulin

It is probable that calmodulin is the calcium binding subunit of all those intracellular enzymes, with the exception of striated muscle actomyosin, which are turned off and on by changes in calcium concentration between about 10^{-7} and 10^{-5} M. A number of these enzymes are summarised together with relevant properties in table 2.3. It can be seen from this list, which is by no means exhaustive, that calmodulin is responsible for the regulation of a wide and diverse range of cellular processes. Lack of space precludes a detailed consideration of these, particularly as many excellent reviews are available elsewhere. The role of calmodulin in the regulation of cAMP-dependent phosphodiesterase, adenylate cyclase, myosin light chain kinase and the Ca^{2+}-Mg^{2+} ATPase has been discussed by Wolff and Brostrom (1979), Klee *et al.* (1980), Cheung and co-workers (Cheung, 1980*a*, 1980*b*) and Klee and Vanaman (1982); similarly, the regulation of phosphorylase kinase and calcium-dependent protein kinases in brain have been reviewed by Cohen (1980*b*) and Schulman *et al.* (1980), respectively. Means *et al.* (1982) have concentrated on the cellular processes regulated by calmodulin rather than specific enzymes, and these authors have drawn together available data on the role of calmodulin in such diverse phenomena as cell motility, cellular transformation, hormone action, cell division and the cell cycle. In section 2.5.2 I have discussed those enzymes for which information is available on the mode of calmodulin–protein interaction – in other words, those enzymes which help us to understand how the calcium binding proteins function. However, it is useful to consider here a few additional points which are of more general interest.

Calmodulin is of central importance in controlling the synthesis and degradation of cyclic AMP in brain in that it acts as the Ca^{2+} binding subunit of both PDE and adenylate cyclase. As these enzymes coexist within the cell (although PDE is in the cytosol and adenylate cyclase in the membrane), it is reasonable to suppose that an influx of calcium would stimulate both enzymes although possibly in a sequential manner (Wolff and Brostrom, 1979; Bradham and Cheung, 1980).

A possible explanation of this apparent contradiction has been offered by Piascik *et al.* (1980) and Potter *et al.* (1980*b*). These workers have observed that in brain, although adenylate cyclase is activated by slight increases in calcium concentration, further rises inhibit the enzyme. In tissues other than brain no calcium-mediated activation can be detected. Thus it has been suggested that calcium modulates cAMP levels in the brain by the following mechanism: after stimulation there is a rise in calcium concentration: in the initial stages of Ca^{2+} influx (0.08–0.2 μM), adenylate cyclase in the membrane is stimulated by the Ca^{2+}-calmodulin complex; however, as the concentration of metal ion increases, adenylate cyclase is inhibited (possibly by Ca^{2+}-calmodulin) but phosphodiesterase is activated by the calcium–calmodulin complex. In this way, levels of cyclic nucleotide in the brain are under the close regulation of calcium ions. Verification of this scheme will have to await further investigation.

Calmodulin also plays an integral role in the regulation of glycogen levels. The importance of calmodulin as δ subunit of muscle phosphorylase kinase has been known for several years, but it has also been shown (Payne and Soderling, 1980; Ahmad *et al.*, 1982) that glycogen synthase kinase from liver is calmodulin dependent.

The importance of calmodulin in the regulation of contractility has been well documented (see, for example, Adelstein and Klee, 1980; Klee *et al.*, 1980; Klee and Vanaman, 1982; Means *et al.*, 1982). It is generally considered that the major point at which this control manifests itself is through myosin light chain kinase. In all mammalian striated muscles the concentration of the enzyme is high, but the function of light chain phosphorylation is still unclear. In fast skeletal muscle, however, there appears to be a correlation between the increase and decrease in light chain phosphate content and the increase and decrease in post-tetanic twitch tension (Manning and Stull, 1979; Stull, 1980). In smooth muscles there is considerable evidence to suggest that the binding of calcium to the calmodulin subunit of MLCK is the triggering mechanism for actin–myosin interaction and therefore contraction (see Small and Sobiesek, 1980; Hartshorne and Siemankowski, 1981, for reviews). In these muscles extent of phosphorylation of the myosin light chain generally correlates with extent of activation of the actomyosin ATPase. Similarly, in non-muscle systems the regulation of acto-

Table 2.3 Calmodulin-dependent enzymes and relevant properties

(a) *Calmodulin-dependent enzymes*

Enzyme	Subunit composition	Calmodulin-binding subunit	References[a]
Cyclic nucleotide phosphodiesterase (E.C. 3.1.4.17)	$\alpha_2 CAM_2$; α = 59 000	α	Cheung (1970); Kakiuchi and Yamazaki (1970); Klee *et al*. (1979*a*); Sharma *et al*. (1980)
Adenylate cyclase (E.C. 4.6.1.1) (from brain)	Catalytic subunit (190 000), G/F subunit (40 000), hormone receptor	Catalytic subunit	Brostrom *et al*. (1975, 1977); Piascik *et al*. (1980); Ross and Gilman (1980); Salter *et al*. (1981)
Guanylate cyclase (E.C. 4.6.1.2)	N.D.	N.D.	Kakiuchi *et al*. (1981*a*)
Myosin light chain kinase	αCAM; (α = 77 000, striated muscle)	α	Pires and Perry (1977); Yazawa *et al*. (1978); Nairn and Perry (1979)
	α = 105 000–130 000 (smooth muscle and non-muscle systems)	α	Dabrowska *et al*. (1977); Hathaway and Adelstein (1979); Yerna *et al*. (1979); Adelstein and Klee (1980)
Phosphorylase kinase (E.C. 2.7.1.38)	$(\alpha\beta\gamma\delta)^{4b}$; α = 145 000; β = 128 000; γ = 45 000 δ = CaM	γ δ' interacts with α and β	Cohen *et al*. (1978); Shenolikar *et al*. (1979); Cohen (1980*a*,*b*)
Glycogen synthase kinase (E.C. 2.7.1.37)	$(\alpha\beta)_n$CaM; α = 51 000; β = 53 000	N.D.	Payne and Soderling (1980); Ahmad *et al*. (1982)
Ca^{2+}–Mg^{2+}ATPase	αCaM; α = 130 000–150 000	α	Gopinath and Vincenzi (1977); Jarrett and Penniston (1977); Niggli *et al*. (1979)
Dynein ATPase	N.D.	N.D.	Jamieson *et al*. (1979); Blum *et al*. (1980; 1981)
NAD^+ kinase (E.C. 2.7.1.23)	N.D.	N.D.	Anderson and Cormier (178); Anderston *et al*. (1980); Cormier *et al*. (1981); Epel *et al*. (1981)
NADPH oxidase (E.C. 1.6.99.1)	N.D.	N.D.	Jones *et al*. (1982)
Phospholipase A2 (E.C. 3.1.1.4)	N.D.	N.D.	Wong and Cheung (1979)
Calmodulin-dependent protein phosphatase (calcineurin)	ABCaM; A = 61 000; B = 15 000	A	Stewart *et al*. (1982)

Tubulin kinase	N.D.	N.D.	Burke and DeLorenzo (1981); DeLorenzo (1981)
Calmodulin-dependent protein kinase	N.D.	N.D.	Schulman and Greengard (1978*a*, *b*); DeLorenzo *et al*. (1979); Landt and McDonald (1980); Schulman *et al*. (1980); Yamauchi and Fujisawa (1980)
Tryptophan 5-mono-oxygenase[d] (E.C. 1.14.16.4)	N.D.	N.D.	Yamauchi and Fujisawa (1979)
Phosphatidylethanolamine methyltransferase[e] (E.C. 2.1.1.17)	N.D.	N.D.	Gil *et al*. (1980); Alemany *et al*. (1982)
cGMP-dependent protein kinase[b]	N.D.	N.D.	Yamaki and Hidaka (1980); Ahrens *et al*. (1982)

(b) *Cellular processes and systems in which calmodulin may play an important role*

Process	Possible point of calmodulin action	References
Motility through the microtubule system	Tubulin dimer, microtubule-associated proteins (MAPs) and τ (tau) factor	Marcum *et al*. (1978); Kakuichi and Sobue (1981); Sobue *et al*. (1981*b*); Lee and Wolff (1982)
Motility through the cytoskeletal system	240 000 molecular weight component of spectrin, fodrin and Tw260/240 (myosin light chain kinase)	Kakuichi *et al*. (1981*b*); Glenney *et al*. (1982*a*,*b*); Palfrey *et al*. (1982)
Mitosis	On the microtubule system (tubulin), the actomyosin system (through MLCK) but possibly calmodulin has another, so far undetermined, effect	Anderson *et al*. (1978); Welsh *et al*. (1978; 1979); Dedman *et al*. (1980)
Flagellum motility	On the microtubule and dynein components of flagella	Blum *et al*. (1980); Means *et al*. (1982)
Cell cycle	Progress from G1 to S phase: possibly calmodulin is required for DNA synthesis	Boynton *et al*. (1980); Chafouleas *et al*. (1982); Means *et al*. (1982); Okumura *et al*. (1982)
Malignant cellular transformation	Increased levels of calmodulin as a result of transformation might affect the cell, possibly through destabilisation of the microtubule system	Watterson *et al*. (1976*b*); Chafouleas *et al*. (1980); MacManus *et al*. (1981); Chafouleas *et al*. (1982); Means *et al*. (1982); Veigl *et al*. (1982)

Table 2.3 continued

Process	Possible point of calmodulin action	References
Exocytosis as exemplified in such processes as:		
(a) *Hormone secretion*		
(i) insulin release from islets of Langerhans	Regulation of cytosolic calmodulin-dependent protein kinase. Also through regulation of cAMP levels	Schubart *et al.* (1980); Gagliardino *et al.* (1980) Krausz *et al.* (1980); Tomlinson *et al.* (1982)
(ii) catecholamine release from chromaffin granules	Possibly a calmodulin-dependent protein kinase	Burgoyne and Geisow (1981)
(b) *Ion secretion*	Phosphorylation by a calmodulin-dependent protein kinase	Ilundain and Naftalin (1979); Taylor *et al.* (1981)
Neurotransmission	Phosphorylation by a calmodulin-dependent protein kinase in synaptosomes, synaptic vesicles and post-synaptic density fractions regulating neurotransmitter release	DeLorenzo *et al.* (1979); DeLorenzo (1980, 1981, 1982)

Notes: CaM = calmodulin; N.D. = not determined.

[a] References given are generally initial observations or those containing data on the calmodulin-binding subunits; the list is representative and certainly not exhaustive.

[b] Phosphorylase kinase has been shown to be activated in the presence of additional calmodulin (δ' subunit) but this is much less tightly bound than the δ subunit (Shenolikar *et al.*, 1979).

[c] Activation of cGMP-dependent protein kinase may be through substrate modification rather than by direct calmodulin binding to the catalytic subunit.

[d] It is probable that tryptophan 5-mono-oxygenase is not directly activated by calmodulin but is the substrate for a calmodulin-dependent protein kinase.

[e] It is probable that phosphatidylethanolamine methyl transferase (E.C. 2.1.1.17) is activated through a calmodulin-dependent protein kinase.

myosin contraction is controlled by MLCK. Antibody studies have shown that in interphase cells most of the calmodulin (Dedman *et al.*, 1978*b*) and MLCK (DeLanerolle *et al.*, 1981) are located on the microfilaments together with actin, myosin and tropomyosin. Calmodulin also appears to be bound to other microfilament-associated proteins, such as spectrin, fodrin and the TW-260/240 proteins from intestinal brush border (Glenney *et al.*, 1982*a*,*b*). Therefore it appears that troponin C is the trigger protein of vertebrate striated muscle systems whereas calmodulin performs a similar function in most, if not all, other contractile systems.

Calmodulin has also been implicated in the regulation of two other contractile systems although its precise role in these processes is still not well understood (see Klee and Vanaman, 1982; Means *et al.*, 1982, for reviews). The assembly and disassembly of microtubules is generally considered to be a calcium-dependent process and this calcium sensitivity has been shown to be dramatically altered in the presence of calmodulin (Marcum *et al.*, 1978). It has been suggested that calmodulin is associated, in the cell, with depolymerised tubulin; whereas little calmodulin is present in those areas where highly structured microtubules exist (Dedman *et al.*, 1980; Salmon and Segall, 1980; Means *et al.*, 1982). Means *et al.* (1982) have therefore suggested that calmodulin is 'involved in the regulated depolymerisation of microtubules that occurs during anaphase chromosome movement'. Some doubts have been expressed about the physiological importance of the observations outlined above for two major reasons: first, troponin C will substitute for calmodulin in the microtubule system (Marcum *et al.*, 1978; Kumagai and Nishida, 1979) and secondly, the affinity of tubulin itself for calmodulin is low. Recently, however, it has been suggested that the actual calmodulin binding protein of the microtubule system is 'tau' (τ) factor – a family of four closely related 55 000–62 000 molecular weight proteins which form part of the microtubule (Kakiuchi and Sobue, 1981; Sobue *et al.*, 1981*b*). It is possible that these proteins mediate the effect of calcium calmodulin on tubulin. It is also possible, in view of the fact that troponin C can be substituted for calmodulin, that an area on the two molecules which is similar in sequence is responsible for binding to the microtubule. This contrasts with most calmodulin-dependent enzymes, which are affected very little by troponin C.

Calmodulin is also probably of importance in the control of the dynein ATPase present in the cilia and flagella of various organisms (see Klee and Vanaman, 1982). Using different extraction procedures, several preparations of dynein ATPases have been made from demembranated cilia of *Tetrahymena*, all of which are activated by the addition of calmodulin although to varying extents (Jamieson *et al.*, 1979, 1980*a*; Blum *et al.*, 1980). Since the addition of the specific calmodulin antagonist W13 caused marked changes in wave pattern of *Tetrahymena* cilia (Means *et al.*, 1982), it seems reasonable to conclude that the motile process as expressed in ciliary movement depends like all other forms of motility on calmodulin or troponin C acting as a regulatory calcium binding subunit.

The other major calmodulin-dependent enzyme system on which most attention has been concentrated is the Ca^{2+}-Mg^{2+} ATPase present in erythrocyte plasma membranes (see Vincenzi and Hinds, 1980; Carafoli, 1981*b*; Klee and Vanaman, 1982, for reviews). This enzyme is responsible for the maintenance of low intracellular concentrations of Ca^{2+} against a high extracellular concentration. It seems probable that at low calcium concentrations calmodulin is present free in the cytoplasm. When there is an increase in calcium within the cytosol of the cell, the cation binds to calmodulin and then the Ca^{2+}–calmodulin complex interacts with the Ca^{2+}-Mg^{2+} ATPase present in the membrane (Graf and Penniston, 1981). Calcium ions are transferred from the cytosol to the outside of the cell. When the calcium concentration in the cytosol is reduced, the calcium--calmodulin and calmodulin–ATPase complexes dissociate, effectively 'turning off' the enzyme.

The Ca^{2+}-Mg^{2+} ATPase has been purified from human erythrocytes using a calmodulin affinity column and appears to comprise a single polypeptide chain of molecular weight about 150 000 (Niggli *et al*., 1979; Gietzen *et al*., 1980), which becomes phosphorylated. *In vivo* crosslinking studies using ^{125}I-calmodulin have confirmed a 1 : 1 ratio for binding of calmodulin to the enzyme, but at present no information is available on sites of interaction.

2.7 CONCLUDING REMARKS

It is now well established that Ca^{2+} acts as a 'second messenger' with a role comparable to that of cyclic AMP. Both of these elicit multiple responses in cells through their actions on a protein receptor molecule. In the case of Ca^{2+} this receptor is generally calmodulin and in the case of cAMP it is cAMP-dependent protein kinase. The two systems are closely intertwined since calmodulin is involved in regulating levels of cAMP in the cell through PDE and adenylate cyclase and acts as a regulatory subunit for some Ca^{2+}-dependent protein kinases. For example, myosin light chain kinase in smooth and non-muscle systems is dependent on Ca^{2+}–calmodulin for activity but that activity is also modified by phosphorylation of the enzymic subunit by cAMP-dependent protein kinase (Adelstein *et al*., 1978).

Although we are now familiar with a large number of calmodulin-regulated enzymes, it is still unclear how they respond to changes in calcium concentration independently of one another. Obviously, as is the case with the various substrates for cAMP-dependent protein kinase, subcellular and cellular localisation is of considerable importance. However, it seems probable that different enzymes are triggered by calmodulin with different numbers of bound calcium atoms; for example, PDE requires $Ca^{2+}_{3\ or\ 4}$–calmodulin whereas MLCK may be stimulated by $Ca^{2+}_{1\ or\ 2}$–calmodulin. (It now appears to be less likely that calmodulin-binding proteins play a central role in controlling calmodulin-regulated enzymes than was once thought to be the case, although such a mechansim cannot be completely ruled out.)

By a combination of these and possibly other mechanisms, almost all eukaryotic cells are able to use a single highly conserved protein both as a receptor for calcium and as a regulatory subunit for a broad spectrum of enzymes. However, in striated muscle, an alternative unique calcium-binding protein – troponin C – is present in very high concentrations (about 70 μM) whereas the concentration of calmodulin (2-5 μM) is much reduced compared with almost all other tissues. It is reasonable to suppose that troponin C has evolved to meet the specialised requirements of controlling muscle contraction. Although calmodulin can substitute for troponin C in reconstituted thin filament preparations (Amphlett *et al.*, 1976; Dedman *et al.*, 1977*b*), the two systems do not behave identically since some of the intricacies of the protein-protein interactions within troponin are not duplicated. Similarly, troponin C cannot act as a calcium binding subunit for the majority of calmodulin-regulated enzymes. The differences in structure in the two groups of proteins must be a function of the different protein–protein interactions in which they are involved.

Our knowledge of how troponin C operates at the molecular level is now considerable yet we know very little of what happens when calmodulin interacts with the calmodulin-regulated enzymes. Hopefully scientists will address this most important aspect of calmodulin biochemistry in the immediate future, and when we are more fully aware of how the different interactions occur the subtleties of the calcium--calmodulin system may become apparent.

ACKNOWLEDGEMENTS

I am most grateful to Barry Levine, Arthur Moir, Valerie Patchell and Michael Wilkinson for invaluable discussion during the preparation of this chapter. Many thanks also to Debbie Williams for excellent secretarial assistance.

REFERENCES

Adelstein, R. S. and Eisenberg, E. (1980) *Ann. Rev. Biochem.* **49**, 921.
Adelstein, R. S. and Klee, C. B. (1980) in *Calcium and Cell Function*, Vol. 1 (Cheung, W. Y., ed.). Academic Press, New York, pp. 167–82.
Adelstein, R. S. and Klee, C. B. (1981) *J. Biol. Chem.* **256**, 7501.
Adelstein, R. S., Conti, M. A., Hathaway, D. R. and Klee, C. B. (1978) *J. Biol. Chem.* **253**, 8347.
Ahmad, Z., De Paoli-Roach, A. A. and Roach, P. J. (1982) *J. Biol. Chem.* **257**, 8348.
Ahrens, H., Paul, A. K., Kuroda, Y. and Sharma, R. K. (1982) *Arch. Biochem. Biophys.* **215**, 597.
Alemany, S., Varela, I., Harper, J. F. and Mato, J. M. (1982) *J. Biol. Chem.* **257**, 9249.
Alexis, M. N. and Gratzer, W. B. (1978) *Biochemistry* **17**, 2319.
Amphlett, G. W., Vanaman, T. C. and Perry, S. V. (1976) *FEBS Lett.* **72**, 163.
Anderson, B., Osborn, M. and Weber, K. (1978) *Eur. J. Cell Biol.* **17**, 354.
Anderson, J. M. and Cormier, M. J. (1978) *Biochem. Biophys. Res. Commun.* **84**, 595.
Anderson, J. M., Charbonneau, H., Jones, H. P., McCann, R. O. and Cormier, M. J. (1980) *Biochemistry* **19**, 3123.
Andersson, A., Forsen, S., Thulin, E. and Vogel, H. J. (1983) *Biochemistry* **22**, 2309.
Andersson, T., Drakenberg, T., Forsen, S. and Thulin, E. (1981) *FEBS Lett.* **125**, 39.

Bagshaw, C. R. and Kendrick-Jones, J. (1979) *J. Mol. Biol.* **130**, 317.
Bailey, K. (1948) *Biochem. J.* **43**, 271.
Barker, W. C., Ketcham, C. K. and Dayhoff, M. O. (1977) in *Calcium Binding Proteins and Calcium Function* (Wasserman, R. H., Corradino, R., Carafoli, E., Kretsinger, R. H., MacLennan, D. H. and Siegel, F. L., eds). Elsevier/North Holland, New York, pp. 73–5.
Bazari, W. L. and Clarke, M. (1981) *J. Biol. Chem.* **256**, 3598.
Blum, J. J., Hayes, A., Jamieson, G. A. and Vanaman, T. C. (1980) *J. Cell Bioi.* 87, 386.
Blum, J. J., Hayes, A., Jamieson, G. A. and Vanaman, T. C. (1981) *Arch. Biochem. Biophys.* **210**, 363.
Blumenthal, D. K. and Stull, J. T. (1980) in *Calcium Binding Proteins: Structure and Function* (Siegel, F. L., Carafoli, E., Kretsinger, R. H., MacLennan, D. H. and Wasserman, R. H., eds), Elsevier/North Holland, New York, pp. 303–4.
Boynton, A. L., Whitfield, J. F. and MacManus, J. P. (1980) *Biochem. Biophys. Res. Commun.* **95**, 745.
Bradham, L. S. and Cheung, W. Y. (1980) in *Calcium and Cell Function*, Vol. 1 (Cheung, W. Y., ed.). Academic Press, New York, pp. 109–26.
Bremel, R. D. and Shaw, M. E. (1978) *FEBS Lett.* **88**, 242.
Brostrom, C. O., Huang, Y-C, Breckinridge, B. M. and Wolff, D. J. (1975) *Proc. Natl Acad. Sci. USA* **72**, 64.
Brostrom, C. O., Brostrom, M. A. and Wolff, D. J. (1977) *J. Biol. Chem.* **252**, 5677.
Brzeska, H., Venyaminov, S. and Drabikowski, W. (1980) in *Calcium Binding Proteins: Structure and Function* (Siegel, F. L., Carafoli, E., Kretsinger, R. H., MacLennan, D. H. and Wasserman, R. H., eds). Elsevier/North Holland, New York, pp. 207–9.
Burgoyne, R. D. and Geisow, M. J. (1981) *FEBS Lett.* **131**, 127.
Burke, B. and DeLorenzo, R. J. (1981) *Proc. Natl Acad. Sci. USA* **78**, 991.
Carafoli, E. (ed.) (1981*a*) *Cell Calcium* **2**, 263–409.
Carafoli, E. (1981*b*) *Cell Calcium* **2**, 353.
Cartaud, A., Ozon, R., Walsh, M. P., Haiech, J. and Demaille, J. G. (1980) *J. Biol. Chem.* **255**, 9404.
Chafouleas, J. G., Pardue, R. L., Brinkley, B. R., Dedman, J. R. and Means, A. R. (1980) in *Calcium Binding Proteins: Structure and Function* (Siegel, F. L., Carafoli, E., Kretsinger, R. H., MacLennan, D. H. and Wasserman, R. H., eds). Elsevier/North Holland, New York, pp. 189–96.
Chafouleas, J. G., Bolton, W. E., Boyd, A. E. and Means, A. R. (1982) in *Mechanisms of Chemical Carcinogenesis*, Vol. 2 (Harris, C. C. and Cerutti, P. A., eds). Alan R. Liss, New York, pp. 545–51.
Chalovitch, J. M. and Eisenberg, E. (1982) *J. Biol. Chem.* **257**, 2432.
Chalovitch, J. M., Chock, P. B. and Eisenberg, E. (1981) *J. Biol. Chem.* **256**, 575.
Charbonneau, H. and Cormier, M. J. (1979) *Biochem. Biophys. Res. Commun.* **90**, 1039.
Cheung, W. Y. (1970) *Biochem. Biophys. Res. Commun.* **38**, 533.
Cheung, W. Y. (1971) *J. Biol. Chem.* **246**, 2859.
Cheung, W. Y. (1980*a*) *Science* **207**, 19.
Cheung, W. Y. (ed.) (1980*b*) *Calcium and Cell Function*, Vol. 1. Academic Press, New York.
Childers, S. R. and Siegel, F. L. (1975) *Biochim. Biophys. Acta* **405**, 99.
Chou, P. Y. and Fasman, G. D. (1974) *Biochemistry* **13**, 222.
Clarke, M., Bazari, W. L. and Kayman, S. C. (1980) *J. Bacteriology* **141**, 397.
Cohen, P. (1980*a*). *Eur. J. Biochem.* **111**, 563.
Cohen, P. (1980*b*). in *Calcium and Cell Function*, Vol. 1 (Cheung, W. Y., ed.). Academic Press, New York, pp. 184–99.
Cohen, P., Burchell, A., Foulkes, J. G., Cohen, P. T. W., Vanaman, T. C. and Nairn, A. C. (1978) *FEBS Lett.* **92**, 287.
Cohen, P., Picton, C. and Klee, C. B. (1979) *FEBS Lett.* **104**, 25–30.
Cole, H. A. and Perry, S. V. (1975) *Biochem. J.* **149**, 525.
Collins, J. H. (1974) *Biochem. Biophys. Res. Commun.* **58**, 301.
Collins, J. H. (1976) in *Calcium in Biological Systems*, Symposium of the Society of Experimental Biology, **30**, 303–34. Cambridge University Press, Cambridge.
Collins, J. H., Potter, J. D., Horn, M. J., Wilshire, G. and Jackman, N. (1973) *FEBS Lett.* **36**, 268.

Collins, J. H., Greaser, M. L., Potter, J. D. and Horn, M. J. (1977) *J. Biol. Chem.* **252**, 6356.

Cook, W. J., Dedman, J. R., Means, A. R. and Bugg, C. E. (1980) *J. Biol. Chem.* **255**, 8152.

Cormier, M. J., Anderson, J. M., Charbonneau, H., Jones, H. P. and McCann, R. O. (1980) in *Calcium and Cell Function*, Vol. 1 (Cheung, W. Y., ed.). Academic Press, New York, pp. 201–18.

Cormier, J. J., Charbonneau, H. and Jarret, H. W. (1981) *Cell Calcium* **2**, 313.

Cox, J. A., Comte, M. and Stein, E. A. (1981*a*) *Biochem. J.* **195**, 205.

Cox, J. A., Kretsinger, R. H. and Stein, E. A. (1981*b*). *Biochim. Biophys. Acta* **670**, 441.

Cox, J. A., Malnoe, A. and Stein, E. A. (1981*c*) *J. Biol. Chem.* **256**, 3218.

Cox, J. A., Fernaz, C., Demaille, J. G., Ortega Perez, R., van Tuinen, D. and Marme, D. (1982) *J. Biol. Chem.* **257**, 10694.

Crouch, T. H. and Klee, C. B. (1980) *Biochemistry* **19**, 3692.

Crouch, T. H., Holroyde, M. J., Collins, J. H., Solaro, R. J. and Potter, J. D. (1981) *Biochemistry* **20**, 6318.

Dabrowska, R., Aromatorio, D., Sherry, J. M. F. and Hartshorne, D. J. (1977) *Biochem. Biophys. Res. Commun.* 78, 1263.

Dalgarno, D. C., Grand, R. J. A., Levine, B. A., Moir, A. J. G., Scott, G. M. M. and Perry, S. V. (1982) *FEBS Lett.* **150**, 54.

Dedman, J. R., Potter, J. D., Jackson, R. L., Johnson, J. D. and Means, A. R. (1977*a*) *J. Biol. Chem.* **252**, 8415.

Dedman, J. R., Potter, J. D. and Means, A. R. (1977*b*) *J. Biol. Chem.* **252**, 2437.

Dedman, J. R., Jackson, R. L., Schrieber, W. E. and Means, A. R. (1978*a*) *J. Biol. Chem.* **253**, 343.

Dedman, J. R., Welsh, M. J. and Means, A. R. (1978*b*) *J. Biol. Chem.* **253**, 7515.

Dedman, J. R., Lin, T., Marcum, J. M., Brinkley, B. R. and Means, A. R. (1980) in *Calcium Binding Proteins: Structure and Function* (Siegel, F. L., Carafoli, E., Kretsinger, R. H., MacLennan, D. H. and Wasserman, R. H., eds). Elsevier/North Holland, New York, pp. 181–8.

DeLanerolle, P., Adelstein, R. S., Feramisco, J. R. and Burridge, K. (1981) *Proc. Natl Acad. Sci. USA* **78**, 4738.

DeLorenzo, R. J. (1980) *Ann. N.Y. Acad. Sci.* **356**, 92.

DeLorenzo, R. J. (1981) *Cell Calcium* **2**, 365.

DeLorenzo, R. J. (1982) *Fed. Proc. Fed. Am. Soc. Exp. Biol.* **41**, 2265.

DeLorenzo, R. J., Freedman, S. D., Yohe, W. B. and Maurer, S. C. (1979) *Proc. Natl Acad. Sci. USA* **76**, 1838.

Delville, A., Grandjean, J., Laszlo, P., Gerday, C., Grabarek, Z. and Drabikowski, W. (1980) *Eur. J. Biochem.* **105**, 289.

Demaille, J., Dutrage, E., Eisenberg, E., Capony, J. P. and Pechere, J-F. (1974) *FEBS Lett.* **42**, 173.

DePaoli-Roach, A. A., Gibbs, J. B. and Roach, P. J. (1979) *FEBS Lett.* **105**, 321.

Dhoot, G. K., Frearson, N. and Perry, S. V. (1979) *Exp. Cell Res.* **122**, 339.

Dieter, P. and Marmé, D. (1980) *Proc. Natl Acad. Sci. USA* **77**, 7311.

Dieter, P. and Marmé, D. (1981) *FEBS Lett.* **125**, 245.

Drabikowski, W., Kuznicki, J. and Grabarek, Z. (1977) *Biochim. Biophys. Acta.* **485**, 124.

Drabikowski, W., Brzeska, H., Kuznicki, J. and Grabarek, Z. (1980) *Ann. N.Y. Acad. Sci.* **356**, 374.

Eaton, B. L. (1976) *Science* **192**, 1337.

Ebashi, S., Kodama, A. and Ebashi, F. (1968) *J. Biochem. (Tokyo)* **64**, 465.

Ebashi, S., Wakabayashi, T. and Ebashi, F. (1971) *J. Biochem. (Tokyo)* **69**, 441–5.

Ebashi, S., Ohtsuki, I. and Mihashi, K. (1972) *Cold Spring Harbor Symp. Quant. Biol.* **37**, 215.

Epel, D. E., Patton, C., Wallace, R. W. and Cheung, W. Y. (1981) *Cell* **23**, 543.

Epstein, P. M., Fiss, K., Hachisu, R. and Adrenyak, D. M. (1982) *Biochem. Biophys. Res. Commun.* **105**, 1142.

Evans, J. S., Levine, B. A., Leavis, P. C., Gergely, J., Grabarek, Z. and Drabikowski, W. (1980*a*) *Biochim. Biophys. Acta.* **623**, 10.

Evans, J. S., Levine, B. A., Leavis, P. C., Gergely, J., Grabarek, Z. and Drabikowski, W. (1980*b*) in *Calcium Binding Proteins: Structure and Function* (Siegel, F. L., Carafoli, E., Kretsinger, R. H., MacLennan, D. H. and Wasserman, R. H., eds). Elsevier/North Holland, New York, pp. 307–8.
Flicker, P. F., Phillips, G. N. and Cohen, C. (1982) *J. Mol. Biol.* **162**, 495.
Gagliardino, J. J., Harrison, D. E., Christie, M. R., Gagliardino, E. E. and Ashcroft, S. J. H. (1980) *Biochem. J.* **192**, 919.
Garfinkel, L. I., Periasamy, M. and Nada-Ginard, B. (1982) *J. Biol. Chem.* **257**, 11078.
Gietzen, K., Tejcka, M. and Wolf, H. U. (1980) *Biochem. J.* **189**, 81.
Gil, M. G., Alemany, S., Cao, D. M., Castano, J. G. and Mato, J. M. (1980) *Biochem. Biophys. Res. Commun.* **94**, 1325.
Giuditta, A., Moore, B. W. and Prozzo, N. (1977) *J. Neurochem.* **29**, 235.
Glenney, J. R. and Weber, K. (1980) *J. Biol. Chem.* **255**, 10551.
Glenney, J. R., Glenney, P., Osborn, M. and Weber, K. (1982*a*) *Cell* **28**, 843.
Glenney, J. R., Glenney, P. and Weber, K. (1982*b*) *Proc. Natl Acad. Sci. USA* **79**, 4002.
Goldberg, A. and Lehman, W. (1978) *Biochem. J.* **171**, 413.
Goodman, M. (1980) in *Calcium Binding Proteins: Structure and Function* (Siegel, F. L., Carafoli, E., Kretsinger, R. H., MacLennan, D. H. and Wasserman, R. H., eds). Elsevier/North Holland, New York, pp. 347–54.
Gopinath, R. M. and Vincenzi, F. F. (1977) *Biochem. Biophys. Res. Commun.* 77, 1203.
Grabarek, Z., Drabikowski, W., Leavis, P. C., Rosenfeld, S. S. and Gergely, J. (1981) *J. Biol. Chem.* **256**, 13121.
Graf, E. and Penniston, J. T. (1981) *Arch. Biochem. Biophys.* **210**, 257.
Grand, R. J. A. (1982) *Life Chemistry Reports* **1**, 105.
Grand, R. J. A. and Perry, S. V. (1978) *FEBS Lett.* **92**, 137.
Grand, R. J. A. and Perry, S. V. (1979) *Biochem. J.* **183**, 285.
Grand, R. J. A. and Perry, S. V. (1980) *Biochem. J.* **189**, 227.
Grand, R. J. A., Perry, S. V. and Weeks, R. A. (1979) *Biochem. J.* **177**, 521.
Grand, R. J. A., Nairn, A. C. and Perry, S. V. (1980*a*) *Biochem. J.* **185**, 755.
Grand, R. J. A., Nairn, A. C., Wall, C. M. and Perry, S. V. (1980*b*) in *Calcium Binding Proteins: Structure and Function* (Siegel, F. L., Carafoli, E., Kretsinger, R. H., MacLennan, D. H. and Wasserman, R. H., eds), Elsevier/North Holland, New York, pp. 213–16.
Grand, R. J. A., Shenolikar, S. and Cohen, P. (1981) *Eur. J. Biochem.* **113**, 359.
Grand, R. J. A., Levine, B. A. and Perry, S. V. (1982) *Biochem. J.* **203**, 61.
Greaser, M. L. and Gergely, J. (1971) *J. Biol. Chem.* **246**, 4226.
Haiech, J., Klee, C. B. and Demaille, J. G. (1981) *Biochemistry* **20**, 3890.
Hartshorne, D. J. and Mueller, H. (1968) *Biochem. Biophys. Res. Commun.* **31**, 647.
Hartshorne, D. J. and Mueller, H. (1969) *Biochem. Biophys. Acta* **175**, 301.
Hartshorne, D. J. and Siemankowski, R. F. (1981) *Ann. Rev. Physiol.* **43**, 519.
Haselgrove, J. C. (1972) *Cold Spring Harbor Symp. Quant. Biol.* **37**, 341.
Hastings, K. E. M. and Emmerson, C. P. (1982) *Proc. Natl Acad. Sci. USA* **79**, 1553.
Hathaway, D. R. and Adelstein, R. S. (1979) *Proc. Natl Acad. Sci. USA* **76**, 1653.
Head, J. F. and Perry, S. V. (1974) *Biochem. J.* **137**, 145.
Head, J. F., Weeks, R. A. and Perry, S. V. (1977) *Biochem. J.* **161**, 465.
Head, J. F., Mader, S. and Kaminer, B. (1979) *J. Cell Biol.* **80**, 211.
Head, J. F., Masure, H. R. and Kaminer, B. (1982) *FEBS Lett.* **137**, 71.
Hidaka, H., Yamaki, T., Asano, M. and Totsuka, T. (1978) *Blood Vessels* **15**, 55.
Hidaka, H., Naka, M. and Yamaki, T. (1979) *Biochem. Biophys. Res. Commun.* **90**, 694.
Hidaka, H., Yamaki, T., Naka, M., Tanaka, T., Hayashi, H. and Kobayashi, R. (1980) *Mol. Pharmacol.* **17**, 66.
Hincke, M. T., McCubbin, W. D. and Kay, C. M. (1978) *Can. J. Biochem.* **56**, 384.
Hitchcock, S. E. (1981) *J. Mol. Biol.* **147**, 153.
Holroyde, M. J., Robertson, S. P., Johnson, J. D., Solaro, R. J. and Potter, J. D. (1980) *J. Biol. Chem.* **255**, 11688.
Hubbard, M., Bardley, M., Sullivan, P., Shepherd, M. and Forrester, I. (1982) *FEBS Lett.* **137**, 85.
Huxley, H. E. (1972) *Cold Spring Harbor Symp. Quant. Biol.* **37**, 361.
Huxley, H. E. and Brown, W. (1967) *J. Mol. Biol.* **30**, 383.

Ilundain, A. and Naftalin, R. J. (1979) *Nature, Lond.* **279**, 446.
Isobe, T. and Okuyama, T. (1978) *Eur. J. Biochem.* **89**, 379.
Itano, T., Itano, R. and Penniston, J. T. (1980) *Biochem. J.* **189**, 455.
Iverson, D. B., Schleicher, M., Zendegui, J. G. and Watterson, D. M. (1981) *Fed. Proc. Fed. Am. Soc. Exp. Biol.* **40**, 1738.
Iwasa, Y., Yoncmitsu, K., Matsui, K., Fukunaga, K. and Miyamoto, E. (1981*a*) *Biochem. Biophys. Res. Commun.* **98**, 656.
Iwasa, Y., Yonemitsu, K. and Miyamoto, E. (1981*b*) *FEBS Lett.* **124**, 207.
Jackson, P., Amphlett, G. W. and Perry, S. V. (1975) *Biochem. J.* **151**, 85.
Jamieson, G. A. and Vanaman, T. C. (1979) *Biochem. Biophys. Res. Commun.* **90**, 1048.
Jamieson, G. A., Vanaman, T. C. and Blum, J. J. (1979) *Proc. Natl Acad. Sci. USA* **76**, 6471.
Jamieson, G. A., Bronson, D. D., Schacht, F. H. and Vanaman, T. C. (1980*a*) *Ann. N.Y. Acad. Sci.* **356**, 1.
Jamieson, G. A., Hayes, A., Blum, J. J. and Vanaman, T. C. (1980*b*) in *Calcium Binding Proteins: Structure and Function* (Siegel, F. L., Carafoli, E., Kretsinger, R. H., MacLennan, D. H. and Wasserman, R. H., eds). Elsevier/North Holland, pp. 165–72.
Jarret, H. W. and Penniston, J. T. (1977) *Biochem. Biophys. Res. Commun.* **77**, 1210.
Jarret, H. W., Charbonneau, H., Anderson, J. M., McCann, R. O. and Cormier, M. J. (1980) *Ann. N.Y. Acad. Sci.* **356**, 119.
Johnson, J. D. and Potter, J. D. (1978) *J. Biol. Chem.* **253**, 3775.
Johnson, J. D., Collins, J. H. and Potter, J. D. (1978) *J. Biol. Chem.* **253**, 6451.
Johnson, J. D., Holroyde, M. J., Crouch, T. H., Solaro, R. J. and Potter, J. D. (1981) *J. Biol. Chem.* **256**, 12194.
Johnson, P., Stockmal, V. B. and Braselton, S. E. H. (1981) *Int. J. Biol. Macromol.* **3**, 267.
Jones, H. P., Ghai, G., Petrone, W. F. and McCord, J. M. (1982) *Biochim. Biophys. Acta* **714**, 152.
Jones, H. P., Matthews, J. C. and Cormier, M. J. (1979) *Biochemistry* **18**, 55.
Kakiuchi, S. and Yamazaki, R. (1970) *Biochem. Biophys. Res. Commun.* **41**, 1104.
Kakiuchi, S. and Sobue, K. (1981) *FEBS Lett.* **132**, 141–143.
Kakiuchi, S., Sobue, K., Yamazaki, R., Nagao, S., Umeki, S., Nozawa, Y., Yazawa, M. and Yagi, K. (1981*a*) *J. Biol. Chem.* **256**, 19–22.
Kakiuchi, S., Sobue, K. and Fujita, M. (1981*b*) *FEBS Lett.* **132**, 144.
Katayama, E. (1979) *J. Biochem. (Tokyo)* **85**, 1379.
Kato, T. and Tonomura, Y. (1975) *J. Biochem. (Tokyo)* **78**, 583.
Keller, C. H., Olwin, B. B., LaPorte, D. C. and Storm, D. R. (1982) *Biochemistry* **21**, 156.
Kendrick-Jones, J. and Jakes, R. (1976) in *International Symposium on Myocardial Failure* (Rieker, G., Weber, A. and Goodwin, J., eds). Tegernsee, Munich, pp. 28–40.
Kendrick-Jones, J., Lehman, W. and Szent-Gyorgyi, A. G. (1970) *J. Mol. Biol.* **54**, 313.
Kilhoffer, M-C., Gerard, D. and Demaille, J. G. (1980*a*) *FEBS Lett.* **120**, 99.
Kilhoffer, M-C., Demaille, J. G. and Gerard, D. (1980*b*) *FEBS Lett.* **116**, 269.
Kilhoffer, M-C., Demaille, J. G. and Gerard, D. (1981) *Biochemistry* **20**, 4407.
Kincaid, R. L., Vaughan, M., Osborne, J. C. and Tkachuk, V. A. (1982) *J. Biol. Chem.* **257**, 10638.
Klee, C. B. (1977) *Biochemistry* **16**, 1017.
Klee, C. B. (1980) in *Calcium and Cell Function*, Vol. 1 (Cheung, W. Y., ed.). Academic Press, New York, pp. 59–77.
Klee, C. B. and Krinks, M. H. (1978) *Biochemistry* **17**, 120.
Klee, C. B. and Vanaman, T. C. (1982) *Adv. Prot. Chem.* **35**, 213.
Klee, C. B., Crouch, T. H. and Krinks, M. H. (1979*a*) *Biochemistry* **18**, 722.
Klee, C. B., Crouch, T. H. and Krinks, M. H. (1979*b*) *Proc. Natl Acad. Sci. USA* **76**, 6270.
Klee, C. B., Crouch, T. H. and Richman, P. G. (1980) *Ann. Rev. Biochem.* **49**, 489.
Klevitt, R. E., Girard, P., Esnouf, M. P. and Williams, R. J. P. (1981*a*) in *Calcium and Phosphate Transport across Biomembranes* (Bronner, F. and Peterlik, M., eds). Academic Press, New York, pp. 25–9.
Klevitt, R. E., Levine, B. A. and Williams, R. J. P. (1981*b*) *FEBS Lett.* **123**, 25.
Kobayashi, R., Tawata, M. and Hidaka, H. (1979) *Biochem. Biophys. Res. Commun.* **88**, 1037.
Kohse, K. P. and Heilmeyer, L. M. G. (1981) *Eur. J. Biochem.* **117**, 507.

Konno, K. (1978) *J. Biochem (Tokyo)* **84**, 1431.
Krausz, Y., Willheim, C. B., Siegel, E. and Sharp, G. W. G. (1980). *J. Clin. Invest.* **66**, 603.
Krebs, J. (1981) *Cell Calcium* **2**, 295.
Kretsinger, R. H. (1972) *Nature New Biol., Lond.* **240**, 85.
Kretsinger, R. H. (1977) in *Calcium Binding Proteins and Calcium Function* (Wasserman, R. H., Corradino, R., Carafoli, E., Kretsinger, R. H., MacLennan, D. H. and Siegel, F. L., eds). Elsevier/North Holland, New York, pp. 63–72.
Kretsinger, R. H. (1980) *Crit. Rev. Biochem.* **8**, 119–74.
Kretsinger, R. H. and Barry, C. D. (1975) *Biochim. Biophys. Acta* **405**, 40.
Kretsinger, R. H. and Nockolds, C. E. (1973) *J. Biol. Chem.* **248**, 3313.
Kretsinger, R. H., Moews, P. C., Coffee, C. J. and Bradshaw, R. A. (1974) in *Calcium Binding Proteins* (Drabikowski, W., Strzelecka-Golaszewska, H. and Carafoli, E., eds). Elsevier, Amsterdam, pp. 703–20.
Kretsinger, R. H., Rudnick, S. E., Sneden, D. A. and Schatz, V. B. (1980) *J. Biol. Chem.* **255**, 8154.
Kumagai, H. and Nishida, E. (1979) *J. Biochem. (Tokyo)* **85**, 1267.
Kuznicki, J., Kuznicki, L. and Drabikowski, W. (1979) *Cell Biol. Int. Rep.* **3**, 17.
Kuznicki, J., Grabarek, Z., Brzeska, H., Drabikowski, W. and Cohen, P. (1981) *FEBS Lett.* **130**, 141.
Lagacé, L., Chandra, T., Woo, S. L. C. and Means, A. R. (1983) *J. Biol. Chem.* **258**, 1684.
Landt, M. and McDonald, J. M. (1980) *Biochem. Biophys. Res. Commun.* **93**, 881.
La Porte, D. C., Toscano, W. A. and Storm, D. R. (1979) *Biochemistry* **18**, 2820.
La Porte, D. C., Wierman, B. M. and Storm, D. R. (1980) *Biochemistry* **19**, 3814.
Leavis, P. C. and Kraft, E. L. (1978) *Arch. Biochem. Biophys.* **186**, 411.
Leavis, P. C., Drabikowski, W., Rosenfeld, S., Grabarek, Z. and Gergely, J. (1977) in *Calcium Binding Proteins and Calcium Function* (Wasserman, R. H., Corradino, R., Carafoli, E., Kretsinger, R. H., MacLennan, D. H. and Siegel, F. L., eds). Elsevier/North Holland, New York, pp. 281–3.
Leavis, P. C., Rosenfeld, S. S., Gergely, J., Grabarek, Z. and Drabikowski, W. (1978) *J. Biol. Chem.* **253**, 5452.
Leavis, P. C., Gergely, J., Grabarek, Z., Drabikowski, W. and Levine, B. A. (1980) in *Calcium Binding Proteins: Structure and Function* (Siegel, F. L., Carafoli, E., Kretsinger, R. H., MacLennan, D. H. and Wasserman, R. H., eds). Elsevier/North Holland, New York, pp. 271–8.
Lee, Y. C. and Wolff, J. (1982) *J. Biol. Chem.* **257**, 6306.
Lehman, W. (1975) *Nature, Lond.* **255**, 424.
Lehman, W. (1977) *Biochem. J.* **163**, 291.
Lehman, W. and Ferrell, M. (1980) *FEBS Lett.* **121**, 273.
Lehman, W. and Szent-Gyorgyi, A. G. (1975) *J. Gen. Physiol.* **66**, 1.
Lehman, W., Regenstein, J. M. and Ransom, A. L. (1976) *Biochim. Biophys. Acta* **434**, 215.
Lehman, W., Head, J. F. and Grant, P. W. (1980) *Biochem. J.* **187**, 447.
Levin, R. M. and Weiss, B. (1977) *Mol. Pharmacol.* **13**, 690.
Levin, R. M. and Weiss, B. (1978) *Biochim. Biophys. Acta* **540**, 197.
Levin, R. M. and Weiss, B. (1979) *J. Pharmacol. Exp. Ther.* **208**, 454.
Levine, B. A. and Williams, R. J. P. (1981) in *The Role of Calcium in Biological Systems*, Vol. 1 (Anghileri, A. M. T. and Anghileri, L. J., eds). CRC Press, Boca Raton, pp. 3–26.
Levine, B. A. and Williams, R. J. P. (1982) in *Calcium and Cell Function*, Vol. II (Cheung, W. Y., ed.). Academic Press, New York, pp. 1–38.
Levine, B. A., Mercola, D., Coffman, D. and Thornton, J. M. (1977) *J. Mol. Biol.* **115**, 743.
Levine, B. A., Thornton, J. M., Fernandes, R., Kelly, C. M. and Mercola, D. (1978) *Biochim. Biophys. Acta* **535**, 11.
Lin, Y. M., Liu, Y. P. and Cheung, W. Y. (1974) *J. Biol. Chem.* **249**. 4943.
MacManus, J. P., Braceland, B. M., Rixon, R. H., Whitfield, J. F. and Morris, H. P. (1981) *FEBS Lett.* **133**, 99.
Maekawa, S. and Abe, T. (1980) *Biochem. Biophys. Res. Commun.* **97**, 621.
Maihle, N. J., Dedman, J. R., Means, A. R., Chafouleas, J. G. and Satir, B. H. (1981) *J. Cell Biol.* **89**, 695.
Mani, R. S., McCubbin, W. D. and Kay, C. M. (1974) *Biochemistry* **13**, 5003.

Manning, D. R. and Stull, J. T. (1979) *Biochem. Biophys. Res. Commun.* **90**, 164.
Marcum, J. M., Dedman, J. R., Brinkley, B. R. and Means, A. R. (1978) *Proc. Natl Acad. Sci. USA* **75**, 3771.
Margossian, S. S. and Cohen, C. (1973) *J. Mol. Biol.* **81**, 409.
Means, A. R. and Dedman, J. R. (1980) *Nature, Lond.* **285**, 73.
Means, A. R., Tash, J. S. and Chafouleas, J. G. (1982) *Physiol. Rev.* **62**, 1.
Moir, A. J. G., Ordidge, M., Grand, R. J. A., Trayer, I. P. and Perry, S. V. (1983) *Biochem. J.* **209**, 417.
Molla, A., Kilhoffer, M-C, Ferraz, C., Audemard, E., Walsh, M. P. and Demaille, J. G. (1981) *J. Biol. Chem.* **256**, 15.
Moore, P. B. and Dedman, J. R. (1982) *J. Biol. Chem.* **257**, 9663.
Munjaal, R. P., Dedman, J. R. and Means, A. R. (1980) *Ann. N.Y. Acad. Sci.* **356**, 110.
Murray, A. C. and Kay, C. M. (1972) *Biochemistry* **11**, 2622.
Nachmias, V. and Asch, A. (1974) *Biochem. Biophys. Res. Commun.* **60**, 656.
Nairn, A. C. and Perry, S. V. (1979) *Biochem. J.* **179**, 89.
Nairn, A. C., Grand, R. J. A., Wall, C. M. and Perry, S. V. (1980) *Ann. N.Y. Acad. Sci.* **356**, 413.
Nairn, A. C., Grand, R. J. A. and Perry, S. V. (1984) *FEBS Lett.* **167**, 215.
Niggli, V., Penniston, J. J. and Carafoli, E. (1979) *J. Biol. Chem.* **254**, 9955.
Norman, J. A., Drummond, A. H. and Moser, P. (1979) *Mol. Pharmacol.* **16**, 1089.
Ohtsuki, I., Yamamoto, K. and Hashimoto, K. (1981) *J. Biochem. (Tokyo)* **90**, 259.
Okumura, K., Kato, T., Ito, J. and Tanaka, R. (1982) *Dev. Brain Res.* **3**, 662.
Ortega Perez, R., Van Tuinen, D., Marme, D., Cox, J. A. and Turian, G. (1981) *FEBS Lett.* **133**, 205
Palfrey, H. C., Schiebler, W. and Greengard, P. (1982) *Proc. Natl Acad. Sci. USA* **79**, 3780.
Parry, D. A. D. and Squire, J. M. (1973) *J. Mol. Biol.* **75**, 33.
Payne, M. E. and Soderling, T. R. (1980) *J. Biol. Chem.* **255**, 8054.
Pearlstone, J. R. and Smillie, L. B. (1978) *Can. J. Biochem.* **56**, 521.
Pearlstone, J. R. and Smillie, L. B. (1982) *J. Biol. Chem.* **257**, 10587.
Pearlstone, J. R., Carpenter, M. R., Johnson, P. and Smillie, L. B. (1976) *Proc. Natl Acad. Sci. USA* **73**, 1902.
Pechere, J-F. (1977) in *Calcium-Binding Proteins and Calcium* (Wasserman, R. H., Corradino, R., Carafoli, E., Kretsinger, R. H., MacLennan, D. H. and Siegel, F. L., eds). Elsevier/North Holland, New York, pp. 213–21.
Perry, S. V. (1979) *Biochem. Soc. Trans.* **7**, 593.
Perry, S. V. and Cole, H. A. (1974) *Biochem. J.* **141**, 733.
Perry, S. V., Grand, R. J. A., Nairn, A. C., Vanaman, T. C. and Wall. C. M. (1979) *Biochem. Soc. Trans.* **7**, 619.
Piascik, M. T., Wisler, P. L., Johnson, C. L. and Potter, J. D. (1980) *J. Biol. Chem.* **255**, 4176.
Picton, C., Klee, C. B. and Cohen, P. (1980) *Eur. J. Biochem.* **111**, 553.
Picton, C., Shenolikar, S., Grand, R. J. A. and Cohen, P. (1983) *Methods in Enzymology* **102**, 219.
Pires, E. M. V. and Perry, S. V. (1977) *Biochem. J.* **167**, 137.
Potter, J. D. and Gergely, J. (1975) *J. Biol. Chem.* **250**, 4628.
Potter, J. D. and Johnson, J. D. (1982) in *Calcium and Cell Function*, Vol. II (Cheung, W. Y., ed.). Academic Press, New York, pp. 145–73.
Potter, J. D., Seidel, J. C., Leavis, P., Lehrer, S. S. and Gergely, J. (1976) *J. Biol. Chem.* **251**, 7551.
Potter, J. D., Johnson, J. D., Dedman, J. R., Schrieber, W. E., Mandel, F., Jackson, R. L. and Means, A. R. (1977) in *Calcium Binding Proteins and Calcium Function* (Wasserman, R. H., Corradino, R., Carafoli, E., Kretsinger, R. H., MacLennan, D. H. and Siegel, F. L., eds). Elsevier/North Holland, New York, pp. 239–50.
Potter, J. D., Robertson, S. P., Collins, J. H. and Johnson, J. D. (1980*a*) in *Calcium Binding Proteins: Structure and Function* (Siegel, F. L., Carafoli, E., Kretsinger, R. H., MacLennan, D. H. and Wasserman, R. H., eds). Elsevier/North Holland, New York, pp. 279–88.
Potter, J. D., Piascik, M. T., Wisler, P. L., Robertson, S. P. and Johnson, C. L. (1980*b*) *Ann. N.Y. Acad. Sci.* **256**, 220.

Prozialeck, W. C. and Weiss, B. (1982) *J. Pharmacol. Exp. Ther.* **222**, 509.
Regenstein, J. M. and Szent-Györgyi, A. G. (1975) *Biochemistry* **14**, 917.
Richman, P. G. (1978) *Biochemistry* **17**, 3001.
Richman, P. G. and Klee, C. B. (1978) *Biochemistry* **17**, 928.
Richman, P. G. and Klee, C. B. (1979) *J. Biol. Chem.* **254**, 5372.
Romero-Herrera, A. E., Castillo, O. and Lehmann, H. (1976) *J. Mol. Evol.* **8**, 251.
Ross, E. M. and Gilman, A. G. (1980) *Ann. Rev. Biochem.* **49**, 533.
Roufogalis, B. D. (1981) *Biochem. Biophys. Res. Commun.* **98**, 607.
Salmon, E. D. and Segall, R. R. (1980) *J. Cell Biol.* **86**, 355.
Salter, R. S., Krinks, M. H., Klee, C. B. and Neer, E. J. (1981) *J. Biol. Chem.* **256**, 9830.
Sasagawa, T., Ericsson, L. H., Walsh, K. A., Schrieber, W. E., Fisher, E. H. and Titani, K. (1982) *Biochemistry* **21**, 2565.
Schubart, U. K., Erlichman, J. and Fleischer, N. (1980) *J. Biol. Chem.* **255**, 4120.
Schulman, H. and Greengard, P. (1978*a*) *Nature, Lond.* **271**, 478.
Schulman, H. and Greengard, P. (1978*b*) *Proc. Natl Acad. Sci. USA* **75**, 5432.
Schulman, H., Huttner, W. B. and Greengard, P. (1980) in *Calcium and Cell Function*, Vol. 1 (Cheung, W. Y., ed.). Academic Press, New York, pp. 220–52.
Seamon, K. B. (1980) *Biochemistry* **19**, 207.
Seamon, K. B. and Moore, B. W. (1980) *J. Biol. Chem.* **255**, 11644.
Seamon, K. B., Hartshorne, D. J. and Bothner-By, A. A. (1977) *Biochemistry* **16**, 4039.
Seymour, J. and O'Brien, E. J. (1980) *Nature, Lond.* **283**, 680.
Sharma, R. K., Wirch, E. and Wang, J. H. (1978) *J. Biol. Chem.* **253**, 3575.
Sharma, R. K., Desai, R., Waisman, D. M. and Wang, J. H. (1979) *J. Biol. Chem.* **254**, 4276.
Sharma, R. K., Wang, T. H., Wirch, E. and Wang, J. H. (1980) *J. Biol. Chem.* **255**, 5916.
Shenolikar, S., Cohen, P. T. W., Cohen, P., Nairn, A. C. and Perry, S. V. (1979) *Eur. J. Biochem.* **100**, 329.
Shimizu, T., Hatano, M., Nagao, S. and Nozawa, Y. (1982) *Biochem. Biophys. Res. Commun.* **106**, 1112.
Siegel, F. L., Carafoli, E., Kretsinger, R. H., MacLennan, D. H. and Wasserman, R. H. (eds) (1980) *Calcium Binding Proteins: Structure and Function*, Elsevier/North Holland, New York.
Sin, I. L., Fernandes, R. and Mercola, D. (1978) *Biochem. Biophys. Res. Commun.* **82**, 1132.
Small, J. V. and Sobiesek, A. (1980) *Int. Rev. Cytology* **64**, 241.
Smoake, J. A., Song, S-Y, and Cheung, W. Y. (1974) *Biochim. Biophys. Acta* **341**, 402.
Sobue, K., Muramoto, Y., Fujita, M. and Kakiuchi, S. (1981*a*) *Biochem. Biophys. Res. Commun.* **100**, 1063.
Sobue, K., Fujita, M., Muramoto, Y. and Kakiuchi, S. (1981*b*) *FEBS Lett.* **132**, 137.
Squire, J. (1981*a*) *The Structural Basis of Muscular Contraction*, Plenum, New York.
Squire, J. (1981*b*) *Nature, Lond.* **291**, 614.
Stewart, A. A., Ingebritsen, T. S., Manalan, A., Klee, C. B. and Cohen, P. (1982) *FEBS Lett.* **137**, 80.
Strasburg, G. M., Greaser, M. L. and Sundaralingam, M. (1980) *J. Biol. Chem.* **255**, 3806.
Stull, J. T. (1980) *Adv. Cyclic Nucleotide Res.* **13**, 39.
Suzuki, Y., Hirabayashi, T. and Watanabe, Y. (1979) *Biochem. Biophys. Res. Commun.* **90**, 253.
Syska, H., Wilkinson, J. M., Grand, R. J. A. and Perry, S. V. (1976) *Biochem. J.* **153**, 375.
Szebenyi, D. M. E., Obendorf, S. K. and Moffat, K. (1981) *Nature, Lond.* **294**, 327.
Takagi, T., Nemoto, T. and Konishi, K. (1980) *Biochem. Biophys. Res. Commun.* **96**, 377.
Talbot, J. A. and Hodges, R. S. (1979) *J. Biol. Chem.* **254**, 3720.
Tanaka, T. and Hidaka, H. (1980) *J. Biol. Chem.* **255**, 11078.
Tanaka, T., Naka, M. and Hidaka, H. (1980) *Biochem. Biophys. Res. Commun.* **92**, 313.
Taylor, K. A. and Amos, L. A. (1981) *J. Mol. Biol.* **147**, 297.
Taylor, L., Guerina, V. J., Donowitz, M., Cohen, M. and Sharp, G. W. G. (1981) *FEBS Lett.* **131**, 322.
Teo, T. S. and Wang, J. H. (1973) *J. Biol. Chem.* **248**, 5950.
Teo, T. S., Wang, T. H. and Wang, J. H. (1973) *J. Biol. Chem.* **248**, 588.
Toda, H., Yazawa, M., Kondo, K., Honma, T., Narita, K. and Yagi, K. (1981) *J. Biochem. (Tokyo)* **90**, 1493.

Tomlinson, S., Walker, S. W. and Brown, B. L. (1982) *Diabetologia* **22**, 1.
Tsuchiya, T., Kaneko, T. and Matsumoto, J. J. (1978) *J. Biochem (Tokyo)* **83**, 1191.
Tsukui, R. and Ebashi, S. (1973) *J. Biochem. (Tokyo)* **73**, 1119.
Tucker, M. M., Robinson, J. B. and Stellwagen, E. (1981) *J. Biol. Chem.* **256**, 9051.
Vanaman, T. C. (1980) in *Calcium and Cell Function*, Vol. 1 (Cheung, W. Y., ed.). Academic Press, New York, pp. 41–58.
Vanaman, T. C. and Perry, S. V. (1978) *Proc. Eur. Conf. Muscle and Motility*, 7th Abstr., p. 37.
van Eerd, J-P and Kawasaki, Y. (1973) *Biochemistry* **12**, 4972.
van Eerd, J-P and Takahashi, K. (1976) *Biochemistry* **15**, 1171.
van Eerd, J-P., Capony, J-P., Ferraz, C. and Pechere, J-F. (1978) *Eur. J. Biochem.* **91**, 231.
Van Eldik, L. J., Grossman, A. R., Iverson, D. B. and Watterson, D. M. (1980) *Proc. Natl Acad. Sci. USA* **77**, 1912.
Veigl, M. L., Sedwick, W. D. and Vanaman, T. C. (1982) *Fed. Proc.* **41**, 2283.
Vincenzi, F. F. and Hinds, T. R. (1980) in *Calcium and Cell Function*, Vol. 1 (Cheung, W. Y., ed.). Academic Press, New York, pp. 128–65.
Waisman, D. M., Stevens, F. C. and Wang, J. H. (1975) *Biochem. Biophys. Res. Commun.* **65**, 975.
Waisman, D. M., Stevens, F. C. and Wang, J. H. (1978) *J. Biol. Chem.* **253**, 1106.
Wall, C., Grand, R. J. A. and Perry, S. V. (1981) *Biochem. J.* **195**, 307.
Wallace, R. W., Lynch, T. J., Tallant, E. A. and Cheung, W. Y. (1978) *J. Biol. Chem.* **254**, 377.
Wallace, R. W., Tallant, E. A. and Cheung, W. Y. (1980) *Biochemistry* **19**, 1831.
Walsh, M., Stevens, F. C., Kuznicki, J. and Drabikowski, W. (1977) *J. Biol. Chem.* **252**, 7440.
Wang, C-L. A., Aquaron, R. R., Leavis, P. C. and Gergely, J. (1982) *Eur. J. Biochem.* **124**, 7.
Wang, J. H. and Desai, R. (1977) *J. Biol. Chem.* **252**, 4175.
Wasserman, R. H. (1980) in *Calcium-Binding Proteins: Structure and Function* (Siegel, F. L., Carafoli, E., Kretsinger, R. H., MacLennan, D. H. and Wasserman, R. H., eds). Elsevier/North Holland, New York, pp. 357–61.
Wasserman, R. H., Fullmer, C. S. and Taylor, A. N. (1978) in *Vitamin D* (Lawson, D. E. M., ed.). Academic Press, London, pp. 136–66.
Watterson, D. M. and Vanaman, T. C. (1976) *Biochem. Biophys. Res. Commun.* **73**, 40–6.
Watterson, D. M. and Vincenzi, F. F. (eds) (1980) Calmodulin and cell functions, *Ann. N.Y. Acad. Sci.* **356**.
Watterson, D. M., Harrelson, W. G., Keller, P. M., Sharief, F. and Vanaman, T. C. (1976*a*) *J. Biol. Chem.* **251**, 4501.
Watterson, D. M., van Eldik, L. J., Smith, R. E. and Vanaman, T. C. (1976*b*) *Proc. Natl Acad. Sci. USA* **73**, 2711.
Watterson, D. M., Sharief, F. and Vanaman, T. C. (1980*a*) *J. Biol. Chem.* **255**, 962.
Watterson, D. M., Iverson, D. B. and van Eldik, L. J. (1980*b*) *Biochemistry* **19**, 5762.
Watterson, D. M., Mendel, P. A. and Vanaman, T. C. (1980*c*) *Biochemistry* **19**, 2672.
Weeds, A. G. and MacLachlan, A. D. (1974) *Nature, Lond.* **252**, 646.
Weeks, R. A. and Perry, S. V. (1978) *Biochem. J.* **173**, 449.
Welsh, M. J., Dedman, J. R., Brinkley, B. R. and Means, A. R. (1978) *Proc. Natl Acad. Sci. USA* **75**, 1867.
Welsh, M. J., Dedman, J. R., Brinkley, B. R. and Means, A. R. (1979) *J. Cell Biol.* **81**, 624.
Wilkinson, J. M. (1974) *Biochim. Biophys. Acta* **359**, 379.
Wilkinson, J. M. (1976) *FEBS Lett.* **70**, 254.
Wilkinson, J. M. (1980) *Eur. J. Biochem.* **103**, 179.
Wilkinson, J. M. and Grand, R. J. A. (1975) in *Proc. 9th FEBS Meeting, 31. Proteins of Contractile Systems* (Biro, E. N. A., ed.). Akademai Kido, Budapest, and Elsevier/North Holland, Amsterdam, pp. 137–44.
Wilkinson, J. M. and Grand, R. J. A. (1978) *Nature, Lond.* **271**, 31.
Wnuk, W., Cox, J. A. and Stein, E. A. (1982) in *Calcium and Cell Function*, Vol. II, (Cheung, W. Y., ed.). Academic Press, New York, pp. 243–78.
Wolff, D. J. and Brostrom, C. O. (1979) *Adv. Cyclic Nucleotide Res.* **11**, 27.

Wolff, D. J., Poirier, P. G., Brostrom, C. O. and Brostrom, M. A. (1977) *J. Biol. Chem.* **252**, 4108.
Wolff, J., Cook, G. H., Goldhammer, A. R. and Berkowitz, S. A. (1980) *Proc. Natl Acad. Sci. USA* 77, 3841.
Wong, P.Y-K. and Cheung, W. Y. (1979) *Biochem. Biophys. Res. Commun.* **90**, 473.
Yamaki, T. and Hidaka, H. (1980) *Biochem. Biophys. Res. Commun.* **94**, 727.
Yamauchi, T. and Fujisawa, H. (1979) *Biochem. Biophys. Res. Commun.* **90**, 28.
Yamauchi, T. and Fujisawa, H. (1980) *FEBS Lett.* **116**, 141.
Yang, S-D., Tallant, E. A. and Cheung, W. Y. (1982) *Biochem. Biophys. Res. Commun.* **106**, 1419.
Yates, L. and Vanaman, T. C. (1981) *Fed. Proc. Fed. Am. Soc. Exp. Biol.* **40**, 1738.
Yazawa, M., Kuwayama, H. and Yagi, K. (1978) *J. Biochem. (Tokyo)* **84**, 1253.
Yazawa, M., Sakuma, M. and Yagi, K. (1980) *J. Biochem. (Tokyo)* **87**, 1313.
Yazawa, M., Yagi, M., Toda, H., Kondo, K., Narita, K., Yamazaki, R., Sobue, K., Kakiuchi, S., Nagao, S. and Nozawa, Y. (1981) *Biochem. Biophys. Res. Commun.* **99**, 1051.
Yerna, M-J., Dabrowska, R., Hartshorne, D. J. and Goldman, R. D. (1979) *Proc. Natl Acad. Sci. USA* **76**, 184.

3
Structure and mechanism of the (Na^+, K^+)-and (Ca^{2+})–ATPases

Michael Forgac and Gilbert Chin

3.1 INTRODUCTION

The (Na^+ and K^+)-stimulated adenosine triphosphatase ((Na^+, K^+)-ATPase) and the (Ca^{2+})-stimulated adenosine triphosphatase ((Ca^{2+})-ATPase) are intrinsic membrane proteins present in eukaryotic cells whose functions are to couple the active transport of either Na^+ and K^+ or Ca^{2+} across the membrane to ATP hydrolysis. The (Na^+, K^+)-ATPase is present in the plasma membrane of essentially all animal cells and is responsible for the maintenance of Na^+ and K^+ gradients. The Na^+ gradient is utilised in the generation of the action potential in nerve cells (Katz, 1966), in the uptake of sugars and amino acids by intestinal epithelia and renal tubules (Crane, 1965, Endou *et al.*, 1975), in the readsorption of Na^+ by the kidney (Jorgensen, P. L., 1980) and to assist in Ca^{2+} and proton efflux from the cell (Miyamato and Racker, 1980; Rindler and Saier, 1981). The K^+ gradient (and K^+ permeability) establish an internally negative membrane potential (Katz, 1966) and a high cytoplasmic K^+ concentration is required for various enzymatic reactions (Evans and Sorges, 1966). The (Ca^{2+})-ATPase is also present in the eukaryotic plasma membrane (Schatzmann and Vincenzi, 1969) as well as various intracellular membranes, such as the sarcoplasmic reticulum of muscle (Hasselbach and Makinose, 1961), and serves to maintain the cytoplasmic Ca^{2+} concentration at a low level. This is essential to the process of muscle contraction, which is activated by an influx of Ca^{2+} into the cytoplasm (Mannherz and Goody, 1976). In addition, Ca^{2+} plays an essential role in the many calmodulin-regulated events in cells (Cheung, 1980) and in the process of synaptic transmission (Miledi, 1973), where it induces fusion of synaptic vesicles to the presynaptic membrane.

Although these proteins transport different ions, they bear a remarkable structural and mechanistic similarity, both to each other (Guidotti, 1979), and to other eukaryotic cation-translocating ATPases, such as the (H^+, K^+)-ATPase of hog stomach (Sachs *et al.*, 1976) and the (H^+)-ATPase of *Neurospora* (Scarborough, 1980). These similarities, which suggest that these proteins may be derived from a common evolutionary ancestor, provide a rationale for discussing the structure and mechanism of the (Na^+, K^+) and Ca^{2+} pumps in a single review. No attempt has been made to present an exhaustive review of the literature on either of these proteins, and the reader is referred to one of the numerous other reviews written on either the (Na^+, K^+)-ATPase (Robinson and Flashner, 1979; Cantley 1981; Jorgensen, P. L., 1982) or the (Ca^{2+})-ATPase (MacLennan and Holland, 1975, de Meis and Vianna, 1979; Carafoli and Zurini, 1982) for additional information.

3.2 STRUCTURE

3.2.1 Purification, subunit molecular weight and reconstitution

Tissues rich in the (Na^+, K^+)-ATPase are mammalian kidney (Kyte, 1971), shark rectal gland (Hokin *et al.*, 1973), eel electric organ (Perrone *et al.*, 1975), and duck nasal salt gland (Hopkins *et al.*, 1976). Membrane preparations, and in some cases solubilised enzymes (Kyte, 1971; Hokin *et al.*, 1973), that catalyse a (Na^+ and K^+)-stimulated hydrolysis of ATP, have been obtained from these sources; the maximal specific activity is 30–40 μmol min^{-1} mg^{-1} protein at 37°C. This activity is specifically and completely inhibited by ouabain and other cardiac glycosides (Glynn, 1957). The particularly simple purification from mammalian kidney consists of a gentle detergent extraction of a microsomal fraction from the homogenate of the outer medulla (Jorgensen, P. L., 1974).

The purified, active enzyme is composed of two discrete polypeptides, a catalytic subunit, α, and a glycoprotein, β. The apparent molecular weights of these peptides have been measured by SDS gel electrophoresis and amino acid analysis (Craig and Kyte, 1980; Peterson and Hokin, 1981) and sedimentation equilibrium and gel filtration (Hastings and Reynolds, 1979; Esmann *et al.*, 1980), and are approximately 110 000 for α and 40 000 for the protein portion of β. The apparent molecular weight of intact β on SDS gels, 55 000–60 000, can be decreased to 50 000 by incubation with a mixture of glycosidases (Craig and Kyte, 1980). Treatment of the membrane-bound enzyme with endoglycosidase D reduces the apparent molecular weight to 42 000 without affecting ATPase activity (G. Chin and M. D. Forgac, unpublished experiments). The *endo* D-treated β subunit still migrates as a diffuse band, and more carbohydrate may still be attached.

Most of the intrinsic protein mass of the sarcoplasmic reticulum, an intracellular membranous network in cardiac and skeletal muscle, is accounted for by a single protein, the (Ca^{2+})-ATPase. Sealed vesicles that possess the same membrane topology as the original reticulum and are competent in active Ca^{2+}

uptake can be prepared. Purification of the enzyme from this membrane fraction by removal of the extrinsic proteins was first accomplished by MacLennan (1970), and resulted in a maximal specific activity of 30–40 μmol min^{-1} mg^{-1} protein at 37°C. The affinity purifications of the calmodulin-binding plasma membrane (Ca^{2+})-ATPases from red blood cells and cardiac sarcolemma have been reported (Caroni and Carafoli, 1981).

The sarcoplasmic reticulum enzyme consists of a single polypeptide with an apparent molecular weight of 100 000 by SDS gel electrophoresis (MacLennan *et al.*, 1971) or of 120 000 by sedimentation equilibrium and gel filtration (Rizzolo *et al.*, 1976). A distinctly higher apparent molecular weight of 150 000 has been obtained by SDS gel electrophoresis of the (Ca^{2+})-ATPases from red blood cells and cardiac sarcolemma (Knauf *et al.*, 1974; Drickamer, 1975; Caroni and Carafoli, 1981). This discrepancy, along with the ability to bind calmodulin (Caroni and Carafoli, 1981; see section 3.4), suggests significant differences in the structures of the plasma membrane and sarcoplasmic reticulum (Ca^{2+})-ATPases, despite functional similarity (see section 3.3.2.1). A recent observation (Steiger and Schatzmann, 1981) has indicated that tryptic removal of a 30 000–40 000 Da peptide from the red blood cell (Ca^{2+})-ATPase produces a 100 000 Da fragment which hydrolyses ATP and is no longer regulated by calmodulin.

Reconstitution into artificial phospholipid vesicles has shown that the α and β subunits of the (Na^+, K^+)-ATPase are sufficient to support coupled hydrolysis and transport (Hilden and Hokin, 1975; Goldin, 1977; see section 3.3.2.1), although an associated proteolipid that is poorly stained on SDS gels has been noticed (Forbush *et al.*, 1978). Similar reconstitution of the ATP-dependent Ca^{2+} transport activity of both classes of (Ca^{2+})-ATPase has been achieved (Zimniak and Racker, 1978; Chiesi and Inesi, 1980; Niggli *et al.*, 1982). A low molecular weight proteolipid of the sarcoplasmic reticulum can be removed without affecting the transport activity of the reconstituted (Ca^{2+})-ATPase (Knowles *et al.*, 1980; MacLennan *et al.*, 1980).

3.2.2 Localisation

Many workers have raised antibodies to the (Na^+, K^+)-ATPase and its subunits; some of the antibodies inhibit activity (Jorgensen, P. L. *et al.*, 1973), and some do not (Kyte, 1974). Electron microscopic examination of immunoferritin deposits in ultrathin frozen sections (Kyte, 1976) is limited by the loss of fine morphology and by the uncertainty of antibody specificity. Other methods for localisation of the (Na^+, K^+)-ATPase are K^+-phosphatase histochemistry (Ernst, 1975) and ouabain autoradiography (Shaver and Stirling, 1978). The problems encountered in these techniques are the specificity of the phosphatase reaction and the removal of nonspecifically bound ouabain, respectively. All of these methods offer only a qualitative description of the presence of the (Na^+, K^+)-ATPase in the plasma membrane.

The structure of the sarcoplasmic reticulum was first observed in the electron microscope by Porter and Palade (1957). A modified staining procedure that included tannic acid as well as glutaraldehyde and osmium tetroxide strikingly revealed the asymmetric distribution of the sarcoplasmic reticulum proteins, with virtually all of the electron-dense stain located on the outer or cytoplasmic side of the vesicles (Saito *et al.*, 1978). A (Ca^{2+})-ATPase-like immunoreactivity has been observed in skeletal muscle cells (Jorgensen, A. O. *et al.*, 1977). Localisation studies in cardiac muscle using ferritin and immunofluorescence have demonstrated that the antibodies raised against the skeletal sarcoplasmic reticulum enzyme cross-react with the cardiac sarcoplasmic reticulum (Ca^{2+})-ATPase; however, no reaction was seen at the cardiac sarcolemma or other cellular membranes (Jorgensen, A. O. *et al.*, 1982).

3.2.3 Quaternary structure

The central questions in this area are still unresolved. What is the state of association of the subunits *in vivo*, and what is the minimal structural unit capable of coupled hydrolysis and transport? Approaches to answering these questions have focused on crosslinking, gel filtration and sedimentation equilibrium, ligand-binding stoichiometry and electron microscopy.

Dimethyl suberimidate, glutaraldehyde and cupric phenanthroline have been used to crosslink the subunits of the (Na^+, K^+)-ATPase (Kyte, 1972, 1975; Giotta, 1976, Liang and Winter, 1977; Craig, 1982*a*). Among the covalent adducts observed are $\alpha\alpha$, $\alpha\beta$, and higher oligomers of α alone. Detergents and ligands appeared to influence the state of association in a complex fashion. A dimer of the β subunit is not seen because of relatively poor staining by Coomassie blue, but has been located by carbohydrate labelling (G. Chin and M. D. Forgac, unpublished experiments). Interpretations of the data are complicated by the varied conditions of crosslinking and restriction to lateral diffusion in the lipid bilayer; however, many workers favour the assignment of $(\alpha\beta)_2$ as the native functional unit (Klingenberg, 1981; Kyte, 1981).

The experimental conditions for hydrodynamic analysis of active unit molecular weight are incompatible with those for assay of activity. A vesicular structure is needed to monitor transport of ions, and solubilisation is required for an accurate size determination. Nevertheless, these experiments have provided measurements of subunit molecular weight and evidence for aggregates (Clarke, 1975; Hastings and Reynolds, 1979; Esmann *et al.*, 1980; Brotherus *et al.*, 1981; Craig, 1982*b*), some of which are able to hydrolyse ATP. Whether the coupling of transport and hydrolysis is maintained in these detergent-dissociated particles is unknown. The stoichiometry of ligand binding has varied, depending on the accuracy of protein assay; a stoichiometry of one high-affinity ATP site per $\alpha\beta$ unit has been found (Moczydlowski and Fortes, 1981; for other ligands see Jorgensen, P. L., 1982), and supports the conclusion that the $\alpha\beta$ monomer is fully capable of ATPase activity.

For the (Ca^{2+})-ATPase, similar experiments using phosphorylation and an

ATP site label (MacLennan *et al.*, 1971; Andersen *et al.*, 1982) produce a stoichiometry of one catalytic site per monomer. Extensive studies on the state of association of the (Ca^{2+})-ATPase in detergents have been carried out by Tanford and co-workers, and they find it possible not only to obtain monomers, but also to replace almost all of the native lipid with detergent while retaining significant ATPase activity (Dean and Tanford, 1978).

Fluorescence resonance energy transfer between mixed populations of (Ca^{2+})-ATPases labelled with donor and acceptor fluorescent groups (Vanderkooi *et al.*, 1977) at varying concentrations has provided evidence for the oligomeric association of monomers in the membrane. This view is consistent with cross-linking of the sarcoplasmic reticulum protein (Murphy, 1976; Hebdon *et al.*, 1979).

Ultrastructural analysis of the (Na^+, K^+)- and (Ca^{2+})-ATPases reconstituted at several different ratios of phospholipid to protein has yielded qualified support for dimeric species (Wang *et al.*, 1979; Skriver *et al.*, 1980). There appears to be sufficient uncertainty in the numbers, dimensions, and identities of the membrane-embedded particles to prevent a finer discrimination between monomer and oligomer. In summary, the available evidence specifies the monomer as the minimum hydrolytic unit, but there are also noncovalent interactions between monomers in the membrane.

3.2.4 Secondary and tertiary structure

Neither the (Na^+, K^+)-ATPase nor the (Ca^{2+})-ATPase has yet been crystallised; therefore their secondary structures are unknown. The membrane-bound (Na^+, K^+)-ATPase will form two-dimensional regular arrays in the presence of phosphate or vanadate (Skriver *et al.*, 1981) and has been analysed by electron microscopy and image reconstruction to a scale of 20 Å (Hebert *et al.*, 1982). This is insufficient resolution for mapping the paths of α-helices and β-sheets. However, based on work with bacteriorhodopsin (Unwin and Henderson, 1975) and thermodynamic calculations (Guidotti, 1977), the lengths of polypeptide that traverse the lipid bilayer are probably α-helices. The peptide backbone would be completely internally hydrogen bonded, and the side chains of nonpolar residues would enjoy favourable interactions with the hydrophobic lipids. Any polar side chains within a largely nonpolar stretch of residues might be exposed to a hydrophilic domain of the protein, or might be stabilised by interaction with polar side chains on a neighbouring helix. The available evidence (tertiary structure discussed subsequently) for the (Na^+, K^+)- and (Ca^{2+})-ATPases is consistent with the structures shown in figure 3.1. The polypeptide chain crosses the bilayer in five to ten helices, and most of the hydrophilic portions of the proteins are exposed on the cytoplasmic side of the membrane, but not grouped in a separate domain as for the red blood cell anion transport protein (Macara and Cantley, 1983).

Impermeant labelling reagents specific for the cytoplasmic or extracellular

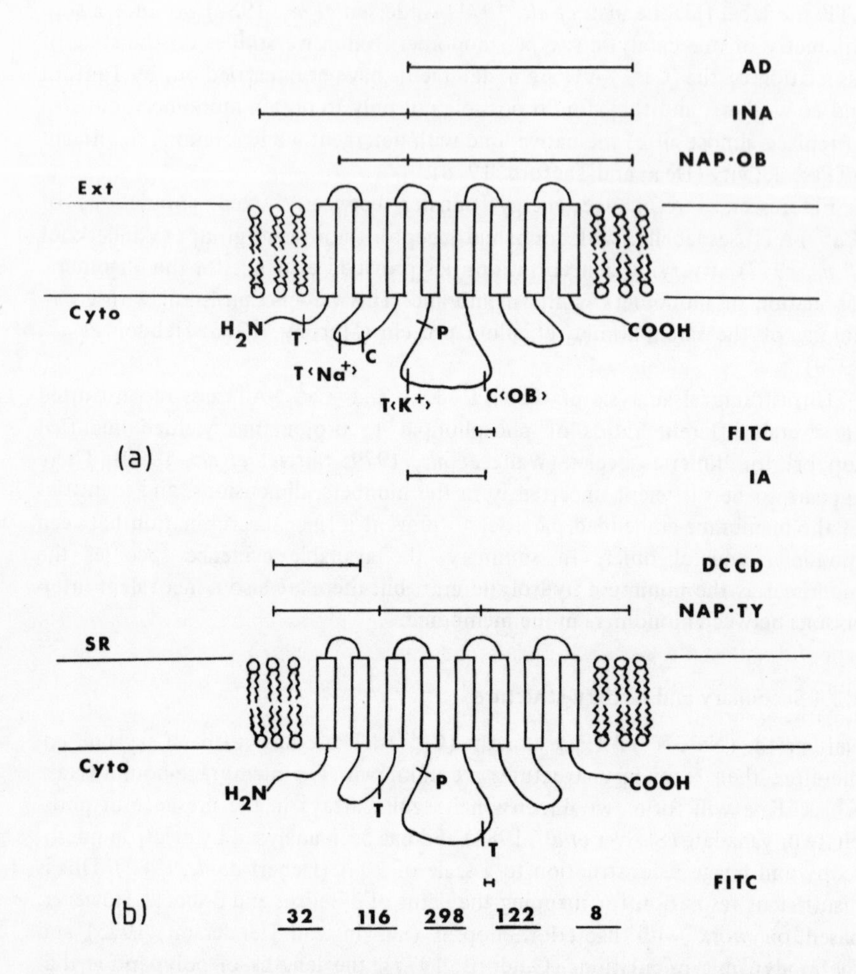

Figure 3.1 Structures of the (Na^+, K^+)-ATPase α subunit (a) and the sarcoplasmic reticulum (Ca^{2+})-ATPase (b). The membrane topology is defined by the cytoplasm (Cyto) and by either the extracellular medium (Ext) or the lumen of the sarcoplasmic reticulum (SR). The intramembranous rectangles represent cross-sections of the cylindrical boundaries of α-helices. Specific sites on the proteins are as follows: T, peptide bonds cleaved by trypsin (in the presence of ligands); C, peptide bonds cleaved by chymotrypsin (in the presence of ligands), P, phosphorylated aspartates. The fragments labelled by extracellular or hydrophobic reagents are placed above the protein, and those labelled by cytoplasmic reagents are placed below. AD, adamantane diazirine (Farley *et al.*, 1980); INA, iodonaphthylazide (Karlish *et al.*, 1977); NAP•OB, *N*-ouabain-*N'*-(2-nitro-4-azidophenyl)ethylenediamine (Jorgensen, P. L. *et al.*, 1982); FITC, fluorescein-5' isothiocyanate (Carilli *et al.*, 1982; Mitchinson *et al.*, 1982); IA, iodoacetate (Castro and Farley, 1979); DCCD, dicyclohexylcarbodiimide (MacLennan and Reithmeier, 1982); NAP•TY, *N*-(2-nitro-4-azidophenyl)-*O*-hexanoyl-2,6-diiodotyramine (Green *et al.*, 1980). For the (Ca^{2+})-ATPase, the numbers of residues which have been sequenced and located are placed beneath the corresponding hydrophilic portions of the protein (MacLennan and Reithmeier, 1982).

surface and a few well-placed proteolytic sites have facilitated the study of the membrane topology of the (Na^+, K^+)-ATPase. Phosphorylation at the active site aspartate residue and covalent reaction with ouabain derivatives have pinpointed regions of the α-subunit that are accessible from the cytoplasmic and extracellular environments (Forbush *et al.*, 1978; Castro and Farley, 1979; Jorgensen, P. L. *et al.*, 1982). The tryptic and chymotryptic fragments of the α-chain have been ordered linearly by protein sequencing (Giotta, 1975; Jorgensen, P. L., 1975, Castro and Farley, 1979). All of the known proteolytic sites have been directly established as cytoplasmically oriented (Chin and Forgac, 1983). The lack of cleavage by proteases attacking the extracellular side suggests that most of the enzyme is within the bilayer or cytoplasm. Furthermore, all of the major proteolytic fragments seem to remain membrane associated; consistent with this is that labelling of intramembranous parts of the α-chain using hydrophobic reagents (Farley *et al.*, 1980; Jorgensen, P. L. *et al.*, 1982) is concentrated in 10 000–15 000 Da segments of the polypeptide that are not solubilised by extensive tryptic digestion (Karlish *et al.*, 1977). The dimensions of these segments are sufficient for two membrane-spanning α-helices connected by a short linker peptide.

The second important aspect of tertiary structure of the (Na^+, K^+)-ATPase is location of the ligand binding sites, that is, which amino acids participate in the formation of a pocket that sequesters cardiac glycosides, nucleotides, or ions. A variety of reactive cardiac glycoside derivatives have been studied. The relatively broad specificity of inhibition by these compounds may reflect a large, vaguely circumscribed binding site that would be difficult to delineate. Fluorescein-5′ isothiocyanate (FITC), which binds competitively with ATP (Karlish, 1980), labels a residue near the ouabain-dependent chymotryptic site (Carilli *et al.*, 1982), and thus connects this residue to the distant active site aspartate (Castro and Farley, 1979) as parts of the nucleotide site. Unfortunately, no specific labels of the ion site structure have been discovered, but two peptide bonds cleaved by trypsin are relevant. One of these bonds is protected, or inaccessible, when high-affinity K^+ binding is seen, and becomes protease-sensitive in synchrony with high-affinity Na^+ binding. The other bond exhibits the converse behaviour of sensitivity only in the presence of the high-affinity K^+ site and inaccessibility when the high-affinity Na^+ site is exposed. The correlation of distinct structural intermediates with those in the mechanistic model (see section 3.3.1) would provide an illuminating stroboscopic description of the structural cycle of transport and hydrolysis.

The β subunit is relatively resistant to structural probes. At least a segment of the peptide appears extracellularly because of the covalently bound carbohydrate. The place or places of attachment of the sugars are unknown, as well as whether the peptide penetrates to the cytoplasmic side of the membrane. Placing a reactive group far from the steroid backbone of digitoxin allows specific though inefficient labelling of the glycoprotein (Ruoho and Hall, 1980), indicating that the core of the cardiac glycoside binding site probably resides solely on the α-chain. The β-chain is embedded in the hydrophobic lipid bilayer, as shown by labelling with a lipid derivative (Montecucco *et al.*, 1981). Investigation of the disposition of the β subunit with respect to the membrane and to the α subunit may reveal some facets of its function. At present, its character as integral to the (Na^+, K^+)-ATPase rests on its stoichiometric presence in all active preparations of the enzyme.

Topological study of the (Ca^{2+})-ATPase is limited by the absence of restrictive labels for the luminally exposed portions of the protein. The sealed, oriented sarcoplasmic reticulum vesicles have permitted workers to identify two trypsin-sensitive peptide bonds, the NH_2-terminus and the active site aspartate, as accessible from the cytoplasm (Thorley-Lawson and Green, 1973; Stewart and MacLennan, 1974; Stewart *et al.*, 1976; Reithmeier and MacLennan, 1981). This one-sided allocation is consistent with the electron microscopic observation that the extramembranous regions of the (Ca^{2+})-ATPase comprise the asymmetrically distributed, protease-sensitive staining on the cytoplasmic side of the sarcoplasmic reticulum vesicles (Saito *et al.*, 1978). The three tryptic fragments have been linearly arranged (Allen *et al.*, 1980; Klip *et al.*, 1980; Tong, 1980), and remain bound to the membrane tightly associated with each other (Thorley-Lawson and Green, 1975). Labelling with hydrophobic reagents provides evidence that all three peptides insert into the lipid bilayer (Green *et al.*, 1980), and, as with the (Na^+, K^+)-ATPase, extensive trypsin digestion results in 10 000–15 000 Da peptides still containing most of the protein-bound label (Gitler and Bercovici, 1980). Thus it seems likely that these segments also represent the membrane-spanning α-helices of the (Ca^{2+})-ATPase.

The Ca^{2+} binding site has tentatively been placed in an NH_2-terminal 13 000 Da fragment (MacLennan *et al.*, 1980). Dicyclohexylcarbodiimide inhibition of (Ca^{2+})-ATPase have been positioned within the polypeptide, and more than half from dicyclohexylcarbodiimide can be found in an NH_2-terminal piece, though not stoichiometrically (Pick and Racker, 1979). Also, black lipid membrane assays of ionophoric properties towards divalent cations implicate this 13 000 Da piece (MacLennan *et al.*, 1980). Finally, the covalent attachment of fluorescein 5′-isothiocyanate close to the NH_2 terminus of the COOH-terminal tryptic fragment of the (Ca^{2+})-ATPase (Mitchinson *et al.*, 1982), the positions of the trypsin-sensitive peptide bonds, and the placement of the active site aspartate demonstrate homologous tertiary structures for the (Na^+, K^+)-ATPase and the sarcoplasmic reticulum (Ca^{2+})-ATPase; the structure of the plasma membrane (Ca^{2+})-ATPase has not been thoroughly examined.

3.2.5 Primary structure and biosynthesis

Little is known about the primary structure of the (Na^+, K^+)-ATPase. The amino acid compositions of the two subunits from several sources have been reported; the NH_2-terminal residue of the α-chain varies while the β-chain seems always to have alanine as its NH_2 terminal amino acid (Kyte, 1972; Perrone *et al.*, 1975; Hopkins *et al.*, 1976). Short sequences at the NH_2- and COOH-termini of the intact subunits and the large tryptic fragments have been obtained (Castro and Farley, 1979, Cantley, 1981). The active site tripeptides containing the phosphorylated aspartates of the (Na^+, K^+)-ATPase and the sarcoplasmic reticulum (Ca^{2+})-ATPase are identical (Bastide *et al.*, 1973). Because of the difficulties in direct sequencing of hydrophobic peptides, the determination of the primary structure will depend on analysis of the corresponding nucleotide sequence.

The absence of the initiator methionines on the mature α- and β-chains signifies processing of the translated peptides. Preliminary studies have employed antibodies and tryptic digestion to identify the products of *in vivo* and *in vitro* translation (Sabatini *et al.*, 1981; McDonough *et al.*, 1982). An interesting result consistent with glycosidase treatment of the purified β subunit (see section 3.2.1) is that inhibition of glycosylation yields an immunoprecipitated peptide with an apparent molecular weight of 38 000 (Sabatini *et al.*, 1981). Biosynthesis has also been studied by purification of the (Na^+, K^+)-ATPase after a pulse of radioactive amino acid (Lo and Edelman, 1976, Churchill and Hokin, 1979), but this approach is restricted to examination of the mature protein.

More than half of the amino acid residues of the sarcoplasmic reticulum (Ca^{2+})-ATPase have been positioned within the polypeptide, and more than half of the remainder have been sequenced (Allen *et al.*, 1980; Green *et al.*, 1980; Klip *et al.*, 1980; MacLennan *et al.*, 1980; Tong, 1980); included are the tryptic sites and the phosphorylated aspartate. Those residues that have been linearly assigned appear to constitute most of the hydrophilic mass of the protein because they are roughly congruent with the peptides released from the membrane by extensive proteolysis (Green *et al.*, 1980). A remarkable finding is that 18 of the 19 tryptophans are contained in hydrophobic peptides which have defied attempts at isolation (Allen *et al.*, 1980). It seems probable that each of the intervening segments (see figure 3.1) crosses the membrane to the lumen, turns, and crosses back to the cytoplasm. The 70–100 residue segments are large enough to form two 50 Å long α-helices, and would anchor the protein in the lipid bilayer (MacLennan and Reithmeier, 1982).

The acetylated NH_2-terminal amino acid (Tong, 1977) is the initiator methionine (Reithmeier *et al.*, 1980). *In vitro* translation experiments have established that the protein is synthesised on membrane-bound polysomes (Greenway and MacLennan, 1978; Chyn *et al.*, 1979). Along with localisation data (see section 3.2.2), developmental studies suggest that biosynthesis of the (Ca^{2+})-ATPase is an early event in formation of the sarcoplasmic reticulum in skeletal muscle (Holland and MacLennan, 1976).

3.3 MECHANISM

3.3.1 A model

Shown in figure 3.2 are models for the mechanism of ion transport by the (Na^+, K^+)- and (Ca^{2+})-ATPases. These models are based largely on those originally proposed by Post *et al.* (1965) and Fahn *et al.* (1966) for the (Na^+, K^+)-

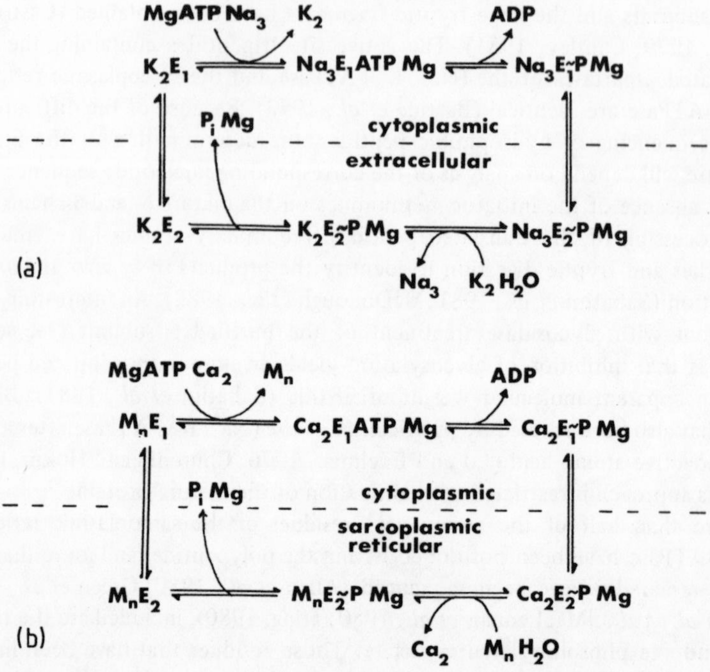

MgATP Na_3 K_2 ADP
K_2E_1 ⇌ Na_3E_1ATP Mg ⇌ $Na_3E_1{\sim}P$ Mg
P_i Mg cytoplasmic
extracellular
K_2E_2 ⇌ $K_2E_2{\sim}P$ Mg ⇌ $Na_3E_2{\sim}P$ Mg
(a)
Na_3 K_2 H_2O

MgATP Ca_2 M_n ADP
M_nE_1 ⇌ Ca_2E_1ATP Mg ⇌ $Ca_2E_1{\sim}P$ Mg
P_i Mg cytoplasmic
sarcoplasmic
reticular
M_nE_2 ⇌ $M_nE_2{\sim}P$ Mg ⇌ $Ca_2E_2{\sim}P$ Mg
(b)
Ca_2 M_n H_2O

Figure 3.2 Models for the mechanism of coupling of ion transport and ATPase activity by the (Na^+, K^+)-ATPase (a) and the (Ca^{2+})-ATPase (b). E_1 (and $E_1{\sim}P$) represent forms of the enzymes possessing high-affinity cytoplasmic sites for either Na^+ (for the Na^+ pump) or Ca^{2+} (for the Ca^{2+} pump), and E_2 (and $E_2{\sim}P$) represent forms of the enzymes possessing high-affinity extracellular sites for K^+ (for the Na^+ pump) or high-affinity sarcoplasmic reticular sites for M^+ (for the Ca^{2+} pump). M^+ denotes the counterion for Ca^{2+}, whose identity and stoichiometry (*n*) are not known for the sarcoplasmic reticulum enzyme.

ATPase, and Kanazawa and Boyer (1973) for the (Ca^{2+})-ATPase. As can be readily seen, the models are identical except for the replacement of Ca^{2+} for Na^+ and M^+ for K^+.

The sequence of steps for the mechanism of the (Na^+, K^+)-ATPase can be described as follows. Binding of Na^+ ions to cytoplasmically oriented high-affinity Na^+ binding sites (on E_1) activates phosphorylation of the enzyme by Mg-ATP to give $E_1{\sim}P$. A conformational change then gives a species ($E_2{\sim}P$) in which the cation binding sites are now extracellularly oriented and have a high affinity for K^+ and a low affinity for Na^+. Replacement of Na^+ by K^+ at these sites activates dephosphorylation of the enzyme and allows reversal of the original conformational change, thus resulting in translocation of K^+ into the cell. Displacement of K^+ by Na^+, Mg^{2+} and ATP then completes the cycle. The following section will describe some of the evidence supporting these models.

3.3.2 Evidence

3.3.2.1 Overall reaction

The overall reactions for the (Na$^+$, K$^+$)- and (Ca^{2+})-ATPases may be described as follows:

$$3Na^+_{cyto} + 2K^+_{ext} + ATP + H_2O \underset{}{\overset{Mg^{2+}}{\rightleftharpoons}} 3Na^+_{ext} + 2K^+_{cyto} + ADP + P_i$$

$$2Ca^{2+}_{cyto} + nM^+_{sarc} + ATP + H_2O \underset{}{\overset{Mg^{2+}}{\rightleftharpoons}} 2Ca^{2+}_{sarc} + nM^+_{cyto} + ADP + P_i$$

The stoichiometry of 3Na$^+$/2K$^+$/ATP was originally obtained for the native (Na$^+$, K$^+$)-ATPase by studies of ATP-driven Na$^+$ and K$^+$ fluxes in the red blood cell (Sen and Post, 1964; Garrahan and Glynn, 1967*b*) and shown to apply to reconstituted preparations of the enzyme isolated from canine kidney (Goldin, 1977) and shark rectal gland (Hilden and Hokin, 1975). A stoichiometry of 2Ca^{2+}/ATP was demonstrated in early studies of ATP-driven Ca^{2+} uptake by sarcoplasmic reticulum vesicles (Hasselbach, 1964; Yamada *et al.*, 1970) but the leakiness of these vesicles to other ions (Meissner and McKinley, 1982) has made difficult the identification of the nature and stoichiometry of the counterion transported. Although (Ca^{2+})-ATPase activity requires Mg^{2+} (Inesi *et al.*, 1970; Kanazawa *et al.*, 1971) and is activated by K$^+$ (Duggan, 1967; de Meis, 1969), no transport of these ions by the (Ca^{2+})-ATPase has been reported. On the other hand, proton efflux has been observed to accompany ATP-dependent Ca^{2+} uptake in a reconstituted preparation of the (Ca^{2+})-ATPase from the red blood cell (Niggli *et al.*, 1982), a result confirmed using sealed inside-out red blood cell vesicles containing a lipid impermeant pH-sensitive fluorescence probe (Forgac and Cantley, 1984). Proton efflux has also been observed during ATP-dependent Ca^{2+} uptake by sarcoplasmic reticulum vesicles (Chiesi and Inesi, 1980), but this proton flux was not abolished by proton ionophores, suggesting that it is the result of passive proton movement in response to a membrane potential generated by Ca^{2+} transport. Other workers have also observed that Ca^{2+} uptake by the sarcoplasmic reticulum is electrogenic (Zimniak and Racker, 1978; Meissner, 1981), in contrast to the red cell Ca^{2+} pump (Niggli *et al.*, 1982), suggesting the possibility that in sarcoplasmic reticulum, Ca^{2+} uptake is not directly coupled to the transport of any other ion but that charge compensation is achieved by fluxes through other ion channels (Meissner, 1981). In summary, protons appear to serve as the counterion for Ca^{2+} in red blood cells, whereas in the sarcoplasmic reticulum the nature and stoichiometry of the counterion (if one exists) remain uncertain. Studies of ion fluxes in tightly sealed reconstituted preparations of the (Ca^{2+})-ATPase purified from sarcoplasmic reticulum should resolve these questions.

That the above reactions are reversible has been demonstrated for both the (Na$^+$, K$^+$)-ATPase (Garrahan and Glynn, 1967*c*; Robinson *et al.*, 1977) and the (Ca^{2+})-ATPase (Kanazawa *et al.*, 1970; Makinose and Hasselbach, 1971). Thus movement of the appropriate ions down their corresponding concentration gradients results in the synthesis of [γ^{32}-P] ATP from ADP and $^{32}P_i$. This observation emphasises the functional similarity between the eukaryotic cation translocating ATPases (i.e. the (Na^+, K^+)- and (Ca^{2+})-ATPases) on the one hand and the prokaryotic ATPases (including the (H^+)-ATPases of mitochondria, chloroplasts and bacteria), which couple ATP synthesis to an oxidatively generated proton gradient (Mitchell, 1961; Racker and Stoeckenius, 1974) on the other.

3.3.2.2 Phosphorylation and dephosphorylation

As predicted by the previously described models, both the (Na^+, K^+)-ATPase (Albers *et al.*, 1963; Post *et al.*, 1965) and the (Ca^{2+})-ATPase (Yamamato and Tonomura, 1967; Inesi *et al.*, 1970) are covalently phosphorylated by [γ-^{32}P] ATP, although a pathway for ATP hydrolysis by the (Na^+, K^+)-ATPase involving no phosphorylated intermediate has been proposed on the basis of steady state kinetic data (Plesner *et al.*, 1981). In the case of the (Na^+, K^+)-ATPase, phosphorylation requires the presence of Na^+ ions ($K_d \sim 8$ mM; Mardh and Post, 1977), which bind to cytoplasmically oriented sites (Blostein, 1979). Rapid kinetic studies indicate that Na^+ and ATP bind in a random order and that pre-equilibration with either ligand enhances the rate of phosphorylation (Mardh and Post, 1977). Similarly, phosphorylation of the (Ca^{2+})-ATPase requires binding of Ca^{2+} ions to high-affinity ($\sim 1\ \mu$M) cytoplasmic sites (Yamamato and Tonomura, 1967; Inesi *et al.*, 1970). In both cases phosphate is attached to an active site aspartate residue to form an acyl phosphate linkage Bastide *et al.*, 1973). Mg^{2+}-ATP appears to serve as the substrate for ATP hydrolysis (Yamamato and Tonomura, 1967; Hexum *et al.*, 1970) and both the substrate and products (i.e. ADP, P_i and Mg^{2+}) bind to and are released from the cytoplasmic side of the protein.

Dephosphorlyation of the (Na^+, K^+)-ATPase is activated by K^+ ions (Post *et al.*, 1969) which bind to high-affinity sites ($K_d \sim 0.1$ mM) located on the extracellular surface (Blostein and Chu, 1977). Using the fluorescent ATP analogue, formycin triphosphate, Karlish *et al.* (1978) have demonstrated that the nucleotide diphosphate product dissociates from the enzyme before the rate-determining step (i.e. K^+-activated dephosphorylation). Although, as discussed earlier, protons can serve as the counterion during Ca^{2+} transport by the red cell (Ca^{2+})-ATPase, dephosphorylation of the sarcoplasmic reticulum enzyme is activated by both K^+ (Yamada and Ikemoto, 1980) and Mg^{2+} (Inesi *et al.*, 1970), as well as protons (Verjovski-Almeida and de Meis, 1977). This result suggests that either K^+ and Mg^{2+} bind to sites independent of the transport sites that activate dephosphorylation or they can bind to the normal transport sites and activate dephosphorylation without being transported. The observation that the Mg^{2+} site responsible for activation of the dephosphorylation of the (Ca^{2+})-ATPase is located on the cytoplasmic side of the protein (Takakuwa and Kanazawa, 1982) suggests that Mg^{2+} is not binding to a counterion transport site.

Reversal of initial phosphorylation by ATP in the presence of ADP results in an ATP/ADP exchange (Hasselbach and Makinose, 1963; Fahn *et al.*, 1966). Evidence for the existence of two forms of the phosphorylated (Na^+, K^+)-ATPase was provided by the observation that certain inhibitors (such as oligomycin and N-ethylmaleimide) prevented the conversion of an ADP-sensitive phosphoenzyme ($E_1{\sim}P$) to a K^+-sensitive phosphoenzyme ($E_2{\sim}P$) and thus activated ATP/ADP exchange (Fahn *et al.*, 1966, 1968). Both the (Na^+K^+)-ATPase (Post *et al.*, 1975) and the (Ca^{2+})-ATPase (Masuda and de Meis, 1973) are phosphorylated by $^{32}P_i$ in the presence of Mg^{2+}, as expected from the reversibility of the overall reaction. This phosphorylation occurs at the same site as phosphorylation by [γ-^{32}P]ATP (Post *et al.*, 1975) and is manifested by a rapid $P_i/H_2{}^{18}O$ exchange (Dahms and Boyer, 1973; Kanazawa and Boyer, 1973). Phosphorylation of the (Na^+, K^+)-ATPase by P_i is activated by K^+ (Dahms and Boyer, 1973; Post *et al.*, 1975), whereas phosphorylation of the (Ca^{2+})-ATPase is promoted by low pH (de Meis and Inesi, 1982). In addition to reversal of the overall reaction in a sided system (see above), synthesis of [γ-^{32}P]ATP in the absence of ion gradients has also been achieved as follows. The enzymes were first phosphorylated with $^{32}P_i$ in the presence of Mg^{2+} and were then exposed to a high concentration of either Na^+ (Taniguchi and Post, 1975) or Ca^{2+} (Knowles and Racker, 1975) in the presence of ADP. Binding of Na^+ or Ca^{2+} to low-affinity sites on $E_2{\sim}P$ (formed from $^{32}P_i$) thus facilitated reversal of the $E_1{\sim}P$ to $E_2{\sim}P$ transition to give a species ($E_1{\sim}P$) which could transfer P_i to ADP. The energy for ATP synthesis is thus provided by the cation-induced conformational change which requires that cations bind to sites from which they are normally released.

3.3.2.3 Changes in conformation and cation affinity

The (Na^+, K^+) and (Ca^{2+})-ATPases exist in two principal conformational states, E_1 and E_2. E_1 (and $E_1{\sim}P$) possess high-affinity cytoplasmic sites for either Na^+ or Ca^{2+}, whereas E_2 (and $E_2{\sim}P$) possess high-affinity extracellular (or sarcoplasmic) sites for either K^+ or M^+. Thus the E_1-E_2 transition presumably represents the actual ion translocation step.

This conformational transition has been detected in the (Na^+, K^+)-ATPase both by changes in the tryptic digestion pattern of the enzyme (Giotta, 1975; Jorgensen, P. L., 1975; Castro and Farley, 1979) and by changes in the intrinsic tryptophan fluorescence (Karlish and Yates, 1978). Jorgensen (1975) found that either Na^+ ions or ATP stabilised a state (E_1) which was cleaved by trypsin to give a 78 000 Da fragment, whereas in the presence of K^+ (E_2), cleavage occurred to give fragments of 60 000 Da and 40 000 Da. The observation that phosphorylation of the enzyme with ATP in the presence of Na^+ and Mg^{2+} resulted in a 'K^+-type' digestion pattern indicates that the initially formed phosphorylated species ($E_1{\sim}P$) is of a higher energy and thus relaxes to the more stable $E_2{\sim}P$ state, so providing at least part of the energy to drive the transport cycle in the forward direction. The enzyme exists predominantly in the E_1 form in the absence of ligands and can be shifted from E_2 to E_1 (even in the presence of K^+) by ATP (Castro and Farley, 1979). Although it has presumably been looked for, no analogous conformationally sensitive cleavage of the (Ca^{2+})-ATPase has been reported.

The E_1-E_2 transition of the (Na^+, K^+)-ATPase can also be detected by an increase in the intrinsic tryptophan fluorescence of the enzyme (Karlish and Yates, 1978). Although the observed changes were small (2–3 per cent), this technique has the advantage of facilitating measurement of the rates of this transition under various conditions. Thus, the rate of the E_2 (K^+) to E_1 (Na^+) transition was shown to be slow ($k \sim 0.2\ s^{-1}$ at 20°C) in the absence of ATP, but was accelerated by binding of ATP at a low-affinity site ($K_d \sim 1$ mM). Both high- and low-affinity ATP sites appear to be involved in supporting (Na^+, K^+)-ATPase activity (Cantley and Josephson, 1976; Robinson, 1976), an observation taken to support an 'alternating site' model for ATP-driven ion transport (Stein *et al.*, 1973; Cantley *et al.*, 1978). As has been pointed out, however, this observation can also be explained by a model in which the high- and low-affinity ATP sites are present sequentially during the catalytic cycle (Smith *et al.*, 1980). Thus at low ATP, the rate of turnover of the enzyme is limited by release of K^+ (which, as shown earlier, is slow), whereas at high concentrations, ATP can bind to $E_2(K^+)$ and accelerate the release of K^+ and hence the overall rate of reaction. That the 'alternating site' model does not apply in this case is suggested by the observation that the entire enzyme can be put into either the $E_1(Na^+)$ or $E_2(K^+)$ form (Jorgensen, P. L., 1975). A Ca^{2+}-induced change in intrinsic fluorescence of the (Ca^{2+})-ATPase has also been observed (Dupont, 1976), and the rate of this process has been followed by stopped-flow fluorimetry (Dupont and Leigh, 1978).

Labelling of both the (Na^+, K^+)-ATPase (Hegyvary and Jorgensen, 1981) and the (Ca^{2+})-ATPase (Pick and Karlish, 1982) with FITC gives species which, although inactive with respect to ATPase activity, still appear capable of undergoing the E_1-E_2 transition, as detected by a change in FITC fluorescence. Thus Ca^{2+} induces a 5–10 per cent decrease in fluorescence of FITC-labelled (Ca^{2+})-ATPase (Pick and Karlish, 1982), an effect which is reversed by both vanadate (a phosphate analogue; Cantley *et al.*, 1977) and low pH. The latter result is consistent with Ca^{2+} and protons stabilising different states of the (Ca^{2+})-ATPase. Although studies have suggested the existence of conformational states in addition to the two described above (Hart and Titus, 1973; Coan and Inesi, 1977; Ikemoto *et al.*, 1978; Forgac, 1980), the E_1 and E_2 states presumably represent the principal conformations with respect to ion translocation.

Extensive evidence has been obtained indicating that the expected changes in cation affinity occur during the catalytic cycle of these enzymes. (See Glynn and Karlish, 1975; MacLennan and Holland, 1975.) Thus the affinity of the cytoplasmic binding sites involved in ion transport and phosphorylation by ATP are the same (K_d for $Na^+ \sim 5$–10 mM; K_d for $Ca^{2+} \sim 1\ \mu$M), and the same concentration of extracellular K^+ activates both dephosphorylation and K^+ influx through the (Na$^+$, K$^+$)-ATPase ($K_d \sim 0.1$ mM). Binding of Na$^+$, K$^+$ or Ca^{2+} to low-affinity sites (representing the sites from which these ions are released) results in inhibition of ATPase activity (Makinose and Hasselbach, 1965; Skou, 1974) and activation of the corresponding exchange reaction. Thus the (Na$^+$, K$^+$)-ATPase catalyses both Na$^+$/Na$^+$ (Garrahan and Glynn, 1967*a*) and K$^+$/K$^+$ (Simons, 1974) exchange while the (Ca^{2+})-ATPase catalyses Ca^{2+}/Ca^{2+} exchange (Takakuwa and Kanazawa, 1981). As discussed previously, high concentrations of Na$^+$ or Ca^{2+} are also required to reverse the overall reactions and synthesise ATP in the absence of ion gradients (Knowles and Racker, 1975; Taniguchi and Post, 1975). Two observations concerning changes in cation affinity are especially noteworthy. First, the (Na$^+$, K$^+$)-ATPase has been shown to be capable of retaining tightly bound K$^+$ after removal of free K$^+$ by exclusion chromatography (Beauge and Glynn, 1979) although in a less than stoichiometric amount. This tightly bound K$^+$ is released on addition of ATP, providing additional support for the role of high ATP concentrations in displacing K$^+$ from the enzyme. Secondly, the affinity of the (Ca^{2+})-ATPase for Ca^{2+} decreased with decreasing pH (Meissner, 1973; Verjovski-Almeida and de Meis, 1977), consistent with a model in which Ca^{2+} and protons compete for the same transport site.

3.3.2.4 Other mechanistic aspects

In addition to the reactions described above, the (Na$^+$, K$^+$)-ATPase can carry out a number of partial reactions, including (Na$^+$)-ATPase activity (Neufeld and Levy, 1969) and (K$^+$)-*p*-nitrophenylphosphatase activity (Judah *et al.*, 1962). Unlike (K$^+$)-*p*NPPase activity, which does not appear to involve ion transport (Drapeau and Blostein, 1980), (Na$^+$)-ATPase activity is associated with a K$^+$-independent Na$^+$ flux which can occur either down (Glynn and Karlish, 1976; Forgac and Chin, 1982) or against (Forgac and Chin, 1981) a Na$^+$ concentration gradient. In the latter case, Na$^+$, at sufficiently high concentrations, was shown to substitute for K$^+$ at the K$^+$ transport sites, whereas in the former case, Na$^+$ transport appears to proceed uncoupled to the movement of any other ion (Dissing and Hoffman, 1983). Such an uncoupled Ca^{2+} flux may be the normal mode of operation of the (Ca^{2+})-ATPase (see section 3.3.2.1), but a 'proton-free' environment can obviously not be obtained under conditions in which the enzyme retains its native structure.

Both the (Na$^+$, K$^+$)-ATPase (Cantley *et al.*, 1977) and the (Ca^{2+})-ATPase (Pick, 1982) are inhibited by vanadate, a compound which binds to the active site and is believed to act as a transition state analogue as a result of its trigonal bipyramidal structure (Cantley *et al.*, 1978). It is also interesting to note that vanadate appears to inhibit only those ATPases whose mechanisms involve a phosphorylated intermediate (L. Cantley, personal communication). Inhibition of the (Na$^+$, K$^+$)-ATPase by cardiac glycosides will be discussed under the section on regulation of activity.

3.3.3 Energetics

Essential to the understanding of the mechanism of the (Na^+, K^+)- and (Ca^{2+})-ATPase is the ability to describe the energetics of the reaction sequence and a knowledge of the precise structural rearrangements of the intermediates involved. With respect to energetics, Jencks (1980) has proposed an energy level diagram emphasising the conditions that must be met in achieving a 'coupled vectorial process'. In line with the model discussed previously, we would like to propose the energy level diagram shown in figure 3.3. This scheme is based on the following observations for the Na^+ pump:

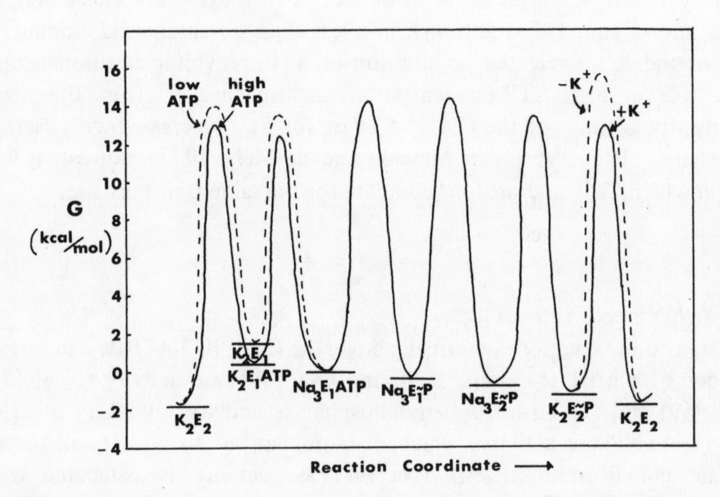

Figure 3.3 Energy level diagram for the (Na^+, K^+)-ATPase. The diagram shown was constructed from the data on the (Na^+, K^+)-ATPase shown in table 3.1. Except where indicated, ligands are present at saturating concentrations. The values of ΔG and $\Delta G\ddagger$ for the step $Na_3E_2{\sim}P \rightarrow K_2E_2{\sim}P$ are not known and the free energies of the intermediates K_2E_1 and K_2E_1ATP are not known. Also not shown is the ΔG of ATP hydrolysis (which would require the inclusion of E_1 + ATP and E_1 + ADP + P_i intermediates), which is -10 to -12 kcal mol^{-1} under physiological conditions.

(a) Since E_1 has a higher affinity for Na^+ than K^+, whereas E_2 has a higher affinity for K^+ than Na^+, then $G(K_2E_1) > G(Na_3E_1ATP\ Mg)$ and $G(Na_3E_2{\sim}P\ Mg) > G(K_2E_2{\sim}P\ Mg)$.

(b) Since, in the presence of Na^+, Mg^{2+} and ATP, the enzyme exists primarily in the $E_2{\sim}P$ state (Fahn *et al.*, 1966; Jorgensen, 1975), then $G(Na_3E_2{\sim}P\ Mg) <$ $G(Na_3E_1ATP\ Mg)$, $G(Na_3E_1{\sim}P\ Mg)$.

(c) Since, on blocking the $E_1{\sim}P$-$E_2{\sim}P$ transition, the enzyme readily carries out ATP/ADP exchange (Fahn *et al.*, 1966), then $G(Na_3E_1ATP\ Mg) \sim$ $G(Na_3E_1{\sim}P\ Mg)$.

(d) Since the enzyme is readily phosphorylated with P_i in the presence of K^+ and Mg^{2+} (Post *et al.*, 1975), then $G(K_2E_2{\sim}P\ Mg) \sim G(K_2E_2)$.

(e) Since the enzyme can be shifted into the E_2 state by the binding of K^+ (Jorgensen, 1975; Karlish and Yates, 1978), then $G(K_2E_2) < G(K_2E_1)$.

(f) Since K^+ facilitates dephosphorylation of the enzyme (Post *et al.*, 1969), then for the $K_2E_2{\sim}P\ Mg \rightarrow K_2E_2$, $\Delta G^{\ddagger}\ (-K^+) > \Delta G^{\ddagger}\ (+K^+)$.

(g) Since high concentrations of ATP activate release of K^+ by the enzyme (Beauge and Glynn, 1979), then, for the step $K_2E_2 \rightarrow Na_3E_1ATP\ Mg$, $\Delta G^{\ddagger}$ (low ATP) $> \Delta G^{\ddagger}$ (high ATP).

Although it is difficult to assign numbers to this scheme because the work done on individual steps has often been done under different conditions and on enzyme isolated from different sources, a partial list of some of the rates and equilibrium constants (and the corresponding values of $\Delta G^{\ddagger}$ and ΔG) for the (Na^+, K^+)-ATPase is shown in table 3.1. It is interesting to note that the intermediates are all of relatively similar energy and that the energy barriers between them are of similar magnitude, suggesting that different steps can readily become rate determining. As can be seen, under steady state conditions, with the cation and nucleotide sites saturated, the enzyme will exist predominantly in the $E_2(K^+)$ state, in agreement with the results of Karlish and Yates (1978). Under physiological conditions, however, the rate of turnover of the pump is limited by the cytoplasmic Na^+ concentration (Sachs, 1970). Since the cytoplasmic ATP concentration is millimolar (sufficient to displace K^+ from the internal sites), the enzyme should exist predominantly in an E_1ATP Mg state, kinetically prevented from undergoing phosphorylation until Na^+ ions bind. A similar description of the energetics of the (Ca^{2+})-ATPase has been presented (Pickart, 1982).

3.3.4 A molecular mechanism

As has been demonstrated in attempts to understand the molecular mechanism of other processes (i.e. O_2 binding by haemoglobin (Perutz, 1970)), an understanding of the mechanism of active transport by the (Na^+, K^+)- and (Ca^{2+})-ATPases will require a detailed knowledge of the structural changes involved in this process. Such detailed structural information has not been obtained for either the (Na^+, K^+)- or (Ca^{2+})-ATPases (see section 3.2). Nevertheless, some information concerning the structural changes involved in ion transport has been obtained. Kyte (1974) demonstrated that the (Na^+, K^+)-ATPase could continue to turn over (at least with respect to ATP hydrolysis) in the presence of an antibody bound to the cytoplasmic surface, suggesting that, as expected from thermodynamic considerations, ion translocation does not require rotation of the entire molecule through the plane of the membrane. A similar result has been obtained for the (Ca^{2+})-ATPase (Dutton *et al.*, 1976).

A number of models have been proposed in which the ion binding sites are present within a cleft in the protein (Jardetsky, 1966; Guidotti, 1979; Cantley,

Table 3.1 Rates, equilibrium constants and values of ΔG and $\Delta G^{\ddagger}$ for individual steps of the (Na^+, K^+)-ATPase reaction mechanism[a]

Step	Rate (s^{-1})	$\Delta G^{\ddagger}$ (kcal mol^{-1})	K_{eq} (k_f/k_r)	ΔG (kcal mol^{-1})	References
$E_1(Na) \underset{k_{-1}}{\overset{k_1}{\rightleftharpoons}} E_2(K)$	k_1 = 290 (low ATP)	14.0	–	–	Glynn and Karlish (1982)
	k_{-1} = 12 (low ATP)	15.9	25	−1.9	Glynn and Karlish (1982)
	k_{-1} = 60 (high ATP)	15.0	–	–	Glynn and Karlish (1982)
$NaE_1ATP \underset{k_{-2}}{\overset{k_2}{\rightleftharpoons}} NaE_1{\sim}P$	k_2 = 220	14.2	–	–	Mardh and Post (1977)
	–	–	2	−0.41	Mardh (1975*b*)
$E_1{\sim}P \underset{k_{-3}}{\overset{k_3}{\rightleftharpoons}} E_2{\sim}P$	k_3 = 77	14.8	–	–	Mardh and Lindahl (1977)
$E_2{\sim}P \underset{k_{-4}}{\overset{k_4}{\rightleftharpoons}} E_2 + P_i$	k_4 = 230 (+K)	14.2	–	–	Mardh (1975*a*)
	k_4 = 3 (−K)	16.8	–	–	Mardh and Post (1977)
	k_{-4} = 80 (+K)	14.8	–	–	Glynn and Karlish (1982)
	–	–	3.0	−0.65	Post *et al.* (1975)

[a]Values for rate and equilibrium constants are given (except where indicated) at saturating ligand concentrations and 20°C, under which conditions the turnover number of the enzyme is 30–60 s^{-1}. Values of ΔG are calculated using the equation $-2.3\,RT \log_{10}(K_{eq})$, and $\Delta G^{\ddagger}$ is calculated using the equation $-2.3\,RT\,(\log_{10}(k_b T/h) - \log(k))$ where $R = 0.002$ kcal mol^{-1} deg, $T = 293$ K, $k_b = 1.38 \times 10^{-16}$ erg deg^{-1} molecule, $h = 6.63 \times 10^{-27}$ erg s, $K_{eq} = k_f/k_r$, and k is the rate constant in s^{-1}.

1981), eliminating the involvement of large structural rearrangements in ion translocation. Guidotti (1979) has pointed out the similarity between such a model and the way in which 2,3-diphosphoglycerate is bound to and released from the haemoglobin molecule. Although fluorescence energy transfer studies indicate a distance of ~ 70 Å between FITC (presumably bound to the internal ATP site) and anthroylouabain (bound to the external ouabain site) (Carilli *et al.*, 1982), no information has been obtained about the distance between the Na^+ and K^+ binding sites (if in fact they are not actually the same site). Whether such a cleft would occur between subunits or within a single subunit is unclear. The β subunit of the (Na^+, K^+)-ATPase probably does not form part of the ion binding site since no comparable subunit is required for Ca^{2+} transport by the (Ca^{2+})-ATPase (see section 3.2.1). The ability of the monomeric form of both the (Na^+, K^+)-ATPase and the (Ca^{2+})-ATPase to catalyse ATP hydrolysis (see section 3.2.3) suggests but does not prove that the monomer is the minimal functional unit. Proof of this proposal will require reconstitution of the monomeric form of these enzymes and demonstration that they retain the ability to transport ions. This has been demonstrated for bacteriorhodopsin (London and Khorana, 1982).

As mentioned in section 3.2.4, little information has been obtained concerning the structure of the ion binding sites. The ability of Ca^{2+} to protect against inhibition of the (Ca^{2+})-ATPase activity by dicyclohexylcarbodiimide (Pick and Racker, 1979) implies that one or more carboxyl groups may be involved in cation binding. Although it has been suggested that the nucleotide and cation binding sites are in close proximity (Grisham *et al.*, 1974), there is extensive evidence for the ability of distant sites on proteins to interact through long-range conformational changes (Perutz, 1970; Zukin *et al.*, 1977). That such changes can occur in the (Na^+, K^+)-ATPase is clear from the observation that binding of cardiac glycosides to an extracellular site on the enzyme causes inhibition of ATP hydrolysis at a site ~ 70 Å away (Carilli *et al.*, 1982). We would therefore like to propose the speculative model for ion transport by the (Na^+, K^+)-ATPase shown in figure 3.4.

The model shows only the E_1 and $E_2{\sim}P$ states and postulates that the cation binding sites (composed of carboxyl side chains) are buried deep within a cleft present in the 100 000 Da subunit. Conversion of internal Na^+ binding sites to external K^+ binding sites involves pivoting of the two domains shown so as to alter the accessibility to these sites and to reduce the number of binding sites from three to two. This change is controlled by ATP binding and phosphorylation at the distant nucleotide site and is affected by the binding of cardiac glycosides to a similarly remote site on the extracellular side of the enzyme. The (Ca^{2+})-ATPase would be expected to have a similar molecular mechanism except with Ca^{2+} replacing Na^+ and M^+ replacing K^+. Protons could readily be accommodated as a counterion for Ca^{2+} in such a model in light of the large changes in pK_a observed for carboxyl groups in different environments on proteins (Hunkapiller *et al.*, 1973).

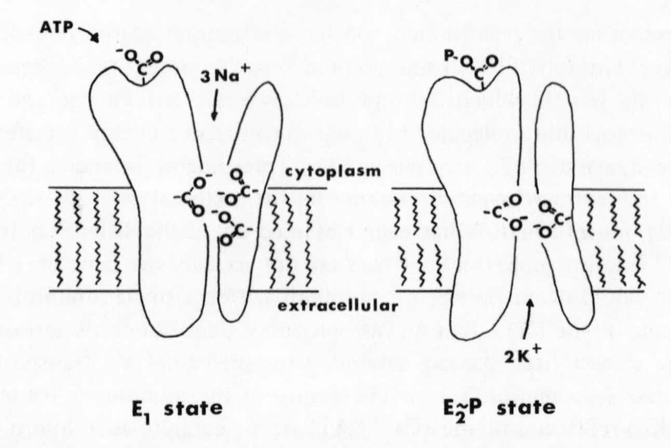

Figure 3.4 Model for the molecular mechanism of the (Na^+, K^+)-ATPase. Shown are the protein conformations corresponding to the E_1 and E_2~P states of the (Na^+, K^+)-ATPase described in figure 3.2. In E_1, the ion binding sites (shown here as carboxyl groups) are accessible from the cytoplasmic side, and in E_2~P, these sites are accessible from the extracellular side (with one fewer site in this state). Changes in the environment of the binding sites convert them from high-affinity Na^+ sites to high-affinity K^+ sites. The conformational change affecting accessibility to these ion binding sites is controlled by ATP binding and phosphorylation at the remote catalytic site indicated by the carboxyl group of the active site Asp residue located on the cytoplasmic domain. The same model could be used to describe the (Ca^{2+})-ATPase with replacement of Na^+ by Ca^{2+} and K^+ by M^+ (as well as the appropriate changes in the number and affinity of the ion binding sites).

3.4 REGULATION OF ACTIVITY

In addition to changes in substrate levels, there appear to exist mechanisms for regulation of the activity of both the (Na^+, K^+)-ATPase and the (Ca^{2+})-ATPase. In the red blood cell, (Ca^{2+})-ATPase activity is directly modulated by the binding of calmodulin (Niggli *et al.*, 1981; Waisman *et al.*, 1981), and in sarcoplasmic reticulum, calmodulin appears to affect activity through phosphorylation of some associated protein (Campbell and MacLennan, 1982, Kirchberger and Antonetz, 1982). In light of the number of processes that are sensitive to the cytoplasmic Ca^{2+} concentration, including muscle contraction (Mannherz and Goody, 1976), calmodulin-regulated reactions (Cheung, 1980), synaptic transmission (Miledi, 1973), oocyte activation (Steinhardt and Epel, 1974, Ridgway *et al.*, 1977) and terminal differentiation of a number of cell types (Hennings *et al.*, 1980; Levenson *et al.*, 1980), it is not surprising that the Ca^{2+} pump should be under tight control. It is clear, however, that there exist other mechanisms in the cell, such as Na^+/Ca^{2+} antiport (Miyamato and Racker, 1980), voltage-sensitive Ca^{2+} channels (Llinas *et al.*, 1976) and Ca^{2+} uptake by mitochondria (Jacobus *et al.*, 1975), for regulating the cytoplasmic Ca^{2+} concentration.

Changes in (Na^+, K^+)-ATPase activity also appear to be associated with many cellular processes, including response to hormones (Clausen and Flatman, 1977; Clausen and Kohn, 1977), cell proliferation (Koch and Leffert, 1979, Rozengurt, 1981) and transformation (Kimelberg and Mayhew, 1975). How activity is regulated in these systems in unclear, although in the case of the insulin response of adipocytes, it does not involve a change in the number of copies of the Na^+ pump (Resh, 1982) necessitating a change in activity of existing Na^+ pumps. A change in the copy number does occur with the glucose transporter in these cells and with the (Na^+, K^+)-ATPase in the response of renal cortex to triiodothyronine (Lo and Edelman, 1976; Cushman and Wardzala, 1980). One possible mechanism of regulating activity involves phosphorylation of the enzyme by a protein kinase. Phosphorylation of the (Na^+, K^+)-ATPase at a noncatalytic site has been observed in murine erythroleukaemia cells (Yeh *et al.*, 1983), where a change in activity of the Na^+ pump appears to be involved in differentiation (Smith *et al.*, 1982). In the context of regulation, it is interesting to note that the (Na^+, K^+)-ATPase is specifically inhibited by micromolar concentrations of ouabain, one of a class of compounds known as cardiac glycosides which bind to an extracellular site on the enzyme (see Glynn, 1964). These compounds are believed to stimulate heart muscle contraction by raising the intracellular Na^+ level, which in turn increases the cytoplasmic Ca^{2+} concentration via a Na^+/Ca^{2+} exchange protein (Miyamoto and Racker, 1980). It does not seem unlikely that, in light of the specificity and affinity of this interaction, there exist one or more naturally occurring serum factors (endogenous glycosides) which bind to the same site and are responsible for regulating the activity of the Na^+ pump *in vivo*. Attempts to isolate and identify such a factor have been reported (see Hoffman and Forbush, 1983).

3.5 RELATION TO OTHER ATPases

In addition to resembling each other, the (Na^+, K^+)- and (Ca^{2+})-ATPases also appear to resemble other eukaryotic cation-translocating ATPases, namely the (H^+, K^+)-ATPase of hog stomach (Sachs *et al.*, 1976) and the (H^+)-ATPase of *Neurospora* (Scarborough, 1980). The (H^+, K^+)-ATPase, which is responsible for acidifying the gastric fluid, is phosphorylated by [γ-^{32}P] ATP to give a K^+-sensitive species of molecular weight ~100 000 Da (Saccomani *et al.*, 1979). The *Neurospora* enzyme, which pumps protons out of the cell, also forms a phosphorylated intermediate (also at an active site Asp; Dame and Scarborough, 1981) of molecular weight ~100 000 Da (Dame and Scarborough, 1980), suggesting that this mechanism of ion translocation has been highly conserved. These eukaryotic ion pumps appear to differ fundamentally from the prokaryotic cation-translocating ATPases (including the mitochondrial and chloroplast (H^+)-ATPases as well as the bacterial (H^+)-ATPase). Like the eukaryotic ATPases, the prokaryotic ATPases can also couple ion transport to ATP hydrolysis (although they normally function in the reverse direction during ATP synthesis), but unlike them do not involve a phosphorylated intermediate (Webb *et al.*, 1980) and are composed of multiple smaller subunits (Pederson, 1975) rather than a single functional polypeptide of molecular weight ~100 000 Da. A (K^+)-ATPase responsible for K^+ transport in *E. coli* has been identified, however, which forms a phosphorylated intermediate of 90 000 Da (Epstein *et al.*, 1979). The prokaryotic and eukaryotic ATPases do resemble each other (and Band 3, the anion channel of the erythrocyte (Guidotti, 1979)) in having most of their mass present on the cytoplasmic side of the membrane (corresponding in mitochondria and chloroplasts to the intraorganellar space). Whether the differences that exist between these species have a functional significance remains to be determined.

ACKNOWLEDGEMENTS

We wish to thank Guido Guidotti, Lewis Cantley and Ian Macara from the Department of Biochemistry and Molecular Biology at Harvard University for many helpful discussions during the preparation of this review, and Nancy Perrone for invaluable secretarial assistance. We also wish to acknowledge support of M. F. by NIH Grant GM 26199 and G. C. by NIH Grants GM 07620 and HL 08893.

REFERENCES

Albers, R. W., Fahn, S. and Koval, G. (1963) *Proc. Natl Acad. Sci. USA* **50**, 474.
Allen, G., Trinnaman, B. J. and Green, N. M. (1980) *Biochem. J.* **187**, 591.
Andersen, J. P., Moller, J. V. and Jorgensen, P. L. (1982) *J. Biol. Chem.* **257**, 8300.
Bastide, F., Meissner, G., Fleischer, S. and Post, R. L. (1973) *J. Biol. Chem.* **248**, 8385.
Beauge, L. A. and Glynn, I. M. (1979) *Nature, Lond.* **280**, 510.
Blostein, R. (1979) *J. Biol. Chem.* **254**, 6673.
Blostein, R. and Chu, L. (1977) *J. Biol. Chem.* **252**, 3035.
Brotherus, J. R., Moller, J. V. and Jorgensen, P. L. (1981) *Biochem. Biophys. Res. Commun.* **100**, 146.
Campbell, K. P. and MacLennan, D. H. (1982) *J. Biol. Chem.* **257**, 1238.
Cantley, L. C. (1981) *Curr. Topics Bioenerg.* **11**, 201.
Cantley, L. C. and Josephson, L. (1976) *Biochemistry* **15**, 5280.
Cantley, L. C., Josephson, L., Warner, R., Yanogasawa, M., Lechene, C. and Guidotti, G. (1977) *J. Biol. Chem.* **252**, 7421.
Cantley, L. C., Cantley, L. G. and Josephson, L. (1978) *J. Biol. Chem.* **253**, 7361–7368.
Carafoli, E. and Zurini, M. (1982) *Biochim. Biophys. Acta* **683**, 279.
Carilli, C. T., Farley, R. A., Perlman, D. M. and Cantley, L. C. (1982) *J. Biol. Chem.* **257**, 5601.
Caroni, P. and Carafoli, E. (1981) *J. Biol. Chem.* **256**, 3263.
Castro, J. and Farley, R. A. (1979) *J. Biol. Chem.* **254**, 2221.
Cheung, W. Y. (1980) *Science* **207**, 19.
Chiesi, M. and Inesi, G. (1980) *Biochemistry* **19**, 2912.
Chin, G. and Forgac, M. (1983) *Biochemistry* **22**, 3405.
Churchill, L. and Hokin, L. E. (1979) *J. Biol. Chem.* **254**, 7388.
Chyn, T. L., Martonosi, A. N., Morimoto, T. and Sabatini, D. D. (1979) *Proc. Natl Acad. Sci. USA* **76**, 1241.
Clarke, S. (1975) *J. Biol. Chem.* **250**, 5459.

Clausen, T. and Flatman, J. A. (1977) *J. Physiol.* **270**, 383.
Clausen, T. and Kohn, P. G. (1977) *J. Physiol.* **265**, 19.
Coan, C. R. and Inesi, G. (1977) *J. Biol. Chem.* **252**, 3044.
Craig, W. S. (1982*a*) *Biochemistry* **21**, 2674.
Craig, W. S. (1982*b*) *Biochemistry* **21**, 5707.
Craig, W. S. and Kyte, J. (1980) *J. Biol. Chem.* **255**, 6262.
Crane, R. K. (1965) *Fed. Proc.* **24**, 1000.
Cushman, S. W. and Wardzala, L. J. (1980) *J. Biol. Chem.* **255**, 4758.
Dahms, A. S. and Boyer, P. D. (1973) *J. Biol. Chem.* **248**, 3155.
Dame, J. B. and Scarborough, G. A. (1980) *Biochemistry* **19**, 2931.
Dame, J. B. and Scarborough, G. A. (1981) *J. Biol. Chem.* **256**, 10724.
Dean, W. L. and Tanford, C. (1978) *Biochemistry* **17**, 1683.
de Meis, L. (1969) *J. Biol. Chem.* **244**, 3733.
de Meis, L. and Inesi, G. (1982) *J. Biol. Chem.* **257**, 1289.
de Meis, L. and Vianna, A. L. (1979) *Ann. Rev. Biochem.* **48**, 275.
Dissing, S. and Hoffman, J. F. (1983) *Curr. Topics Memb. Trans.* **19.**
Drapeau, P. and Blostein, R. (1980) *J. Biol. Chem.* **255**, 7827.
Drickamer, L. K. (1975) *J. Biol. Chem.* **250**, 1952.
Duggan, P. F. (1967) *Life Sci.* **6**, 561.
Dupont, Y. (1976) *Biochem. Biophys. Res. Commun.* **71**, 544.
Dupont, Y. and Leigh, J. B. (1978) *Nature, Lond.* **273**, 396.
Dutton, A., Rees, E. D. and Singer, S. J. (1976) *Proc. Natl Acad. Sci. USA* **73**, 1532.
Endou, H., Reuter, E. and Weber, H. (1975) *Biochim. Biophys. Acta* **389**, 516.
Epstein, W., Laimins, L. and Hesse, J. (1979) *Eleventh Int. Cong. Biochem.* (Toronto), p. 449.
Ernst, S. A. (1975) *J. Cell Biol.* **66**, 586.
Esmann, M., Christiansen, C., Karlsson, K., Hansson, G. C. and Skou, J. C. (1980) *Biochim. Biophys. Acta* **603**, 1.
Evans, H. J. and Sorges, G. J. (1966) *Ann. Rev. Plant Physiol.* **17**, 47.
Fahn, S., Koval, G. J. and Albers, R. W. (1966) *J. Biol. Chem.* **241**, 1882.
Fahn, S., Koval, G. J. and Albers, R. W. (1968) *J. Biol. Chem.* **243**, 1993.
Farley, R. A., Goldman, D. and Bayley, H. (1980) *J. Biol. Chem.* **255**, 860.
Forbush, B., III, Kaplan, J. H. and Hoffman, J. F. (1978) *Biochemistry* **17**, 3667.
Forgac, M. D. (1980) *J. Biol. Chem.* **255**, 1547.
Forgac, M. and Cantley, L. (1984) *J. Memb. Biol.* (in press).
Forgac, M. and Chin, G. (1981) *J. Biol. Chem.* **256**, 3645.
Forgac, M. and Chin, G. (1982) *J. Biol. Chem.* **257**, 5652.
Garrahan, P. J. and Glynn, I. M. (1967*a*) *J. Physiol.* **192**, 159.
Garrahan, P. J. and Glynn, I. M. (1967*b*) *J. Physiol.* **192**, 217.
Garrahan, P. J. and Glynn, I. M. (1967*c*) *J. Physiol.* **192**, 237.
Giotta, G. J. (1975) *J. Biol. Chem.* **250**, 5159.
Giotta, G. J. (1976) *J. Biol. Chem.* **251**, 1247.
Gitler, C. and Bercovici, T. (1980) *Ann. N. Y. Acad. Sci.* **346**, 199.
Glynn, I. M. (1957) *J. Physiol.* **136**, 148.
Glynn, I. M. (1964) *Pharm. Rev.* **16**, 381.
Glynn, I. M. and Karlish, S. J. D. (1975) *Ann. Rev. Biochem.* **44**, 13.
Glynn, I. M. and Karlish, S. J. D. (1976) *J. Physiol.* **256**, 465.
Glynn, I. M. and Karlish, S. J. D. (1982) in *Membranes and Transport*, Vol. 1 (Martonosi, A. N., ed.) Plenum, New York, pp. 529–36.
Goldin, S. M. (1977). *J. Biol. Chem.* **252**, 5630.
Green, N. M., Allen, G. and Hebdon, G. M. (1980) *Ann. N. Y. Acad. Sci.* **358**, 149.
Greenway, D. C. and MacLennan, D. H. (1978) *Can. J. Biochem.* **56**, 452.
Grisham, C. M., Gupta, R. K., Barnett, R. E. and Mildvan, A. S. (1974) *J. Biol. Chem.* **249**, 6738.
Guidotti, G. (1977) *J. Supra. Struct.* **7**, 489.
Guidotti, G. (1979) *The Neurosciences* (Schmitt, F. O. and Warden, F. G., eds). MIT Press, Cambridge, Mass., pp. 831–40.
Hart, W. M. and Titus, E. O. (1973) *J. Biol. Chem.* **248**, 4674.
Hasselbach, W. (1964) *Prog. Biophys. Mol. Biol.* **14**, 167.

Hasselbach, W. and Makinose, M. (1961) *Biochem. Z.* **333**, 518.
Hasselbach, W. and Makinose, M. (1963) *Biochem. Z.* **339**, 94.
Hastings, D. F. and Reynolds, J. A. (1979) *Biochemistry* **18**, 817.
Hebdon, G. M., Cunningham, L. W. and Green, N. M. (1979) *Biochem. J.* **179**, 135.
Hebert, H., Jorgensen, P. L., Skriver, E. and Maunsbach, A. B. (1982) *Biochim. Biophys. Acta* **689**, 571.
Hegyvary, C. and Jorgensen, P. L. (1981) *J. Biol. Chem.* **256**, 6296.
Hennings, H., Michael, D., Cheng, C., Steinert, P., Holbrook, K. and Yuspa, S. H. (1980) *Cell* **19**, 245.
Hexum, T., Samson, F. E. and Himes, R. H. (1970) *Biochim. Biophys. Acta* **212**, 322.
Hilden, S. and Hokin, L. E. (1975) *J. Biol. Chem.* **250**, 6292.
Hoffman, J. F. and Forbush, B. (eds) (1983) *Current Topics in Membranes and Transport* Vol. 19. Academic Press, New York.
Hokin, L. E., Dahl, J. L., Deupree, J. D., Dixon, J. F., Hackney, J. F. and Perdue, J. F. (1973) *J. Biol. Chem.* **248**, 2593.
Holland, P. C. and MacLennan, D. H. (1976) *J. Biol. Chem.* **251**, 2030.
Hopkins, B. E., Wagner, H., Jr and Smith, T. W. (1976) *J. Biol. Chem.* **251**, 4365.
Hunkapiller, M. W., Smallcombe, S. H., Whitaker, D. R. and Richards, J. H. (1973) *J. Biol. Chem.* **248**, 8306.
Ikemoto, N., Morgan, J. F. and Shinpei, Y. (1978) *J. Biol. Chem.* **253**, 7827.
Inesi, G., Maring, E., Murphy, A. J. and McFarland, B. H. (1970) *Arch. Biochem. Biophys.* **138**, 285.
Jacobus, W. E., Tiozzo, R., Lugli, G., Lehninger, A. L. and Carafoli, E. (1975) *J. Biol. Chem.* **250**, 7863.
Jardetsky, O. (1966) *Nature, Lond.* **211**, 969.
Jencks, W. P. (1980) *Adv. Enzym.* **51**, 75.
Jorgensen, A. O., Kalnins, V. I., Zubrzycka, E. and MacLennan, D. H. (1977) *J. Cell Biol.* **74**, 287.
Jorgensen, A. O., Shen, A. C. Y., Daly, P. and MacLennan, D. H. (1982) *J. Cell Biol.* **93**, 883.
Jorgensen, P. L. (1974) *Biochim. Biophys. Acta* **356**, 36.
Jorgensen, P. L. (1975) *Biochim. Biophys. Acta* **401**, 399.
Jorgensen, P. L. (1980) *Phys. Rev.* **60**, 864.
Jorgensen, P. L. (1982) *Biochim. Biophys. Acta* **694**, 27.
Jorgensen, P. L., Hansen, O., Glynn, I. M. and Cavieres, J. D. (1973) *Biochim. Biophys. Acta* **291**, 795.
Jorgensen, P. L., Karlish, S. J. D. and Gitler, C. (1982) *J. Biol. Chem.* **257**, 7435.
Judah, J. D., Ahmed, K. and McLean, A. E. (1962) *Biochim. Biophys. Acta* **65**, 472.
Kanazawa, T. and Boyer, P. D. (1973) *J. Biol. Chem.* **248**, 3163.
Kanazawa, T., Yamada, S. and Tonomura, Y. (1970) *J. Biochem.* **68**, 593.
Kanazawa, T., Yamada, S., Yamamoto, T. and Tonomura, Y. (1971) *J. Biochem.* **70**, 95.
Karlish, S. J. D. (1980) *J. Bioenerg. Biomem.* **12**, 111.
Karlish, S. J. D. and Yates, D. W. (1978) *Biochim. Biophys. Acta* **527**, 115.
Karlish, S. J. D., Yates, D. W. and Glynn, I. M. (1976) *Nature, Lond.* **263**, 251.
Karlish, S. J. D., Jorgensen, P. L. and Gitler, C. (1977) *Nature, Lond.* **269**, 715.
Katz, B. (1966) *Nerve, Muscle and Synapse*, McGraw-Hill, New York.
Kimelberg, H. K. and Mayhew, E. (1975) *J. Biol. Chem.* **250**, 100.
Kirchberger, M. A. and Antonetz, T. (1982) *J. Biol. Chem.* **257**, 5685.
Klingenberg, M. (1981) *Nature, Lond.* **290**, 449.
Klip, A., Reithmeier, R. A. F. and MacLennan, D. H. (1980) *J. Biol. Chem.* **255**, 6562.
Knauf, P. A., Proverbio, F. and Hoffman, J. F. (1974) *J. Gen. Phys.* **63**, 324.
Knowles, A. F. and Racker, E. (1975) *J. Biol. Chem.* **250**, 1949.
Knowles, A. F., Zimniak, P., Alfonzo, M., Zimniak, A. and Racker, E. (1980) *J. Mem. Biol.* **55**, 233.
Koch, K. S. and Leffert, H. L. (1979) *Cell* **18**, 153.
Kyte, J. (1971) *J. Biol. Chem.* **246**, 4157.
Kyte, J. (1972) *J. Biol. Chem.* **247**, 7642.
Kyte, J. (1974) *J. Biol. Chem.* **249**, 3652.

Kyte, J. (1975) *J. Biol. Chem.* **250**, 7443.
Kyte, J. (1976) *J. Cell Biol.* **68**, 304.
Kyte, J. (1981) *Nature, Lond.* **292**, 201.
Levenson, R., Housman, D. and Cantley, L. (1980) *Proc. Natl Acad. Sci. USA* **77**, 5948.
Liang, S. M. and Winter, C. G. (1977) *J. Biol. Chem.* **252**, 8278.
Llinas, R., Steinberg, I. Z. and Walton, K. (1976) *Proc. Natl Acad. Sci. USA* **73**, 2919.
Lo, C. S. and Edelman, I. S. (1976) *J. Biol. Chem.* **251**, 7834.
London, E. and Khorana, H. G. (1982) *J. Biol. Chem.* **257**, 7003.
Macara, I. and Cantley, L. C. (1983) in *Cell Membranes* (Elson, E. L., Frazier, W. A. and Glaser, L., eds). Plenum, New York, pp. 41–87.
McDonough, A. A., Hiatt, A. and Edelman, I. S. (1982) *J. Mem. Biol.* **69**, 13.
MacLennan, D. H. (1970) *J. Biol. Chem.* **245**, 4508.
MacLennan, D. H. and Holland, P. C. (1975) *Ann. Rev. Biophys. Bioeng.* **4**, 377.
MacLennan, D. H. and Reithmeier, R. A. F. (1982) *Membranes and Transport*, Vol. 1. (Martonosi, A. N. ed.). Plenum, New York, pp. 567–71.
MacLennan, D. H., Seeman, P., Iles, G. H. and Yip, C. C. (1971) *J. Biol. Chem.* **246**, 2702.
MacLennan, D. H., Reithmeier, R. A. F., Shoshan, V., Campbell, K. P. and LeBel, D. (1980) *Ann. N.Y. Acad. Sci.* **358**, 138.
Makinose, M. and Hasselbach, W. (1965) *Biochem. Z.* **343**, 360.
Makinose, M. and Hasselbach, W. (1971) *FEBS Lett.* **12**, 271.
Mannherz, H. G. and Goody, R. S. (1976) *Ann. Rev. Biochem.* **45**, 427.
Mardh, S. (1975*a*) *Biochim. Biophys. Acta* **391**, 448.
Mardh, S. (1975*b*) *Biochim. Biophys. Acta* **391**, 464.
Mardh, S. and Lindahl, S. (1977) *J. Biol. Chem.* **252**, 8058.
Mardh, S. and Post, R. L. (1977) *J. Biol. Chem.* **252**, 633.
Masuda, H. and de Meis, L. (1973) *Biochemistry* **12**, 4581.
Meissner, G. (1973) *Biochim. Biophys. Acta* **298**, 906.
Meissner, G. (1981) *J. Biol. Chem.* **256**, 636.
Meissner, G. and McKinley, D. (1982) *J. Biol. Chem.* **257**, 7704.
Miledi, R. (1973) *Proc. Roy. Soc. B.* **183**, 421.
Mitchell, P. (1961) *Nature, Lond.* **191**, 144.
Mitchinson, C., Wilderspin, A. F., Trinnaman, B. J. and Green, N. M. (1982) *FEBS Lett.* **146**, 87.
Miyamoto, H. and Racker, E. (1980) *J. Biol. Chem.* **255**, 2656.
Moczydlowski, E. G. and Fortes, P. A. G. (1981) *J. Biol. Chem.* **256**, 2346.
Montecucco, C., Bisson, R., Gach, C. and Johansson, A. (1981) *FEBS Lett.* **128**, 17.
Murphy, A. J. (1976) *Biochem. Biophys. Res. Commun.* **70**, 160.
Neufeld, A. H. and Levy, H. M. (1969) *J. Biol. Chem.* **244**, 6493.
Niggli, V., Adunyah, E. S., Penniston, J. T. and Carafoli, E. (1981) *J. Biol. Chem.* **256**, 395.
Niggli, V., Siegel, E. and Carafoli, E. (1982) *J. Biol. Chem.* **257**, 2350.
Pederson, P. L. (1975) *Bioenerg.* **6**, 243.
Perrone, J. R., Hackney, J. F., Dixon, J. F. and Hokin, L. E. (1975) *J. Biol. Chem.* **250**, 4178.
Perutz, M. F. (1970) *Nature, Lond.* **228**, 726.
Peterson, G. L. and Hokin, L. E. (1981) *J. Biol. Chem.* **256**, 3751.
Pick, U. (1982) *J. Biol. Chem.* **257**, 6111.
Pick, U. and Racker, E. (1979) *Biochemistry*, **18**, 109.
Pick, U. and Karlish, S. J. D. (1982) *J. Biol. Chem.* **257**, 6120.
Pickart, C. M. (1982) Ph.D. Thesis, Brandeis University.
Plesner, I. W., Plesner, L., Norby, J. G. and Klodos, I. (1981) *Biochim. Biophys. Acta* **643**, 483.
Porter, K. R. and Palade, G. E. (1957) *J. Biophys. Biochem. Cytol.* **3**, 269.
Post, R. L., Sen, A. K. and Rosenthal, A. S. (1965) *J. Biol. Chem.* **240**, 1437.
Post, R. L., Kume, S., Tobin, T., Orcutt, B. and Sen, A. K. (1969) *J. Gen. Physiol.* **54**, 306.
Post, R. L., Toda, G. and Rogers, F. N. (1975) *J. Biol. Chem.* **250**, 691.
Racker, E. and Stoeckenius, W. (1974) *J. Biol. Chem.* **249**, 662.
Reithmeier, R. A. F. and MacLennan, D. H. (1981) *J. Biol. Chem.* **256**, 5957.
Reithmeier, R. A. F., de Leon, S. and MacLennan, D. H. (1980) *J. Biol. Chem.* **255**, 11839.

Resh, M. D. (1982) *J. Biol. Chem.* **257**, 11946.
Ridgway, E. B., Gilkey, J. C. and Jaffe, L. F. (1977) *Proc. Natl Acad. Sci. USA*, **74**, 623.
Rindler, M. J. and Saier, M. H. (1981) *J. Biol. Chem.* **256**, 10820.
Rizzolo, L. J., LeMaire, M., Reynolds, J. A. and Tanford, C. (1976) *Biochemistry* **15**, 3433.
Robinson, J. D. (1976) *Biochim. Biophys. Acta* **429**, 1006.
Robinson, J. D. and Flashner, M. S. (1979) *Biochim. Biophys. Acta* **549**, 145.
Robinson, J. D., Hall, E. S. and Dunham, P. B. (1977) *Nature, Lond.* **269**, 165.
Rozengurt, E. (1981) *Adv. Enzym. Reg.* **19**, 61.
Ruoho, A. E. and Hall, C. C. (1980) *Ann. N.Y. Acad. Sci.* **346**, 90.
Sabatini, D., Colman, D., Sabban, E., Sherman, J., Morimoto, T., Kreibich, G. and Adesnik, M. (1981) *Cold Spring Harbor Symp. Quant. Biol.* **66**, 807.
Saccomani, G., Dailey, D. W. and Sachs, G. (1979) *J. Biol. Chem.* **254**, 2821.
Sachs, J. R. (1970) *J. Gen. Physiol.* **56**, 322.
Sachs, G., Chang, H. H., Rabon, E., Shackman, R., Lewin, M. and Saccomani, G. (1976) *J. Biol. Chem.* **251**, 7690.
Saito, A., Wang, C. T. and Fleischer, S. (1978) *J. Cell Biol.* **79**, 601.
Scarborough, G. A. (1980) *Biochemistry* **19**, 2925.
Schatzmann, H. J. and Vincenzi, F. F. (1969) *J. Physiol.* **201**, 369.
Sen, A. K. and Post, R. L. (1964) *J. Biol. Chem.* **239**, 345.
Shaver, J. L. F. and Stirling, C. (1978) *J. Cell Biol.* **76**, 278.
Simons, T. J. B. (1974) *J. Physiol.* **237**, 123.
Skou, J. C. (1974) *Biochim. Biophys. Acta* **339**, 234.
Skriver, E., Maunsbach, A. B. and Jorgensen, P. L. (1980) *J. Cell Biol.* **86**, 746.
Skriver, E., Maunsbach, A. B. and Jorgensen, P. L. (1981) *FEBS Lett.* **131**, 219.
Smith, R. L., Zinn, K. and Cantley, L. C. (1980) *J. Biol. Chem.* **255**, 9852.
Smith, R. L., Macara, I. G., Levenson, R., Housman, D. and Cantley, L. C. (1982) *J. Biol. Chem.* **257**, 773.
Steiger, J. and Schatzmann, H. J. (1981) *Cell Calcium* **2**, 601.
Stein, W. D., Lieb, W. R., Karlish, S. J. D. and Eilam, Y. (1973) *Proc. Natl Acad. Sci. USA* **70**, 275.
Steinhardt, R. A. and Epel, D. (1974) *Proc. Natl Acad. Sci. USA* **71**, 1915.
Stewart, P. S. and MacLennan, D. H. (1974) *J. Biol. Chem.* **249**, 985.
Stewart, P. S., MacLennan, D. H. and Shamoo, A. E. (1976) *J. Biol. Chem.* **251**, 712.
Takakuwa, Y. and Kanazawa, T. (1981) *J. Biol. Chem.* **256**, 2696.
Takakuwa, Y. and Kanazawa, T. (1982) *J. Biol. Chem.* **257**, 426.
Taniguchi, K. and Post, R. L. (1975) *J. Biol. Chem.* **250**, 3010.
Thorley-Lawson, D. A. and Green, N. M. (1973) *Eur. J. Biochem.* **40**, 403.
Thorley-Lawson, D. A. and Green, N. M. (1975) *Eur. J. Biochem.* **59**, 193.
Tong, S. W. (1977) *Biochem. Biophys. Res. Commun.* **74**, 1242.
Tong, S. W. (1980) *Arch. Biochem. Biophys.* **203**, 780.
Unwin, P. N. T. and Henderson, R. (1975) *J. Mol. Biol.* **94**, 425.
Vanderkooi, J. M., Ierokomas, A., Nakamura, H. and Martonosi, A. (1977) *Biochemistry* **16**, 1262.
Verjovski-Almeida, S. and de Meis, L. (1977) *Biochemistry* **16**, 329.
Waisman, D. M., Gimble, J. M., Goodman, D. B. and Rasmussen, H. (1981) *J. Biol. Chem.* **256**, 409.
Wang, C. T., Saito, A. and Fleischer, S. (1979) *J. Biol. Chem.* **254**, 9209.
Webb, M. R., Grubmeyer, C., Penefsky, H. S. and Trentham, D. R. (1980) *J. Biol. Chem.* **255**, 11637.
Yamada, S. and Ikemoto, N. (1980) *J. Biol. Chem.* **255**, 3108.
Yamada, S., Yamamoto, T. and Tonomura, Y. (1970) *J. Biochem.* **67**, 789.
Yamamoto, T. and Tonomura, Y. (1967) *J. Biochem.* **62**, 558.
Yeh, L.-A., Ling, L., English, L. and Cantley, L. (1983) *J. Biol. Chem.* **258**, 6567.
Zimniak, P. and Racker, E. (1978) *J. Biol. Chem.* **253**, 4631.
Zukin, R. S., Hartig, P. R. and Koshland, D. E. (1977) *Proc. Natl Acad. Sci. USA* **74**, 1932.

4
Metallothionein

Peter E. Hunziker and Jeremias H. R. Kägi

4.1 INTRODUCTION

Metallothioneins are nonenzymic low molecular weight proteins of extremely high sulphur and metal content which have recently become the object of intensive interest in many branches of the life sciences (Kägi and Nordberg, 1979; Webb, 1979). They are chiefly found in parenchymatous tissues of vertebrates and invertebrates but have also been isolated from some microorganisms and plants. They were initially discovered by Margoshes and Vallee (1957) in equine kidney in their search for the biological role of cadmium, and they are still the only biological compound known to contain this metal. It has already been shown in the earliest studies that cadmium is only one of several optional metallic constituents of the metallothioneins, the others being most commonly zinc and copper (Kägi and Vallee, 1960, 1961). Hence, it is believed that these proteins play a general role in the metabolism and the detoxification of a number of essential and nonessential trace metals. In mammals, their principal physiological purpose appears to be exerting a homeostatic function for zinc and copper, both by dispensing these metal ions when they are needed for the synthesis of metal-dependent cellular components such as metalloenzymes, nucleic acids and supramolecular structures and by sequestering them in a chemically innocuous form when they exceed a critical intracellular concentration (Richards and Cousins, 1976; Brady, 1982; Bremner, 1982; Webb and Cain, 1982; Whanger and Ridlington, 1982). The latter purpose extends to nonessential trace elements such as cadmium, mercury, silver, gold, and bismuth which bind to the protein with high affinity, thereby being prevented from reacting with other intracellular targets (Cherian and Goyer, 1978; cit. in Nordberg and Kojima, 1979). The ability of a number of transition and post-transition metals to induce the synthesis of metallothionein by gene activation (Piscator, 1964; Suzuki and Yoshikawa, 1976; Durnam and Palmiter, 1981) further underscores the suggestion that these proteins are part of a fast-responding and efficient biological feedback mechanism controlling the intracellular concentration of chemically reactive free metal ions.

The covalent structure of the metallothioneins is highly adapted to form metal complexes. This property is linked to the presence of a large number of cysteine residues and to their arrangement in the amino acid sequence to accommodate the metal ions in tetrahedrally structured metal–thiolate clusters (Vašák and Kägi, 1983*b*). In the present chapter, an up-to-date account is given of the current knowledge concerning the occurrence and the structural, evolutional and spectroscopic features of these proteins.

4.2 DEFINITION AND IDENTIFICATION

Historically, the cadmium-, zinc- and copper-containing sulphur-rich protein from equine renal cortex was designated as metallothionein (Kägi and Vallee, 1960). This protein has the following characteristic properties (Kojima and Kägi, 1978):

molecular weight 6000–7000 (i.e. 10 000 when measured by gel filtration);
high metal content;
characteristic amino acid composition (high cysteine content, no aromatic amino acids);
optical features characteristic of metal-thiolates (mercaptides);
unique amino acid sequence (fixed distribution of cysteine residues).

At the First International Meeting on Metallothionein and other Low Molecular Weight Metal-binding Proteins in 1978 (Nordberg and Kojima, 1979), the definition was adopted that any protein resembling equine renal metallothionein in several of these criteria can be classified as a metallothionein.

4.3 OCCURRENCE

Metallothioneins are now known to occur in various animal phyla as well as in certain eukaryotic microorganisms and in some plants. There are also some reports on their occurrence in prokaryotes (Maclean *et al.*, 1972; Olafson *et al.*, 1979*a*). An exhaustive list of the species and tissues in which metallothioneins were found until 1978 was given by Nordberg and Kojima (1979). Thus far, most studies have been concentrated on mammals. The protein is most abundant in the parenchymatous tissues of liver, kidney and intestines; however, its occurrence is also documented in many other tissues and organs (Żelazowski and Piotrowski, 1977). The amount of metallothioneins in different species and tissues is highly variable. Thus, human liver, often a rich source, may contain several hundred milligrams of metallothionein. There are also large differences in function of age. Very high concentrations occur in the livers of newborn pigs, fetal and newborn rats, and in human and lamb fetuses (Bakka and Webb, 1981). On the other hand, the tissues of most adult laboratory animals have a relatively low natural abundance of metallothionein, yet their content can be increased as much as 40 times following administation of large doses of 'inducer metals' (Piotrowski *et al*., 1973). During the last few years, many studies have also been reported on the formation of metallothionein in cultured mammalian cells (table 4.1). In most cultured cells the protein can be induced either by Cd(II) or by glucocorticoid hormones. In some cadmium-resistant mutant strains, the development of tolerance was shown to be due to selective amplification of the genes coding for metallothionein (Beach and Palmiter, 1981).

Metallothioneins strongly resembling the mammalian forms have also been isolated from birds and from fish (cit. in Nordberg and Kojima, 1979; Overnell

Table 4.1 Metallothionein in cultured cells

Cell type	Inducing agent	Reference
Human		
Embryonic fibroblasts	Cd	Lucis *et al.* (1970)
Skin fibroblasts	Cd	Chan *et al.* (1979)
Fetal lung fibroblast c.l.[a]	Cd	Hart and Keating (1980)
Lymphocytes	Zn	Phillips (1979)
Skin epithelial c.l.	Cd	Rugstad and Norseth (1975)
HeLa c.l.	Cd, Zn	Rudd and Herschman (1979)
HeLa c.l.	Dexamethasone	Karin and Herschman (1979)
Prostatic carcinoma c.l.	Dexamethasone	Giles and Cousins (1982)
Breast tumour c.l.	Cd	Bazzell *et al.* (1979)
Monkey		
Kidney c.l.	Cd	Kimura *et al.* (1979*b*)
Pig		
Liver c.l.	Cd	Daniel *et al.* (1977)
Kidney c.l.	Cd	Webb and Daniel (1975)
Rabbit		
Kidney c.l.	Cd	Kobayashi and Kimura (1980)
Rat		
Liver c.l.	Cd	Rudd and Herschman (1978)
Hepatocytes	Dexamethasone	Failla and Cousins (1978)
Hepatocytes	Dexamethasone	Karin *et al.* (1980)
Kidney epithelial cells	Cd	Cherian (1980)
Chinese hamster		
CHO c.l.	Cd	Hildebrand *et al.* (1979)
Mouse		
Fibroblast c.l.	Cd	Rugstad and Norseth (1978)
Friend leukaemia c.l.	Cd	Beach and Palmiter (1981)
Tumour c.l.	Dexamethasone Cd	Mayo and Palmiter (1981)
Tumour c.l.	Cd	Kobayashi and Kimura (1980)
Ehrlich ascites tumour c.l.	Cd, Zn	Koch *et al.* (1980)

[a]c.l. = cell line.

and Coombs, 1979). During the past few years, many reports have appeared also on the occurrence of such proteins in invertebrate species (cit. in Roesijadi, 1981), yet only a few of them have been characterised completely. The forms most similar to vertebrate metallothioneins were isolated from the hepatopancreas of the crab *Scylla serrata* (Olafson *et al.*, 1979*b*) and *Cancer pagurus* (Overnell and Trewhella, 1979). The proteins from molluscs are reported to contain in general less cysteine and a lower metal concentration. As in vertebrates, their synthesis is induced following metal exposure. Proteins resembling metallothioneins both in metal and amino acid composition have also been identified in earthworms (Suzuki *et al.*, 1980) and in a marine oligochaete (Thompson *et al.*, 1982).

A number of proteins from microorganisms and plants were classified as metallothioneins based on their metal and cysteine composition. The first microbial form reported is a copper-containing, cysteine-rich protein isolated from a culture of *Saccharomyces cerevisiae* grown in the presence of 0.2 mM $CuSO_4$. Its molecular weight measured by gel filtration is 9500, similar to that of vertebrate metallothioneins (Prinz and Weser, 1975). However, subsequent amino acid sequence studies failed to reveal a significant primary structure relationship (Kimura *et al.*, 1981). Another copper-containing metallothionein was isolated from *Neurospora crassa* grown on a medium supplemented with 0.5 mM $CuSO_4$ (Lerch, 1979). It consists of only 25 amino acid residues but shows a remarkable structural similarity to the N-terminal portion of vertebrate metallothioneins (see section 4.6). A protein resembling mammalian metallothioneins in amino acid composition has also been isolated from *Tetrahymena pyriformis* (Nakamura *et al.*, 1981). The occurrence of metallothionein in plants is documented by the recent characterisation of an extremely cysteine-rich copper protein from the roots of a metal-resistant strain of the grass *Agrostis gigantea* growing on a copper-enriched nutrient solution (Rauser and Curvetto, 1980) and by the isolation of a cadmium-binding metallothionein-like protein from the roots of tomato plants exposed for several weeks to 1.8×10^{-5} M $CdCl_2$ (Bartolf *et al.*, 1980).

4.4 ISOMETALLOTHIONEINS

Isolation procedures based on ion exchange properties and electrophoresis have led to the recognition of a variety of forms of metallothionein which were subsequently shown to differ in amino acid composition and sequence (*cit.* in Nordberg and Kojima, 1979). In all mammals examined thus far, there are at least two major isometallothioneins that appear to be coded by different genes which are thought to have arisen by gene duplication during evolution. They are usually numbered according to the sequence in which they elute from anion exchangers as metallothionein-1 (MT-1), metallothionein-2 (MT-2), etc. (Nordberg and Kojima, 1979). Their separation has been greatly aided by the introduction of high performance liquid chromatography (HPLC) combined with on-line atomic absorption measurement of metals (Suzuki, 1980) or by reversed-phase HPLC

(Klauser *et al.*, 1983). These methods resolve not only differently charged isometallothioneins but also other polymorphic variants not separable by classical chromatographic procedures. Thus, by Suzuki's procedure it was recently possible to resolve four different isometallothioneins from the renal cortex of monkeys exposed to $CdCl_2$ in their diet (Nomiyama and Nomiyama, 1982). Similarly, by reversed-phase HPLC, the two principal differently charged forms of rabbit liver metallothionein were subfractionated further to yield a total of four isometallothioneins (Klauser *et al.*, 1983). The same method has also allowed the isolation of four different isoproteins from the hitherto unresolved metallothionein-1 fraction of adult human liver. As shown in figure 4.1, these

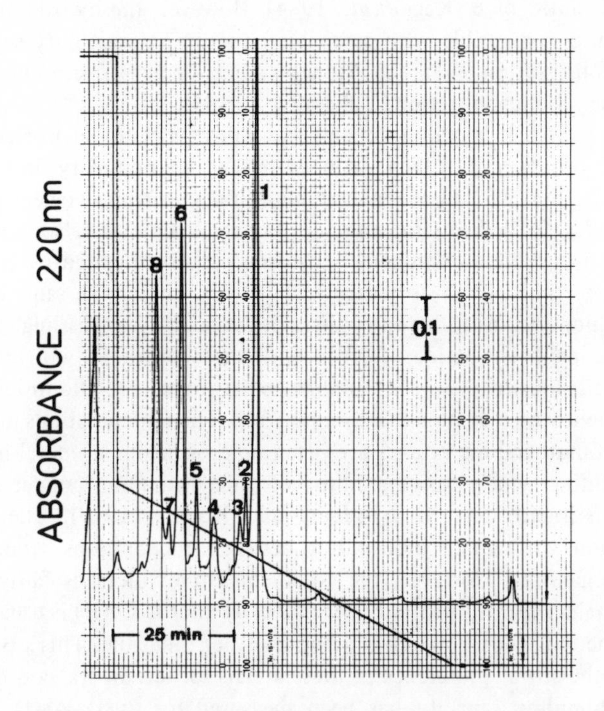

Figure 4.1 Metallothionein isoprotein HPLC patterns from crude extract of human liver. Crude metallothionein, 5 mg, prepared by gel filtration (Kägi *et al.*, 1974) was taken up in 0.5 ml 0.02 M Tris-HCl and chromatographed on semipreparative LiChrosorb RP 18 (10 μm) using a linear acetonitrile gradient (straight line) in 0.05 M Tris-HCl, pH 7.5, at a flow rate of 2 ml min^{-1}. Isometallothioneins were identified by amino acid analysis in the following peaks: 1 (MT-2), 3 (MT-1A), 5 (MT-1B), 6 (MT-1C) and 8 (MT-1D). Peaks 2, 4 and 7 were not identified (Hunziker and Kägi, 1983).

four subforms of metallothionein-1 and the single form contained in metallothionein-2 can now be resolved and quantified in a single step starting from a crude extract of human liver (Hunziker and Kägi, 1983).

4.5 PHYSICAL AND CHEMICAL CHARACTERISATION

All mammalian metallothioneins characterised thus far are small single-chain proteins. By gel filtration in 0.1 M salt solutions most forms are eluted at a position corresponding to that of a globular protein of molecular weight of about 10 000. This is considerably higher than the values measured by equilibrium centrifugation of equine renal metallothionein (Kägi *et al.*, 1974) or calculated from the chain weight and the metal composition, i.e. 6500–6800. The discrepancy was taken as evidence of a nonglobular shape for these proteins. Assuming a degree of hydration of 0.2, the molecular mass and the Stokes radius (16.1 Å) are consistent with the dimensions of a rigid prolate ellipsoid with an axial ratio of 6 (Kägi *et al.*, 1974). However, the hydrodynamic parameters are also compatible with a flexible structure, a possibility suggested by the recent observation that the Stokes radius decreases substantially upon increasing the salt concentration (Vašák *et al.*, 1984).

Native mammalian metallothioneins are proteins with an isoelectric point close to 4 (Nordberg *et al.*, 1972). As measured by free boundary electrophoresis, between pH 7.5 and 9.5 equine renal metallothionein carries two negative charges (Bethune *et al.*, 1979). The same number of charges is present also in the human, rabbit and murine metallothionein-1 isoforms, as documented by comparative electrophoretic mobility measurements at pH 8.6. Under the same conditions, the metallothionein-2 isoforms carry three negative charges (Kissling, 1979).

Metallothionein is, besides ferritin, the metalloprotein with the highest metal content known. However, unlike most other metalloproteins, native metallothionein varies in its metal composition. Thus, preparations judged to be pure by chromatographic criteria are often heterogeneous in metal ion composition, exhibiting widely varying ratios of zinc, cadmium, copper and minor metallic constituents (*cit.* in Nordberg and Kojima, 1979). There are also large variations in metal composition between preparations from different individual organisms. However, the total content of metal is fairly constant. In all mammalian forms, a total of seven Zn(II) and/or Cd(II) is usually bound, indicating the existence of seven binding sites per molecule. This stoichiometry is related to the strong preference of these metals for tetrahedral coordination. A larger metal-binding capacity has been measured for Cu(I), Ag(I) and Hg(II) metal ions, which can also accept other types of coordination (Kägi and Vallee, 1961; Lerch, 1981; Bernhard *et al.*, 1983).

In part, the metal ion composition is determined by the extent of the exposure of the organism to the different metals. Thus, exposure to salts of zinc, cadmium, copper, mercury, silver, gold and bismuth yields metallothioneins rich in these metals (*cit.* in Nordberg and Kojima, 1979). In organisms not subjected to experimental pretreatment with metal salts, thus far only Zn(II), Cd(II), Cu(I) and Hg(II) have been detected (table 4.2). In most mammals, Zn(II) is the major component; however, depending on the species, the tissue, the stage of development and the age, the proportions of Cd(II) and Cu(I) can be substantial. Also very characteristic are the variations in the metal compo-

Table 4.2 Occurrence and metal composition of metallothionein (all data refer to metallothionein obtained from organisms not subjected to experimental pretreatment with metals)

Species	Organ	Metal composition				Reference
		Zn	Cd	Cu	Hg	
Man	Liver	++++++	±	±	—	Bühler and Kägi (1974)
	Kidney	+++	+++	+	±	Pulido *et al.* (1966)
	Fetal liver	++++		++		Riordan and Richards (1980)
	Neonatal liver	++++		++		Bakka and Webb (1981)
Horse	Liver	+++++	+	±		Kägi *et al.* (1974)
	Kidney	+++	+++	±		Kägi and Vallee (1960)
	Intestine	++++	++	±		Kägi *et al.* (1979)
Sea lion	Kidney	+++			+++	Lee *et al.* (1977)
Sheep	Fetal liver	++++		++		Bremner *et al.* (1977)
Pig	Liver	++++		++		Bremner (1976)
Rabbit	Neonatal liver	+++++		+		Bakka and Webb (1981)
Syrian hamster	Neonatal liver	++		++++		Bakka and Webb (1981)
Chinese hamster	Neonatal liver	++++		++		Bakka and Webb (1981)
Rat	Neonatal liver	+++++		+		Mason *et al.* (1980)
	Neonatal kidney	+++++		+		Brady and Webb (1981)
	Adult liver	++++		++		Kägi *et al.* (1979)
	Adult kidney	++		++++		Brady and Webb (1981)
	Adult testis	+++++		+		Brady and Webb (1981)
Mouse	Neonatal liver	+++		+++		Bakka and Webb (1981)
Duck	Liver	++	+	+++		Brown *et al.* (1977)
Flounder	Liver	+	±	+++++		Brown (1977)
Skipjack	Liver	+	+++	+++		Takeda and Shimizu (1982)
Eel	Liver	+++	±	+++		Noël-Lambot *et al.* (1978)
Crab	Hepatopancreas	±	+++	+++		Overnell and Trewhella (1979)
Limpet	Soft tissue	+++	++	+		Howard and Nickless (1975)

sition of metallothioneins isolated from different tissues. Metallothioneins from equine and human kidney cortex contain more cadmium and less zinc than those from the liver of these species (Pulido *et al.*, 1966; Bühler and Kägi, 1974; Kojima *et al.*, 1979). Analogous differences exist with respect to the zinc and copper content between hepatic and renal metallothioneins of rodents (Suzuki, 1979; Żelazowski and Szymańska, 1980). Since the metallothioneins occurring in equine liver and kidney were found to be the same proteins (Kojima *et al.*, 1979), such differences in metal content cannot have any structural basis. They must be explained by local physiological circumstances and by the mechanisms controlling the flow of the various metals through the organism.

The other exceptional compositional feature of the metallothioneins is the abundance of sulphur. With a total of 11 per cent they have the largest sulphur content of any known protein (Kägi *et al.*, 1974). It is accounted for by the extreme preponderance of cysteine residues, which comprise about one-third of all amino acid residues (table 4.3) yielding for cadmium- and/or zinc-metallothionein a cysteine-to-metal stoichiometry of about 3. All cysteine residues participate in metal binding. There are no disulphide bonds and no labile sulphur. Beside cysteine, in most metallothioneins there are also relatively large amounts of serine and of the basic amino acids lysine and arginine. The remaining amino acids are inconspicuous. Characteristically, there are no aromatic amino acids. Histidine has been found thus far only in avian metallothionein (Weser *et al.*, 1973; Oh *et al.*, 1979; Klauser *et al.*, 1983).

4.6 AMINO ACID SEQUENCES

Amino acid sequence data provide essential information (1) on the nature and the extent of the polymorphism of metallothionein, (2) on their evolution, and (3) on the structural basis of the metal–protein interaction. Thus far, sequences have been determined either completely or partially for 15 different metallothioneins including two from an invertebrate species and one from a fungus (figure 4.2). For one of the human isometallothioneins (metallothionein-2) (Karin and Richards, 1982*a,b*) and one of the mouse isometallothioneins (metallothionein-1) (Durnam *et al.*, 1980; Glanville *et al.*, 1981; Mbikay *et al.*, 1981), the nucleotide sequence of cDNA and of genomic DNA has also been determined. The sequences of the Chinese hamster isometallothioneins (Griffith *et al.*, 1983) and of rat metallothionein-1 (Andersen *et al.*, 1983) were derived from the nucleotide sequence of cDNA only. There is no known homology to other proteins, thus establishing that the metallothioneins belong to a novel superfamily of proteins.

4.6.1 Sequence homology

Based on the alignment made in figure 4.2, all metallothioneins show readily recognisable homology, indicating their relatedness. All mammalian forms contain 61 residues, and the two crab isometallothioneins 57 and 58 residues,

Table 4.3 Amino acid composition of some metallothioneins (MT) whose complete sequences are known[a]

Amino acid	Human MT-2	Equine MT-1A	Equine MT-1B	Mouse MT-1	Mouse MT-2	*Scylla* MT-1	*Scylla* MT-2	*Neurospora* MT
Ala	7	5	7	5	6	2	1	1
Arg	—	2	1	—	—	1	2	—
Asn	1	1	1	2	1	2	2	2
Asp	3	2	2	2	3	1	4	1
Cys	20	20	20	20	20	18	18	7
Gln	1	1	2	1	3	1	1	—
Glu	1	1	1	—	—	5	4	—
Gly	5	7	5	5	4	5	3	6
Ile	1	—	—	—	1	—	—	—
Leu	—	—	—	—	—	—	—	—
Lys	8	6	7	7	8	8	8	1
Met	1	1	1	1	1	—	—	—
Pro	2	3	2	2	2	4	6	—
Ser	8	8	8	9	10	7	5	7
Thr	2	3	1	5	1	3	3	—
Val	1	1	3	2	1	1	—	—
Total residues	61	61	61	61	61	58	57	25

[a]For references, see figure 4.2.

```
                        1                  10                   20                    30                  40                  50                     60
                                           A     S   T
Human      MT-1        Ac-M D P N C S C A  P G G S C T C A G S C K C K E · C K C T S C K K S C C S C C P V G C A K C A Q G C I C K G(T,S)D K · C S C C A-OH
                                           T V   V   A                                                                                     A   E
Human      MT-2        Ac-M D P N C S C A A G D S C T C A G S C K C K E · C K C T S C K K S C C S C C P V G C A K C A Q G C I C K G A S D K · C S C C A-OH
Equine     MT-1A       Ac-M D P N C S C P T G G S C T C A G S C K C K E · C R C T S C K K S C C S C C P G G C A R C A Q G C V C K G A S D K · C S C C A-OH
Equine     MT-1B       Ac-M D P N C S C V A G E S C T C A G S C K C K Q · C R C A S C K K S C C S C C P V G C A K C A Q G C V C K G A S D K · C S C C A-OH
                                                                          A
Rabbit     MT-2         X-M D P N C S C A A D G(S,C,T,C,A,T,S,C)K C K   · C K C T S C K K S C C S C C P S G C A K C A Q G C I C K G A S D K · C S C C A-OH
                                                                          E
Hamster    MT-1         X-M D P N C S C S T G S T C T C S S S C G C K D · C K C T S C K K S C C S C C P V G C S K C A Q G C V C K G A S D K · C T C C A-OH
Hamster    MT-2         X-M D P N C S C A T D G S C S C A G S C K C K E · C K C T T C K K S C C S C C P V G C A K C S Q G C V C K E A S D K · C S C C A-OH
Mouse      MT-1        Ac-M D P N C S C S T G G S C T C S S C A C K N · C K C T S C K K S C C S C C P V G C S K C A Q G C V C K G A A D K · C T C C A-OH
Mouse      MT-2        Ac-M D P N C S C A S D G S C S C A G A C K C K Q · C K C T S C K K S C C S C C P V G C A K C S Q G C I C K Q A S D K · C S C C A-OH
Rat        MT-1         X-M D P N C S C S T G G S C T C S S S C G C K N · C K C T S C K K S C C S C C P V G C S K C A Q G C V C K G A S D K · C T C C A-OH
Rat        MT-2        Ac-M D P N C S C A T D G S C S C A G S C K C K Q · C K C T S C K K S C C S C C P.... .(Partial sequence)
Plaice                 Ac-M D P · C E C S K T G T C N C G G(S)(C)T(C)K N · (C)G(C)T.....(Partial sequence)... ............................K ·(C,T,C)C Q-OH
Scylla     MT-1             P G P C · C · · N D K C V C K E G · G C K E G C Q C T S C R C S P C E K C S S G C · K C A N K E E C S K T C S K A C S C C P T-OH
"          MT-2             P D P C · C · · N D K C D C K E G · E C K T G C K C T S C R C P P C E Q C S S G C · K C A N K E D C R K T C S K P C S C C P-OH
Neurospora MT               H-G D C G C S G A S S C N C G S G C S C S N · C G S K-OH
```

One-letter symbols:

A = Alanine	M = Methionine
C = Cysteine	N = Asparagine
D = Aspartic Acid	P = Proline
E = Glutamic acid	Q = Glutamine
G = Glycine	R = Arginine
I = Isoleucine	S = Serine
K = Lysine	T = Threonine
L = Leucine	V = Valine

X = Undetermined

Other symbols:

Ac = Acetyl
H = Free amino terminus
OH = Free carboxyl terminus

References:

Human MT-1	Kissling and Kägi, 1979
Human MT-2	Kissling and Kägi, 1977
Equine MT-1A	Kojima et al., 1979
Equine MT-1B	Kojima et al., 1976
Rabbit MT-2	Kimura et al., 1979a
Hamster MT-1	Griffith et al., 1983
Hamster MT-2	Griffith et al., 1983
Mouse MT-1	Huang et al., 1977
Mouse MT-2	Huang et al., 1981
Rat MT-1	Andersen et al., 1983
Rat MT-2	Kissling et al., 1979
Plaice MT	Overnell et al., 1981
Scylla MT-1	Lerch et al., 1982
Scylla MT-2	Lerch et al., 1982
Neurospora MT	Lerch, 1979

Figure 4.2 Amino acid sequences of metallothioneins (MT). (The numeration refers to the sequence of mammalian metallothioneins. The residues enclosed within parentheses require further identification. Dots between adjacent residues denote deletions introduced for optimal alignment. Residues 58 and 59 of the human and equine metallothioneins were reassigned following sequence re-examination (M. Kimura and J. H. R. Kägi, unpublished data). Residue 23 of mouse metallothionein-1 was reassigned on the basis of the cDNA sequence (Mbikay *et al.*, 1981).)

respectively. The copper-containing metallothionein from the fungus *Neurospora crassa* containing only 25 residues is homologous to the N-terminal portion of the proteins derived from animals. The data also document the structural differences between the isometallothioneins. Thus, the two forms isolated from equine liver differ in a total of seven amino acid positions. Most of the replacements are explicable by single base changes. However, the number of substituents exceeds that to be expected for allelic proteins. It suggests instead that they are coded by different cistrons that have evolved by duplication from an ancestral gene. The same interpretation pertains to the human isometallothioneins, which also differ from each other in seven positions. The sequence determined for human metallothionein-1 is still equivocal in some of the assignments. This can be attributed to the heterogeneity of this chromatographic fraction documented by recent HPLC studies (see section 4.4 and figure 4.1). As shown previously for the horse proteins (Kojima and Kägi, 1978; Kojima *et al*., 1979), some of these additional human isoproteins may represent minor variants which differ from the major isometallothioneins by single amino acid substitutions and, hence, may originate from allelic genes. The pronounced heterogeneity of metallothionein from human tissues is also consistent with the recent report that the human genome contains as many as 11 genes and/or pseudogenes coding for metallothionein (Karin and Richards, 1982*b*). In marked contrast to the minor differences between the principal equine and human isoproteins, the two Chinese hamster and mouse isometallothioneins differ from each other in 20 and 25 per cent of all residues, respectively. Judged from the partial sequence data available, the same applies to the rat isometallothioneins. The larger differences in these metallothioneins could indicate either that the murine forms have evolved at substantially higher rates or else that their genes have separated earlier than those expressed in the horse and in man and, hence, that the isoproteins of different mammals did not arise from a single common gene duplication. The latter suggestion receives some support from the observation that each of the murine sequences differs to about the same extent from the two equine sequences, indicating comparable evolutionary distances (table 4.4).

The functional significance of the emergence of multiple isometallothioneins is still unknown. One possible explanation for the development of quite different isoproteins in the murine could be a certain specialisation towards the binding of copper. This metal is much more abundant in the rodent metallothioneins than in the equine and human proteins, where the metals are more restricted to zinc and cadmium (table 4.2). Other factors that may also have led to the evolution of metallothionein variants could be the need for differential regulation of the expression of the various genes. The fact that after partial hepatectomy the two major isometallothioneins in rat liver are induced to a different degree offers some support to this view (Ohtake and Koga, 1979; Webb and Cain, 1982).

Table 4.4 Amino acid sequence identity of metallothioneins[a]

	Human MT-2	Equine MT-1A	Equine MT-1B	Mouse MT-1	Mouse MT-2	*Scylla* MT-1	*Scylla* MT-2	*Neurospora* MT
Human MT-2	100							
Equine MT-1A	89	100						
Equine MT-1B	90	89	100					
Mouse MT-1	82	82	80	100				
Mouse MT-2	87	80	82	76	100			
Scylla MT-1	47	43	41	41	41	100		
Scylla MT-2	46	40	40	41	42	82	100	
Neurospora MT	32	32	32	44	32	28	24	100

[a] The figures indicate observed percentage identity (I_{obs}) based on alignment in figure 4.2.

4.6.2 Evolutionary rate

With the still very limited number of sequences available and with the uncertainties brought about the the expression of a number of different metallothionein genes in mammals (Karin and Richards, 1982*b*), it would be premature to construct a phylogenetic tree for this protein. However, from the known vertebrate and invertebrate structures, it is possible to gain an estimate of the average rate of its evolution. The relevant data are summarised in table 4.4 in the form of an amino acid sequence identity matrix. From the percentage of observed identity (I_{obs}) the percentage of amino acid substitution events that have taken place can be calculated by correcting for repeated changes at the same amino acid site by the expression $N_{corr} = 100 \ln (100/I_{obs})$ (adapted from Margoliash and Fitch, 1968). By making the reasonable assumption that some of the genes coding for the equine and human isometallothioneins are orthologous and accepting that 75 million years have elapsed since the last common ancestor of these euplacental mammals, one obtains from the 89.5 per cent average structural identity of equine metallothionein-1A and metallothionein-1B with human metallothionein-2 a rate of 7.5×10^{-10} amino acid substitutions per codon per year. A comparable rate of about 7.2×10^{-10} amino acid substitutions per codon per year is calculated from the average 42 per cent sequence identity between the crab metallothioneins and all mammalian forms sequenced and by using a figure of about 600 million years since the separation of vertebrates and invertebrates. Similarly, from the 32 per cent average sequence identity between *Neurospora* metallothionein and the corresponding segments of the appropriately aligned animal proteins and from an estimate of 1150 million years elapsed since the divergence of fungi and animals, an evolutionary rate of about 5×10^{-10} amino acid substitutions per codon per year is obtained. These admittedly very tentative figures indicate an overall rate of metallothionein evolution ranging between that of cytochrome *c*, i.e. 4.2×10^{-10} amino acid substitutions per codon per year, and that of the haemoglobin chains, i.e. 10.1×10^{-10} amino acid substitutions per codon per year (King and Jukes, 1969).

4.6.3 Variable and invariable residues

A detailed comparison of the mammalian sequences shows that the amino acid substitutions that occurred in evolution are limited to certain regions in the chain. The major hotspots are the positions 8, 9, 10, 11, 14 and 23 in the amino-terminal half and the positions 39 and 54 in the carboxyl-terminal half. Nearly completely invariant are segments at the amino-terminus (positions 1–7), in the centre (positions 28–38) and at the carboxyl-terminus (positions 56–61) of the chain. The significance of this preservation is as yet unknown, but it implies the existence of stringent structural requirements for metal binding and chain folding. All vertebrate metallothioneins have a blocked amino-terminal methionine which is the initiator residue (Durnam *et al.*, 1980; Berger *et al.*, 1981). Its persistence after *in vivo* protein synthesis has been attributed to its juxtaposition with aspartic acid, whose negative charge renders the Met–Asp bond resistant to the post-translational cleavage by methionine aminopeptidase (Schechter *et al.*, 1978).

The most striking structural feature of all metallothioneins sequenced thus far is the preservation of the cysteine residues (figure 4.3). Their positions are

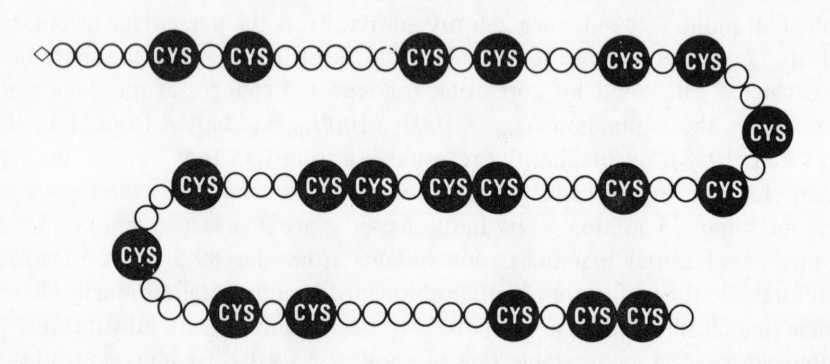

Figure 4.3 Mode of distribution of cysteine residues in mammalian metallothioneins (adapted from Kojima and Kägi, 1978).

identical in all mammalian metallothioneins. There is also a complete correspondence between all cysteines of *Neurospora* metallothionein and the seven first cysteines of all vertebrate chains. Similarly, depending on the alignment made, there is correspondence of 80-90 per cent of the cysteine residue of the crab metallothioneins with those of the mammalian forms.

Highly typical also is the relative arrangement of the cysteine residues within the chain. In mammals, 14 of the 20 cysteine residues form part of seven Cys–*X*–Cys tripeptide sequences where *X* is an amino acid other than cysteine. In the metallothioneins from crab and *Neurospora*, these sequences occur three times. In addition, there are Cys–Cys and Cys–*X*–*Y*–Cys sequences. Another interesting feature of the vertebrate and invertebrate metallothioneins is the close association of the basic amino acids lysine and arginine and of serine (or the homologous threonine) with the cysteine residues, the central portion of the chain consisting almost exclusively of these residues. Although a possible role of the hydroxyl amino acid residues in conjunction with the metal-binding cysteine residues is still not obvious, it has been suggested that the basic residues may serve to balance the excess negative charge generated on formation of the metal-thiolate complexes with bivalent metal ions (Kojima *et al.*, 1976). In this regard it is noteworthy that *Neurospora* metallothionein, which *in vivo* is thought to bind monovalent copper in a cluster, contains but a single carboxyl-terminal lysine (Lerch, 1981). That the lysine residues serve a structural role in Cd(II)- and Zn(II)-binding metallothioneins is also underscored by the recent finding that in the native protein these residues are partially protected from chemical modification and that their deprotonation produces marked changes in the polypeptide chain conformation (J. Pande, M. Vašák and J. H. R. Kägi, unpublished observation).

4.7 METAL-BINDING SITES

The abundance of cysteine residues and their arrangement in the chelating –Cys–*X*–Cys– and –Cys–*X*–*Y*–Cys– oligopeptide sequences predispose metallothionein for metal binding. Similar sequences are known to form the metal-binding sites in the iron–sulphur proteins (Palmer, 1975) and in the zinc-containing horse liver alcohol dehydrogenase (Eklund *et al*., 1976) and aspartate transcarbamylase (Nelbach *et al*., 1972). In each of these cases the metals are bound to the protein by coordination to the deprotonated cysteine side chains through thiolate (or mercaptide) bonds. For mammalian metallomeins it has been suggested that the –Cys–*X*–Cys- dithiol sites function as nucleation centres in the assembling of the ultimate multidentate metal coordination structure that is determined by the ligand–metal stoichiometry and by the constraints imposed by the nature of the metal ions and the structure of the polypeptide chain (Kojima *et al*., 1976). Very recent studies have shown that hexapeptides containing such a dithiol sequence form both mononuclear and oligonuclear metal-thiolate complexes with Cd(II), Zn(II) and Co(II), which exhibit spectra features that are indistinguishable from those of metallothionein (Willner *et al.*, 1983).

The details of the mode of metal binding have been clarified both by chemical and spectroscopic means (for references see Nordberg and Kojima, 1979). It was shown by a variety of methods that in native metallothionein all cysteine residues participate in metal binding and that the affinity of the different metal ions for the protein increases in the order typical for thiolate model complexes, i.e. Zn(II) < Pb(II) < Cd(II) < Cu(I), Ag(I), Hg(II), Bi(III). Accordingly, the least firmly bound Zn(II) can be displaced by the other metal ions. Zinc, lead and cadmium are also readily removed on exposure to low pH, yielding the metal-free protein apometallothionein. By adding to the latter appropriate amounts of metal salts, well-defined forms of (mammalian) metallothionein containing exactly 7 mol of Zn(II), Cd(II), Hg(II), Pb(II), Bi(III), Co(II) or Ni(II) can be prepared (Bernhard *et al*., 1983; Vašák and Kägi, 1983*b*).

The stability of the metal-protein complexes has been measured for Zn(II) and Cd(II) by following spectrophotometrically the displacement of metals as a function of increasing proton concentration (Kägi and Vallee, 1961; Vasák and Kägi, 1983*b*). The values obtained for the binding constants, i.e. $K_{Cd} = 1 \times 10^{22}\ M^{-1}$ and $K_{Zn} = 1 \times 10^{18}\ M^{-1}$ are in reasonable agreement with the range expected for the cumulative association constants of complexes with several thiolate ligands. If recalculated for pH 7, the apparent stability constants are still about $2 \times 10^{16}\ M^{-1}$ and $2 \times 10^{12}\ M^{-1}$ for Cd(II) and Zn(II), respectively. The 10 000-fold higher affinity for Cd(II) in these complexes must be contrasted with the about ten-fold difference in the first stepwise stability constants of complexes of Cd(II) and Zn(II) with monodentate thiols (Gurd and Wilcox, 1956). It illustrates the potentiation of metal ion selectivity by chelation to binding sites containing several thiolate groups. This effect is responsible for the preferential accumulation of Cd(II) peculiar to metallothionein.

4.7.1 Coordination geometry

Information on the geometry of metal coordination in metallothionein has been derived from a large variety of spectroscopic studies (Vašák and Kägi, 1983*b*). The electronic absorption spectra of metallothionein containing a full complement of either Zn(II), Cd(II), Hg(II), Pb(II) or Bi(III) and of the metal-free apometallothionein are shown in figure 4.4. Lacking aromatic amino acids,

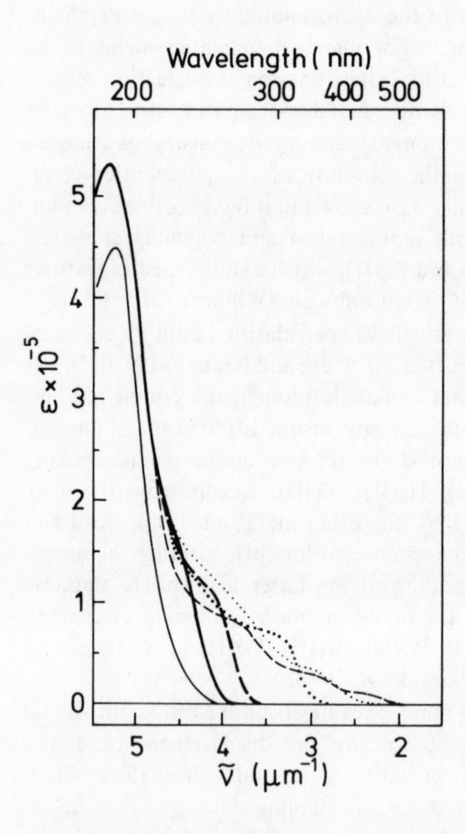

Figure 4.4 UV-absorption spectra of apometallothionein at pH 2 (thin full line) and of metallothionein at pH 8 containing a full complement (7 mol per mole) of zinc (thick full line), cadmium (thick broken line), mercury (thick dotted line), bismuth (thin broken line) and lead (thin dotted line) (Bernhard *et al.*, 1983).

apometallothionein exhibits a plain absorption with a single maximum at 190 nm arising mainly from the transitions of the secondary amide chromophores of the polypeptide chain and of the thiol side chains of the cysteine residues (Bühler and Kägi, 1979). In the metal-containing forms, the far-UV absorption is intensified, and at the low-energy side, new absorption bands appear whose positions indicated by absorption shoulders are specific for the particular metal. Spectra very similar to those of Zn(II)- and Cd(II)-metallothionein are also displayed by complexes of these metals with simple thiols (Kägi and Vallee, 1961; Vašák *et al.*, 1981*a*) and oligopeptides containing –Cys–*X*–Cys– sequences (Willner *et al.*, 1983), thereby documenting their origin in the metal–thiolate transitions. The very strong metal-dependence of these spectra is diagnostic of an electron transfer nature for the transitions involved. According to a semiempirical theory of Jørgensen (1970), the frequency of the lowest energy band ν (μm^{-1}) of such spectra is related to the difference of the optical electronegativities of the ligand $\chi_{opt}(L)$, and of the metal, $\chi_{opt}(M)$, by the expression

$$\nu(\mu m^{-1}) = (3.0\,\mu m^{-1})\,[\chi_{opt}(L) - \chi_{opt}(M)]$$

As shown in table 4.5, there is excellent agreement between the spectral location calculated from the values of the geometry-dependent $\chi_{opt}(M)$ derived from

Table 4.5 Location of first metal–thiolate electron transfer transition in monosubstituted metal derivatives of metallothionein

Derivative	Observed		Calculated[a]	
	μm^{-1}	(nm)	μm^{-1}	(nm)
$Zn(II)_7$-metallothionein	4.33	(231)	4.31	(232)
$Cd(II)_7$-metallothionein	4.00	(250)	4.01	(249)
$Hg(II)_7$-metallothionein	3.25	(308)	3.21	(312)
$Pb(II)_7$-metallothionein	2.50	(400)	2.49	(401)

[a]Calculated from the difference of the optical electronegativities of the bonded ions (Jørgensen, 1970). The values for the optical electronegativities employed were 2.6 for RS^-, 1.15 for Zn(II), 1.27 for Cd(II), 1.50 for Hg(II) and 1.77 for Pb(II). The values of the metals were deduced from absorption spectra of tetrahedralide complexes (Day and Seal, 1972).

tetrahedral–tetrahalide model complexes and the resolved lowest energy transition for four different monosubstituted metallothioneins, indicating that this band is indeed the first Laporte-allowed electron transfer transition and that the coordination geometry is also tetrahedral (Vašák *et al.*, 1981*a*).

Very strong support for this coordination geometry comes also from spectroscopic studies of metallothionein reconstituted with paramagnetic Co(II), a derivative whose optical and magneto-optical features are much more sensitive to the mode of metal coordination than those of the metallothioneins containing d^{10} metal ions (Vašák *et al.*, 1981*b*). The absorption spectrum of this intensely green-coloured complex displays, besides the Co(II)–thiolate electron transfer bands in the high-energy region, characteristic *d–d* bands in the visible and near-IR region with maxima at 600, 690 and 743 nm and at 1150 and 1275 nm, respectively (figure 4.5). These features closely resemble those of inorganic tetrahedral tetrathiolate complexes and of Co(II) derivatives of crystallographic-

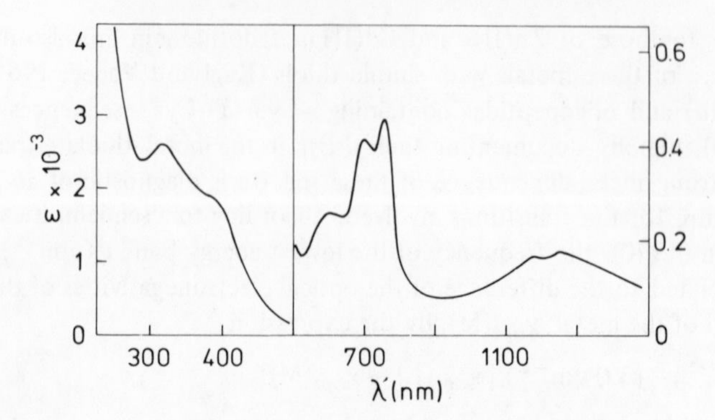

Figure 4.5 Absorption spectrum of rabbit liver $Co(II)_7$-metallothionein in 0.05 M Tris-HCl, pH 7. The molar absorbancy refers to the metal (from Vašák and Kägi, 1983*a*).

ally defined metalloproteins where the metal is known to be bound tetrahedrally to four cysteine residues, e.g. in rubredoxin (May and Kuo, 1978) and in the structural metal sites of horse liver alcohol dehydrogenase (Maret *et al.*, 1979). Hence it follows that in Co(II)-metallothionein too, the cobalt ions are coordinated to groups of four thiolate ligands forming high-spin tetrahedral complexes. The simplicity of the absorption spectrum of Co(II)-metallothionein suggests, furthermore, that the various metal-binding sites of metallothionein are both chemically and structurally similar. This is also supported by the results gained from the application of more specialised methods to the study of this metalloprotein, such as perturbed angular correlation of gamma-ray spectroscopy (PAC) (Vašák and Bauer, 1982) and of extended X-ray absorption fine structure (EXAFS) measurements (Garner *et al.*, 1982). Both methods independently confirmed that the metal-binding sites have a geometry that is close to that of a tetrahedron. The EXAFS study carried out on a 6 per cent zinc-containing form of sheep liver metallothionein gave unambiguous evidence that all zinc sites are essentially equivalent and that the metal ion is attached to four sulphur atoms with a Zn–S distance of approximately 2.29 Å. Comparable results were obtained with solely zinc-containing and with both zinc- and cadmium-containing preparations of rabbit liver metallothionein yielding an average Zn–S distance of 2.28 Å (Ross *et al.*, 1983). The PAC measurements monitoring nuclear quadrupole interactions in excited cadmium nuclei (^{111m}Cd) incorporated into metallothionein revealed in addition to the predominant tetrahedral tetrathiolate microsymmetry a minor contribution from another symmetry type also present in the overall framework of the metal-binding site.

4.7.2 Metal–thiolate clusters

The abundance of thiolate ligands in metallothionein and the unusual stoichiometry of only about three thiolate ligands per bivalent metal ion have prompted the suggestion that in this protein the metal complexes may exist as clusters in which neighbouring metal ions are linked by one or more bridging thiolate ligands (Weser and Rupp, 1979*a*). This view has now received compelling support from the spectroscopic evidence that in metallothionein each bivalent metal ion is linked to four thiolate ligands. Since, in contrast to the iron sulphur proteins, metallothionein contains no significant amounts of nonprotein sulphur, such an allocation of thiolate ligands is feasible only if eight of the 20 cysteine residues of the polypeptide chain provide sulphur bridges between adjacent metal ions, thereby forming thiolate-metal clusters. The occurrence of such oligonuclear aggregates is now also documented by various spectroscopic methods, among them most directly magnetic resonance measurements monitoring metal–metal interactions.

In elegant studies, Sadler *et al.* (1978) and Otvos and Armitage (1979) used the potential of ^{113}Cd NMR spectroscopy as a highly specific and natural probe of its environment in ^{113}Cd-enriched metallothionein. The usefulness of this method relies both on the sensitivity of the ^{113}Cd chemical shift to small differences in coordination, which allows the resolution of the signals from the different sites, and on the observation that the resonances are split into multiplets as a consequence of ^{113}Cd–^{113}Cd scalar coupling of neighbouring nuclei via thiolate ligands (Otvos and Armitage, 1980). This latter feature offers the possibility of identifying interacting nuclei by homonuclear decoupling studies. As shown in figure 4.6, a solely ^{113}Cd(II)-containing form of rabbit-liver metallothionein-1 displays a total of eight resonances. The occurrence of eight peaks in the spectrum of a protein with seven metal-binding sites was tentatively attributed to residual heterogeneity in the metallothionein sample employed in this study. By homonuclear decoupling studies the eight multiplets could be shown to arise from two separate thiolate clusters designated A and B and containing four and three metal ions, respectively. Cluster A gives rise to five signals. Two of them, 7 and 7′, are half as intensive as the remaining signals (1 and 1′, 5 and 5′, and 6 and 6′) (table 4.6). This was explained by the existence of a variant form of cluster A, designated as A′, which is thought to exist in equal abundance as cluster A. The remaining signals (2, 3 and 4) are allocated to cluster B. From these and other data Otvos and Armitage proposed a spatial model for the clusters in metallothionein. It depicts each metal ion as tetrahedrally linked to four thiolate ligands, and locates 11 cysteine residues in cluster A/A′ and the remaining nine in cluster B. The large variation in ^{113}Cd chemical shift ranging from 604 to 670 ppm is attributed in part to the unequal coordination environment created by the different numbers of bridging ligands in the various sites.

Recently, these studies were also extended to the metallothioneins from human and calf liver and from the hepatopancreas of the crab *Scylla serrata*. The ^{113}Cd NMR spectra of the two ^{113}Cd(II)-reconstituted human liver isometallothioneins are similar to those of the two rabbit proteins, and by homonuclear decoupling measurements the same two-cluster arrangement was confirmed (Armitage *et al.*, 1982). These studies also documented nicely the sensitivity

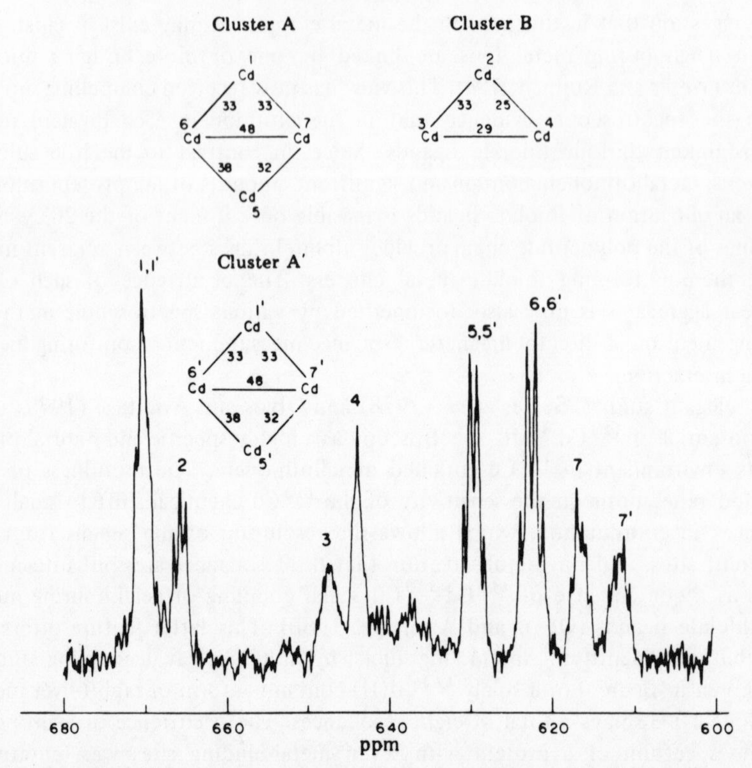

Figure 4.6 A 'fully relaxed' proton decoupled ^{113}Cd NMR spectrum of rabbit liver Cd(II)-metallothionein-1 (~ 8 mM) and schematic representations of the metal cluster structures in the protein, established by homonuclear decoupling experiments (see also table 4.6). The spin-coupling connections between adjacent metal ions in the clusters are indicated by the lines connecting the Cd in the schematic structures. The number beside each Cd refers to the corresponding resonance in the ^{113}Cd spectrum and the numbers appearing on the lines connecting the metals are measured two-bond coupling constants (± 3 Hz). The cysteine thiolate ligands which bridge the adjacent metals have been omitted from the drawings for clarity (from Otvos and Armitage, 1980).

of the chemical shift of the ^{113}Cd NMR resonance to the difference in primary structure of the two rabbit isoproteins and to the heterogeneity of human metallothionein-1 (Boulanger and Armitage, 1982). The ^{113}Cd NMR spectra of the two isometallothioneins from calf liver, which were prepared to contain 3.9 mol ^{113}Cd and 2.6 mol Cu, displayed only the four major multiplets previously assigned to the four-metal cluster A (Briggs and Armitage, 1982). Hence, it was concluded that copper is bound selectively to the three-metal cluster B site. Very well resolved ^{113}Cd NMR spectra with five and six multiplets in the chemical shift range of 620 to 660 ppm were also observed with crab metallothionein-1 and 2, respectively. Homonuclear decoupling established that the total of six metals bound per mole of protein are arranged in two separate three-metal clusters (Otvos *et al*., 1982).

Table 4.6 Chemical shifts and integrated areas of the ^{113}Cd resonances in Cd-metallothionein[a]

Resonance[b]	Chemical shift (ppm)		Relative integrated area[c]
1, 1′	670.4,	670.1	1.03
2	665.1		0.52
3	647.5		0.48
4	643.5		0.65
5, 5′	629.8,	628.8	0.93
6, 6′	622.1,	622.0	1.04
7	615.9		0.48
7′	611.2		0.42
Cluster A[d]	–		3.90
Cluster B[e]	–		1.65
Cluster A plus cluster B	–		5.55

[a] From Otvos and Armitage (1980).
[b] The resonance numbering scheme corresponds to that which appears in figure 4.6.
[c] The areas of the resonances in the rabbit Cd-metallothionein-1 spectrum in figure 4.6 are reported relative to the average area of the overlapping resonances 1 and 1′, 5 and 5′, and 6 and 6′. These three multiplets are each assumed to represent single equivalents of ^{113}Cd.
[d] The resonances assigned to ^{113}Cd in cluster A are 1, 1′, 5, 5′, 6, 6′, 7 and 7′.
[e] The resonances assigned to ^{113}Cd in cluster B are 2, 3 and 4.

Independent evidence for the existence of clusters in metallothionein comes also from the study of the electron spin resonance (ESR) and magnetic properties of preparations of Co(II)-metallothionein made by the addition of varying amounts of $CoCl_2$ to the apoprotein (Vašák and Kägi, 1981). The ESR spectrum of the fully reconstituted $Co(II)_7$-metallothionein resembles that of a rhombically distorted Co(II) complex ($S = 3/2$) with parameters of $g_x \simeq 5.9$, $g_y \simeq 4.2$, and $g_z \simeq 2.0$ typical of tetrahedral coordination (figure 4.7, left). Essentially the same ESR profile is seen in Co(II)-metallothionein of lower cobalt-to-protein ratios. However, its amplitude varies remarkably in function of stoichiometry (figure 4.7, right). At low metal-to-protein ratios – up to about four equivalents of Co(II) added – the intensity of the signal at $g_x \simeq 5.9$ is proportional to the metal incorporated and is comparable to that obtained at corresponding concentrations of the high-spin $[CoCl_4]^{2-}$ model complex. Beyond four equivalents of Co(II) per protein, further additions of metal progressively reduce the signal size, yielding at saturation a nearly diamagnetic complex with an amplitude of only about 5 per cent of that of the tetrahalide model complex. This loss of paramagnetism on approaching saturation can be accounted for by antiferromagnetic coupling of vicinal paramagnetic metal ions. The efficiency of the spin cancelling implies the existence of clustered structures in which the metal ions are interacting through bonds via a shared ligand. The sharp transition from the

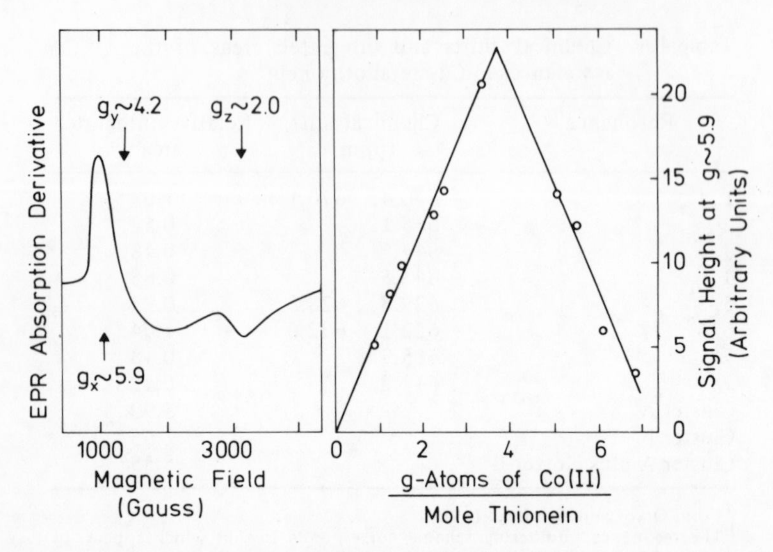

Figure 4.7 Electron spin resonance (ESR) titration of rabbit liver apometallothionein with Co(II). Left: ESR spectrum of $Co(II)_7$-metallothionein at 4 K. Right: dependency of ESR signal height at $g_x \approx 5.9$ on Co(II)-to-protein ratio (from Vasák and Kägi, 1981).

paramagnetic nonclustered to the diamagnetic clustered structure upon filling-up of the metal-binding sites of metallothionein indicates that under the conditions employed the noninteracting complexes are formed preferentially.

The loss of paramagnetism in fully reconstituted Co(II)-metallothionein was also confirmed by magnetic susceptibility measurements (Vašák and Kägi, 1981). On approaching saturation of the metal-binding sites of metallothionein by Co(II) the effective magnetic moment decreases from 5.2 Bohr magnetons initially to an average value of about 2.56 Bohr magnetons per Co(II) bound. This is much lower than the spin-only value of high-spin Co(II) (3.86 Bohr magnetons) but it is comparable to the value of 2.64 Bohr magnetons reported for the crystallographically defined tetranuclear Co(II)-benzene-thiolate (SPh) complex, $[Co_4(SPh)_{10}]^{2-}$ of Dance (1979). In both cases the reduction in paramagnetic moment is thought to be mediated by a superexchange interaction via s- and p-orbitals of the bridging thiolate ligands.

The combined effect of two adjacent metal ions on these bridging thiolate ligands was also documented by X-ray photoelectron spectroscopy (ESCA). This method yields information on the charge distribution in the complex by measuring the core electron binding energy of the bonded atoms in the complex. As pointed out first by Weser and co-workers (1973) and Weser and Rupp (1979*b*), the chemical shift of the sulfur $2p_{1/2,3/2}$ electron binding energy profile of the various metal forms of metallothionein is conditioned by the loss of the proton on metal binding and by the extent of polarisation of the sulphur valence shell electrons on binding of Zn(II), Cd(II) or Hg(II). The most pronounced feature of the ESCA spectra of metallothionein is, however, the loss of the doublet structure and the very substantial broadening of the sulphur core electron binding profile as compared with that of the thiol model compound cysteamine hydrochloride (figure 4.8). This broadening is consistent

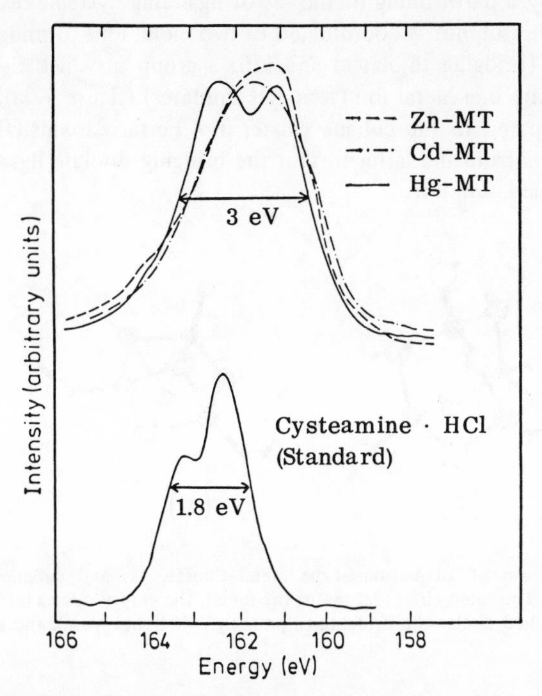

Figure 4.8 Sulphur $2p_{1/2,3/2}$ electron binding energy profiles (ESCA) of different metal forms of rabbit liver metallothionein (top) and of cysteamine-HCl (bottom).

with the view that metallothionein contains two forms of thiolate ligand which are located in a different charge environment and, hence, differ in core electron binding energy. Measurements of model complexes containing bridging and terminal chloride ligands (Hamer and Walton, 1974) allow the inference that, on coordination of a thiolate ligand to a second metal ion, the sulphur $2p_{1/2,3/2}$ core electron binding energy is shifted upwards by 1–2 eV.

The detailed organisation of the cluster structures in metallothionein is still unknown. However, the fixed sulphur-to-metal ratio, the rules of stereochemistry and the well-established spectroscopic features restrict the potential models considerably. One of the few defined compounds whose crystallographic structure is known and whose spectroscopic properties closely resemble those of metallothionein is the complex of Zn(II) or Co(II) with benzenethiol studied by Dance (1979, 1980). In appropriate solvents, these substances form one-dimensional polymers in which groups of four T_d coordinated bivalent metal ions (Me(II)) are joined via six bridging thiolate ligands to form a regular $Me(II)_4S_6$ adamantane cluster. This 10-vertices cage-like polyhedral structure is built up entirely of metal–thiolate units of tetrahedral symmetry. The same arrangement of the metal and sulphur atoms is known to occur in nonmolecular form in zinc blende (ZnS). In the molecular complex under discussion, such a structure is characterised by a partitioning of the set of liganding cysteine residues into a group in which the sulphur is coordinated to two metal ions forming an integral part of the cage (bridging thiolates) and into a group in which the sulphur is coordinated to only one metal ion (terminal thiolates) (figure 4.9a). This structure can be compared to the cubane cluster in 4 Fe ferredoxins (figure 4.9b). However, it differs from the latter in that the bridging thiolate ligands take the place of the inorganic sulphur.

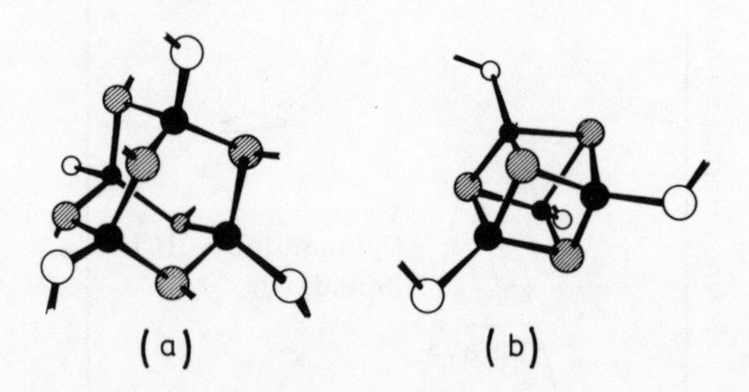

Figure 4.9 Comparison of adamantane-type metal–thiolate (a) and cubane-type metal–sulphur (b) clusters. The filled circles represent the metal, the empty circles terminal sulphur ligands and the hatched circles the bridging sulphur ligands (from Vašák and Kägi, 1983*b*).

The two metal–thiolate clusters proposed by Otvos and Armitage (1980) on the basis of ^{113}Cd NMR homonuclear decoupling studies on metallothionein can be viewed as somewhat modified adamantane structures, one of them containing four bivalent metal ions and 11 cysteine residues, the other three bivalent metal ions and nine cysteine residues. These structures are also compatible with all other spectroscopic data currently available. Recently, this model has received additional strong support from the observation that the two clusters are part of two different domains in metallothionein and that the four-metal cluster can be separated from the remainder of the protein by proteolytic cleavage in the middle of the chain (Boulanger *et al.*, 1982; Winge and Miklossy, 1982). The four-metal cluster designated cluster A/A′ by Otvos and Armitage (1980) is made up of the carboxyl-terminal half of the polypeptide chain (figure 4.10).

4.8 CONFORMATION OF THE POLYPEPTIDE CHAIN

The X-ray structure of metallothionein remains to be determined (Melis *et al.*, 1983). Thus, except for some of its gross physical features derived from hydrodynamic measurement (see section 4.5) and the recent discovery of its domain

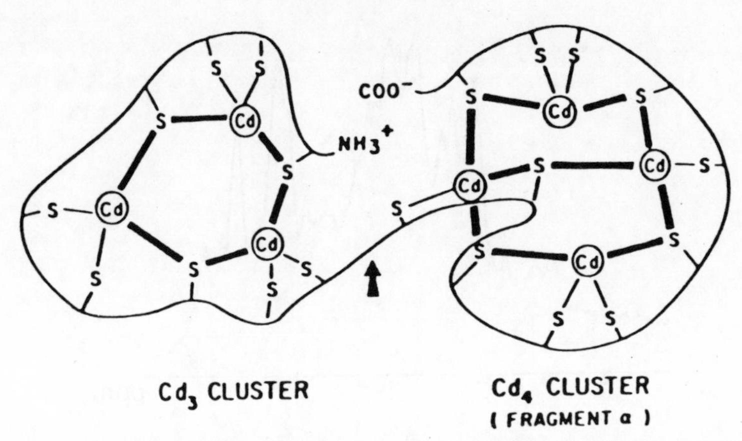

Figure 4.10 Proposed domain structure of metallothionein (from Winge and Miklossy, 1982).

structure (Winge and Miklossy, 1982), little is known as yet about its overall spatial organisation. Nonetheless, based on the observation that native metallothionein is heat-stable and highly resistant to proteolysis (Webb, 1972) and that some of its peptide hydrogens are shielded from exchanging with the solvent (Ulmer and Vallee, 1971), it appears that the protein has an ordered and compact tertiary structure and that metal binding contributes to the stability of the folded conformation. This is also supported by comparative ^1H NMR studies conducted on metallothionein and apometallothionein from different sources (Vašák *et al*., 1980). The differences between the two forms are particularly striking in the low field region (6.5 to 9.5 ppm), where, in the absence of aromatic amino acid side chains and histidine, only resonances of the secondary amide groups of the polypeptide chain and of the nitrogen-bound protons of the side chains occur. In the apoprotein at low pH in ^1H_2O the secondary amide resonances are collected to a relatively narrow band between 8.6 and 8.1 ppm (figure 4.11, bottom). The resonances at 7.51 and 7.18 ppm originate from the labile side-chain protons of lysine and arginine, respectively. Judged from the temperature dependence of the chemical shifts of the maximum of the amide resonances (6 ppb/°C) and from the $^2H/^1H$-exchange rate measured in ^2H_2O, it was concluded that most of the amide protons of the metal-free protein are freely exposed to the solvent. In marked contrast, the metal-containing protein displays in ^1H_2O, pH 7.5, a broad spectrum of distinct secondary amide resonances that extend between 6.8 and 9.6 ppm (figure 4.11, top). The various amide protons contributing to the envelope of resonances also differ in their rate of isotopic exchange. Thus, upon dissolution of lyophilised native metallothionein in ^2H_2O, the intensity of 12 amide resonances decreases with a rate substantially lower ($t_{1/2} \approx$ 20–320 min) than that of the other amide hydrogens ($t_{1/2} < 2.4$ min). The shielding of some of the amide protons of the native protein from the solvent was also documented by the lower temperature

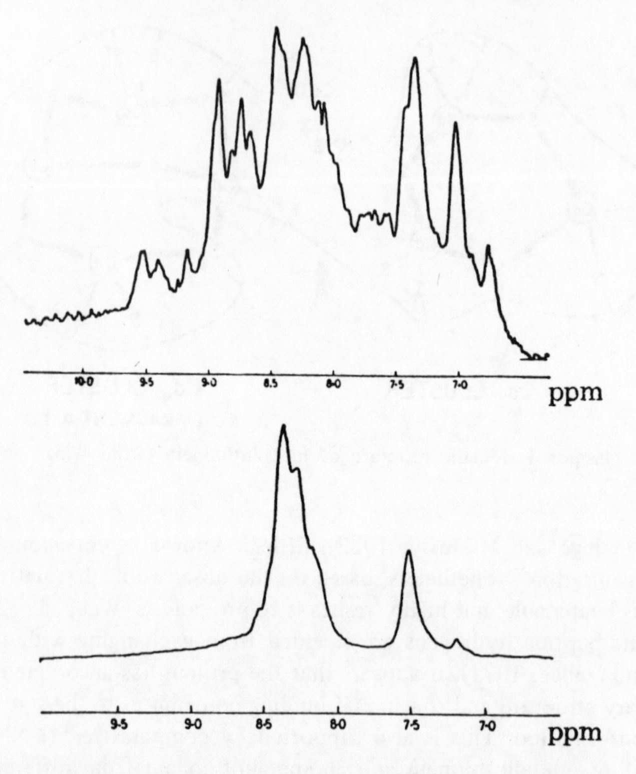

Figure 4.11 Low-field region of the 270 MHz ^{1}H NMR spectra of equine metallothionein-1B in ^1H_2O, pH 7.5 (top), and of apometallothionein-1B (bottom) in ^1H_2O, pH 1. Chemical shift is expressed in parts per million downfield from the internal standard DSS. The solvent ^1H_2O signal was suppressed by a gated pulse. (Reprinted with permission from Vašák *et al.* (1980). Copyright 1982 American Chemical Society.)

dependence of their chemical shifts (0.1–4 ppb/°C). Both the retardation of peptide hydrogen exchange and the distribution of distinct secondary amide resonances over a wide chemical shift span are taken as a strong indication of the existence of a well-defined and discrete tertiary structure.

The structural differences between the metal-free and the metal-containing protein are also manifested in the high-field region (Rupp *et al.*, 1974; Vašák *et al.*, 1980). However, because of the absence of aromatic residues in the molecule and, hence, of ring current effects, shifts arising from changes in the proton environments are generally much smaller and have not as yet been amenable to detailed structural interpretations. It seems likely, though, that these limitations will be overcome by two-dimensional NMR techniques applied successfully to a number of small proteins (Nagayama, 1981; Wider *et al.*, 1984). Studies on metallothionein in progress have already allowed the identification of many individual spin systems and have permitted some sequential assignments (Neuhaus *et al.*, 1983).

Some information on the secondary folding of the polypeptide chain in metallothionein has also been obtained from the measurement of conformation-dependent circular dichroism and of infrared absorption (IR) properties. Typical far-UV circular dichroism spectra of the metal-free and the zinc-containing form of human isometallothionein-2 are shown in figure 4.12. Both spectra display

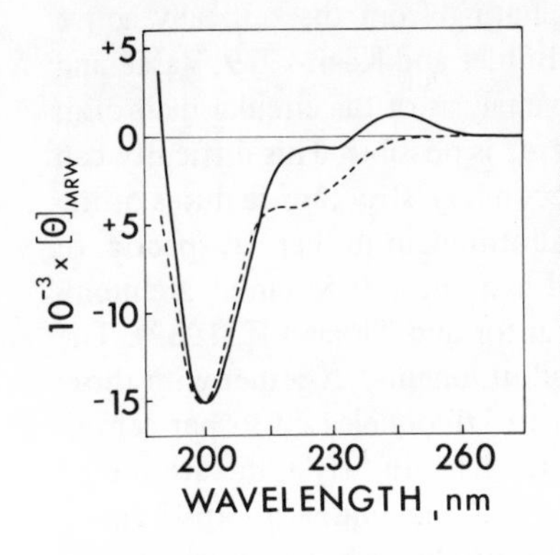

Figure 4.12 Circular dichroism spectra of human liver Zn(II)-metallothionein (MT-2) in 0.05 M sodium phosphate buffer, pH 7.0 (solid line) of human apo-metallothionein in 0.02 M perchloric acid, pH 1.7 (broken line); ordinate; mean residue ellipticity ($[\theta]_{MRW}$) refers to a mean residue weight of 99 (adapted from Bühler and Kägr, 1979).

their strongest ellipticity band at 200 nm, suggesting a large proportion of unordered structure (Bühler and Kägi, 1979). The preponderance of this conformation is typical of all metallothioneins examined thus far (see Nordberg and Kojima, 1979). This was also verified by a computer fitting analysis of the circular dichroism spectrum of the metal-free form of equine liver metallothionein-1A using parameters derived from crystallographically defined proteins (Provencher and Glöckner, 1981). The remaining secondary structure features are attributed to β-sheet and β-turn conformations and to a small amount of α-helix (D. Gilg and J. H. R. Kägi, in preparation). The quantitative estimates are given in table 4.7. It is interesting that these figures are in fairly good agreement with the values calculated from the amino acid sequence using the secondary structure prediction method of Chou and Fasman (1971).

Table 4.7 Secondary structure analysis of equine liver metallothionein-1A

Type	Secondary structure estimated from circular dichroism spectrum[a] (%)	Secondary structure prediction from sequence[b] (%)
α-Helix	6	10
β-Sheet	18	16
β-Turn	21	26
Disordered structure	55	48

[a]Evaluated according to Provencher and Glöckner (1981).
[b]Calculated according to Chou and Fasman (1971).

Considering the pronounced effect of metal binding on the ^{1}H NMR features of metallothionein, it is probable that some of the differences between the circular dichroism spectra of holo- and apometallothionein (figure 4.12) arise from changes in secondary structure induced by metal complexation. However, because in the metal-containing form the conformation-dependent polypeptide ellipticity bands are superimposed by strong bands originating from the optically active metal–thiolate transitions (Weser *et al.*, 1973; Bühler and Kägi, 1979; Vašák and Kägi, 1983*b*), no quantitative and comparative analysis of the circular dichroism spectrum in terms of secondary structure elements is possible. This difficulty can be circumvented by comparing, instead, the secondary structure features of the metal-free and metal-containing forms of metallothionein in their IR spectra. In this method, position and shape of the amide I band near 1650 cm^{-1} are monitored in a solution of the protein in $^{2}H_2O$ (Cantor and Timasheff, 1982). The IR spectra of the two forms of rabbit liver metallothionein-2, together with those of myoglobin (76 per cent α-helix) and pancreatic ribonuclease (36 per cent β-structure) as reference proteins, are shown in figure 4.13. In $^{2}H_2O$, the disordered polypeptide conformation in proteins displays a maximum at 1645 cm^{-1} and a weak shoulder at 1670 cm^{-1}. β-Structure produces both a maximum at 1634 cm^{-1} and a strong and a weak shoulder at 1660 cm^{-1} and 1680 cm^{-1}, respectively. α-Helix conformation has a single maximum at 1650 cm^{-1} (Eckert *et al.*, 1977). The position of the maximum of the metal-free form of metallothionein at 1647 cm^{-1} is consistent with the preponderance of unordered polypeptide structure mentioned previously. The displacement of the amide I band to 1643 cm^{-1}, the intensification of the shoulder at 1660 cm^{-1}, and the increase in total amplitude are suggestive of a substantial increase in β-structure content on metal binding (D. Gilg and J. H. R. Kägi, unpublished observation).

4.9 CONCLUSION

The large body of chemical and spectroscopic data summarised in this chapter reflects the considerable progress made in recent years in the understanding of the structure of this highly unusual protein. Among the major findings are (1) the remarkable conservation of its primary structure manifesting itself both in the nearly complete invariance of the positions of the metal-binding cysteine residues and in the rather low rate of molecular evolution ranging between that of cytochrome *c* and of the globins, and (2) the detailed knowledge gained from physical studies on the structural organization of the metal-binding sites. Without the benefit of crystallographic data, it has been established unambiguously that bivalent metal ions are bound in tetrahedral microsymmetry to four thiolate ligands and that these complexes are joined to form discrete metal–thiolate clusters. In mammalian metallothioneins there are two separate metal–thiolate clusters that are located in different segments of the molecule, one containing four metals, the other three. In the crustacean metallothioneins there are two three-metal clusters. These oligonuclear metal–thiolate aggregates consti-

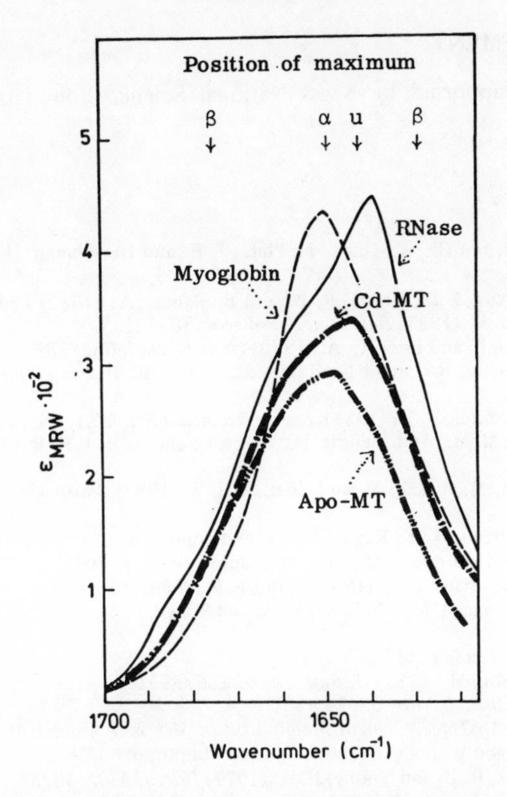

Figure 4.13 Amide I region in IR spectra of rabbit liver apometallothionein-2 (apo-MT) at pH 1.6, metallothionein (Cd-MT), and of myoglobin and ribonuclease (RNase) at pH 7.5. All spectra are recorded in ^2H_2O. The vertical arrows indicate the positions of the maxima of protein amide I transitions typical for pleated sheet (β), α-helix (α) and unordered (u) conformations in ^2H_2O (adapted from Vašák and Kägi, 1983*b*).

tute to our knowledge the first example of clusters of group-2B ions in a biological system. They differ clearly both in composition and structure from those found in the iron–sulphur proteins, and imply by their unique occurrence in metallothionein that their spontaneous formation and their stability are dependent on the very special distribution of the cysteine residues that is typical of these proteins. The stringency of the stereochemical requirements of bioinorganic structures may explain, in fact, the preservation of the positions of most of these residues in evolution. The persistence of the two-cluster arrangement both in vertebrates and invertebrates suggests that they may serve an as yet unknown biological function. Clearly, further studies on the evolution of these proteins and on the chemical and biochemical properties of these novel clusters are indispensable to a full understanding of the role played by this protein in cellular processes.

ACKNOWLEDGEMENT

This work was supported by Swiss National Science Foundation Grant No. 3.207-0.82.

REFERENCES

Andersen, R. D., Birren, B. W., Ganz, T., Pletz, J. E. and Herschman, H. R. (1983) *DNA* **2**, 15.

Armitage, I. M., Otvos, J. D., Briggs, R. W. and Boulanger, Y. (1982) *Fed. Proc.* **41**, 2974.

Bakka, A. and Webb, M. (1981) *Biochem. Pharmacol.* **30**, 721.

Bartolf, M., Brennan, E. and Price, C. A. (1980) *Plant Physiol.* **66**, 438.

Bazzell, K. L., Coleman, R. L. and Nordquist, R. E. (1979) *Toxicol. Appl. Pharmacol.* **50**, 199.

Beach, L. R. and Palmiter, R. D. (1981). *Proc. Natl Acad. Sci. USA,* **78**, 2110.

Berger, C., Kissling, M. M., Andersen, R. D., Weser, U. and Kägi, J. H. R. (1981) *Experientia* **37**, 619.

Bernhard, W., Good, M., Vašák, M. and Kägi, J. H. R. (1983). *Inorg. Chim. Acta, Bioinorg. Chem.* **79**, 154.

Bethune, J. L., Budreau, A. J., Kägi, J. H. R. and Vallee, B. L. (1979) In *Metallothionein* (Kägi, J. H. R. and Nordberg, M., eds). Birkhäuser, Basel, p. 207

Boulanger, Y. and Armitage, I. M. (1982) *J. Inorg. Biochem.* **17**, 147.

Boulanger, Y., Armitage, I. M., Miklossy, K.-A. and Winge, D. R. (1982) *J. Biol. Chem.* **257**, 13717.

Brady, F. O. (1982) *TIBS,* **7**, 143.

Brady, F. O. and Webb, M. (1981). *J. Biol. Chem.* **256**, 3931.

Bremner, I. (1976). *Brit. J. Nutr.* **35**, 245.

Bremner, I. (1982) In *Trace Element Metabolism in Man and Animals* (Gawthorne, J. M., Howell, J. McC. and White, C. L., eds). Springer, Berlin, p. 637.

Bremner, I., Williams, R. B. and Young, B. W. (1977) *Brit. J. Nutr.* **38**, 87.

Briggs, R. W. and Armitage, I. M. (1982) *J. Biol. Chem.* **257**, 1259.

Brown, D. A. (1977) *Mar. Biol.* **44**, 203.

Brown, D. A., Bawden, C. A., Chatel, K. W. and Parsons, T. R. (1977) *Environ. Conserv.* **4**, 213.

Bühler, R. H. O. and Kägi, J. H. R. (1974) *FEBS Lett.* **39**, 229.

Bühler, R. H. O. and Kägi, J. H. R. (1979) In *Metallothionein* (Kägi, J. H. R. and Nordberg, M., eds). Birkhäuser, Basel, p. 211.

Cantor, C. R. and Timasheff, S. N. (1982) in *The Proteins* (Neurath, H. and Hill, R. L., eds). 3rd Edn, Vol. V. Academic Press, New York, p. 145.

Chan, W.-Y., Garnica, A. D. and Rennert, O. M. (1979) *Pediat. Res.* **13**, 197.

Cherian, M. G. (1980) *Toxicology* **17**, 225.

Cherian, M. G. and Goyer, R. A. (1978) *Life Sci.* **23**, 1.

Chou, P. Y. and Fasman, G. D. (1971) *Biochemistry* **13**, 211.

Dance, I. G. (1979) *J. Am. Chem. Soc.* **101**, 6264.

Dance, I. G. (1980) *J. Am. Chem. Soc.* **102**, 3445.

Daniel, M. R., Webb, M. and Cempel, M. (1977) *Chem.-Biol. Interact.* **16**, 101.

Day, P. and Seal, R. H. (1972) *J. Chem. Soc., Dalton Trans.* 2054.

Durnam, D. M. and Palmiter, R. D. (1981) *J. Biol. Chem.* **256**, 5712.

Durnam, D. M., Perrin, F., Gannon, F. and Palmiter, R. D. (1980) *Proc. Natl Acad. Sci. USA* **77**, 6511.

Eckert, K., Grosse, R., Malur, J. and Repke, K. R. H. (1977) *Biopolymers* **16**, 2549.

Eklund, H., Nordström, B., Zeppezauer, E., Söderlund, G., Ohlsson, I., Boiwe, T., Söderberg, B.-O., Tapia, O., Brändén, C.-I. and Åkeson, Å. (1976). *J. Mol. Biol.* **102**, 27.

Failla, M. L. and Cousins, R. J. (1978) *Biochim. Biophys. Acta* **543**, 293.

Garner, C. D., Hasnain, S. S., Bremner, I. and Bordas, J. (1982) *J. Inorg. Biochem.* **16**, 253.

Giles, P. J. and Cousins, R. J. (1982) *Cancer Res.* **42**, 2.

Glanville, N., Durnam, D. M. and Palmiter, R. D. (1981). *Nature, Lond.* **292**, 267.

Griffith, B. B., Walters, R. A., Enger, M. D., Hildebrand, C. E. and Griffith, J. K. (1983) *Nucleic Acids Res.* **11**, 901.
Gurd, F. R. N. and Wilcox, P. E. (1956) in *Advances in Protein Chemistry* (Anson, M. L., Bailey, K. and Edsall, J. T., eds), Vol. XI. Academic Press, New York, p. 311.
Hamer, A. D. and Walton, R. A. (1974) *Inorg. Chem.* **13**, 1446.
Hart, B. A. and Keating, R. F. (1980) *Chem.-Biol. Interact.* **29**, 67.
Hildebrand, C. E., Tobey, R. A., Campbell, E. W. and Enger, M. D. (1979) *Exp. Cell Res.* **124**, 237.
Howard, A. G. and Nickless, G. (1975) *J. Chromat.* **104**, 457.
Huang, I-Y., Yoshida, A., Tsunoo, H. and Nakajima, H. (1977) *J. Biol. Chem.* **252**, 8217.
Huang, I-Y., Kimura, M., Hata, A., Tsunoo, H. and Yoshida, A. (1981) *J. Biochem*, **89**, 1839.
Hunziker, P. E. and Kägi, J. H. R. (1983) Abstract, *15th Meeting of the Federation of European Biochemical Societies*, Brussels, p. 217.
Jørgensen, C. K. (1970) *Prog. Inorg. Chem.* **12**, 101.
Kägi, J. H. R. and Nordberg, M. (1979) *Metallothionein.* Birkhäuser, Basel.
Kägi, J. H. R. and Vallee, B. L. (1960) *J. Biol. Chem.* **235**, 3460.
Kägi, J. H. R. and Vallee, B. L. (1961) *J. Biol. Chem.* **236**, 2435.
Kägi, J. H. R., Himmelhoch, S. R., Whanger, P. D., Bethune, J. L. and Vallee, B. L. (1974) *J. Biol. Chem.* **249**, 3537.
Kägi, J. H. R., Kojima, Y., Berger, C., Kissling, M. M., Lerch, K. and Vašák, M. (1979) in *Metalloproteins* (Weser, U., ed.). Georg Thieme, Stuttgart, p. 194.
Karin, M. and Herschman, H. R. (1979) *Science,* **204**, 176.
Karin, M. and Richards, R. I. (1982*a*) *Nucleic Acids Res.* **10**, 3165.
Karin, M. and Richards, R. I. (1982*b*) *Nature, Lond.* **299**, 797.
Karin, M., Herschman, H. R. and Weinstein, D. (1980) *Biochem. Biophys. Res. Commun.* **92**, 1052.
Kimura, M., Otaki, N. and Imano, M. (1979*a*) in *Metallothionein* (Kägi, J. H. R. and Nordberg, M., eds). Birkhäuser, Basel, p. 163.
Kimura, M., Otaki, N. and Kakefuda, T. (1979*b*) in *Metallothionein* (Kägi, J. H. R. and Nordberg, M., eds). Birkhäuser, Basel, p. 187.
Kimura, M., Otaki, N., Hartmann, H.-J. and Weser, U. (1981) *Regard sur lar Biochimie,* **3**, 101.
King, J. L. and Jukes, T. H. (1969) *Science,* **164**, 788.
Kissling, M. M. (1979) Ph.D. Thesis, University of Zürich.
Kissling, M. M. and Kägi, J. H. R. (1977) *FEBS Lett.* **82**, 247.
Kissling, M. M. and Kägi, J. H. R. (1979) in *Metallothionein* (Kägi, J. H. R. and Nordberg, M., eds). Birkhäuser, Basel, p. 145.
Kissling, M. M., Berger, C., Kägi, J. H. R., Andersen, R. D. and Weser, U. (1979) in *Metallothionein* (Kägi, J. H. R. and Nordberg, M., eds). Birkhäuser, Basel, p. 181.
Klauser, S., Kägi, J. H. R. and Wilson, K. J. (1983) *Biochem. J.* **209**, 71.
Kobayashi, S. and Kimura, M. (1980) *Toxicol. Lett.* **5**, 357.
Koch, J., Wielgus, S., Shankara, B., Saryan, L. A., Shaw, F. and Petering, D. H. (1980) *Biochem. J.* **189**, 95.
Kojima, Y. and Kägi, J. H. R. (1978) *TIBS*, **3**, 90.
Kojima, Y., Berger, C., Vallee, B. L. and Kägi, J. H. R. (1976) *Proc. Natl Acad. Sci. USA* **73**, 3413.
Kojima, Y., Berger, C. and Kägi, J. H. R. (1979) in *Metallothionein* (Kägi, J. H. R. and Nordberg, M., eds). Birkhäuser, Basel, p. 153.
Lee, S. S., Mate, B. R., von der Trenck, K. T., Rimerman, R. A. and Bühler, D. R. (1977) *Comp. Biochem. Physiol.* **57C**, 45.
Lerch, K. (1979) in *Metallothionein* (Kägi, J. H. R. and Nordberg, M., eds). Birkhäuser, Basel, p. 173.
Lerch, K. (1981) in *Metal Ions in Biological Systems* (Sigel, H., ed.), Vol. 13. Marcel Dekker, New York, p. 299.
Lerch, K., Ammer, D. and Olafson, R. W. (1982). *J. Biol. Chem.* **257**, 2420.
Lucis, O. J., Shaikh, Z. A. and Embil, J. A., Jr (1970) *Experientia* **26**, 1109.
Maclean, F. I., Lucis, O. I., Shaikh, Z. A. and Jansz, E. R. (1972) *Fed. Proc.* **31**, 699.
Maret, W., Anderson, I., Dietrich, H., Schneider-Bernlöhr, H., Einarsson, R. and Zeppezauer, M. (1979) *Eur. J. Biochem.* **98**, 501.

Margoliash, E. and Fitch, W. M. (1968) *Ann. N.Y. Acad. Sci.* **151**, 359.
Margoshes, M. and Vallee, B. L. (1957) *J. Am. Chem. Soc.* **79**, 4813.
Mason, R., Bakka, A., Samarawickrama, G. P. and Webb, M. (1980) *Brit. J. Nutr.* **45**, 375.
May, S. W. and Kuo, J.-Y. (1978) *Biochemistry* **17**, 3333.
Mayo, K. E. and Palmiter, R. D. (1981) *J. Biol. Chem.* **256**, 2621.
Mbikay, M., Maiti, I. B. and Thirion, J.-P. (1981) *Biochem. Biophys. Res. Commun.* **103**, 825.
Melis, K. A., Carter, D. C., Stout, C. D. and Winge, D. R. (1983) *J. Biol. Chem.* **258**, 6255.
Nagayama, K. (1981) *Adv. Biophys.* **14**, 139.
Nakamura, Y., Katayama, S., Okada, Y., Suzuki, F. and Nagata, Y. (1981) *Agric. Biol. Chem.* **45**, 1167.
Nelbach, M. E., Pigiet, V. P., Gerhart, J. C. and Schachman, H. K. (1972) *Biochemistry* **11**, 315.
Neuhaus, D., Wagner, G., Vašák, M., Kägi, J. and Wüthrich, K. (1983) Abstract, *VIth International Meeting of the Royal Society of Chemistry*, Edinburgh.
Noël-Lambot, F., Gerday, Ch. and Distèche, A. (1978) *Comp. Biochem. Physiol.* **61C**, 177.
Nomiyama, K. and Nomiyama, H. (1982) *J. Chromatogr.* **228**, 285.
Nordberg, G. F., Nordberg, M., Piscator, M. and Vesterberg, O. (1972) *Biochem. J.* **126**, 491.
Nordberg, M. and Kojima, Y. (1979) in *Metallothionein* (Kägi, J. H. R. and Nordberg, M., eds). Birkhäuser, Basel, p. 41.
Oh, S. H., Nakaue, H., Deagen, J. T., Whanger, P. D. and Arscott, G. H. (1979) *J. Nutr.* **109**, 1720.
Ohtake, H. and Koga, M. (1979) *Biochem. J.* **183**, 683.
Olafson, R. W., Abel, K. and Sim, R. G. (1979*a*) *Biochem. Biophys. Res. Commun.* **89**, 36.
Olafson, R. W., Sim, R. G. and Boto, K. G. (1979*b*) *Comp. Biochem. Physiol.* **62B**, 407.
Otvos, J. D. and Armitage, I. M. (1979) in *Metallothionein* (Kägi, J. H. R. and Nordberg, M., eds). Birkhäuser, Basel, p. 249.
Otvos, J. D. and Armitage, I. M. (1980) *Proc. Natl Acad. Sci. USA* **77**, 7094.
Otvos, J. D., Olafson, R. W. and Armitage, I. M. (1982) *J. Biol. Chem.* **257**, 2427.
Overnell, J. and Coombs, T. L. (1979) *Biochem. J.* **183**, 277.
Overnell, J. and Trewhella, E. (1979) *Comp. Biochem. Physiol.* **64C**, 69.
Overnell, J., Berger, C. and Wilson, K. J. (1981) *Biochem. Soc. Trans.* **9**, 217.
Palmer, G. (1975) in *The Enzymes* (Boyer, P. D., ed.), Vol. 12. Academic Press, New York, p. 1.
Phillips, J. L. (1979) *Biol. Trace Elem. Res.* **1**, 359.
Piotrowski, J. K., Bolanowska, W. and Sapota, A. (1973) *Acta Biochim. Pol.* **20**, 207.
Piscator, M. (1964) *Nord. Hyg. Tidskr.* **45**, 76.
Prinz, R. and Weser, U. (1975) *Hoppe-Seyler's Z. Physiol. Chem.* **356**, 767.
Provencher, S. W. and Glöckner, J. (1981) *Biochemistry* **20**, 33.
Pulido, R., Kägi, J. H. R. and Vallee, B. L. (1966) *Biochemistry* **5**, 1768.
Rauser, W. E. and Curvetto, N. R. (1980) *Nature, Lond.* **287**, 563.
Richards, M. P. and Cousins, R. J. (1976) *Proc. Soc. Exp. Biol. Med.* **153**, 52.
Riordan, J. R. and Richards, V. (1980) *J. Biol. Chem.* **255**, 5380.
Roesijadi, G. (1981). *Marine Environ. Res.* **4**, 167.
Ross, I., Binstead, N., Blackburn, N. J., Bremner, I., Diakun, G. P., Hasnain, S. S., Knowles, P. F., Vašák, M. and Garner, C. D. (1983) in *EXAFS and Near Edge Structure* (Bianconi, A., Incoccia, L. and Stipcich, S., eds). Springer-Verlag, Berlin, p. 337.
Rudd, C. J. and Herschman, H. R. (1978) *Toxicol. Appl. Pharmacol.* **44**, 511.
Rudd, C. J. and Herschman, H. R. (1979) *Toxicol. Appl. Pharmacol.* **47**, 273.
Rugstad, H. E. and Norseth, T. (1975) *Nature, Lond.* **257**, 136.
Rugstad, H. E. and Norseth, T. (1978) *Biochem. Pharmacol.* **27**, 647.
Rupp, H., Voelter, W. and Weser, U. (1974) *FEBS Lett.* **40**, 176.
Sadler, P. J., Bakka, A. and Beynon, P. J. (1978) *FEBS Lett.* **94**, 315.
Schechter, I., Zemell, R. and Burstein, Y. (1978) Abstracts, *11th Meeting of the Federation of European Biochemical Societies*, Copenhagen, p. 103.
Suzuki, K. T. (1979) *Arch. Environ. Contam. Toxicol.* **8**, 255.
Suzuki, K. T. (1980) *Anal. Biochem.* **102**, 31.
Suzuki, K. T., Yamamura, M. and Mori, T. (1980) *Arch. Environ. Contam. Toxicol.* **9**, 415.

Suzuki, Y. and Yoshikawa, H. (1976) *Ind. Health* **14**, 25.
Takeda, H. and Shimizu, C. (1982) *Nippon Suisan Gakkaishi* **48**, 717.
Thompson, K. A., Brown, D. A., Chapman, P. M. and Brinkhurst, R. O. (1982) *Trans. Am. Microsc. Soc.* **101**, 10.
Ulmer, D. D. and Vallee, B. L. (1971) *Adv. Chem. Ser.* **100**, 187.
Vašák, M. and Bauer, R. (1982) *J. Am. Chem. Soc.* **104**, 3236.
Vašák, M. and Kägi, J. H. R. (1981) *Proc. Natl Acad. Sci. USA* **78**, 6709.
Vašák, M. and Kägi, J. H. R. (1983*a*) in *Biomineralization and Biological Metal Accumulation* (Westbroek, P. and de Jong, E. W., eds). D. Reidel, Dordrecht, p. 429.
Vašák, M. and Kägi, J. H. R. (1983*b*) in *Metal Ions in Biological Systems* (Sigel, H., ed.), Vol. 15, Marcel Dekker, New York, p. 213.
Vašák, M., Galdes, A., Hill, H. A. O., Kägi, J. H. R., Bremner, I. and Young, B. W. (1980) *Biochemistry* **19**, 416.
Vašák, M., Kägi, J. H. R. and Hill, H. A. O. (1981*a*) *Biochemistry* **20**, 2852.
Vašák, M., Kägi, J. H. R., Holmquist, B. and Vallee, B. L. (1981*b*). *Biochemistry* **20**, 6659.
Vašák, M., Berger, C.and Kägi, J. H. R. (1984) *FEBS Lett.* **168**, 174.
Webb, M. (1972) *Biochem. Pharmacol.* **21**, 2751.
Webb, M. (1979) in *The Chemistry, Biochemistry and Biology of Cadmium* (Webb, M., ed.), Vol. 2. Elsevier/North-Holland, Amsterdam, p. 195.
Webb, M. and Cain, K. (1982) *Biochem. Pharmacol.* **31**, 137.
Webb, M. and Daniel, M. (1975) *Chem.-Biol. Interact.* **10**, 269.
Weser, U. and Rupp, H. (1979*a*) in *Metallothionein* (Kägi, J. H. R. and Nordberg, M., eds). Birkhäuser, Basel, p. 221.
Weser, U. and Rupp, H. (1979*b*) in *The Chemistry, Biochemistry and Biology of Cadmium* (Webb, M., ed.), Vol. 2. Elsevier/North-Holland, Amsterdam, p. 267.
Weser, U., Rupp, H., Donay, F., Linnemann, F., Voelter, W., Voetsch, W. and Jung, G. (1973) *Eur. J. Biochem.* **39**, 127.
Whanger, P. D. and Ridlington, J. W. (1982) in *Biological Roles of Metallothionein* (Foulkes, E. C., ed.), Vol. 9, Elsevier/North-Holland, New York, p. 263.
Wider, G., Macura, S., Kumar, A., Ernst, R. R. and Wüthrich, K. (1984) *J. Magn. Reson.* **56**, 207.
Willner, H., Vašák, M. and Kägi, J. H. R. (1983) *Inorg. Chim. Acta, Bioinorg. Chem.* **79**, 106.
Winge, D. H. and Miklossy, K.-A. (1982) *J. Biol. Chem.* **257**, 3471.
Żelazowski, A. J. and Piotrowski, J. K. (1977) *Experientia* **33**, 1624.
Żelazowski, A. J. and Szymańska, J. A. (1980) *Biol. Trace Elem. Res.* **2**, 137.

5
Transferrins

Jeremy H. Brock

5.1 INTRODUCTION

Transferrin is a serum β-globulin with the property of reversibly binding iron. It was originally isolated by Schade and Caroline (1946), although its existence had been inferred from earlier studies which had demonstrated that a small amount of the iron in blood was associated with the plasma protein fraction (Fontes and Thivolle, 1925). There are two other proteins which resemble transferrin in many of their properties: ovotransferrin, present in avian egg white and first isolated by Alderton *et al*. (1946), and lactoferrin, originally isolated from milk by Johansson *et al*. (1958). This review will deal with the properties and function of all three proteins, and in addition will mention briefly a few other iron-binding proteins which share some affinity with transferrin. Emphasis will be on molecular aspects of structure and function, and the chapter will deal only briefly with synthesis, catabolism, genetic polymorphism and clinical aspects of the transferrins, which are described in more detail in the comprehensive review by Morgan (1981*a*). Other recent reviews dealing with the structure and function of transferrin are those by Aisen (1980), Aisen and Listowsky (1980) and Chasteen (1983*a*), and the properties of the transferrins in the wider context of iron metabolism are dealt with by Bezkorovainy (1980).

5.2 NOMENCLATURE

The name transferrin is now universally used for the iron-binding protein of plasma, in preference to the earlier name of siderophilin (Schade *et al*., 1949). The name lactoferrin is now almost always used for the iron-binding protein of milk and other secretions, and neutrophil granules: the alternative name lactotransferrin is more cumbersome and implies an iron transport function which is at best only doubtfully established. For the protein of avian egg white the name ovotransferrin is preferable to the alternative name of conalbumin, which erroneously suggests affinity with albumin. The names transferrin, lactoferrin and ovotransferrin will therefore be used in this review, and their universal adoption is urged.

Frieden and Aisen (1980) have suggested that transferrin molecules containing different numbers of bound iron atoms be referred to as follows: Fe_2Tf for the diferric (saturated) protein, Fe_NTf for the monoferric species containing iron bound to the N-terminal binding site, $TfFe_C$ for the monoferric transferrin with iron bound to the C-terminal site, and Tf for the iron-free (apo) protein. This last abbreviation is, however, frequently used when referring to transferrin without specifying its degree of saturation, and in this review the abbreviation apoTf will be used for iron-free transferrin. Otherwise, the nomenclature proposed by Frieden and Aisen (1980) will be adopted, and is to be recommended.

5.3 OCCURRENCE, SYNTHESIS AND CATABOLISM

Transferrin is essentially a protein of serum and extravascular fluid. Its existence in a wide variety of vertebrates has been established, and transferrin or transferrin-like proteins have been found in moths (Palmour and Sutton, 1971), crabs (Martin *et al*., 1982) and spiders (Lee, M. Y. *et al*., 1978). In man, serum normally contains 2–4 mg ml^{-1} of transferrin, at an iron saturation of about 30 per cent. Both these figures vary considerably in pathological states (Morgan, 1981*a*). Lactoferrin, in contrast, is normally found in serum only at microgram levels (Bennett and Mohla, 1976), but is more plentiful in a wide range of external secretions such as milk, seminal fluid, tears, sweat, and nasal and genital secretions (Masson *et al*., 1966; Masson, 1970). It is also found in bile (Van Vugt *et al*., 1975), probably originating in the pancreas (Figarella and Sarles, 1975), and in the secondary granules of neutrophils (Baggiolini *et al*., 1970). In most cases it appears to be largely devoid of iron, and it has recently been shown that most of the iron in human milk is not, as was originally assumed, bound to lactoferrin, but present in the lipid and low molecular weight fractions (Fransson and Lönnerdal, 1980). In some species, such as the rabbit, lactoferrin is virtually absent from milk and is replaced by a variant of serum transferrin with modified sialic acid content (Baker *et al*., 1968).

Ovotransferrin has been identified in egg white from a variety of birds, and constitutes between 2 and 16 per cent of the dry weight (Feeney *et al*., 1960). Only hen ovotransferrin has been studied to any extent, and it differs from hen serum transferrin only in its carbohydrate moiety (Williams, 1962; Lee, D. C. *et al*., 1978).

Transferrin is synthesised mainly in the liver (Morgan, 1981*a*). Lactoferrin is synthesised in many of the secretory organs in which it is found (Masson, 1970), and local production may be increased as a result of inflammation of the mammary gland (Harmon *et al.*, 1975), pancreas (Multigner *et al.*, 1981) and parotid salivary gland (Tabak *et al.*, 1978). Ovotransferrin is synthesised in the oviduct (Palmiter, 1972). Newly synthesised transferrins contain an additional leader sequence of 19 amino acids which is lost prior to secretion of the protein (Schreiber *et al.*, 1979; Jeltsch and Chambon, 1982). Precursors of rat and human transferrins lacking sialic acid can also be detected in the liver (Schreiber *et al.*, 1979, 1981).

The liver is the major site of catabolism of transferrin and circulating lactoferrin, though the latter is removed more rapidly than the former (section 5.4.5). Synthesis and catabolism of the transferrins are reviewed in detail by Bezkorovainy (1980) and Morgan (1981*a*).

5.4 STRUCTURE OF THE TRANSFERRINS

5.4.1 Molecular weight

Measurements of molecular weights by gel filtration, polyacrylamide gel electrophoresis in sodium dodecyl sulphate, and ultracentrifugation have given values in the range 74 000-80 000 for transferrins (summarised by Bezkorovainy (1980)) and in the range 80 000-92 000 for the lactoferrins. The recent elucidation of the complete amino acid sequences of human transferrin (MacGillivray *et al.*, 1982, 1983) and ovotransferrin (Jeltsch and Chambon, 1982; Williams *et al.*, 1982*a*) give unequivocal values of 79 570 and 77 700 respectively, including the carbohydrate moieties. Some species variation may occur, for example bovine transferrin migrated slightly ahead of human transferrin when a mixture of both proteins was subjected to polyacrylamide gel electrophoresis in sodium dodecyl sulphate (Hatton *et al.*, 1977), and its molecular weight appears to be in the range 74 000-77 000 (Stratil and Spooner, 1971; Richardson *et al.*, 1973).

5.4.2 Isoelectric point

Human transferrin has an isoelectric point in the range 5.4-5.9, Fe_2Tf having a slightly higher pI than apoTf (Keller and Pennell, 1959; Bearn and Parker, 1966; Hovanessian and Awdeh, 1976). Most other serum transferrins have fairly similar pI values, although shark transferrin has an exceptionally high pI of 9.2 (Got *et al.*, 1967). Ovotransferrin has an isoelectric point of 6.0 (Bain and Deutsch, 1948). Lactoferrins have a much higher pI than most serum transferrins, values in the range 8.4-9.0 being obtained (Roberts *et al.*, 1973; Weiner and Szuchet, 1975; Kinkade *et al.*, 1976; Van Snick and Masson, 1976). Earlier studies had reported a pI below 7 for human lactoferrin (Masson *et al.*, 1969; Masson, 1970), but this was later shown to be an artifact arising from binding of ampholytes to the protein (Roberts *et al.*, 1973). Lactoferrin, perhaps because of its high isoelectric point, displays a marked tendency to complex with other molecules; these may include spermatozoal surface components (Roberts and Boettcher,

1969), agar and trypan blue (Malmquist and Johansson, 1971), DNA (Loisillier *et al*., 1968; Bennett and Davis, 1982), serum albumin (Hekman, 1971; Lampreave, 1976) and β-lactoglobulin (Brock *et al*., 1976; Lampreave, 1976). Lactoferrin also polymerises readily in the presence of calcium ions (Bennett *et al*., 1981). These properties make it essential that the purity of lactoferrin preparations be carefully checked, since binding of other substances can lead to spurious physicochemical data (Masson, 1970) and may affect biological properties.

5.4.3 Molecular size and shape

Rosseneu-Motref *et al.* (1971) showed, using dielectric dispersion and viscosity measurements, that human Fe_2Tf has the form of a prolate ellipsoid of dimensions 28 × 55 Å. For apoTf the dimensions were 25 × 62 Å, and these dimensions agree well with values of 21 × 68 Å reported by Yeh *et al.* (1979) for apo-ovotransferrin, using light scattering and transient electric birefringence. Martel *et al.* (1980) used the technique of small-angle neutron scattering to study the shape of human apotransferrin and concluded that the molecule had the form of an oblate spheroid with the dimensions 46.6 × 46.6 × 15.8 Å.

The most valuable information regarding the molecular size and shape of transferrin has come from X-ray diffraction studies. Extensive investigations of rabbit transferrin at 6 Å resolution have shown it to possess a bilobal structure joined by a bridging region (Gorinsky *et al*., 1979) (figure 5.1). There is a cleft in

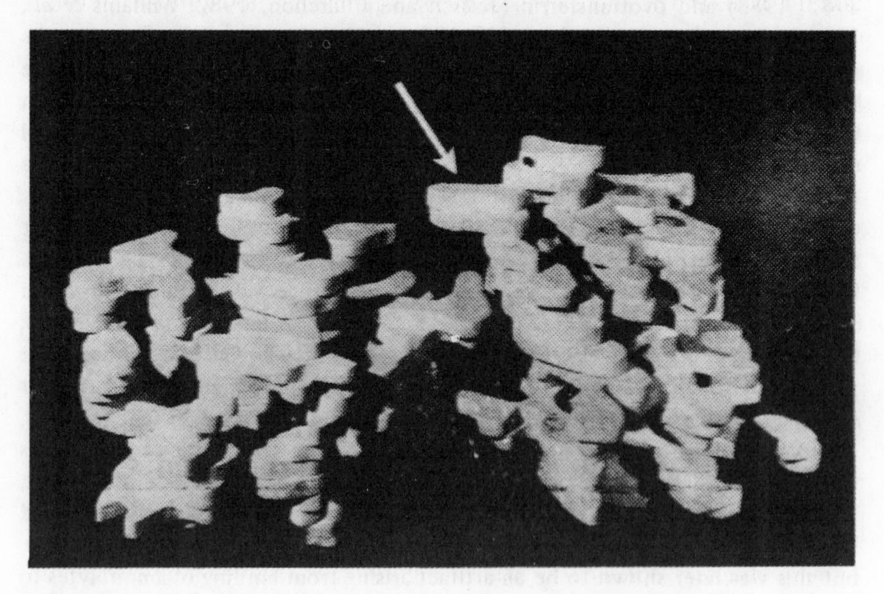

Figure 5.1 A balsa-wood model of the rabbit transferrin molecule derived from X-ray crystallographic studies. The bilobal nature of the molecule is clearly seen, and the arrow indicates a region of high-electron density which may be an α-helix. Reprinted by permission from *Nature, Lond.* **287**, 157. Copyright © 1979, Macmillan Journals Ltd.

each lobe, which may represent the iron-binding site of each region. The dimensions of the molecule, 95 × 60 × 50 Å, are somewhat larger than those obtained by hydrodynamic measurements. A similar study at 5 Å resolution on ovotransferrin (Abola *et al.*, 1982) has also revealed a bilobal structure of dimensions 98 × 60 × 40 Å. The individual lobes have dimensions of 56 × 40 Å and 60 × 36 Å. There is some evidence that each globular domain consists of two subdomains. These studies support the conclusions derived from amino acid sequencing and fragmentation studies (sections 5.4.4 and 5.4.6) that transferrin consists of two largely independent iron-binding domains possessing a certain degree of internal homology with each other. Preliminary X-ray crystallographic data have also been reported for human transferrin (DeLucas *et al.*, 1978) and lactoferrin (Baker and Rumball, 1977).

5.4.4 Primary structure

The complete amino acid sequence of human transferrin has recently been elucidated by MacGillivray *et al.* (1982, 1983). The protein was found to contain 679 amino acids, and 42 per cent homology was found between residues 1–336 and 337–679 when these were aligned, with gaps inserted at appropriate positions (figure 5.2). This suggests that transferrin may have evolved by gene duplication from an ancestral protein similar to a single domain. The N-terminal domain contained eight disulphide bridges, whereas the C-terminal domain contained eleven. The positions of those in the N-terminal region were established as connecting cystines 9–48, 19–39, 117–194, 137–331 and 227–241. The positions of the remaining bridges were not established, although Williams *et al.* (1982*a*) have inferred the positions of some of them from homologies between human transferrin and ovotransferrin, and between the two domains of human transferrin. It is of interest that only seven residues separate the last cystine in the N-terminal domain (331) and the first in the C-terminal domain (339). This implies that the bridging peptide connecting the domains is quite short and may account for the fact that this protein cannot be readily cleaved by proteolytic enzymes to yield the two isolated domains simultaneously, unlike some other transferrins (Esparza and Brock, 1980*a*). Both glycan chains were linked to asparagines in the C-terminal region, at positions 413 and 611.

The complete amino acid sequence of hen ovotransferrin has also been reported, but in this case the sequence was derived from the nucleotide sequence of cDNA transcribed from ovotransferrin mRNA (Jeltsch and Chambon, 1982). The protein, excluding the leader sequence, consisted of 686 amino acids, the single glycan chain being located in the C-terminal region attached to Asn-473. Close agreement was found between this sequence and the almost complete primary structure obtained by Williams *et al.* (1982*a*) by conventional amino-acid sequencing. Homology between the N- and C- terminal domains was 37 per cent, a figure similar to the 40 per cent for human transferrin. Sequencing of cystine-containing peptides (Elleman and Williams, 1970), together with the high

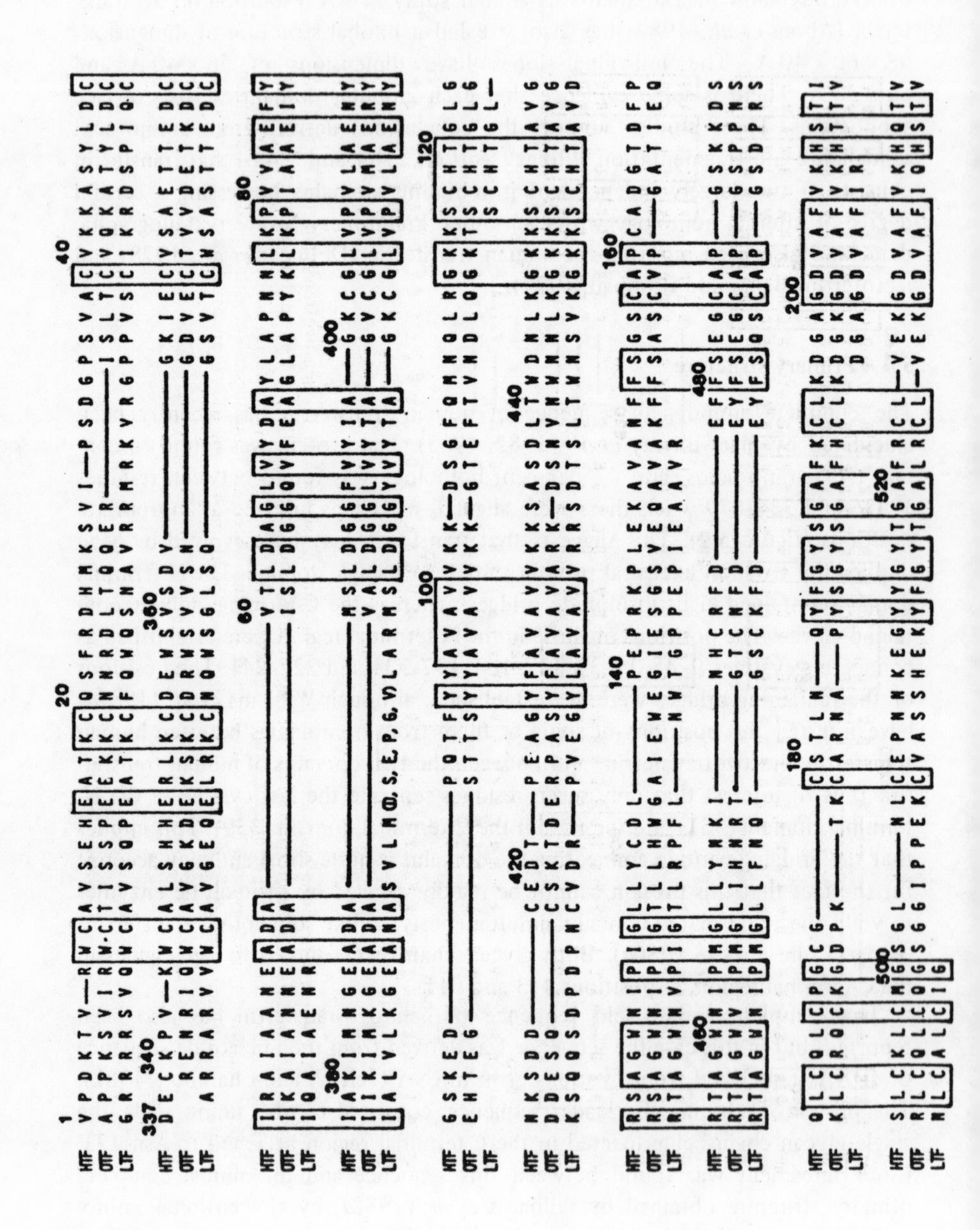

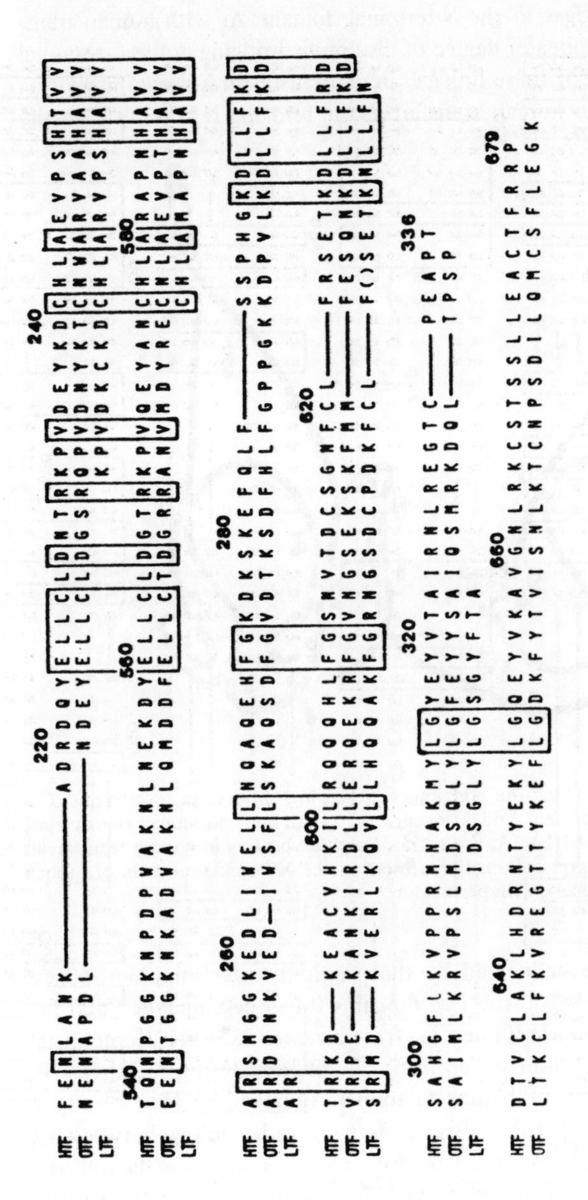

Figure 5.2 A comparison of the complete sequences of human transferrin (HFT), hen ovotransferrin (OTF) and the partial sequence of human lactoferrin (LTF). The upper three lines represent the N-terminal domain of each protein and the lower lines the carboxyl-terminal domain. Numbers refer to the human transferrin sequence. Gaps have been placed to maximise homology, and boxes contain those residues which are homologous in both domains of all three proteins. Glycosylated asparagines in human transferrin are at positions 413 and 611. From MacGillivray *et al.* (1983).

degree of homology (86 per cent) of cystine residues between the two domains, allowed the positions of all the disulphide bridges to be located. There are six in the N-terminal domain and nine in the C-terminal domain, of which six are homologous with the bridges in the N-terminal domain. As with human transferrin, therefore, there is a greater degree of disulphide bridging in the C-terminal domain. The arrangement of these linkages in ovotransferrin is shown in a string model (figure 5.3). Unlike human transferrin, the bridging region between the

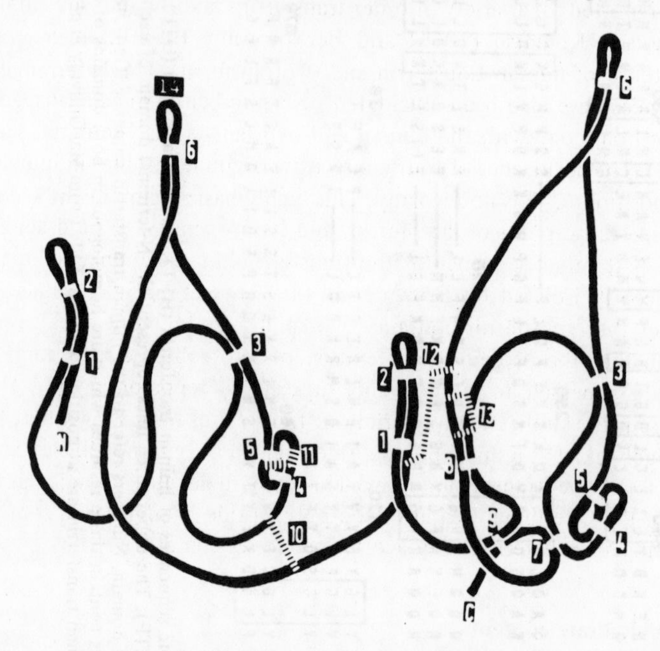

Figure 5.3 A string model showing the positions of disulphide bridges in transferrins. The white bars, numbered 1–9, represent bridges occurring in both hen and human transferrins, and the hatched bars, numbered 10–13, the extra disulphide bridges in human transferrin. The figure 14 represents an extra half-cystine present in the N-terminal domain of human lactoferrin. From Williams (1982*a*), with permission.

domains cannot readily be delineated by the proximity of disulphide bridges, but from the homology between the two domains it can be concluded that the bridge consists of a nine-amino-acid peptide from residues 333–341. Comparison of the primary structures of human transferrin and ovotransferrin, together with the partial primary structure of human lactoferrin obtained by Metz-Boutigue *et al.* (1981), show (figure 5.2) that there is 50 per cent homology between the human transferrin sequence and those of both ovotransferrin and lactoferrin.

In an earlier report MacGillivray *et al.* (1977) had suggested, on the basis of the partial primary structure then available, that human transferrin possessed a degree of fourfold homology. This proposal has not been supported by the complete sequence now available, nor by that of ovotransferrin. However, some evidence for an ancient fourfold homology can be derived from the base sequence of ovotransferrin cDNA (Jeltsch and Chambon, 1982), and this would be consistent with the X-ray crystallographic evidence indicating that each domain may consist of two globular subdomains (Abola *et al.*, 1982). The possible evolutionary significance of homologies within transferrins has recently been reviewed by Williams (1982*a*).

Although complete sequence information is lacking, amino acid compositions have been reported for a variety of other transferrins and the data are summarised by Feeney and Komatsu (1966) and Bezkorovainy (1980). In general, they resemble those of human transferrin and ovotransferrin. The N-terminal amino acid sequences have also been elucidated for a number of transferrins, and these show close homology with the human and ovotransferrin N-terminal sequences (figure 5.4). The occurrence of four consecutive arginine residues in human lactoferrin at positions 2–5 is noteworthy. This highly basic region might account for the high isoelectric point of lactoferrin, and its propensity to bind acidic molecules. The N-terminal sequence of the transferrin-like melanoma cell membrane protein recently isolated by Brown *et al.* (1982) (section 5.7.2) is also shown.

The fact that transferrin contains two similar iron-binding sites gave rise to speculation that the protein might consist of two subunits, and some experimental data appeared to support this suggestion (Jeppsson, 1967). However, many subsequent studies with a variety of transferrins have shown that transferrins invariably exist as a single polypeptide chain. The only possible exception to this rule is bovine transferrin, in which some molecules contain an internal cleavage in the C-terminal region of the polypeptide chain (Maeda *et al.*, 1980). The resulting microheterogeneity is discussed in section 5.4.7.

5.4.5 Carbohydrate content

The transferrins are glycoproteins. The percentage of carbohydrate varies according to species, and also within different tissues of the same species. Furthermore, transferrin molecules from the same tissue of the same species may exhibit differences in the structure of their glycan chains, particularly with respect to sialic acid content.

The structure of the two glycan chains of human transferrin has been elucidated. The most frequent structure is a biantennary glycan of the N-acetyllactosamine type (Spik *et al.*, 1975; Dorland *et al.*, 1977). However, the presence of two types of triantennary glycan has now been established (Montreuil and Spik, 1975; Wong *et al.*, 1978; Krusius and Finne, 1981), though the relative distribution of the bi- and triantennary glycan chains within transferrin molecules is controversial. Montreuil and Spik (1975) concluded that the triantennary glycan, when present, was always linked to the asparagine residue in the peptide Asn-Val-Thr: comparison of this sequence with the primary structure reported by MacGillivray *et al.* (1983) identifies Asn-611 as the linkage point. The other glycan linkage point, Asn-413, would therefore possess only biantennary glycan

Protein	*Sequence*	*Reference*
Human transferrin	Val – Pro – Asp – Lys – – – Thr – Val – – – Arg – Trp – Cys – Ala – Val – Ser	MacGillivray *et al.* (1982)
Rabbit transferrin	Val – Thr – Glu – Lys – – – Thr – Val – – – Asn – Trp – ? – Ala – Val – Ser	Heaphy and Williams (1982*a*)
Rat transferrin	Val – Pro – Asp – Lys – – – Thr – Val – – – Lys – Trp – Cys – Ala – Val – Ser	Schreiber *et al.* (1979)
Pig transferrin	Val – Pro – Glu – Lys – – – Thr	Graham and Williams (1975)
Sheep transferrin	Ser – Pro – Glu – Lys – – – Thr – Val – – – Arg – Trp	Guérin *et al.* (1976)
Bovine transferrin	Asp – Pro – Glu – Arg – – – Thr – Val – – – Arg	Brock *et al.* (1980); Maeda *et al.* (1980)
Hen transferrin	Ala – Pro – Pro – Lys – – – Ser – Val – Ile – Arg – Trp – Cys – Thr – Ile – Ser	Jeltsch and Chambon (1982)
Duck transferrin	Ala – Pro – Pro – Lys – – – Thr	Graham and Williams (1975)
Human lactoferrin	Gly – Arg – Arg – Arg – Arg – Ser – Val – – – Glu – Trp – Cys – Ala – Val – Ser	Jollès *et al.* (1976)
*p*97 protein	– – – – – – – Gly – Met – – – Glu – Val – – – Arg – Trp – Cys – Ala – Thr – Ser	Brown *et al.* (1982)

Figure 5.4 The N-terminal amino acid sequence of several transferrins and human melanoma cell surface protein *p*97 (see section 5.7.2). Homologous residues are boxed.

chains. However, Hatton *et al.* (1979), by chromatography of peptides on concanavalin-A–Sepharose, found that triantennary glycans could be linked at either of the two asparagine residues. More than one type of triantennary glycan was also reported (Hatton *et al.*, 1979; Regoeczi *et al.*, 1979). Kerckaert and Bayard (1982) have recently examined fractions reactive and unreactive to concanavalin-A by sodium dodecyl sulphate–polyacrylamide gel electrophoresis, and concluded that transferrin molecules possess either two triantennary or two biantennary glycans, thus excluding molecules containing one chain of each type, as in the proposal of Montreuil and Spik (1975). Further studies of the glycans from human and other transferrins may help to resolve these differences. There is now also evidence for the presence of tetra-antennary glycans on a small proportion of transferrin molecules (März *et al.*, 1982).

The structure of the carbohydrate moeity of ovotransferrin has also been investigated. Only a single glycan chain is present, but there are two variant structures, though whether these differ in the structure of the glycan itself or in the linkage point is controversial (Graham and Williams, 1975; Iwase and Hotta, 1977; Dorland *et al.*, 1979). Unlike serum transferrins, the glycan does not contain galactose or sialic acid. Assuming that each carbohydrate chain contained two galactose residues, Hudson *et al.* (1973) calculated that hen serum transferrin and bovine transferrin each contained a single glycan chain; rabbit, human and equine transferrins contained two; and porcine transferrin four. However, Graham and Williams (1975) isolated two different glycopeptides from digests of bovine transferrin and thus concluded that there should be two glycan chains per molecule. They also concluded that porcine transferrin contained only a single glycan chain. Controversy also exists regarding the number of glycan chains in rabbit transferrin: Leger *et al.* (1978) found evidence of only a single type of glycopeptide but further studies by Hudson and co-workers indicated that two identical glycopeptides were present in each transferrin molecule (Strickland and Hudson, 1978; Strickland *et al.*, 1979). Finally, Heaphy and Williams (1982*a*) found that the carbohydrate content of rabbit transferrin, and of the isolated C-terminal domain which contained the carbohydrate moiety, was similar to that of ovotransferrin, which definitely contains only a single chain, and about half that of human transferrin, which contains two. On the basis of these findings they concluded that rabbit transferrin contains only a single glycan chain.

A number of factors may contribute to the discrepancies in the results obtained by different groups of workers. A potential source of error is contamination of transferrin preparations with haemopexin, which frequently cochromatographs with transferrin and has a higher carbohydrate content (Hatton *et al.*, 1977; Leger *et al.*, 1978). Conclusions based on the number of glycopeptides isolated from proteolytic digests are only reliable if it can be assumed that multiple linkage points with identical peptide structures do not exist, and that heterogeneity with respect to the linkage point of one or more glycan chains is not present among different transferrin molecules. A more detailed knowledge of the primary structures of the transferrins involved should help to resolve some of these questions. Elucidation of the primary structures of human transferrin and ovotransferrin has established that the glycan chain(s) of both proteins are located in the C-terminal domain, and analysis of isolated domains has shown that the same is true for rabbit (Heaphy and Williams, 1982*a*) and bovine (Brock *et al.*, 1978*a*) transferrins. Although this might suggest that the carbohydrate moiety of transferrins is always located in the C-terminal domain, this is not the case with human lactoferrin, in which one glycan chain is found in each domain (Bluard-Deconinck *et al.*, 1978). The same may also be true of porcine transferrin (Esparza, 1979; Esparza and Brock, 1980*a*). The glycan moieties of human and bovine lactoferrins contain fucose (Spik *et al.*, 1974; Van Halbeek *et al.*, 1981; Matsumoto *et al.*, 1982), a sugar not found normally in the glycans of serum transferrins, although it does appear to be present in porcine serum transferrin (Hudson *et al.*, 1973; Graham and Williams, 1975). Although human lactoferrin possesses only two glycan chains per molecule, four or five different types of biantennary glycan chain occur (Matsumoto *et al.*, 1982; Spik *et al.*, 1982).

Though the glycan chains of transferrins appear to play little part in their iron-binding and iron-donating properties (sections 5.5 and 5.6), they are important in determining the rate of elimination of the proteins from the circulation. Transferrin is catabolised in the liver, and its removal from the circulation is effected by interaction of the carbohydrate moiety with the hepatic asialoglycoprotein receptor. This has been studied extensively by Regoeczi and his colleagues by measuring the rate of elimination of labelled transferrins from the circulation. Although human transferrin, like other glycoproteins, is eliminated more rapidly when the sialic acid content is low (Wong and Regoeczi, 1977), elimination is much slower than with most other asialoglycoproteins (Regoeczi *et al.*, 1974). With other transferrins, particularly when homologous systems are used, there is virtually no interaction with the asialoglycoprotein receptor (Hatton *et al.*, 1974, 1977, 1978; Regoeczi and Hatton, 1974; Regoeczi *et al.*, 1975). However, some of the human transferrin molecules bearing triantennary glycans are eliminated more rapidly (Wong *et al.*, 1978; Hatton *et al.*, 1979; Debanne *et al.*, 1981). Lactoferrin is rapidly removed from the circulation by the liver due to its possession of a fucosyl α-1-3-N-acetyl glucosaminyl group in the glycan chain, which is much more readily recognised by the hepatic receptor than is the non-fucose-containing glycan of transferrin (Prieels *et al.*, 1978).

5.4.6 Isolation of single domains of transferrin

The concept of a structure consisting of two largely independent domains, each carrying its own iron-binding site, has received strong support from studies demonstrating that cleavage of some transferrins, usually by proteases, could under certain conditions yield half-molecules capable of binding just a single iron atom.

Williams (1974, 1975) treated ovotransferrin containing a single iron atom in either the N- or C-terminal site with trypsin or chymotrypsin. The unoccupied domain was cleaved to small peptides, leaving a single domain carrying one iron atom. Peptide maps of the two fragments were quite different, and the fragments did not cross-react immunologically. Tsao *et al.* (1974*a*) also obtained an iron-binding fragment from ovotransferrin by cleavage with cyanogen bromide, and comparison of this fragment with those obtained by Williams led the latter to conclude that it corresponded to the C-terminal domain.

It was not possible to isolate both fragments simultaneously by cleaving Fe_2-ovotransferrin; trypsin and chymotrypsin were unable to split the saturated protein, and subtilisin destroyed the N-terminal domain, leaving the C-terminal fragment only. Simultaneous production of both the N- and C-terminal fragments was, however, achieved by tryptic cleavage of bovine transferrin (Brock *et al.*, 1976, 1978*a*; Brock and Arzabe, 1976), thus clearly indicating the independence of the two iron-binding domains. In addition, brief tryptic digestion of bovine apotransferrin yielded the C-terminal fragment, but the N-terminal fragment was destroyed (Brock *et al.*, 1978*a*). As with ovotransferrin, peptide maps and immunological studies showed no similarity between the fragments (figure 5.5) and carbohydrate was found only in the C-terminal fragment.

In view of the different behaviour of iron-saturated bovine and ovotransferrins when treated with trypsin, the effect of this enzyme on several other transferrins was investigated (Esparza and Brock, 1980*a*). Porcine Fe_2Tf also yielded two monoferric fragments, though a variety of chromatographic procedures failed to separate them (Esparza, 1979). Human and rabbit transferrins underwent some internal cleavage, but this had little effect on their iron-binding properties and did not result in any monoferric fragments being liberated. Equine transferrin showed the presence of two major fragments of molecular weights 32 000 and 51 000 on unreduced sodium dodecyl sulphate–polyacrylamide gel electrophoresis, but these appeared to remain noncovalently bound together in the absence of detergent. The reasons for these differences among transferrins from various species may reflect dissimilarities in the structure and accessibility of the bridge region between domains. The lack of arginine or lysine residues in the bridge region of human transferrin (MacGillivray *et al.*, 1982) may, for example, account for the failure of trypsin to cleave this protein to monoferric fragments. The conditions employed for digestion may also affect the result, as is emphasised by the recent study of Keung *et al.* (1982), who found that when Fe_2-ovotransferrin was treated with immobilised subtilisin, direct cleavage to two monoferric fragments occurred and the N-terminal domain was not destroyed, as happens when free subtilisin is used.

Monoferric fragments have been obtained by tryptic or chymotryptic digestion of the appropriate monoferric human (Evans and Williams, 1978) or rabbit (Heaphy and Williams, 1982*a*) transferrins. The N-terminal fragments have also been obtained by digesting diferric human or rabbit transferrins with thermolysin or subtilisin, respectively (Lineback-Zins and Brew, 1980; Heaphy and Williams, 1982*a*; Zak *et al.*, 1983).

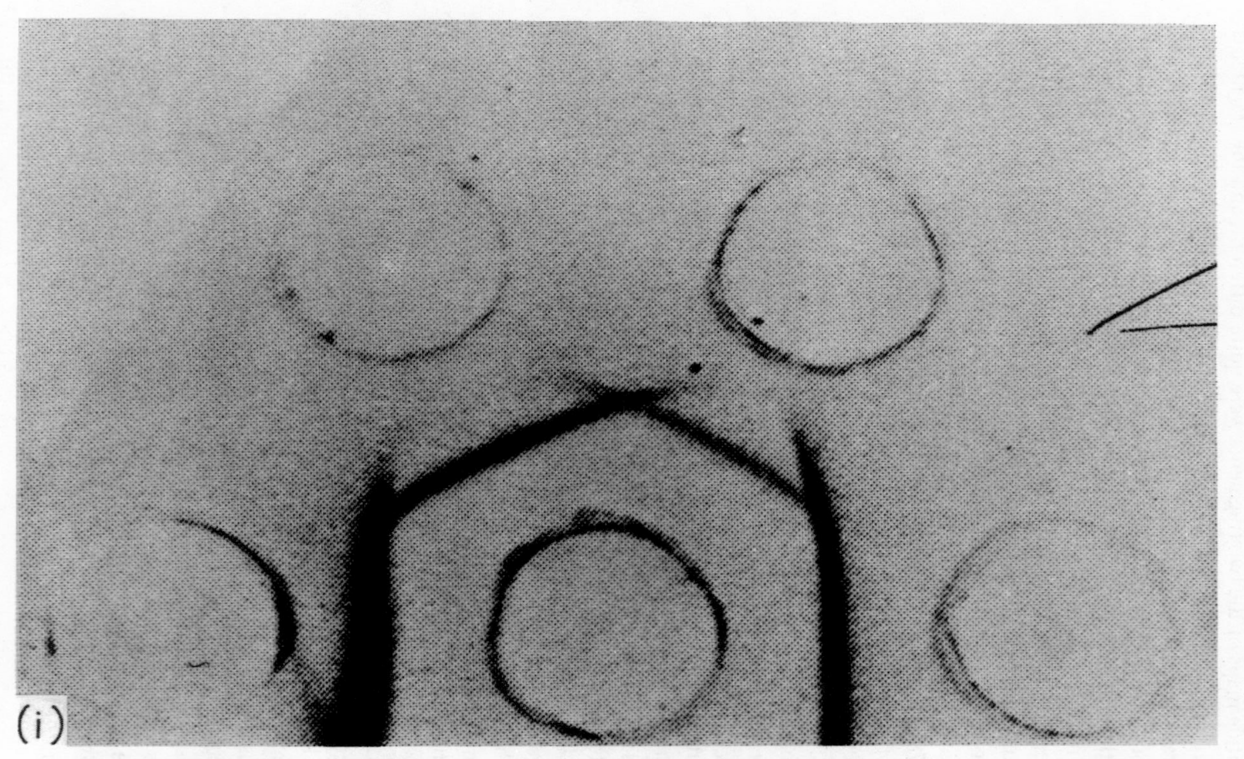

Figure 5.5 Antigenic properties (i) and peptide maps (ii) of monoferric fragments from bovine transferrin. In (i), the N-terminal (upper left well) and C-terminal (upper right well) monoferric fragments, and undigested bovine transferrin (lower left and right wells) are reacted against antiserum to bovine transferrin. Diagram (ii) shows peptide maps of bovine transferrin (a) and the N-terminal (b) and C-terminal (c) monoferric fragments. (i) From Brock and Arzabe (1976) with permission; (ii) from Brock *et al.* (1978*a*) with permission.

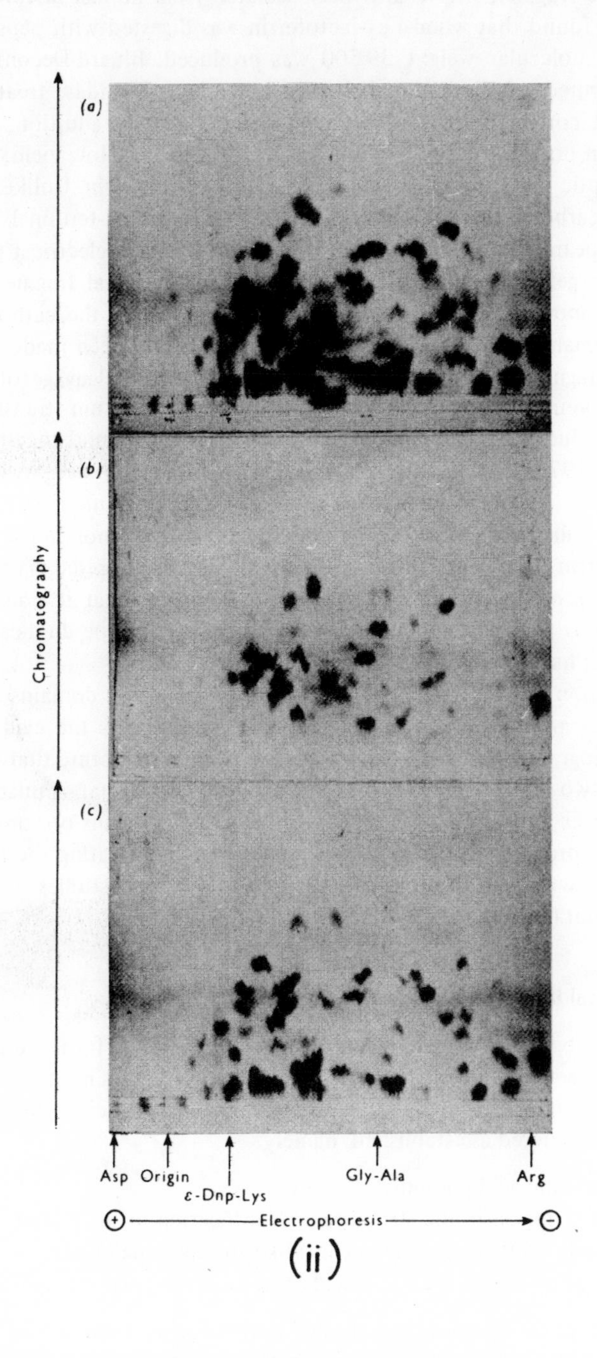

(a)
(b)
(c)
Chromatography
Asp
Origin
ε-Dnp-Lys
Gly-Ala
Arg
⊕ Electrophoresis ⊖
(ii)

Monoferric fragments have also been isolated from human lactoferrin. Line *et al.* (1976) found that when Fe_2-lactoferrin was digested with pepsin a single fragment of molecular weight 39 500 was produced. Bluard-Deconinck *et al.* (1978) confirmed this finding and showed by carboxypeptidase treatment that this fragment corresponded to the C-terminal domain. In addition, they were able to obtain both N- and C-terminal fragments, albeit in low yield, by tryptic or chymotryptic digestion of 30 per cent saturated lactoferrin. Unlike the serum transferrins, carbohydrate was found in both the N- and C-terminal fragments. They also appeared to differ markedly from each other in electrical properties, as in agarose gel electrophoresis at pH 8.6 the C-terminal fragments moved towards the anode and the N-terminal fragment towards the cathode. Intact lactoferrin remained near the origin. Attempts have also been made to prepare monoferric fragments from bovine lactoferrin by direct cleavage of the iron-saturated protein with trypsin, but, as with equine serum transferrin, the fragments produced appeared to remain noncovalently bound together (Brock *et al.*, 1976, 1978*b*). A lactoferrin fragment of molecular weight 45 000 was identified in bovine colostrum, though it was not isolated (Butler, 1973).

Recently, Williams *et al.* (1982*b*) have shown that when monoferric fragments of ovotransferrin are injected into mice, they are rapidly excreted via the kidney, whereas intact ovotransferrin is retained. They suggest that if transferrin did indeed evolve from a single-domain ancestral protein by gene duplication, such a protein must have been membrane bound.

In conclusion, the isolation of individual iron-binding domains of various transferrins as monoferric half-molecules strongly supports the evidence from X-ray crystallography and internal homology of primary structure that transferrin consists of two independent domains which are somewhat similar, but not identical to each other. These fragments offer a useful means of comparing the metal-binding properties of individual sites, and of investigating the interaction of transferrin with cell membrane receptors. Some such studies have already been carried out (sections 5.5.2, 5.6.2.2 and 5.6.3).

5.4.7 Structural heterogeneity of the transferrins

It has been recognised for many years that transferrins may form multiple bands in starch gel electrophoresis which are not due to variations in the iron content. A number of studies have investigated transferrin heterogeneity, and four causes can now be considered as established, namely:

(1) variable sialic acid content;
(2) other differences in the structure and/or location of the glycan chains;
(3) genetically controlled amino acid substitutions or deletions;
(4) certain other miscellaneous, and in some cases unexplained, heterogeneities.

The sialic acid content has been studied in transferrins from several species. In human transferrin, two or three components are normally detected in starch gel electrophoresis (Heide and Haupt, 1964; Graham and Williams, 1975; Regoeczi *et al.*, 1977), but these are reduced to a single band after neuraminidase treatment. Wong and Regoeczi (1977) fractionated purified human transferrin (homozygous type C) on DEAE cellulose into five fractions, containing 0.46, 1.4, 2.4, 3.2 and 3.9 residues of sialic acid per molecule. They also observed that some loss of sialic acid occurred during storage. Similar studies have shown that bovine transferrin may contain between zero and five sialic acid residues per molecule (Stratil and Spooner, 1971), those with two or three residues being the most abundant. Transferrin in a few individual animals showed fewer bands than normal in starch gel electrophoresis, due to the absence of the trisialotransferrin, which apparently is caused by a genetically controlled abnormality in sialotransferase (Spooner and Baxter, 1969; Stratil and Spooner, 1971; Spooner *et al.*, 1977). In the rat, two major bands containing two and three residues of sialic acid respectively are present (Schreiber *et al.*, 1979) and these probably correspond to the two 'isotransferrins' detected by electrophoresis (Beaton *et al.*, 1961; Gordon and Louis, 1963; Huebers *et al.*, 1977; Okada *et al.*, 1979). Variable sialic acid content was also found to be the cause of electrophoretic heterogeneity in human (Wolfson and Robbins, 1971) and porcine (Roberts and Boursnell, 1975) lactoferrins.

Different degrees of sialylation of human transferrin are found in various biological fluids (Van Eijk *et al.*, 1983), and Stibler *et al.* (1978, 1979) have shown that an abnormal transferrin band observed when sera of alcoholics are subjected to isoelectric focussing is caused by the presence of molecules with an unusually low sialic acid content. Detection of this band could possibly be of use as a clinical test for alcoholism (Stibler *et al.*, 1979).

Variability in the structure of the glycan chain appears to be involved in the electrophoretic heterogeneity of hen ovotransferrin, as no differences have been found in amino acid composition (Wenn and Williams, 1968) or peptide maps (Williams and Wenn, 1970) between the two major components. Since sialic acid is absent, other sugars must be involved. Iwase and Hotta (1977) analysed ovotransferrin glycopeptides by sequential glycosidase treatment and found that the glycan chain of the minor component had the structure Asn-$[GlcNAc]_3$-$[Man]_4$-$[GlcNAc]_2$, and that of the major component the structure Asn-$[GlcNAc]_2$-$[Man]_3$-$[GlcNAc]_3$. The two variants also differed in the point of attachment of the glycan chain to the protein. The major electrophoretic band contained a further variant which bound to concanavalin-A and was thought to possess more than one type of glycan chain. It is not certain whether the heterogeneity of human transferrin with respect to bi- and triantennary glycan chains (section 5.4.5) causes electrophoretic microheterogeneity, other than as a result of the presence of additional sialic acid residues on the triantennary glycans (Wong *et al.*, 1978).

It is now well known that transferrins from many species show genetically controlled deletions or substitutions in their polypeptide chains, leading to genetic variants which may be distinguished in starch gel electrophoresis or isoelectric focussing. In man more than 90 per cent of the population is homozygous for transferrin C, though at least 22 further alleles have been identified (Kühnl and Spielmann, 1978). Human lactoferrin apparently does not show genetic polymorphism (Wolfson and Robbins, 1971). Full details of the occurrence and frequency of transferrin genetic variants in man and other species can be found elsewhere (Lush, 1966; Giblett, 1969; Manwell and Baker, 1970) and only structural aspects will be considered here. By comparing the complete amino acid sequence of human transferrin C (MacGillivray *et al.*, 1982) with peptide sequences from other variants (Wang and Sutton, 1965; Wang *et al.*, 1966, 1967; Howard *et al.*, 1968) it can be deduced that in TfD_1 Gly-227 is substituted by Asp, in TfD_{chi} His-300 is substituted by Arg and in TfB_2 Gly-651 is substituted by Glu. The first two substitutions will therefore affect the N-terminal domain and the last the C-terminal domain. In bovine transferrin, electrophoresis of monoferric fragments of the four common variants (A, D_1, D_2 and E) has shown that all except D_2 have electrophoretically indistinguishable C-terminal domains, and A and D_2 have indistinguishable N-terminal domains (Brock *et al.*, 1980). It is of interest that variants D_1 and D_2, which display only a marginal difference in mobility when the intact proteins are subjected to electrophoresis, showed substantial differences when the individual domains were examined. This approach may be of value in confirming differences between doubtfully distinguishable polymorphic variants in other species. Genetic variants do not normally differ in their antigenic properties, but equine transferrin appears to be an exception to this rule, since genetic variants could be classified into two subgroups displaying slight antigenic differences (Kaminski *et al.*, 1981). There is little evidence for any functional differences between genetic variants, although a slight difference in antimicrobial activity between human transferrins C and D has been reported (Lawrence *et al.*, 1977).

A heterogeneity that causes bovine transferrin to run as two bands in electrophoresis but due neither to variation in sialic acid content nor to genetic polymorphism has been recognised by several workers (Makarechian and Howell, 1966; Spooner and Baxter, 1969; Hatton *et al.*, 1977), and was shown by Brock *et al.* (1980) to be located in the C-terminal domain of the molecule. Maeda *et al.* (1980) found that when the two components were reduced with 2-mercaptoethanol and subjected to electrophoresis in 7 M urea, one of them showed evidence of internal cleavage of the polypeptide chain. Under these conditions a polypeptide of molecular weight about 6000 was released from the C-terminus of the variant molecule. Interestingly, the band corresponding to the cleaved variant is not seen in fetal bovine transferrin (Spooner *et al.*, 1970) until near term, suggesting that the development of some plasma proteolytic activity may be responsible. Pairs of bands in electrophoresis have also been observed in porcine transferrin (Stratil and Kubek, 1974), and in homozygous carp transferrin (which does not contain sialic acid) four bands are seen (Valenta *et al.*, 1976). Recently, multiple heterogeneity has also been detected in human transferrin by high-resolution two-dimensional electrophoresis (Anderson and Anderson, 1979).

A number of minor components were present which differed slightly from the major fraction in either molecular weight or charge, even when heterogeneity due to sialic acid had been eliminated. This heterogeneity may reflect differences between molecules with bi- or triantennary glycan chains or cleavage of small peptides from some molecules, analogous to the findings with bovine transferrin.

All the above-mentioned heterogeneities represent minor structural differences in what appear to be functionally normal transferrin molecules. However, a rare variant of human transferrin with abnormal iron-binding properties has been isolated by Evans *et al.* (1982) following the screening of 5000 individual sera by electrophoresis in 6 M urea according to the method of Makey and Seal (1976). Iron bound at the C-terminal site was found to be abnormally labile in the presence of urea, and gave a visible spectrum with λ_{max} at 435 nm instead of the usual 465 nm. This variant also showed a diminished ability to deliver iron to cells (Young *et al.*, 1984).

5.5 METAL BINDING

5.5.1 General aspects

All transferrins are capable of binding two ferric ions, or certain other metal ions, per molecule. These ions occupy specific sites and, as discussed previously, each occupies a separate domain of the molecule. Metal binding requires the synergistic binding of an anion, normally (bi)carbonate (the form of the anion that is bound is controversial, see section 5.5.6). Apotransferrins are colourless, but develop a red-brown colour giving λ_{max} at 465-470 nm when iron is bound. Under physiological conditions, the effective association constant for the Fe_2N transferrin complex is of the order of 10^{24} M^{-1} (Aisen and Leibman, 1968), but this decreases as the pH is lowered (table 5.1), so that at pH 4.5 it is less than that of iron citrate, and citrate will then remove iron quantitatively from transferrin. Although it is frequently stated that iron dissociates from transferrin at

Table 5.1 Relative stability constants K_a, of iron citrate and iron transferrin complexes at different pH values and atmospheric pCO_2 (from Aisen (1977))

pH	Transferrin	Citrate
7.4	21.3	16.9
6.6	18.1	16.0
4.5	9.7	12.4

pH 5-6, such statements are meaningless unless conditions with respect to competing chelators are defined. For example, at one extreme, human transferrin will lose its iron at pH values as high as 7.4 in the presence of pyrophosphate (Egyed, 1975; Pollack *et al.*, 1977; Carver and Frieden, 1978; Morgan, 1979), whereas, at the other extreme, bovine transferrin in a nonchelating buffer such as acetate will form the characteristic red-brown complex when iron is added at pH values as low as 4.3 (J. H. Brock and I. Esparza, unpublished observations).

Lactoferrin has a higher affinity for iron than transferrin, the apparent affinity constant at pH 6.4 being some 26 times greater than that of transferrin (Aisen and Leibman, 1972), and as a consequence citrate cannot effectively compete for lactoferrin-bound iron until pH 2 is reached (Masson and Heremans, 1968). Transferrin will also bind iron as Fe^{2+}, though with low affinity, and the complex is rapidly converted to the Fe^{3+} form in the presence of oxygen (Kojima and Bates, 1981).

Binding of iron, or other metals, causes a significant conformational change to occur in the transferrin molecule. The isoelectric point increases (Keller and Pennell, 1959; Bearn and Parker, 1966; Hovanessian and Awdeh, 1976), and this causes molecules of differing iron content to elute sequentially from DEAE cellulose columns with a gradient of increasing ionic strength (Lane, 1971). The structure becomes more compact when iron is bound, as is evident from a decrease in the effective hydrodynamic volume (Bezkorovainy, 1966; Rosseneu-Motref *et al.*, 1971). As a consequence, the protein becomes markedly less susceptible to proteolytic degradation (Azari and Feeney, 1958; Williams, 1974; Brock *et al.*, 1976; Esparza and Brock, 1980*a*) and denaturation (Bezkorovainy and Grohlich, 1967; Makey and Seal, 1976). Conversely, apotransferrins possess antigenic sites which become hidden when iron is bound (Tengerdy *et al.*, 1966; Kourilsky and Burtin, 1968).

5.5.2 Differences and cooperativity between the sites

In the past there has been much controversy as to whether cooperativity exists between the metal binding sites of transferrin, and whether binding is random, or favours one of the two sites. The first study to address this problem was carried out by Warner and Weber (1953), who concluded, on the basis of equilibrium dialysis studies against citrate as a competing chelator, that strong positive cooperativity existed between the sites. The obvious corollary of this observation was that Fe^{3+} was taken up pairwise, and that monoferric transferrins were unlikely to be able to exist. These conclusions were challenged by Aasa *et al.* (1963), who pointed out that true equilibrium had probably not been reached in the experiments of Warner and Weber (1953), and that when equilibrium was obtained, dialysis studies indicated that the sites were independent and functionally equivalent. Electron paramagnetic resonance spectroscopic studies carried out at the same time also indicated that the sites were similar. From the first of these conclusions it follows that monoferric transferrins should exist, and this was subsequently confirmed electrophoretically for human transferrin by Aisen *et al.* (1966) and for ovotransferrin (Williams *et al.*, 1970) by isoelectric focussing. From the second conclusion it follows that iron binding should occur randomly between the sites, the proportions of apo-, monoferric and diferric transferrins in partially saturated transferrin solutions obeying the rules of mathematical probability. Again, confirmatory experimental evidence was forthcoming when Wenn and Williams (1968) showed, by isoelectric focussing, that the proportion of the three species found in solutions of ovotransferrin saturated to varying degrees agreed closely with the predicted values.

As evidence began to accumulate that transferrin did not consist of two structurally identical units, and suggestions were made (see section 5.6.6) that the two sites might have different physiological functions, a re-evaluation of the concept of identical iron-binding sites became necessary. Line *et al.* (1967) provided an early challenge to this view when they showed that tyrosine residues at the two sites differed in their ability to react with tetranitromethane. Modification by diethylpyrocarbonate of the histidine residues involved in metal-binding (Mazurier *et al.*, 1977, 1981) and spectroscopic studies, particularly the use of EPR, also began to reveal differences between the sites. Aisen *et al.* (1969) showed that when iron was added to a chromium–transferrin complex, it displaced chromium in a nonrandom manner. Aasa (1972) reinvestigated the EPR spectrum of Fe_2-transferrin and concluded that, contrary to earlier observations (Aasa *et al.*, 1963), two different signals could be detected. Similar conclusions were reached when the EPR spectra of the vanadyl (Cannon and Chasteen, 1975; Chasteen *et al.*, 1977) and copper (Mazurier *et al.*, 1977; Zweier and Aisen, 1977; Zweier, 1978; Froncisz and Aisen, 1982) complexes were investigated. In addition, EPR spectroscopy was used to demonstrate that when Fe_2Tf is treated with the chaotropic agent $NaClO_4$ one site undergoes a conformational change whereas the other does not (Price and Gibson, 1972*a*; Baldwin *et al.*, 1982). The two monoferric human transferrins and ovotransferrins also show slight but definite differences in their EPR spectra (Butterworth *et al.*, 1975; Aisen *et al.*, 1978) (figure 5.6) and differences between the sites of human lactoferrin have also been detected (Ainscough *et al.*, 1980).

Although it was originally thought that the optical spectra in the visible region of the two iron-binding sites were identical (Aasa *et al.*, 1963), more recent studies have shown slight differences in the visible spectra of the two sites of ovotransferrin (Aisen *et al.*, 1973*a*), bovine transferrin (Brock and Arzabe, 1976) and human transferrin (Zak *et al.*, 1983). An equimolar mixture of the two monoferric fragments of bovine transferrin gave a spectrum indistinguishable from that of intact bovine transferrin (figure 5.7). Differences between the iron binding sites of ovotransferrin with respect to thermal denaturation have also been detected using differential scanning calorimetry (Donovan and Ross, 1975; Donovan *et al.*, 1976; Evans *et al.*, 1977).

In view of the large body of data indicating non-identity of the metal binding sites, Aisen *et al.* (1978) reinvestigated the binding constants by equilibrium dialysis against citrate as originally described by Aasa *et al.* (1963). However, they allowed for the possible formation of ferric dicitrate complexes, and as a further precaution maintained the citrate concentration at a constant value and varied the iron concentration, rather than vice versa, as used by Aasa *et al.*

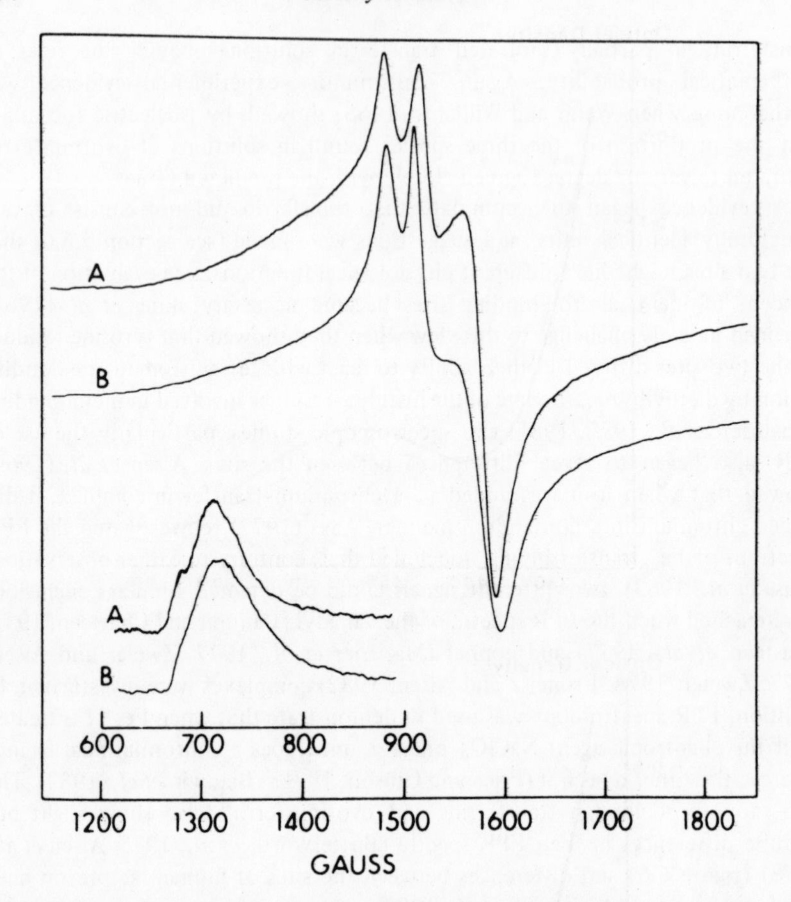

Figure 5.6 The EPR spectra of monoferric human transferrins in 0.2 M KCl–0.1 M 4-(2-hydroxyethyl)-1-piperazine ethanesulphonic acid – 0.05 M $KHCO_3$, pH 7.5. Microwave power, 10 mW, modulation 10 G; microwave frequency 9.114 GHz; temperature 77 K. A, spectrum of $TfFe_C$; B, spectrum of Fe_NTf. Inset shows low field gain increased six-fold. From Aisen *et al.* (1978) with permission.

(1963). Under these conditions they confirmed the difference in the apparent stability constants for the two sites, the values obtained for the binding of iron at pH 7.4 and atmospheric pCO_2 being 4.7×10^{20} M^{-1} and 2.4×10^{19} M^{-1}.

Since the most recent studies have indicated a difference between the sites in their affinity for iron, nonrandom binding of iron – or other metals – is a likely possibility. An early indication of such behaviour was provided by Luk (1971) who studied the binding of lanthanides to transferrin by observing changes in the fluorescence intensity of the ions. He found that whereas transferrin would bind two Tb^{3+}, Eu^{3+}, Er^{3+} and Ho^{3+} ions, only a single Pr^{3+} or Nd^{3+} ion was bound, suggesting that only one site was able to accommodate these latter two ions. Princiotto and Zapolski (1975) made the important observation that when human

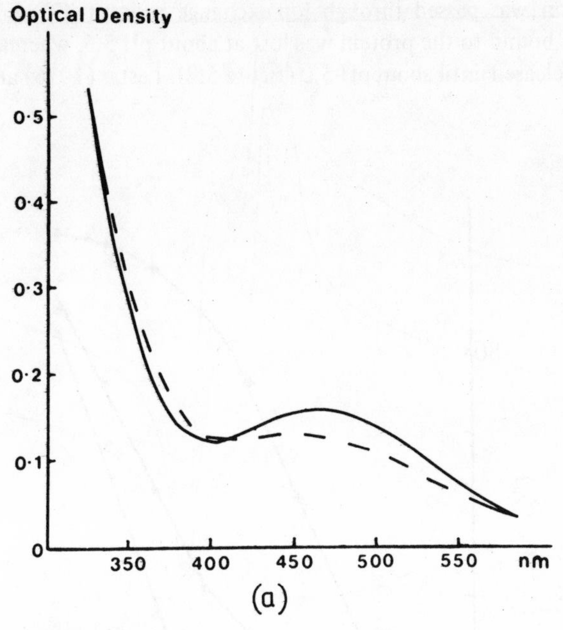

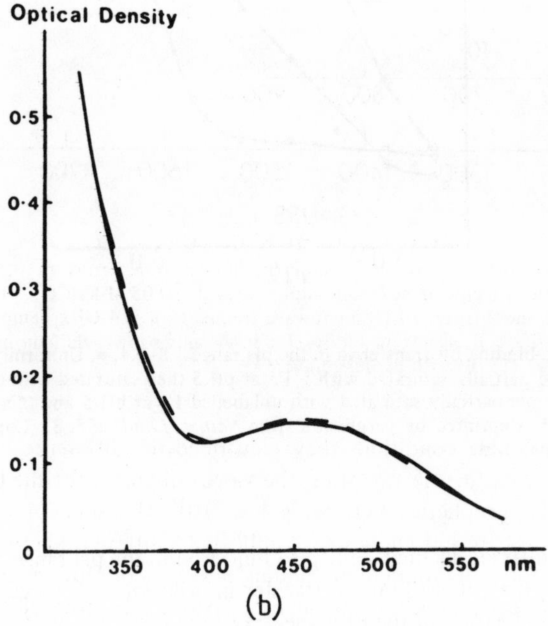

Figure 5.7 (a) Visible spectra of N-terminal (- - -) and C-terminal (—) monoferric fragments of bovine transferrin. (b) Visible spectra of bovine Fe_2-transferrin (—) and an equimolar mixture of the N- and C-terminal fragments (- - -). From Brock and Arzabe (1976) with permission.

Fe_2-transferrin was passed through ion-exchange resin at different pH values, half the iron bound to the protein was lost at about pH 5.5, whereas the remainder was not released until about pH 5.0 (figure 5.8). Lestas (1976) also concluded

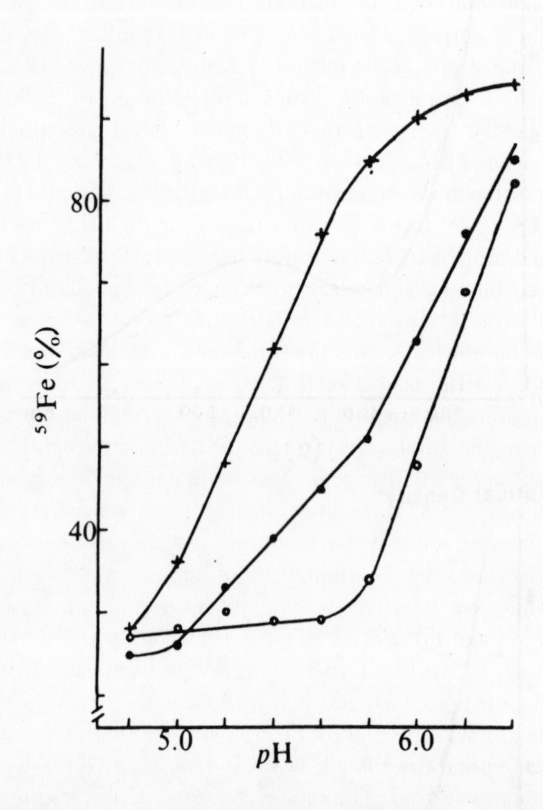

Figure 5.8 Iron-binding by transferrin in the pH range 4.8–6.4. •, Uniformly labelled transferrin: +, sample partially saturated with ^{59}Fe at pH 5 then saturated with unlabelled Fe at pH 7.45. ○, Sample partially saturated with unlabelled Fe at pH 5 and then saturated with ^{59}Fe at pH 7.45. Reprinted by permission from *Nature, Lond.* **255**, 87. Copyright © 1975, Macmillan Journals Ltd.

that the sites differed in their ability to bind iron in the pH range 5–6 by studying the absorption at 470nm of transferrin solutions of varying degrees of saturation. Investigation of the EPR spectra of the Cu^{2+} and VO^{2+} complexes of transferrin also revealed that only a single site was capable of binding the metal ions at pH 6.0, and that in the case of the Cu^{2+} complex, the second site only became occupied at pH > 7.2 (Cannon and Chasteen, 1975; Zweier, 1978). Harris (1977*a*) showed that when one equivalent of ^{59}Fe, as the nitrilotriacetate complex, was added to human transferrin at pH 6.0, and a second equivalent of ^{55}Fe was added after raising the pH to 7.5, subsequent lowering of the pH to 6.3 labilised predominantly ^{55}Fe. Harris (1977*a*) and Cannon and Chasteen (1975) referred to the acid-stable and acid-labile sites as site A and site B, respectively: subsequently, Evans and Williams (1978) showed by analysis of monoferric fragments that these corresponded to the C- and N-terminal sites, respectively. Iron added to ovotransferrin as the nitrilotriacetate complex at pH 6.0 also binds preferentially to the C-terminal site (Evans and Holbrook, 1975; Williams *et al.*, 1978), but with bovine transferrin the N-terminal site is preferentially occupied under these conditions (Brock and Arzabe, 1976; Brock *et al.*, 1978*a*). One of the sites of rat transferrin also binds iron preferentially at low pH (Morgan *et al.*, 1978; Okada *et al.*, 1978) but with rabbit transferrin no difference between the sites was observed during dissociation of iron at acid pH (Princiotto and Zapolski, 1978). Differential binding and release of iron with respect to pH also occurs with human lactoferrin (Mazurier and Spik, 1980).

Further studies have revealed that the situation is more complex than a simple difference in binding affinities between the sites of some transferrins at low pH. Harris (1977*a*) found that preferential binding of iron to the C-terminal site occurred even when the addition was made at physiological pH. Iron nitrilotriacetate also reacts preferentially with the C-terminal site of rabbit transferrin (Heaphy and Williams, 1982*a*) despite the lack of difference between the sites when iron dissociates at acid pH. Furthermore, if ferric citrate was used instead of ferric nitrilotriacetate, the N-terminal site of human transferrin was occupied preferentially, while use of ferrous ascorbate as iron donor revealed no site preference (Zapolski and Princiotto, 1977*a*). These observations have been confirmed by a number of workers (Aisen *et al.*, 1978; Evans and Williams, 1978; Zapolski and Princiotto, 1980*a*), and it is now clear that when nitrilotriacetate or analogous chelators are used as iron donors the C-terminal site of human transferrin is preferentially occupied, whereas a variety of other donors (ferric citrate, ferric oxalate, ferrous ammonium sulphate or ferric chloride) donate iron preferentially to the N-terminal site. However, if transferrin partially saturated with one of the latter group of donors is left for several days, some of the iron migrates to the C-terminal site, this being the site with the higher affinity constant (Aisen *et al.*, 1978; Van Eijk *et al.*, 1978). It is thus evident that although one site may be favoured thermodynamically as a result of its greater affinity constant, it may not necessarily be favoured kinetically, this factor depending upon the nature of the iron donor.

Recently, the influence of pH and salt concentration on the distribution of iron between the sites of human transferrin has been carefully investigated (Chasteen and Williams, 1981; Williams *et al.*, 1982*c*), using the urea-polyacrylamide gel electrophoresis system of Makey and Seal (1976) to separate and quantify the four forms of transferrin, i.e. apoTf, $Fe_N Tf$, $TfFe_C$, and $Fe_2 Tf$. Metal binding to the two sites can be described by four site constants (Aisen *et al.*, 1978) and the equilibria can be represented by:

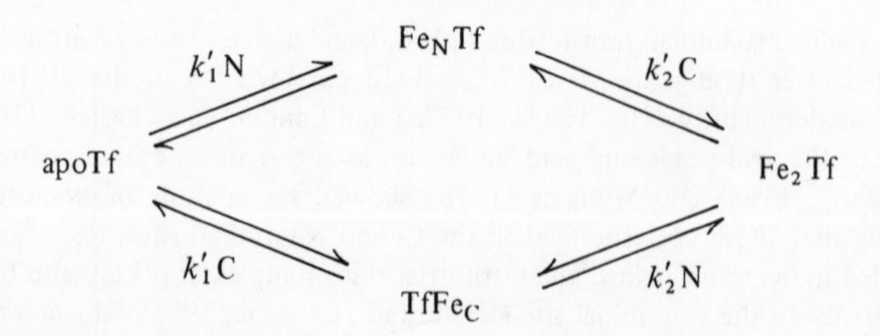

The constants $k_1'N$, $k_1'C$, $k_2'N$ and $k_2'C$ are dependent upon temperature, pH, ionic strength, concentration of anion(s) capable of forming a ternary complex with iron and transferrin, and other factors influencing iron binding. If $k_1'N/k_1'C > 1$, then iron added to apotransferrin will bind preferentially to the N-terminal site *provided thermodynamic equilibrium is obtained* and vice versa. Kinetic considerations may, nevertheless, result in iron binding initially to the opposite site. At low pH, i.e. within the range 6.1-6.9, the value of $k_1'N/k_1'C$ is approximately 0.06, in both the presence and absence of 0.5 M NaCl, but at high pH (> 8.2), the figures become 1.03 and 1.85 in the absence or presence of 0.5 M NaCl, respectively. Thus the C-terminal site becomes thermodynamically less favoured as the pH increases, and this is accentuated in the presence of high salt concentration (figure 5.9). Higher salt concentrations thus facilitate release from the C-terminal site and hinder release from the N-terminal site, confirming the earlier conclusions of Van Eijk *et al.* (1978). These effects of salt are due to the chloride ion and other anions such as thiocyanate and perchlorate, and cations such as Li^+ have similar effects (Baldwin, 1980; Baldwin and De Sousa, 1981; Baldwin *et al.*, 1982; Folajtar and Chasteen, 1982). The factor $k_2'N/k_1'N$ is a measure of cooperativity between the sites (it can be shown that $k_2'N/k_1'N = k_2'C/k_1'C$). If this factor is greater than unity, then there is positive cooperativity, and vice versa. In the absence of NaCl, there is negative cooperativity below pH 8.0, and positive cooperativity above this value. In the presence of 0.5 M NaCl, negative cooperativity exists below pH 6.9, and positive cooperativity at pH > 7.5. Between these two values binding is non-cooperative (figure 5.9). These studies should help to provide a basis for explaining possible functional differences between the sites.

It should be noted that although most studies have concentrated on binding of iron at physiological pH and its loss at lower pH values, metal binding capacity is also lost at high pH. Iron is lost, apparently in random fashion, between pH 9 and 10 (Zapolski and Princiotto, 1980*a*), but loss of Cu^{2+} occurs preferentially from the C-terminal site between pH 9.2 and 9.6 (Zweier, 1978).

Thus it can now be accepted, in contrast to earlier assumptions, that the iron-binding sites of transferrin may differ in their binding affinities and spectroscopic properties. The extent of these differences may vary among transferrins of different species, and appear to be particularly evident in human transferrin. Recent studies demonstrating the extent to which salt and pH influence the

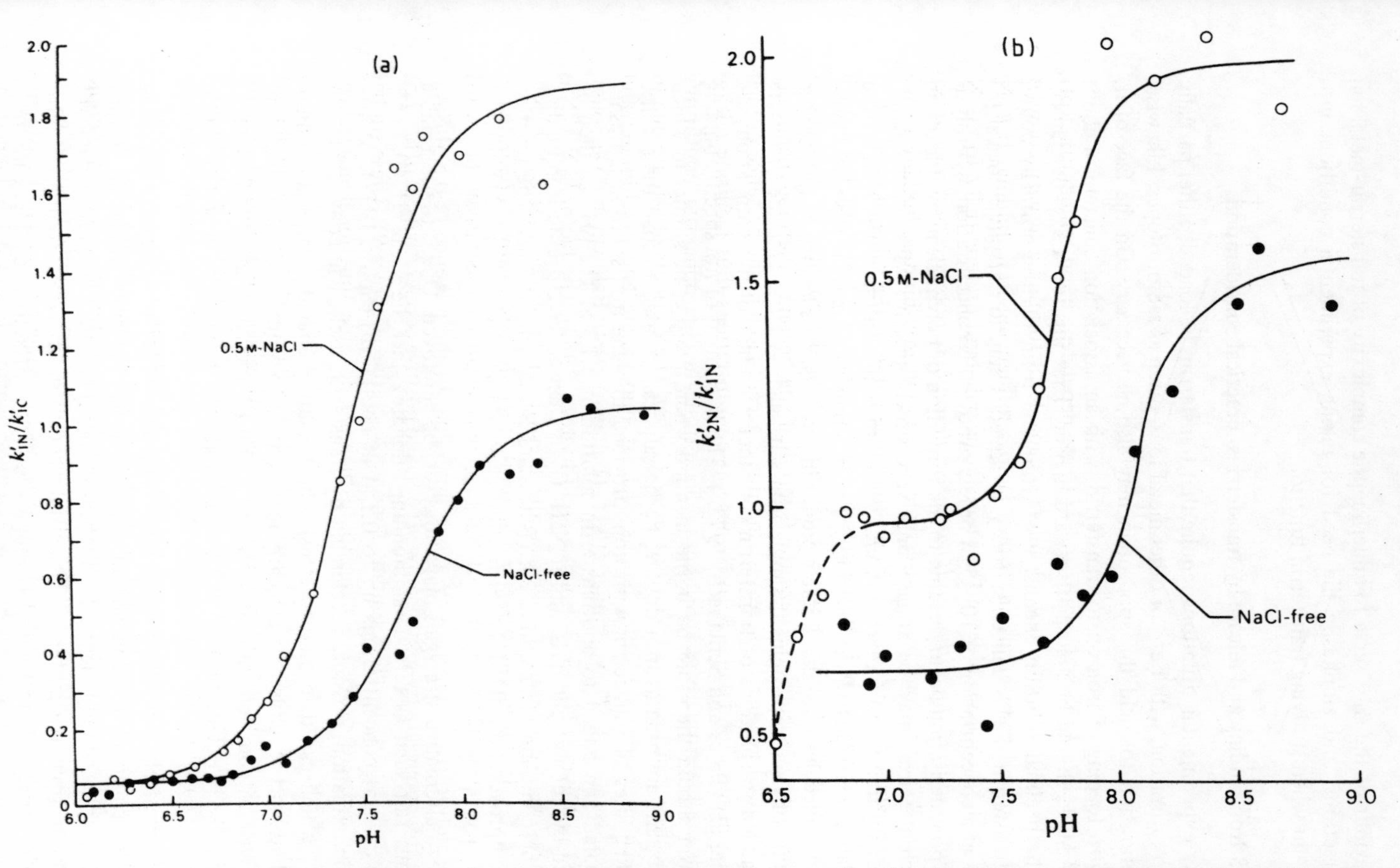

Figure 5.9 Variation of site preference factor, $k'_1{}_N/k'_1{}_C$ (a) and cooperativity factor, $k'_2{}_N/k'_1{}_N$ (b) of human transferrin with pH. Conditions: 1 mM $NaHCO_3$, 1 mM sodium citrate, 50 mM 4-morpholine ethanesulphonic acid, 50 mM 4-(2-hydroxyethyl)-1-piperazine ethanesulphonic acid, 50 mM 4-2(hydroxyethyl)-1-piperazine propane sulphonic acid at pH values shown at 37°C, with and without 0.5 M NaCl. From Chasteen and Williams (1981) with permission.

behaviour of the sites serve to underline the complexity of transferrin-metal ion interactions, and emphasise the need for taking experimental conditions into account when studying transferrin function.

5.5.3 Iron binding and release by transferrins: practical considerations

Many experimental situations require that transferrins be loaded, either partially or to saturation, with Fe^{3+} or other metal ions, often as radioisotopes. Likewise, desaturation to yield the apoprotein may also be necessary. On the face of it, *in vitro* loading of iron on to transferrin, with an association constant of about 10^{20} M^{-1}, should be a straightforward matter. That this is not so is due largely to the peculiar solution chemistry of iron. The form of iron normally bound by transferrin, Fe^{3+}, can exist freely in aqueous solution at physiological pH only at a concentration of 10^{-18} M or less. Above this concentration, hydrolysis and polymerisation occur, leading to the formation of polynuclear complexes, in which the iron may be at best only very slowly reactive with other ligands, including transferrin (Allerton *et al.*, 1966; Spiro and Saltman, 1969; Bates and Schlabach, 1973*a*). More alarmingly, such complexes may involve not only water molecules, but also suitable groups in proteins, thus leading to nonspecific binding of iron to proteins if simple iron salts such as ferric chloride or nitrate are added to solutions of transferrin (or other proteins), or to serum, at physiological pH (Bates and Schlabach, 1973*a*). The likelihood of nonspecific binding of iron should therefore be borne in mind when interpreting studies in which ferric salts are used as iron donors. Ferric salts can be used to load the specific binding sites of transferrin with iron, provided that the iron solution is freshly prepared and added to apotransferrin solution at pH 3.5. The pH can then be raised by adding solid Tris, and if $NaHCO_3$ is then added, specific iron-binding will be complete after about 2 h (Bates and Schlabach, 1973*a*). The reaction rate improves if the pH of the ferric chloride solution is lowered to 1.2 (Workman *et al.*, 1975).

Two alternative strategies may, however, be employed. One is to use a stable ferrous salt such as ferrous ammonium sulphate, since Fe^{2+}, as $(H_2O)_6Fe^{2+}$, may exist at concentrations up to 0.1 M at neutral pH (Spiro, 1977). Atmospheric oxidation, possibly accelerated by transferrin itself, will convert Fe^{2+} to Fe^{3+}, which can then bind to transferrin. Freshly prepared ferrous ammonium sulphate, pH 2, added to a bicarbonate-containing solution of apotransferrin, will lead to complete saturation within 3 min (Workman *et al.*, 1975; Huebers *et al.*, 1981*a*). This reagent will also saturate the transferrin in serum within 1.5 min.

Alternatively, Fe^{3+} may be stabilised at neutral pH by a suitable chelating agent: such an agent should be able to form a stable complex at neutral pH, but this complex should be both kinetically and thermodynamically capable of delivering iron to transferrin. The citrate and nitrilotriacetate complexes fulfil both criteria, and are frequently employed. For satisfactory loading, care must be taken to ensure that only reactive low molecular weight species are present in the chelate solution. Bates *et al.* (1967) have shown that in order to ensure the absence of polymers in neutral ferric citrate solutions a 20-fold excess of citrate ions should be present, and additional citrate should therefore be added to solutions of commercial ferric citrate preparations. Such preparations will otherwise polymerise in solution, leading to slow reaction with the protein and the possibility of nonspecific binding. Even when the presence of the low molecular weight complex is assured, solutions are best left overnight to ensure completion of the reaction. Nitrilotriacetate forms stable low molecular weight chelates at neutral pH at a 4:1 chelate to iron ratio (Schlabach and Bates, 1975), and donates its iron rapidly, though the N-terminal site of human transferrin takes about 30 min to become fully loaded (Zapolski and Princiotto, 1977*a*). Nitrilotriacetate may also occupy the anion-binding site of transferrin if bicarbonate is absent from the solution (Bates and Wernicke, 1971), and this is best avoided by the use of bicarbonate-containing buffers (Bates and Schlabach, 1973*a*). Chelates may also remain associated with the protein, and if necessary these can be removed by dialysis against $NaClO_4$ (Price and Gibson, 1972*b*). It has recently been reported that even when nitrilotriacetate is used as donor, some undialysable polynuclear iron complexes may form, and that these are best removed by gel filtration (Smit *et al.*, 1981).

It is important to realise that when subsaturating amounts of iron are added to apotransferrin the initial distribution of iron between the sites will reflect the kinetically rather than thermodynamically favoured interaction, so that redistribution may occur with time (Van Eijk *et al.*, 1978). The degree to which this occurs is affected by pH and salt concentration (Chasteen and Williams, 1981; Williams *et al.*, 1982*c*). It is therefore necessary to control experimental conditions if information on the individual iron-binding sites is sought.

Removal of iron from transferrin is normally achieved by dialysis against citrate at pH 5. This procedure is satisfactory, though it may be necessary subsequently to remove bound citrate as described previously. With lactoferrin, the higher binding affinity means that the dialysis must be carried out using citric acid at pH 2 to remove the iron (Masson and Heremans, 1968). However, it is now apparent that this procedure causes some irreversible changes in the structure and properties of the protein (Van Snick and Masson, 1976; Ainscough *et al.*, 1980). Dialysis against EDTA-phosphate-acetate buffer at pH 4 (Mazurier and Spik, 1980) offers a less drastic method of removing iron from lactoferrin.

5.5.4 Binding of metals other than iron

Although iron is the most important (and perhaps the only) metal bound by transferrin *in vivo*, transferrins are capable of binding a large number of other metal ions. In many cases it has been shown that these are bound at the same sites as iron. Aisen (1980) has proposed that four criteria should be met if a metal is to be considered as specifically bound by transferrin:

(1) The metal should be displaced by iron when the latter is added.

(2) Binding of (bi)carbonate, or another anion, should accompany metal binding.
(3) Not more than two metal ions should be bound per transferrin molecule.
(4) The metal ion should not bind to iron-saturated transferrin.

Metals which bind to transferrin or ovotransferrin and have been shown to fulfil most or all of these criteria include the divalent and trivalent transition metal ions Cr^{3+}, Cu^{2+}, Mn^{2+}, Co^{3+}, Co^{2+}, Cd^{2+}, Zn^{2+}, VO^{2+}, Sc^{2+}, Ga^{3+} and Ni^{2+} (Feeney and Komatsu, 1966; Aisen *et al.*, 1969; Tan and Woodworth, 1969; Woodworth *et al.*, 1970; Ford-Hutchinson and Perkins, 1971; Harris *et al.*, 1974; Cannon and Chasteen, 1975; Zweier, 1978; Harris, 1983), the trivalent lanthanides Er^{3+}, Tb^{3+}, Ho^{3+}, Nd^{3+} and Pr^{3+}(Luk, 1971; Meares and Ledbetter, 1977; O'Hara *et al.*, 1981), the tetravalent actinides Th^{4+} and Pu^{4+} (Stevens *et al.*, 1968; Pecoraro *et al.*, 1981; Harris *et al.*, 1981) and Pt^{2+} (Stjernholm *et al.*, 1978). Several of the transition metal ions have also been shown to bind to lactoferrin (Ainscough *et al.*, 1979, 1980; Carmichael and Vincent, 1979). In most cases two metal ions are bound per transferrin molecule but only single Nd^{3+} or Pr^{3+} ions are bound (Luk, 1971), presumably due to the inability of one of the sites to accommodate these large ions, which have atomic ionic radii of 0.98 and 0.99, respectively (Pecoraro *et al.*, 1981). On the other hand, more than two Zn^{2+} or Cu^{2+} ions may be bound per protein molecule, due to binding at sites other than the specific metal binding site (Jones and Perkins, 1965; Charlwood, 1979). Tan and Woodworth (1969) compared the relative binding affinities by spectroscopic displacement studies, and showed that the order of binding was $Fe^{3+} > Cr^{3+}$ and $Cu^{2+} > Mn^{3+}$, Co^{2+} and $Cd^{2+} > Zn^{2+} > Ni^{2+}$. Feeney and Komatsu (1966), quoting unpublished work of Inman (1956), reported that human transferrin bound Co^{3+} with greater affinity than Fe^{3+}, but this is not in accordance with more recent displacement studies using NMR spectroscopy (Zweier *et al.*, 1981) or visible spectroscopy (T. Mainou-Fowler and J. H. Brock, unpublished observations), both of which indicated that Co^{3+} is displaced by Fe^{3+}. In the case of lactoferrin, however, there is some suggestion that the affinity for other metals may approach that for iron, since addition of Fe^{3+} to Cr_2Lf results in the liberation of only one Cr^{3+} ion (Ainscough *et al.*, 1980). The affinity of bovine lactoferrin for Fe^{3+} and VO^{2+} must also be similar for both ions, since a large excess of one will displace the other from the protein (Carmichael and Vincent, 1979). Thus conditions 1 and 4 for specific binding listed earlier may not always apply in the case of lactoferrin. In accord with its ability to retain Fe^{3+} down to low pH values, bovine lactoferrin will also retain VO^{2+} down to pH 4–4.2 (Carmichael and Vincent, 1979), while removal of Ga^{3+} from transferrin by lactoferrin becomes more rapid at lower pH (Hoffer *et al.*, 1977; Weiner *et al.*, 1981).

5.5.5 Location of the metal-binding sites

From electron spin resonance spectral data it was concluded that the iron-binding sites of transferrin and ovotransferrin must be at least 10 Å apart (Aasa *et al.*,

1963; Windle *et al.*, 1963). More recently, fluorescent emission from excited bound Tb^{3+} has been employed. O'Hara *et al.* (1981) measured energy transfer from Tb^{3+} bound at one site to Mn^{3+} or Fe^{3+} bound at the other. By comparing the resultant reduction in emission lifetime relative to Tb_2Tf, they concluded that the intersite distance was 35.5 ± 4.5 Å. An earlier figure of 25 Å, based on fluorescence quenching (Meares and Ledbetter, 1977), was probably too low due to overestimation of quenching, and the figure of 43 Å obtained by Luk (1971), using the same technique as O'Hara *et al.* (1981), was probably erroneous due to the use of $FeCl_3$ to prepare monoferric transferrin, which could have led to non-specific binding of iron and excessive fluorescence lifetime reduction. A recent investigation of the intersite distance using NMR spectroscopy gave a value of $>$ 16 Å for human transferrin (Zweier *et al.*, 1981). Unfortunately, X-ray crystallographic studies, which offer great promise for localising the iron-binding sites, have not yet reached the stage where this has been achieved. However, Abola *et al.* (1982) have tentatively located one of the iron-binding sites of ovotransferrin in what appears to be a shallow groove about 12 Å deep. This agrees with an estimate of 17 Å for the distance of the binding sites from the surface of the transferrin molecule by using terbium fluorescence (Yeh and Meares, 1980). Studies on the binding of synergistic anions (Schlabach and Bates, 1975; Harris and Gelb, 1980), accessibility of water molecules (Koenig and Schillinger, 1969), and using circular dichroism (Mazurier *et al.*, 1976) also suggest that the sites must be near the surface of the molecule.

5.5.6 The requirement for binding of a synergistic anion

It was observed many years ago that addition of iron to apotransferrin only resulted in the formation of the characteristic red complex if bicarbonate were available (Schade *et al.*, 1949), and it was thus concluded that metal binding only occurred if synergistic binding of bicarbonate also took place. This observation has been amply confirmed (Warner and Weber, 1953; Aisen *et al.*, 1967; Price and Gibson, 1972*b*; Harris *et al.*, 1974; Bates and Schlabach, 1975; Tsang *et al.*, 1975; Gelb and Harris, 1980). The results of EPR studies suggesting that iron–transferrin complexes could be formed in the absence of bicarbonate (or another anion) (Aisen *et al.*, 1967; Aasa and Aisen, 1968) were later shown to be due to contamination with citrate (Price and Gibson, 1972*b*), and it is now clear that only nonspecific binding of iron can occur without the participation of a synergistic anion (Bates and Schlabach, 1975; Tsang *et al.*, 1975). Bicarbonate is also required for iron binding to lactoferrin (Masson and Heremans, 1968). Some studies have indicated that carbonate rather than bicarbonate is the form of the anion that is bound (Bates and Schlabach, 1973*b*; Schlabach and Bates, 1975), but recent proton-release studies favour bicarbonate (Gelb and Harris, 1980). This controversial point has recently been investigated by Zweier *et al.* (1981) using ^{13}C-NMR spectroscopy of the Co^{3+}–transferrin complex. They concluded that the anion was present as carbonate at the N-terminal site, but probably as bicarbonate at the C-terminal site. It therefore seems likely that the form of the anion bound varies with the site involved, and may also vary according to the metal ion bound and perhaps with salt concentration or pH. Such variability could explain the earlier conflicting results.

A variety of other anions may substitute for (bi)carbonate at the anion-binding site (Young and Perkins, 1968; Price and Gibson, 1972*b*; Aisen *et al.*, 1973*b*; Egyed, 1973; Schlabach and Bates, 1975), though almost all such complexes are less stable than those containing (bi)carbonate. Other anions will therefore be displaced unless CO_2 is excluded from the system. A definitive study of the anion-binding requirements of transferrin was carried out by Schlabach and Bates (1975) who examined the ability of 25 organic and six inorganic anions to promote iron binding. None of the inorganic anions (other than (bi)carbonate) was active, and from the structures of the 22 organic anions that did promote iron-binding, it was proposed that the anion must possess a carboxylic acid group, and another functional group capable of acting as an Fe-ligand within 6.3 Å. These must be capable of achieving a carbonate-like configuration, and their Van der Waals radii must be able to fit into a site 3 Å deep, 6 Å wide and 4–6 Å or more long. The close proximity of metal and anion has been confirmed in EPR (Najarian *et al.*, 1978; Zweier *et al.*, 1979) and NMR (Harris *et al.*, 1974) studies.

The work of Schlabach and Bates (1975) was extended by Campbell and Chasteen (1977), who took advantage of the fact that between pH 7.5 and 9 the N- and C-terminal sites of transferrin exist in different conformations, designated B and A respectively, when VO^{2+} is bound, and these can be distinguished by their EPR spectra (Chasteen *et al.*, 1977). Nitrilotriacetate, malonate, oxalate and (bi)carbonate could be bound with VO^{2+} in both the A and B conformations; maleate only in the A conformation, and a variety of monocarboxylic acids with weakly acidic proximal groups allowed binding only in the B conformation. When Cu^{2+} is bound, however, oxalate will bind only at the B (N-terminal) site (Zweier and Aisen, 1977). Bromopyruvate can act as a synergistic anion, but can also bind covalently at the metal-binding site, thus providing an affinity label for the site (Patch and Carrano, 1982). Under these conditions the ternary complex is stabilised to the extent that some iron can remain bound to transferrin at pH as low as 4. From these studies it is now generally accepted that transferrin binds the metal and anion at interlocking sites, in which the carboxyl group(s) interact with the protein and the proximal functional group with the metal. The structure of the anion may be represented as:

```
        R
        |
  L – CH – C = O
           |
           O⁻
```

where L is the iron-binding group. The substituent R may vary considerably and may project from the surface of the transferrin molecule, since even a molecule as large as xylenol orange can act as a synergistic anion (Harris and Gelb, 1980).

Some iron chelates, notably nitrilotriacetate, can also act as synergistic anions, and consequently addition of such chelates to apotransferrin may result in the formation of the ternary transferrin-iron-chelate complex. The iron in this complex is more labile than when (bi)carbonate acts as synergistic anion (Bates and Schlabach, 1973*b*). It was originally thought that bicarbonate would, if present, displace the chelate anion directly, but more recent data suggest that transient quaternary Fe-transferrin-anion-bicarbonate complexes are formed (Kojima and Bates, 1979; Bates, 1982; Cowart *et al.*, 1982), from which the chelate anion is subsequently released. The proposed mechanism can be represented by the following equations (Bates, 1982):

(1) Fe^{3+}—Tf—Chel + HCO_3^- = HCO_3^-—Fe^{3+}—Tf—Chel
(2) HCO_3^-—Fe^{3+}—Tf—Chel = Fe^{3+}—Chel + apoTf + HCO_3^-
(3) apoTf + HCO_3^- = apoTf—HCO_3^-
(4) apoTf—HCO_3^- + Fe^{3+}—Chel = Chel—Fe^{3+}—Tf—HCO_3^-
(3) Chel—Fe^{3+}—Tf—HCO_3^- = Chel + Fe^{3+}—Tf—HCO_3^-

Citrate, although possessing many of the characteristics of a synergistic anion, does not form a coloured complex with iron and transferrin (Bates and Schlabach, 1975; Phelps and Antonini, 1975). Its ability to destabilise the iron-(bi)carbonate-transferrin complex (Aisen and Leibman, 1968) does not necessarily imply an ability to compete with the anion, since borate and phosphate, which cannot act as synergistic anions (Schlabach and Bates, 1975) also destabilise the ternary complex (Rogers *et al.*, 1977*a*).

5.5.7 Ligands at the metal-binding sites

Ferric iron bound to transferrin possesses an octahedral ligand field with six coordinating ligands. Studies on the binding of synergistic anions described in the previous section indicate that one ligand is almost certainly coordinated with the anion. A water molecule also appears to occupy one ligand (Koenig and Schillinger, 1969; Zweier and Aisen, 1977). The remaining ligands must therefore be coordinated to the protein, and considerable effort has been devoted to elucidating the groups involved. Three different approaches have been used: hydrogen ion titration, spectroscopic studies, and chemical modification.

Early titration studies with human transferrin or ovotransferrin established that three protons were released per Fe^{3+} bound, and that groups with a p$K > 10$ were involved (Warner and Weber, 1953; Wishnia *et al.*, 1961; Windle *et al.*, 1963). These studies strongly suggested the involvement of three tyrosines at each metal-binding site, and have been confirmed in a more recent investigation (Gelb and Harris, 1980). Various spectroscopic studies have also implicated tyrosine in the metal-binding site; the techniques used include Raman resonance spectroscopy (Carey and Young, 1974; Gaber *et al.*, 1974; Tomimatsu *et al.*, 1976), proton magnetic resonance (Woodworth *et al.*, 1970) and ultraviolet difference spectroscopy (Tan and Woodworth, 1969; Krysteva *et al.*, 1976;

Tomimatsu and Donovan, 1976; Gelb and Harris, 1980; Pecoraro *et al*., 1981). Most workers concluded that three tyrosines were involved at each site, though Tan and Woodworth (1969) considered two more likely, and one less tyrosine was thought to be involved when divalent ions such as Cu^{2+} were bound (Windle *et al*., 1963; Tan and Woodworth, 1969; Zweier and Aisen, 1977). However, interpretation of ultraviolet difference spectra has generally been based on the assumption that the change in absorbance when tyrosine coordinates to a metal ion is constant regardless of the metal involved. By studying the binding of various metal ions to ethylene-bis[(*o*-hydroxyphenyl)glycine] (EHPG), a chelate analogue of transferrin, Pecoraro *et al*. (1981) have shown that this assumption is unjustified, since different Δe values were obtained for chelation by EHPG, and that both trivalent and divalent cations were coordinated to only two tyrosines. When tetravalent Th^{4+} was bound, two tyrosines were involved at one site and only one at the other site (Harris *et al*., 1981). To reconcile these results with proton release studies they propose that a hydroxo ion, rather than a water molecule, is coordinated to Fe^{3+}, but that water itself is coordinated to Cu^{2+}.

Chemical modification studies implicating tyrosine residues have included nitration (Line *et al*., 1967; Tsao *et al*., 1974*b*; Williams, 1982*b*), acetylation (Komatsu and Feeney, 1967), iodination (Phillips and Azari, 1972), bromoacetylation (Line *et al*., 1967), ethoxyformylation (Krysteva *et al*., 1975; Rogers *et al*., 1977*b*), photooxidation (Rogers *et al*., 1977*b*) and periodate oxidation Geoghegan *et al*., 1980). Again, some authors have concluded that three residues are involved, and others only two. A possible pitfall in chemical modification studies is that some residues may be protected when iron is bound, not because they are directly involved in metal binding but as a result of concomitant conformational changes in the structure of the protein. The difficulty of interpreting results was illustrated by Williams (1982*b*), who reported that when iron was bound by ovotransferrin, 7 mol of tyrosine per mole of protein were protected against nitration. However, subsequent sequence analysis of the nitrated peptides showed that ten different tyrosine residues were involved, indicating that some residues must have undergone partial nitration. By taking into account the homology in primary structure between the two domains of ovotransferrin, it was possible to assign eight of the ten protected tyrosines to four conserved pairs at positions 82, 92, 188 and 191 in the N-terminal domain and at positions 415, 431, 521 and 524 in the C-terminal domain. Clearly, some of these residues are likely candidates for involvement in the iron-binding site.

The other amino acid residue that is firmly implicated in the metal-binding site is histidine. Early ESR and EPR studies indicated that the binding of both Fe^{3+} and Cu^{2+} to transferrin, ovotransferrin and lactoferrin involved two histidyl residues per metal ion (Aasa *et al*., 1963; Windle *et al*., 1963). More recently, Mazurier *et al*. (1977) reported that two independent subspectra were obtained from dicupric transferrin and lactoferrin, and from these they concluded that at the B [= N-terminal] site only one histidyl residue was involved whereas at least three were involved at the A [= C-terminal] site. However, Froncisz and

Aisen (1982) have pointed out that in these studies the complexes were formed at pH 8.6, which causes one site to form an abnormal complex that does not involve binding of a synergistic anion. Comparable studies carried out at lower pH revealed only one tyrosine per metal-binding site (Zweier and Aisen, 1977).

Spectroscopic and circular dichroism studies have implicated tryptophan as an iron-binding ligand (Lehrer, 1969; Tomimatsu and Donovan, 1976; Mazurier *et al*., 1976), but it seems likely that the observed changes were due to modification of the protein conformation by metal binding rather than to actual involvement of tryptophan as a ligand, since sulphenylation of all the tryptophan residues of transferrin did not cause loss of iron binding (Ford-Hutchinson and Perkins, 1972).

The involvement of arginine and lysine residues in iron binding has been reported, but this probably reflects their role as ligands for the synergistic anion rather than the metal itself (Zweier and Aisen, 1977; Rogers *et al*., 1978). Zweier *et al*. (1981), using NMR spectroscopy, have concluded that arginine is involved at the N-terminal site, and either lysine or arginine at the C-terminal site. However, Shewale and Brew (1982) have reported that differential kinetic labelling of the lysine residues in human transferrin implicates both Lys-206, from the N-terminal domain, and Lys-534, from the C-terminal domain, in the iron-binding sites. This also suggests that the neighbouring histidine residues (207 and 535) are strong candidates for involvement in metal binding.

In conclusion, therefore, the most recent data tend to favour the coordination of two tyrosines and either one or two histidines to each bound metal ion, the remaining metal ligands being occupied by a water molecule or hydroxo ion, and the synergistic anion. Binding of the anion appears to involve arginine and perhaps lysine. The results of chemical modification have been summarised by Feeney *et al.* (1983) and Chasteen (1983*b*).

5.6 FUNCTION OF THE TRANSFERRINS

5.6.1 Introduction

The major function of serum transferrin is the transport of iron from sites of storage and absorption and of erythrocyte catabolism, to haematopoietic tissue. Most of this iron will be destined for erythroid precursors, though a small but significant proportion is also taken up by non-erythroid cells, an activity that appears to play an important part in cell transformation and division. Some iron may also be delivered to or taken up from the hepatic stores, and in pregnancy delivered to the placenta for transport to the fetus. Although the mechanism by which transferrin donates its iron to cells has been extensively investigated, many features of the reaction are still not well understood. Much less work has been carried out on the acquisition of iron by transferrin *in vivo*.

A second function of transferrin is defence against systemic infection. The ability of transferrin to bind iron with high affinity makes it difficult for microorganisms to acquire this metal, thus impeding their growth. This activity of transferrin can be considered part of the body's nonspecific immunity against infection, though its importance relative to other mechanisms is difficult to assess. Lactoferrin and ovotransferrin are also thought to play roles as antimicrobial agents, protecting secretory surfaces and the developing embryo, respectively. Unlike serum transferrin, however, neither protein appears to play any significant role in iron transport, although lactoferrin may play a minor part in certain circumstances. It may also act as a regulator of the immune system. Several other functions have been proposed for both lactoferrin and transferrin, and these will be discussed below.

5.6.2 Cellular uptake of transferrin-bound iron

The uptake of transferrin-bound iron by erythroid precursors was first demonstrated by Walsh *et al.* (1949), and it is now accepted that transferrin is the only iron donor of significance to the erythroid marrow. It is evident that the high affinity of transferrin for iron means that a specialised mechanism must exist to enable iron to be released from the protein and incorporated into the cell. The experimental system that served as a model for the vast majority of investigations up to the late 1970s was first introduced by Jandl *et al.* (1959). These workers used reticulocyte-rich blood as a source of immature erythroid cells, which were incubated with $^{59}FeCl_3$ in the presence or absence of transferrin and other proteins. Only when transferrin was present were significant amounts of ^{59}Fe incorporated into haem; in all other cases any iron taken up by the cells was found to be associated with the membrane, due to what would now be recognised as nonspecific binding resulting from the use of $FeCl_3$.

From current knowledge, the iron uptake process may be split into four stages:

(1) interaction of transferrin with a specific membrane receptor;
(2) internalisation of the transferrin-receptor complex;
(3) removal of iron from transferrin;
(4) release of apotransferrin by the cell.

It should be noted, however, that the importance of step (2) is not universally accepted, and step (4) may not apply in all cases.

5.6.2.1 Interaction of transferrin with cell membrane receptors

The existence of a specific receptor for transferrin in the reticulocyte membrane was first suggested from the initial work of Jandl *et al.* (1959). Further progress was made by Garrett *et al.* (1973), who demonstrated that ^{125}I-transferrin was bound by reticulocyte membranes, and that the iron content of the bound transferrin was lower than that of the transferrin originally added. The transferrin could be extracted from the membrane with detergents, bound to another macromolecule which was thought to represent the transferrin receptor. These studies provided the basis for a large number of subsequent investigations, commencing with Speyer and Fielding (1974), concerned with the isolation and characterisation of the transferrin receptor.

Almost all attempts to isolate and characterise the transferrin receptor up to the end of the 1970s used reticulocyte-rich blood (commonly referred to simply as 'reticulocytes') as the starting material. The results obtained were frequently discordant, particularly with respect to molecular weight, with estimates ranging from 30 000 (Light, 1977) to 400 000 (Leibman and Aisen, 1977). Although the variability in molecular weight values, which have been summarised by Bezkorovainy (1980), may be partly attributed to methodological differences, the difficulties associated with working with detergent-solubilised material and the heterogeneity of the cell suspension used as starting material may also be responsible. It is perhaps for the latter reason that more consistent results have been obtained in recent studies of receptors isolated from non-erythroid cells, usually from cultured cell lines. These studies have generally indicated that the receptor has a molecular weight of 170 000–200 000, and is composed of two subunits, each of molecular weight about 95 000 (Hamilton *et al.*, 1979; Seligman *et al.*, 1979; Wada *et al.*, 1979; Fernandez-Pol and Klos, 1980; Enns and Sussman, 1981*a*; Goding and Burns, 1981; Goding and Harris, 1981; Shindelman *et al.*, 1981; Sutherland *et al.*, 1981; Trowbridge and Omary, 1981; Bleil and Bretscher, 1982), a conclusion which has also been reached by at least some of the investigators working with reticulocytes (Leibman and Aisen, 1977; Witt and Woodworth, 1978; Ecarot-Charrier *et al.*, 1980). The placental transferrin receptor also appears to be similar (Enns and Sussman, 1981*a*).

The receptor is glycosylated (Leibman and Aisen, 1977; Seligman *et al.*, 1979; Wada *et al.*, 1979; Núñez *et al.*, 1981; Lebman *et al.*, 1982; Newman *et al.*, 1982; Schneider *et al.*, 1982). Each subunit is thought to contain three glycan chains composed of galactose, N-acetyl glucosamine and sialic acid, two of which are also unusually rich in mannose (Newman *et al.*, 1982). The receptor also contains covalently bound palmitic acid which may anchor the receptor to the membrane (Omary and Trowbridge, 1981*a, b*) and is phosphorylated (Schneider *et al.*, 1982). A fragment of about 70 000 molecular weight which retains its transferrin-binding ability can be proteolytically cleaved from the receptor (Bleil and Bretscher, 1982; Schneider *et al.*, 1982), indicating that a major portion of the receptor probably protrudes from the membrane. A proposed model of the structure of the receptor is shown in figure 5.10. Interestingly, studies with mouse × human somatic cell hybrids indicate that the genes which code for both transferrin and the receptor are located on chromosome 3 (Enns *et al.*, 1982; Goodfellow *et al.*, 1982).

Many workers have used Scatchard analysis of saturation binding data to estimate the number and affinity of transferrin receptors on a wide variety of cell types, and some of the more recent data are summarised in table 5.2. The number of receptors on erythroid precursors decreases as the cell matures (Núñez *et al.*, 1977; Van Bockxmeer and Morgan, 1979; Frazier *et al.*, 1982; Iacopetta *et al.*, 1982). Muller and Shinitzky (1979), by studying the effect of cholesterol

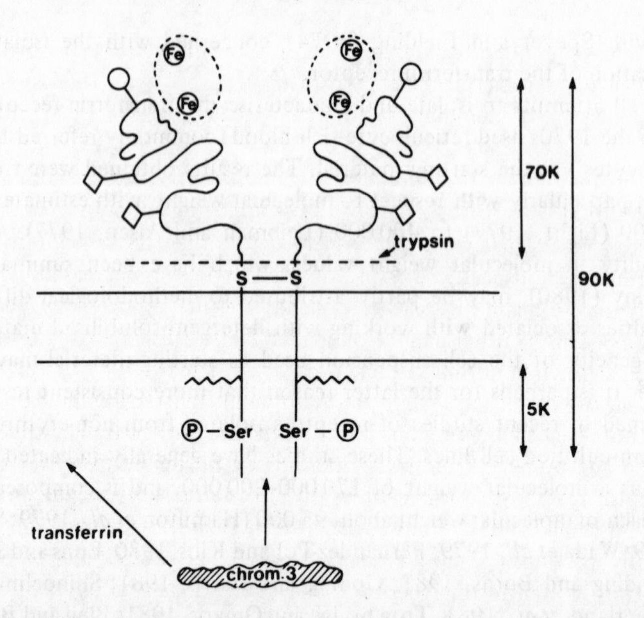

Figure 5.10 Schematic representation of the cell surface transferrin receptor. The positions of fatty acids, phosphoserine and disulphide bonds are deduced by analogy with the structure of other transmembrane proteins. Key: ◇, High-mannose oligosaccharide chain; ○, complex-type chain; zigzag line, covalently bound fatty acid. From Newman *et al.* (1982) with permission.

on the number of transferrin receptors on bone-marrow cells, proposed that decreased membrane lipid fluidity may result in the expression of progressively fewer receptors, but Pan and Johnstone (1983) have shown that receptors are shed as the cell matures. Although it has been suggested that the affinity of the receptor might also decrease during maturation (Light, 1978), more recent studies have not revealed any evidence of this, and while there is some variation in the affinity constants shown in table 5.2 these seem more likely to be due to variations in experimental technique.

In non-erythroid cells the possession of transferrin receptors is associated with cell transformation and division (Larrick and Cresswell, 1979*a*; Galbraith *et al.*, 1980*a*; Larrick and Logue, 1980; Brock and Rankin, 1981; Shindelman *et al.*, 1981; Tei *et al.*, 1982, Yeh *et al.*, 1982), and it is now known that the proliferation-associated antigen recognised by monoclonal antibody OKT9 is in fact the transferrin receptor (Goding and Burns, 1981; Sutherland *et al.*, 1981; Trowbridge and Omary, 1981). Some of the cell types in which transferrin receptors have been identified are shown in table 5.2. Differentiation of transformed cells normally results in a decrease in the number of receptors expressed (Tei *et al.*, 1982; Yeh *et al.*, 1982) and possibly also a decrease in their affinity (Tei *et al.*, 1982). The Friend erythroleukaemic cell is an exception to this rule, however, because differentiation is accompanied by initiation of haemoglobin synthesis, and consequently the number of transferrin receptors expressed is maintained or increased (Hu *et al*., 1977; Glass *et al*., 1978; Yeoh and Morgan, 1979; Wilczynska and Schulman, 1980; Iacopetta *et al*., 1982; Trowbridge *et al*., 1982).

Iron status may affect the number of receptors expressed on both erythroid and non-erythroid cells. Black *et al*. (1979) found that reticulocytes from iron-deficient rats possessed fewer transferrin receptors than those from iron-replete animals, although nonspecific binding of transferrin was increased. This latter observation may explain the earlier contrary results reported by Okada and Brown (1977). On the other hand, iron-loaded HeLa cells expressed fewer receptors than cells grown in normal media (Ward *et al*., 1982).

Studies on the molecular weight, structure and transferrin-binding properties have so far not revealed any obvious differences between the transferrin receptors of erythroid and non-erythroid cells, and antisera to the human placental transferrin receptor react with the receptors on both reticulocytes and tumour cells (Enns and Sussman, 1981*b*; Enns *et al*., 1981). However, Lebman *et al*. (1982) have recently produced a monoclonal antibody that reacts preferentially with human erythroid transferrin receptors, and Trowbridge *et al*. (1982) have produced another monoclonal antibody that reacts with transformed murine cells but not with erythroid precursors. Whether these antigenic differences reflect genuine structural differences, or simply the relative accessibility of antigenic sites is not yet known.

The receptor shows greatest affinity for iron-saturated transferrin, that for apotransferrin being about two orders of magnitude lower at physiological pH (Jandl and Katz, 1963; Kornfeld, 1969, Van Bockxmeer *et al*., 1978; Verhoef *et al*., 1979; Young and Aisen, 1981). Some workers have nevertheless reported no difference in the binding of apo- and iron-containing transferrin (Hamilton *et al*., 1979; Ecarot-Charrier *et al*., 1980; Karin and Mintz, 1981; Ward *et al*., 1982), but in these studies the concentrations of transferrin used were of the order of a few micrograms per millilitre. Since culture media can contain sufficient iron to saturate up to 10 $\mu g\ ml^{-1}$ of transferrin (Brock and Rankin, 1981), it seems likely that the apotransferrin was converted to Fe_2-transferrin.

The binding of transferrin to the membrane receptor is pH-dependent. Optimal binding appears to occur at pH 7–8 (Van Bockxmeer *et al*., 1978; Shisheva *et al*., 1982). However, if the pH is reduced to about 5, bound ^{125}I-transferrin can no longer be rapidly displaced by excess unlabelled transferrin, suggesting that a conformational change effectively locks the transferrin to the receptor (Seligman *et al*., 1979; Wada *et al*., 1979; Ecarot-Charrier *et al*., 1980). Moreover, the affinity of the receptor for apotransferrin is greater at about pH 5 than at pH 7 (Dautry-Varsat *et al*., 1983; Klausner *et al*., 1983*a*).

The receptor is specific for transferrin. Neither lactoferrin nor ovotransferrin will bind to or donate iron to mammalian reticulocytes (Zapolski and Princiotto, 1976; Brock and Esparza, 1979; Cox *et al*., 1979; Van Bockxmeer and Morgan, 1982), earlier contrary results (Jandl *et al*., 1959; Williams, 1975; Line *et al*.,

Table 5.2 Transferrin receptors in various cell types

Cell	Species	Number of receptors per cell	Binding affinity for Fe_2Tf (litres/M)	Comments	Reference
Early normoblasts	rat	300 000	4.2×10^6		Iacopetta *et al.* (1982)
Intermediate normoblasts	rat	800 000	3.3×10^6		Iacopetta *et al.* (1982)
Pronormoblasts	mouse	14 600 000	–	Not corrected for nonspecific binding	Núñez *et al.* (1977)
Normoblasts	mouse	1 000 000–3 570 000	–	Not corrected for nonspecific binding	Núñez *et al.* (1977)
Nucleated red blood cells and orthochromatophilic normoblasts	mouse	720 000	–	Not corrected for nonspecific binding	Núñez *et al.* (1977)
Bone-marrow cells	rat	330 000	2.5×10^6	20–50% saturated transferrin used	Verhoef and Noordeloos, (1977)
Reticulocytes	rabbit	800 000	3×10^7		Schulman *et al.* (1981)
Reticulocytes	rabbit	100 000	4.1×10^7		Van Bockxmeer *et al.* (1978)
Reticulocytes	rat	189 000	4.1×10^6	20–50% saturated transferrin used	Verhoef and Noordeloos (1977)
Reticulocytes	rat	105 000	4.2×10^6		Iacopetta *et al.* (1982)
Reticulocytes	mouse	86 000	4.2×10^6		Iacopetta *et al.* (1982)
Reticulocytes	mouse	61 000	5.9×10^8	Not corrected for nonspecific binding	De Abrew (1981)
Reticulocytes	human	383 000	$1\text{–}3 \times 10^7$	Relatively immature cells	Frazier *et al.* (1982)
		46 000		Relatively mature cells	
Reticulocytes	chick	230 000	1.2×10^6	Aerobic conditions	Woodworth *et al.* (1982)
	chick	192 000	1.3×10^6	Anaerobic conditions	

Teratocarcinoma cells	mouse	5 700	1.5×10^8	Human transferrin used. Affinity calculated from authors' data	Karin and Mintz (1981)
Myeloid leukaemic cells	mouse	450 000	3.73×10^7	Undifferentiated	Tei *et al.* (1982)
	mouse	285 000	1.89×10^7	Differentiated	
Thymic lymphoma	mouse	120 000	5.1×10^8	Calculated from mean of three experiments: not corrected for non-specific binding	De Abrew (1981)
Embryo fibroblasts	rat	90 000	1.1×10^7	High-affinity sites	Octave *et al.* (1981)
	rat	950 000	1.6×10^5	Low-affinity sites	
Lymphoblastoid line (Raji)	human	500 000	5×10^7		Schulman *et al.* (1981)
T-lymphoblastoid cells	human	60 000	1×10^8	Figure given for affinity in original publication (1×10^{12}) is wrong (P. Cresswell, pers. comm.)	Larrick and Cresswell (1979*b*)
B-lymphoblastoid cells	human	30 000	1×10^8		
K562 leukaemic cell line	human	1 000 000	8×10^7		Schulman *et al.* (1981)
K562 leukaemic cell line	human	198 000	5–8 $\times 10^8$	Proliferating	Frazier *et al.* (1982)
		44 000		Non proliferating	
K562 leukaemic cell line	human		5.3×10^8	Fe_2-transferrin pH 7.3	Klausner *et al.* (1983*a*)
K562 leukaemic cell line	human	150 000	2.1×10^7	Apotranferrin pH 7.3	
			4.8×10^7	Apotransferrin pH 4.8	

Table 5.2 (*cont'd.*)

Cell	Species	Number of receptors per cell	Binding affinity for Fe_2Tf (litres/M)	Comments	Reference
Choriocarcinoma cell line	human	370 000	4.25×10^8		Hamilton *et al.* (1979)
HeLa cells	human	2 800 000	3.7×10^7	Normal medium: calculated from authors' data	Ward *et al.* (1982)
	human	1 300 000	5.2×10^7	Fe-supplemented medium: figures calculated from authors' data	Ward *et al.* (1982)
Hepatoma HepG2 cell line	human	60 000	1.4×10^8	Fe_2-transferrin pH 7.3	Dautry-Varsat *et al.* (1982)
	human	–	1.4×10^7	Apotransferrin pH 7.3	
	human	35 000	7.7×10^7	Fe_2-transferrin pH 5.4	
	human	58 000	7.7×10^7	Apotransferrin pH 5.4	
Fibroblasts	human	59 000	$1–2.5 \times 10^7$	Proliferating	Frazier *et al.* (1982)
	human	18 000		Nonproliferating	
Transformed peripheral blood lymphocytes	human	80 000	1.8×10^8	Corrected for presence of nontransformed cells	Galbraith *et al.* (1980*c*)
Hepatocytes	rat	63 000	1.6×10^7	Fe_2-transferrin	Young and Aisen (1980, 1981)
	rat	63 000	7.2×10^5	Apotransferrin	
Mammary gland cells	rabbit	700 000	3.2×10^{11}	High-affinity sites	Shisheva *et al.* (1982)
		700 000	4.2×10^{10}	Low-affinity sites	
Spleen cells	mouse	50 000	4.3×10^8	Not corrected for nonspecific binding or non-reactive cells	De Abrew (1981)
Thymus cells	mouse	< 2 000	–		De Abrew (1981)

1976) being due probably to the presence of nonspecifically bound or chelate-bound iron. Lactoferrin also fails to inhibit the binding of transferrin to HeLa cells (Ward *et al.*, 1982). Ovotransferrin will, however, deliver iron to chick embryo red cells (Williams and Woodworth, 1973; Keung and Azari, 1982), whereas mammalian transferrins will not (Keung and Azari, 1982). Within mammalian species, however, reticulocytes can take up iron from heterologous transferrins, though the rate is usually lower than with the homologous protein, and nonspecific binding is increased (Morgan, 1964; Verhoef *et al.*, 1973, 1979; Esparza and Brock, 1980*a*; Van Bockxmeer and Morgan, 1982).

The carbohydrate moiety of the receptor appears to be important in the binding of transferrin; Steiner (1980) found that treatment of the receptor with β-galactosidase caused a decrease in transferrin binding, and that binding was inhibited by galactose or N-acetyl galactosamine. Neuraminidase treatment of reticulocytes gave conflicting results (Steiner, 1980; Loh, 1982). Leibman and Aisen (1977) reported that the 176 000 molecular weight component identified as the transferrin receptor in reticulocyte membranes was also present in mature erythrocytes, but without its carbohydrate moiety, thus implying that the carbohydrate was important for transferrin binding. The binding of transferrin to rat reticulocyte receptors, and cellular iron uptake, were shown by Van der Heul *et al.* (1978) to be inhibited by polyclonal antibodies and Fab fragments, but more recent studies using monoclonal antibodies have given variable results. Monoclonals OKT9 and B3/25 to the human transferrin receptor do not affect transferrin binding or iron uptake *in vitro* (Sutherland *et al.*, 1981; Trowbridge and Lopez, 1982), although B3/25 could inhibit the growth of human melanoma M21 in nude mice (Trowbridge and Domingo, 1981). Conversely, bound transferrin does not inhibit the binding of OKT9 to the receptor (Sutherland *et al.*, 1981). However, other monoclonal antibodies to human and mouse transferrin receptors have been obtained which do interfere with transferrin binding (Lebman *et al.*, 1982; Trowbridge and Lopez, 1982; Trowbridge *et al.*, 1982). These results suggest that monoclonal antibodies may be a useful tool in helping to understand the molecular events in the transferrin–cell receptor interaction.

There is some evidence that macrophages, although nonproliferating cells, may possess transferrin receptors. Though Van Snick *et al.* (1977) reported that macrophages bound only very small amounts of transferrin, binding data of Wyllie (1977) and of Nishisato and Aisen (1982) are indicative of specific receptors. However, Wyllie (1977) reported that 80 per cent saturated transferrin was bound more readily than transferrin of low saturation, but the opposite was found by Nishisato and Aisen (1982). Since macrophages can bind a large variety of macromolecules to a greater or lesser extent, it seems desirable to characterise the putative transferrin receptor. MacSween and MacDonald (1969) reported that macrophages took up transferrin-bound iron, but this may have occurred via fluid-phase endocytosis (R. N. M. MacSween, pers. comm.). Wilkins *et al.* (1977) found that when ^{59}Fe-transferrin was injected into the synovial fluid of patients with rheumatoid arthritis, ^{59}Fe could be found after 2 h in synovial macrophages, and interpreted this as indicative of uptake of transferrin-bound iron. However, the intervention of lactoferrin in this inflammatory condition cannot be excluded (section 5.6.11).

In contrast to the voluminous literature on the nature of the transferrin receptor, few studies have attempted to determine which structural features of the transferrin molecule are important in binding to the cell membrane. Kornfeld (1968) reported that chemical modifications of the transferrin molecule causing blockage of free amino groups or changes in net charge tended to decrease specific binding and increase nonspecific binding to reticulocytes, but iron transfer to the cells was unchanged or even increased. He also showed that enzymatic cleavage of sialic acid and some of the neutral sugars of the carbohydrate moiety of transferrin had no effect on its iron-donating properties. The non-involvement of sialic acid has been confirmed in more recent studies (Van Eijk *et al.*, 1982 Young *et al.*, 1983). The fact that chick embryo red cells take up iron at the same rate from both ovotransferrin and chicken serum transferrin (Keung and Azari, 1982), which have different carbohydrate moieties (Williams, 1962), also suggests that the carbohydrate content of transferrin plays no part in its interaction with cell membrane receptors. Significantly, monoferric fragments of bovine transferrin and ovotransferrin showed relatively little ability to donate iron to rabbit reticulocytes (figure 5.11) or chick embryo red cells, respectively

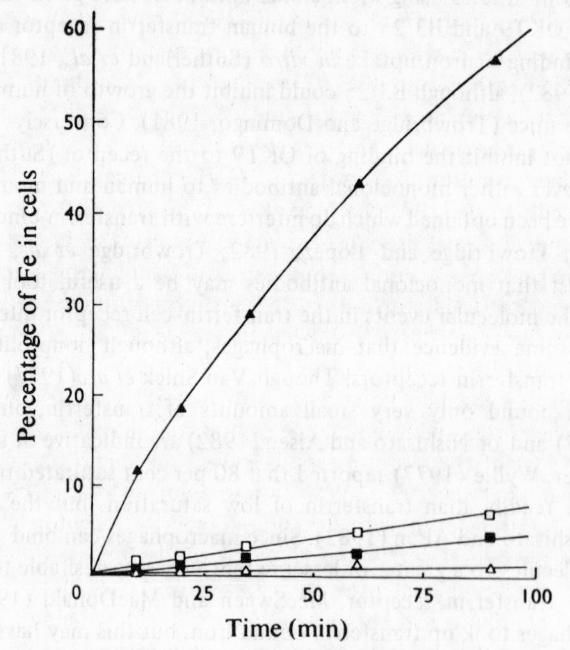

Figure 5.11 Incorporation of iron into rabbit reticulocytes from: bovine transferrin (▲), N-terminal monoferric fragment (△), C-terminal monoferric fragment (■) and C-terminal monoferric fragment, lacking internal cleavages in the polypeptide chain, derived by brief tryptic digestion of bovine apotransferrin (□). From Brock *et al.* (1978*a*) with permission.

(Brock *et al.*, 1978*a*; Esparza and Brock, 1980*b*; Brown-Mason and Woodworth, 1982; Keung and Azari, 1982) even though the intact proteins donated iron efficiently. In both cases significant binding of the fragments to the cells was nevertheless observed, and, in the case of ovotransferrin, a mixture of N- and C-terminal fragments could also donate iron to chick reticulocytes (Brown-Mason and Woodworth, 1984). This suggests that the poor iron-donating ability of fragments is due not to cleavage of the receptor-binding region of the transferrin molecule, but to a requirement for the binding of both domains of the protein. This idea is compatible with several observations indicating that transferrin uptake is a two-stage process and requires a divalent ligand, the second stage being temperature dependent and probably involving internalisation of the transferrin (Baker and Morgan, 1969; Martinez-Medellin *et al.*, 1977; Van Bockxmeer *et al.*, 1978; Takahashi and Tavassoli, 1982). This aspect is described in more detail below. Preliminary studies with a panel of antitransferrin monoclonal antibodies have shown that some, but not all, could inhibit binding to the transferrin receptor (Bartek *et al.*, 1983), and this approach may help to define the receptor-binding region of the transferrin molecule.

5.6.2.2 Internalisation of transferrin

Autoradiographic evidence for internalisation of transferrin by reticulocytes was first presented by Morgan and Appleton (1969), and this was supported by further morphological studies, mainly by Morgan and his associates (summarised by Morgan (1981*a*)) during the following decade. Furthermore, transferrin bound to reticulocytes at 4°C was found to be susceptible to proteolysis or iodination by lactoperoxidase, whereas much of the transferrin became inaccessible when the temperature was raised to 37°C (Hemmaplardh and Morgan, 1976; Martinez-Medellin *et al.*, 1977; Harding *et al.*, 1983). These studies imply a temperature-dependent internalisation of transferrin previously bound in a temperature-independent step. A similar approach was adopted by Karin and Mintz (1981) to demonstrate internalisation of transferrin by mouse teratocarcinoma cells. Bleil and Bretscher (1982) showed that in HeLa cells the receptor itself could be iodinated by lactoperoxidase at 0°C, and that on warming to 37°C the labelled receptors ceased to be accessible to trypsin. They concluded that the receptor recycles over a period of about 21 min, and that at any given time only about 25 per cent of the receptors are expressed on the cell surface. In support of this mechanism, Frazier *et al.* (1982) have compared the number of surface binding sites for transferrin with the total number of receptors determined by radioimmunoassay. In a variety of cell types the latter assay always gave the higher figure, suggesting that many receptors are unavailable to transferrin. In reticulocytes only about half the receptors were unable to bind transferrin, but in fibroblasts or leukaemic cells the proportion increased to 80-90 per cent. Binding of transferrin to the receptor causes capping or patching (Hemmaplardh and Morgan, 1977; Galbraith and Galbraith, 1980; Ekblom *et al.*, 1983; Enns *et al.*, 1983), a phenomenon which is known to lead to internalisation when, for example, antigens are bound to the surface immunoglobulin of B-lymphocytes. Internalisation is more rapid when 95 per cent of the receptors are occupied than when only 5 per cent are occupied, indicating that transferrin binding promotes internalisation (Klausner *et al.*, 1983*b*). Such a process seems to require a multi-subunit membrane receptor, and it is noteworthy that in transformed T and B lymphocytic cell lines Goding and Harris (1981) demonstrated by two-dimensional electrophoresis that the transferrin receptor is the major multi-subunit membrane component. The studies with monoferric fragments (section 5.6.2.1) that suggest that transferrin needs to act as a divalent ligand also support this idea. Similarly, antibodies to the transferrin receptor can, like transferrin itself, cause receptor aggregation and internalisation, but monovalent Fab fragments do not (Enns *et al.*, 1983). Finally, it seems likely that cells of the placenta also internalise transferrin as the protein can be demonstrated in coated vesicles (Booth and Wilson, 1981; Pearse, 1982).

Although there is much convincing evidence for the internalisation of transferrin, it does not necessarily follow that this process is essential for cellular iron uptake, and a number of studies have yielded data indicating that iron may be acquired without internalisation. Several groups of workers have reported that membrane preparations from reticulocytes which had been incubated with ^{59}Fe–^{125}I-transferrin contained relatively more ^{59}Fe than ^{125}I-transferrin, suggesting that a membrane component acted as an iron recipient (Garrett *et al.*, 1973; Fielding and Speyer, 1974; Verhoef and Noordeloos, 1977; Glass *et al.*, 1980). However, the putative iron-binding component was not investigated in any detail, and Morgan (1981*a*) has drawn attention to the difficulty of ensuring that membrane preparations are free of intracellular organelles. Nevertheless, Brown *et al.* (1982) have isolated an iron-binding protein of indisputable membrane origin from human melanoma cells which is distinct from the transferrin receptor; this might correspond to the membrane iron-binding component identified in reticulocytes by Glass *et al.* (1980). A further argument against the obligatory requirement for endocytosis comes from studies showing that reticulocytes can remove iron from matrix-bound transferrin, albeit at a much slower rate than from free transferrin (Glass *et al.*, 1977; Hemmaplardh and Morgan, 1977; Loh *et al.*, 1977; Zaman *et al.*, 1980). Although all four groups reported essentially similar findings, i.e. that uptake from insolubilised transferrin was about 3–10 per cent of that from free transferrin, Hemmaplardh and Morgan (1977) considered that the uptake could be explained by experimental artifacts such as exchange of iron with membrane-bound transferrin or leaching of transferrin from the matrix, whereas the other three groups considered the uptake to be significant, given that steric factors would undoubtedly reduce the rate of uptake from insolubilised transferrin.

Recently, Woodworth *et al.* (1982) have reinterpreted some of the earlier data which were thought to support the obligatory need for endocytosis and have proposed a novel mechanism in which the first, temperature-independent, stage of transferrin binding represents binding to free receptors, and the second, temperature-sensitive, step, binding to receptors already occupied by transferrin, the rate of which depends upon the rate of release of the transferrin already in residence. A somewhat similar proposal has been made by Núñez and Glass (1983). The highly specific microtubule inhibitor nocodazole did not affect the rate of iron uptake by reticulocytes, indicating that endocytosis was not essential.

Overall, therefore, there is now little doubt that receptor-mediated endocytosis of transferrin occurs during cellular uptake of transferrin-bound iron. Whether endocytosis is *essential* for iron uptake is, however, still not entirely clear. Since Frazier *et al*. (1982) have shown that non-erythroid cells internalise a greater proportion of their transferrin receptors than do reticulocytes, it is possible that iron release at the cell membrane, if it occurs at all, may be a relatively more important process in reticulocytes. Indeed, both mechanisms may operate simultaneously (Zaman *et al.*, 1980; Van der Heul *et al.*, 1984).

5.6.2.3 Release of iron from transferrin

The extremely high affinity of transferrin for iron implies that a specific mechanism must exist to enable the protein to surrender its iron to the cell. The nature of this mechanism is still uncertain, and its elucidation has been complicated by the controversy over whether or not it is necessary for endocytosis of transferrin to occur. At one extreme, it has been proposed that transferrin may undergo a conformational change while bound to the cell membrane, resulting in iron release to a membrane iron-binding component (section 5.6.2.2), whereas at the other extreme it has been suggested that transferrin may actually deliver iron directly to the mitochondria.

In vitro, the iron–transferrin complex may be destabilised by protonation, attack upon the synergistic anion, or by a chelator of higher affinity (sections 5.5.2 and 5.5.3), and all these mechanisms are possible candidates for *in vivo* release. Removal of iron by reduction, which has also been proposed, seems less likely in view of the fact that very strong reducing agents such as dithionite are needed to liberate iron from transferrin *in vitro* (Harris *et al*., 1977; Kojima and Bates, 1979).

An initial attack upon the (bi)carbonate ion is suggested by studies in which substitution of oxalate or malonate at the anion-binding site inhibits iron acquisition by reticulocytes but does not prevent transferrin binding (Aisen and Leibman, 1973; Egyed, 1973; Williams and Woodworth, 1973). Transferrin-bound gallium is also taken up less readily by tumour cells from complexes containing oxalate or malonate (Terner *et al*., 1981). However, this may simply be a reflection of the fact that competing chelators remove iron less readily from transferrin when the anion-binding site is occupied by oxalate (Morgan, 1977).

Evidence for the importance of protonation comes from studies in which lysosomotropic agents such as NH_4Cl, which cause an increase in intravacuolar pH, inhibit iron uptake but do not affect transferrin binding or internalisation in both erythroid and non-erythroid cells (Paterson and Morgan, 1980; Morgan, 1981*b*; Octave *et al*., 1982*a, b*; Karin and Mintz, 1981; Ciechanover *et al*., 1983;

Harding and Stahl, 1983). Removal of iron within an endocytotic vesicle seems reasonable, since the intravesicular pH could be lowered either by the action of a proton pump, which would account for the requirement for metabolic energy, or by fusion with a lysosome, which would reduce the intravesicular pH to about 4.6 (Ohkuma and Poole, 1978). Recent evidence favours incorporation into a non-lysosomal acidic vacuole (Van Renswoude *et al.*, 1982; Dautry-Varsat *et al.*, 1983; Harding *et al.*, 1983; Klausner *et al.*, 1983*a*). It remains to be explained how iron leaves the vesicle before the latter becomes reincorporated into the cell membrane. However, in fibroblasts it has been proposed that both receptor-mediated and fluid-phase endocytosis of transferrin occur, and that fusion with lysosomes results in degradation of the transferrin incorporated by fluid-phase endocytosis, whereas receptor-bound transferrin is protected and recycled (Octave *et al.*, 1979, 1981, 1983).

Removal of iron by competing chelates, perhaps in conjunction with protonation, may also occur. Pyrophosphate and organic phosphates can remove iron from transferrin *in vitro* even at neutral pH (Egyed, 1975; Morgan, 1977, 1979; Pollack *et al.*, 1977; Carver and Frieden, 1978) and it is possible that ATP fulfils a dual role as energy source and iron chelator (Egyed, 1977; Kailis and Morgan, 1977; Morgan, 1979). Haemoglobin itself seems to facilitate the ATP-mediated release of iron (Egyed *et al.*, 1980), though neither the mechanism nor the physiological significance of this effect has been established.

Iron released by any of the above mechanisms probably enters an intracellular labile pool (Jacobs, 1977*a*) from which it can be taken up by mitochondria for haem synthesis. However, the nature and to some extent the existence of such a pool remains speculative (Romslo, 1980) and it has been suggested that transferrin may deliver iron directly to the mitochondria. Transferrin has been found in the mitochondrial fraction of disrupted reticulocytes (Neuwirt *et al.*, 1975) and association of transferrin with the mitochondria has been demonstrated morphologically (Isobe *et al.*, 1981). Several studies have shown that mitochondria can remove iron from transferrin *in vitro* (Koller *et al.*, 1976; Ulvik *et al.*, 1976; Konopka, 1978; Konopka and Turska, 1979). This uptake was oxygen dependent and enhanced by pyrophosphate (Konopka and Romslo, 1980, 1981).

Clearly, further studies are needed to elucidate the mechanism of iron release from transferrin *in vivo*, and once again more than one mechanism may operate.

5.6.2.4 Release of apotransferrin

Early studies with reticulocytes established that, following delivery of iron, undegraded apotransferrin was released (Jandl and Katz, 1963), and this also occurs with non-erythroid cells (Brock and Rankin, 1981; Karin and Mintz, 1981; Klausner *et al.*, 1983*b*; Octave *et al.*, 1983). Since, at physiological pH, apotransferrin binds to the receptor with lower affinity than iron-containing transferrin (section 5.6.2.1), the latter will displace the former. Dissociation within the endocytotic vesicle as a result of low pH, as suggested by Karin and Mintz (1981), seems unlikely, as discussed earlier. In fibroblasts, however, at

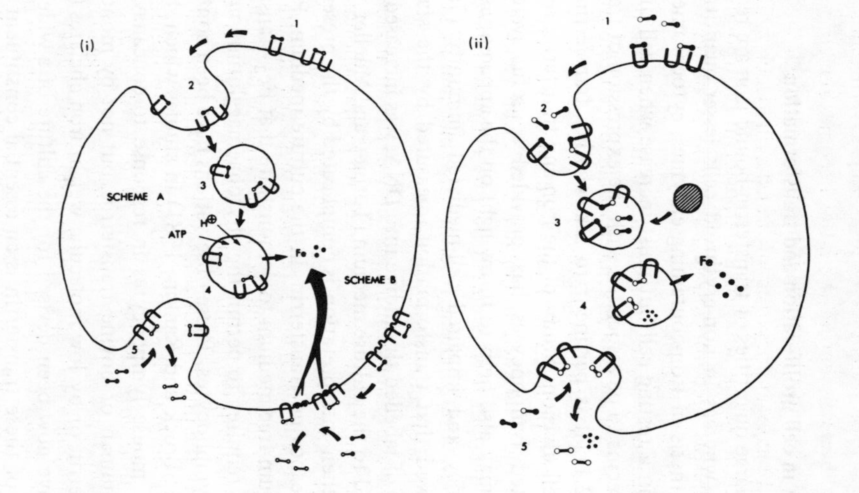

Fig. 5.12 Proposed mechanisms for cellular uptake of transferrin-bound iron.

(i) Mechanisms which may operate in reticulocytes and non-erythroid cells: *Scheme A*, Mechanism involving endocytosis: (1) Binding of mono- and diferric transferrin to membrane receptors; bridging and trigger 'patching' may occur. (2) Endocytosis of transferrin–receptor complexes. (3) Formation of endocytotic vacuole. (4) Action of proton pump and/or ATP causes release of iron from receptor-bound transferrin. (5) Vacuole fuses with cell membrane. Receptors re-expressed. Apotransferrin displaced by fresh iron-containing transferrin molecules. *Scheme B*, Mechanism without endocytosis. Transferrin binds to receptor as before. Iron removed at membrane and taken up by membrane iron-binding component which passes iron to the cytoplasm. Apotransferrin displaced as before.

(ii) Mechanism involving lysosomes which may operate in some non-erythroid cells. Essentially as scheme A in (i) but a variable amount of transferrin may also be taken up by fluid-phase endocytosis at step 2. At step 3, fusion with a lysosome occurs. Iron is released due to lowered intravacuolar pH, free transferrin is digested by proteases, receptor-bound transferrin is protected due to stabilisation of the complex at low pH. At step 5, some degraded transferrin may be released as well as intact apotransferrin. This mechanism closely resembles that proposed by Octave *et al.* (1981).

least some of the transferrin is degraded (Hemmaplardh and Morgan, 1974; Octave *et al.*, 1981), probably due to uptake by fluid-phase endocytosis, and, in carcinoma cells, receptor-bound transferrin may follow either a short cycle involving only peripheral endosomes, or a longer cycle which additionally involves a juxtanuclear compartment (Hopkins, 1983; Hopkins and Trowbridge, 1983). Various stages in the process of cellular uptake of transferrin-bound iron, taking into account different alternative mechanisms, are shown in figure 5.12.

5.6.3 The role of transferrin in cell proliferation and transformation

In the erythroid precursor, large quantities of transferrin-bound iron are required for haemoglobin synthesis, whereas in non-erythroid cells lesser quantities are needed, mainly for the synthesis of iron-containing enzymes, cytochromes, etc. Since such needs are low in a resting cell but must increase when cell division occurs, and since division seems to be associated with the expression of transferrin receptors (section 5.6.2.1), it is pertinent to ask whether transferrin is an absolute requirement for cell division. Study of the iron and transferrin requirements of mitogen-stimulated lymphocytes has provided a useful system to explore this question, and may also help to throw light on the apparent association between iron deficiency and impaired cell-mediated immunity (Jacobs, 1977*b*). Early studies showed that transformation, measured by the standard technique of incorporation of labelled thymidine into DNA, was increased when human transferrin was added to the culture medium (Tormey and Mueller, 1972; Tormey *et al.*, 1972), but their interpretation is complicated by the presence of fetal calf serum, and hence bovine transferrin, in the culture medium. Phillips and Azari (1975) used a serum-free medium to demonstrate that Fe_2-transferrin, but not apotransferrin, was required to permit the phytohaemagglutinin-induced transformation of human lymphocytes. These observations have been confirmed and extended (Brock, 1981; Brock and Rankin, 1981) in studies which showed that the transformation of mouse lymphocytes in response to concanavalin-A was enhanced by mouse, human or bovine transferrin, but not by monoferric fragments of bovine transferrin or by low molecular weight iron chelates (figure 5.13). Serum-free media have now been devised for the culture of a wide range of cells, and in almost all of these transferrin is an essential constituent (table 5.3). Monoclonal antibodies which block the binding of transferrin to the membrane receptor also inhibit cell growth (Trowbridge and Lopez, 1982, Trowbridge *et al.*, 1982; Mendelsohn *et al.*, 1983). It thus seems clear that transferrin is an essential requirement for cell proliferation, and that, as with reticulocytes, transferrins from both homologous and heterologous species can interact with the cell. Some workers have nevertheless reported that bovine transferrin does not interact with human transferrin receptors (Galbraith *et al.*, 1980*b*; Ward *et al.*, 1982). It is possible that this may be due to loss of bound bovine transferrin during experimental procedures, as the affinity of bovine transferrin for the rat and hamster transferrin receptors is reported to be considerably lower than that

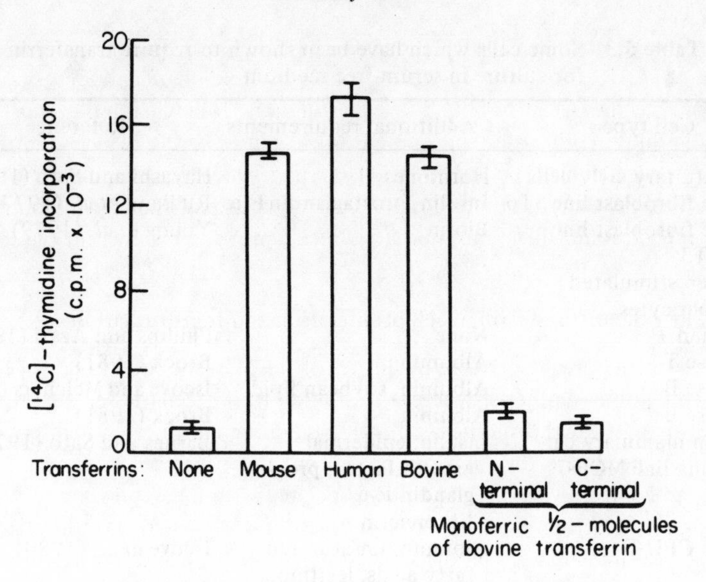

Figure 5.13 ^{14}C-thymidine incorporation into mouse lymphocytes cultured for 72 h in serum-free medium in the presence of concanavalin-A (1 μg ml^{-1}) and transferrins and monoferric transferrin fragments (50 μg ml^{-1}; 30 per cent iron saturation). Mean ± standard deviation (n = 3). From Brock (1981) with permission.

of human transferrin (Messmer, 1973; Verhoef *et al.*, 1979), and indeed it is a less efficient donor of iron to rabbit and rat reticulocytes than many other transferrins (Verhoef *et al.*, 1979; Esparza and Brock, 1980*a*). That it should be totally ineffective as an iron donor to non-erythroid cells, as has been reported (Sephton and Kraft, 1978), seems unlikely, given the universally recognised growth-promoting effect of fetal calf serum, in which the abundant supply of bovine transferrin may well compensate for its lower efficiency as an iron donor. Uptake of iron by lymphocytes precedes DNA synthesis (Brock and Rankin, 1981), in agreement with the observations that iron is required for DNA synthesis (Hoffbrand *et al.*, 1976).

Although lymphocytes will not transform in serum-free medium if iron is added as a low molecular weight chelate, iron is nevertheless taken up from such compounds (Brock and Rankin, 1981). This raises the question of whether transferrin might fulfil some function additional to that of supplying iron. It is possible that the apparently divalent nature of the interaction between transferrin and the membrane receptor may in itself be a necessary signal for triggering lymphocyte proliferation (Brock and Mainou-Fowler, 1983), although recent evidence indicates that, for other cell types, proliferation may occur in certain cases in the absence of transferrin (Block and Bothwell, 1983; Fernandez-Pol *et al.*, 1983; Landschulz *et al.*, 1984).

Table 5.3 Some cells which have been shown to require transferrin for culture in serum-free medium

Cell type	Additional requirements	Reference
Rat pituitary GH_3 cells	Hormones	Hayashi and Sato (1976)
Mouse fibroblast line 3T6	Insulin, prostaglandin $F_2\alpha$	Rudland *et al.* (1977)
Mouse fibroblast line SV3T3	Biotin	Young *et al.* (1979)
Mitogen stimulated lymphocytes:		
Human T	None	Phillips and Azari (1975)
Mouse T	Albumin	Brock (1981)
Mouse B	Albumin, soybean lipid	Iscove and Melchers (1978)
Mouse B	Albumin	Brock (1981)
Human mammary carcinoma line MCF-7	Insulin, epidermal growth factor, prostaglandin $F_2\alpha$ fibronectin	Barnes and Sato (1979)
Mouse CFU-E	Albumin, unsaturated fatty acids, lecithin, cholesterol	Iscove *et al.* (1980)
Human promyelocytic line HL-60	Insulin	Breitman *et al.* (1980)
Rat cerebellar cells	Insulin, progesterone, putrescine, selenium	Messer *et al.* (1981)
Rat neuroblastoma line	Insulin, progesterone, putrescine, selenium	Bottenstein and Sato (1979)
Chick dorsal root ganglia	Insulin, progesterone, putrescine, selenium, nerve growth factor	Bottenstein *et al.* (1980)
Various chick and rat neuron-like cells	Insulin, progesterone, putrescine, selenium, nerve growth factor	Skaper *et al.* (1979)

5.6.4 Iron acquisition by transferrin *in vivo*

In contrast to the large amount of work carried out on the mechanism of iron donation by transferrin, the manner in which iron is acquired by transferrin *in vivo* has received scant attention. The main sources of iron are those cells of the mononuclear phagocyte system (MPS) (this term is preferred to the older and less precise reticuloendothelial system (RES)) involved in erythrocyte catabolism, principally Kupffer cells and splenic macrophages, the intestinal mucosa, and hepatocytes. Since the last of these can both acquire and release iron, they will be dealt with separately (section 5.6.5).

The most fundamental question is whether transferrin is actively involved in the control of iron release, or whether it merely acts as a passive recipient of iron whose release is controlled by other, perhaps intracellular, events. At the present time most evidence favours the latter alternative. In general, studies in man and experimental animals have failed to detect any correlation between transferrin concentration or saturation and the rate of iron absorption or release from the MPS (Schade *et al*., 1969; Levine *et al*., 1972; Fillet *et al*., 1974; Rosenmund *et al*., 1980; Finch *et al*., 1982). This is supported by cell culture studies in which it was found that release of iron by macrophages following ingestion of radiolabelled transferrin-antitransferrin immune complexes was unaffected by the availability of transferrin iron-binding sites in the culture medium (Esparza and Brock, 1981). Contrary results in an earlier study by Fedorko (1974) may have been due to the use of ferritin-antiferritin immune complexes to load the cells, which would have given a much greater loading with iron. An involvement of transferrin in the control of iron release by cells of the MPS would probably require direct interaction between transferrin and the macrophages, and the few reports in this area have yielded controversial results (section 5.6.2.1). Until new evidence is forthcoming it seems safest to conclude that transferrin acts as a passive recipient for iron released by cells, and that a specific interaction is not required.

5.6.5 Interaction of transferrin with hepatocytes

Since hepatocytes can both remove iron from and donate iron to transferrin, any attempt to analyse the mechanisms involved must take into account the fact that both processes may occur simultaneously. Young and Aisen (1980, 1981), by using desferrioxamine to sequester iron released by the cells, were able to show that uptake of iron took place by a receptor-mediated process analogous to that occurring in other cells, but that release occurred independently of the extracellular availability of transferrin iron-binding sites, as appears also to be the case with macrophages (see above). Since uptake was proportional to transferrin saturation, the net flux of iron is dependent upon this factor, as has been confirmed by Baker *et al*. (1981). An earlier report suggesting that transferrin-bound iron entered the hepatocytes by a process of simple diffusion (Grohlich *et al*., 1979) may have been due to the poor viability of the cells used in the study. A further complication arises from the fact that hepatocytes are also involved in transferrin catabolism via the asialoglycoprotein receptor (section 5.4.5). Young *et al*. (1983) have recently shown that asialotransferrin can bind to both the asialoglycoprotein receptor and to the transferrin receptor, but only protein bound to the latter is returned intact after delivery of iron. Transferrin entering the cell via the asialoglycoprotein receptor is catabolised. This probably accounts for the so-called diacytosis of asialotransferrins (Tolleshaug *et al*., 1981) in which some asialotransferrin taken up by the liver was released intact rather than catabolised.

5.6.6 Functional heterogeneity of the iron-binding sites

Some 18 years ago Fletcher and Huehns (1967, 1968) reported that when transferrin in plasma was partially depleted of iron by *in vitro* incubation with reticulocytes, the remaining iron was relatively less available to erythroid precursors than was iron on transferrin which had been brought to the same percentage saturation by chemical means. From these and similar experiments they proposed, first, that the two sites of transferrin were not identical in their ability to donate and accept iron to and from different tissues; iron released from the gut mucosa or iron stores would be preferentially bound to one site of transferrin, which would also donate its iron preferentially to erythroid precursors, and the other site would preferentially donate iron to the storage tissues. Secondly, it was proposed that diferric transferrin was a more efficient iron donor than monoferric transferrin. Until this time the metal-binding sites of transferrin were generally thought to be chemically identical (section 5.5.2) and transferrin-bound iron was considered to exist as a homogeneous pool. The Fletcher–Huehns hypothesis, as these proposals came to be known, obviously implied chemical differences between the sites, and generated a large number of biochemical and biophysical studies from which it is now clear (section 5.5.2) that the two sites do indeed differ chemically. Simultaneously, other workers investigated the proposed functional differences between the sites, either by refinements of the *in vitro* techniques originally used by Fletcher and Huehns, or by *in vivo* experiments. These experiments, to be described in this section, have not yielded such clear-cut results. Princiotto and Zapolski (1976) took advantage of the non-random uptake of iron by human transferrin *in vitro* to radiolabel either one or other site, and found that rabbit reticulocytes did indeed acquire iron preferentially from one site. Nevertheless, Harris and Aisen (1975) and Harris (1977*b*) reported that when homologous rabbit or human reticulocyte–transferrin systems were used, no functional difference was apparent, and attributed the earlier results to the use of a heterologous system. However, no difference was found in the ability of the sites of bovine transferrin to donate iron to rabbit reticulocytes (Esparza and Brock, 1980*b*), or of those of rat and rabbit transferrins to donate iron to human reticulocytes (Huebers *et al.*, 1981*b*). Further studies have generally failed to show differences between the abilities of the two sites of human (Morgan *et al.*, 1968; Huebers *et al.*, 1981*b*), rabbit (Delaney *et al.*, 1982; Heaphy and Williams, 1982*b*) or rat (Beamish *et al.*, 1975; Morgan *et al.*, 1978; Huebers *et al.*, 1981*a*; Young, 1982) transferrins to donate iron to homologous reticulocytes, nor could any difference be found in the iron-donating properties of the two monoferric human transferrins (Van der Heul *et al.*, 1981). However, some further contrary results have also been reported (Verhoef *et al.*, 1978; Van Baarlen *et al.*, 1980). A number of variables may influence the results of this type of experiment, such as the presence of nonspecifically bound iron, contamination with labelled haemoglobin derived from initial incubation with reticulocytes and, in the case of heterologous systems, exchange of iron with homologous transferrin bound to the cells. Also, as pointed out by Martinez-Medellin and Benavides (1979), a difference between the rate of release of iron from the two sites will only manifest itself if iron detachment from transferrin is the rate-limiting step of the donation mechanism, and this appears to be the case only when the medium used provides the cells with a suboptimal energy supply.

In vivo experiments have also been used to investigate possible functional differences between the sites. In studies by Hahn and Ganzoni (1975) and by Brown and co-workers (Awai *et al.*, 1975*a, b*; Okada *et al.*, 1977) it was found that injection into rats of transferrin which had been labelled with radioiron following *in vitro* incubation with reticulocytes resulted in nonuniform uptake of the label into different organs. In support of the Fletcher–Huehns hypothesis, relatively more label was incorporated into erythrocytes than into the liver. These results were later reinterpreted as being due to differences between the two rat isotransferrins, rather than between the sites of individual molecules (Okada *et al.*, 1979). However, other workers, using similar techniques, have failed to detect differences between either the two sites or the two isotransferrins in the rat (Pootrakul *et al.*, 1977; Huebers *et al.*, 1978, 1981*a*; Zapolski and Princiotto, 1980*b*). Other *in vivo* studies have also failed to detect differences between the binding sites in man (Huebers *et al.*, 1981*b, c*) or ruminants (Gibbons *et al.*, 1976).

There has been more support for Fletcher and Huehns' other proposal, namely, that diferric transferrin is a better iron donor than monoferric transferrin. Several workers have demonstrated that reticulocytes, liver cells and embryo cells took up more iron from diferric transferrin than from a similar quantity of iron offered as predominantly monoferric species (Beamish *et al.*, 1975; Zapolski and Princiotto, 1977*b*; Huebers *et al.*, 1978, 1981*b, c*). *In vivo*, iron is taken up from the circulation by erythroid cells and the liver more rapidly when present as diferric transferrin (Brown *et al.*, 1975; Hahn *et al.*, 1975; Christensen *et al.*, 1978; Skarberg *et al.*, 1978; Huebers *et al.*, 1981*a, b*). This may be explained, at least in part, by the fact that iron atoms are generally removed pairwise from diferric transferrin (Huebers *et al.*, 1978, 1981*a*; Groen *et al.*, 1982; Young, 1982). Each cycle of diferric transferrin will thus yield two iron atoms, as against only one from each cycle of monoferric transferrin.

Finally, there remains the question of whether iron is acquired nonrandomly by transferrin *in vivo*. Zapolski and Princiotto (1980*b*) showed that, in the rat, radioiron released by cells of the MPS, which had been loaded with heat-damaged labelled erythrocytes, was preferentially bound to the acid-stable (presumably C-terminal) site of transferrin. However, this does not distinguish between preferential loading of the C-terminal site and preferential utilisation of iron bound to the N-terminal site. Marx *et al.* (1982) studied the site distribution of transferrin-bound iron following intragastric administration of ^{59}Fe to iron-deficient rabbits. In these animals the low serum iron levels meant that most iron taken up converted apotransferrin to monoferric transferrin, rather than monoferric transferrin to diferric transferrin. In most of the animals radioiron was found mainly as Fe_NTf, suggesting preferential uptake by the N-terminal site. Young (1982) examined the site distribution of iron taken up by apotransferrin incubated with isolated rat hepatocytes and found that both monoferric transferrins were generated in equal amounts, indicating that no site preference existed, and similar results were obtained by Nishisato and Aisen (1983) when apotransferrin was incubated with iron-loaded macrophages. Clearly, further studies in this area are required to solve these apparently conflicting results.

In conclusion, it seems fair to say that although the Fletcher–Huehns hypothesis correctly predicted the preferential utilisation of iron bound to diferric transferrin, the bulk of evidence currently available does not favour functional heterogeneity of the sites with respect to iron donation to reticulocytes. Further study is required regarding nonrandom uptake of iron *in vivo*. Like all good hypotheses, it remains controversial and has stimulated much valuable work on the chemistry and biology of transferrin which has established beyond question the chemical non-identity of the sites (section 5.5.2). It is perhaps curious that despite this chemical difference no clear evidence for functional heterogeneity has been unequivocally presented.

5.6.7 Transport of metals other than iron by transferrin

Although transferrin can bind a wide variety of metal ions *in vitro* (section 5.5.4), there is only rather limited evidence that this property is exploited *in vivo*, despite the fact that since transferrin is normally far from fully saturated with iron, ample binding capacity would be available. There is some evidence that small amounts of chromium (Hopkins and Schwartz, 1964) and manganese (Hancock *et al.*, 1973) are bound to transferrin *in vivo*, but the destination of these metals has not been ascertained. Vanadium is transported by transferrin *in vivo*, and incorporated into ferritin in the liver (Sabbioni and Marafante, 1981). Although transferrin does not bind copper in plasma, mucosal transferrin may act as a vehicle for copper absorption (El-Shobaki and Rummel, 1979) and the same may be true for chromium (Wollenberg *et al.*, 1982). Transferrin has also been reported to bind a small but relatively constant proportion of serum zinc (Boyett and Sullivan, 1970) and to mediate the transfer of absorbed zinc to portal plasma (Evans and Winter, 1975; Evans, 1976). However, Charlwood (1979) showed by chelator competition studies that practically all zinc will be bound by albumin rather than to transferrin. This does not rule out the possibility that transferrin might be kinetically more accessible to absorbed zinc and physiological pCO_2 levels may favour binding by transferrin (Harris, 1983), but a re-evaluation of the work of Boyett and Sullivan (1970) and of Evans' group by Smith *et al.* (1978) and Chesters and Will (1981) has cast doubt on the identity of the zinc-binding protein detected by the earlier workers. Transferrin-bound zinc has been reported to enhance the transformation of mitogen-stimulated lymphocytes in serum-free medium (Phillips and Azari, 1974; Phillips, 1976, 1978, 1980) but this does not necessarily imply that transferrin acts as a donor of zinc to lymphocytes *in vivo*.

There has been some interest in the role of transferrin in transporting metals not normally present in the body. Plutonium-238 is bound by transferrin *in vivo* (Stevens *et al.*, 1968) and its metabolism shares many features in common with iron (Stover *et al.*, 1968; Priest and Haines, 1982). Metastable indium-113, used for organ scanning, is also transported by transferrin (Hosain *et al.*, 1969). Of particular interest has been the role of transferrin in the transport of gallium-67, which is used to detect tumours. Injected transferrin-bound gallium localises in tumours with a greater degree of specificity than does gallium citrate (Wong *et al.*, 1980). This is probably due to the low association constant for gallium-transferrin of 2.5×10^5 litres mol^{-1} (Larson *et al.*, 1978), which would allow significant nonspecific binding of gallium citrate to other proteins. Transferrin-bound gallium, on the other hand, is thought to be taken up via transferrin receptors (Larson *et al.*, 1980) which are found on tumour cells (section 5.6.2.1). The role of transferrin in gallium uptake by tumour cells is supported by several studies showing that ^{67}Ga uptake *in vitro* is enhanced by transferrin (Harris and Sephton, 1977; Sephton and Kraft, 1978; Larson *et al.*, 1979; Noujaim *et al.*, 1979; Rasey *et al.*, 1982). However, Gams *et al.* (1975) and Vallabhajosula *et al.* (1981) reported that uptake was lower from Ga-transferrin than from Ga-citrate, and it has been proposed that the low pH within tumours would cause dissociation of gallium from transferrin, and that while transferrin may transport gallium to the site of the tumour, free gallium is actually involved in cellular uptake mechanisms (Sephton, 1981; Vallabhajosula *et al.*, 1981). It should be noted that little gallium is taken up by erythroid precursors, despite their abundant transferrin receptors (Logan *et al.*, 1981). The role of transferrin in gallium uptake is summarised in a symposium edited by Noujaim and Weibe (1981).

5.6.8 Antimicrobial properties of the transferrins

All microorganisms, with the possible exception of lactobacilli (Archibald, 1983), require iron for growth. In consequence, transferrins, by virtue of their high affinity for iron, can retard microbial growth by making this element relatively unavailable. The role of transferrin and ovotransferrin as mediators of the bacteriostatic properties of serum and egg white respectively has been recognised since the discovery of the proteins (Schade and Caroline, 1944, 1946; Alderton *et al.*, 1946). Similar properties were also established for lactoferrin soon after its characterisation (Masson *et al.*, 1966; Oram and Reiter, 1968). Several comprehensive reviews of the possible role of transferrins in protection against infection have appeared (Weinberg, 1978; Bullen, 1981; Finkelstein *et al.*, 1983), hence this review will concentrate on mechanistic aspects.

The bacteriostatic activity of transferrin in serum, or other biological fluids, and its reversal by addition of sufficient iron to saturate the protein, have been demonstrated against a wide variety of organisms (see review by Weinberg 1978)). In many cases these observations correlate with an increased virulence of the organisms when they are introduced into an iron-loaded host (Bullen, 1981). Nevertheless, bacteriostasis *in vitro* is often transient and it is now known that many microorganisms produce low molecular weight iron chelators of high affinity known as siderophores, which are potentially capable of removing iron from transferrins. Most siderophores are derivatives of catechol or hydroxamic acid; their structure and properties have been reviewed by Bezkorovainy (1980) and by Neilands (1980, 1981). Synthesis of siderophores occurs in response to a low-iron environment, and the differing abilities of various microorganisms to do so may relate to their virulence (Rogers, 1973; Miles *et al.*, 1979). Concurrent synthesis of receptors for the iron–siderophore complex also occurs (Braun *et al.*, 1976; Ichihara and Mizushima, 1978). The siderophore enterochelin (also known as enterobactin), produced by many Enterobacteriaceae, has been extensively studied and is a cyclic trimer of 2,3-dihydroxy-N-benzoyl-L-serine (O'Brien and Gibson, 1970; Pollack and Neilands, 1970). It can remove iron from transferrin *in vitro* (Rogers, 1973; Rogers *et al.*, 1977*c*; Carrano and Raymond, 1979), and can abolish the bacteriostatic activity of transferrin and lactoferrin on *E. coli* (Rogers, 1973; Rogers *et al.*, 1977*c*; Rogers and Synge, 1978, Brock *et al.*, 1983) (figure 5.14). The dividing organisms acquire lactoferrin- or transferrin-bound iron without degrading the protein (Tidmarsh and Rosenberg, 1981;

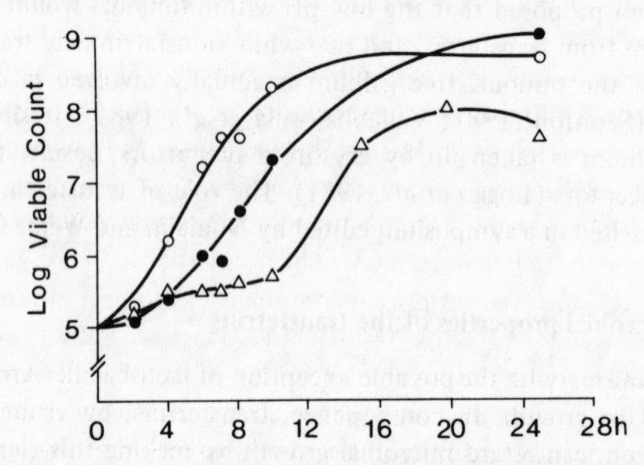

Figure 5.14 Reversal by iron and enterochelin of the lactoferrin-mediated inhibition of *Escherichia coli* by human milk: Δ, milk only; •, mild + 50 μM Fe-nitrilotriacetate complex; ○, milk + 3.2 μM enterochelin. From Brock *et al.* (1983) with permission.

Brock *et al.*, 1983). It has been reported, on the basis of the nondialysable nature of at least part of the enterochelin in enterochelin–transferrin mixtures, that iron may be acquired as an iron–siderophore–transferrin complex (Kvach *et al.*, 1977) but this conclusion may be erroneous, as synthetic analogues of enterochelin bind to dialysis tubing (Carrano and Raymond, 1979). Moreover, growth of *Salmonella paratyphi* can occur when the organisms and transferrin are separated by a dialysis sack (Tidmarsh and Rosenberg, 1981).

Antibodies may enhance the antimicrobial properties of transferrin and lactoferrin (Bullen *et al.*, 1967, 1972; Rogers, 1976; Stephens *et al.*, 1980), apparently by preventing the secretion of siderophores (Rogers, 1973). However, they may also react with the siderophores themselves (Moore *et al.*, 1980). Some workers have also reported that the bacteriostatic activity against *E. coli* is diminished if antibody to the O-antigen is removed by absorption (Fitzgerald and Rogers, 1980; Stephens *et al.*, 1980) but others have found no such correlation (Dolby and Honour, 1979; Brock *et al*., 1983). Furthermore, some have reported no enhancement of bacteriostasis by antibodies (Law and Reiter, 1977; Samson *et al*., 1979). Clearly, the role of antibodies in what appears to be an essentially nutritional mechanism requires further investigation.

The neisseriae are unusual organisms in that they appear to be able to grow in the presence of transferrin, lactoferrin, or synthetic iron chelators, yet siderophore production cannot be detected (Norrod and Williams, 1978*a, b*; Archibald and DeVoe, 1980; Mickelesen *et al*., 1982). Direct contact between the iron-binding protein and the organisms seems to be necessary (Archibald and DeVoe, 1979; Holbein, 1981). Moreover, *N. meningitidis* grows well in the presence of transferrin, but poorly in the presence of ovotransferrin, suggesting a degree of specificity in the interaction (Archibald and DeVoe, 1979; Mickelesen and Sparling, 1981). The mechanism appears to be highly effective, since growth of *N. meningitidis* in a low-iron defined medium is actually enhanced by the addition of normal human serum (Holbein, 1981). Further details of this novel mechanism are awaited with interest.

Although transferrins generally produce a bacteriostatic rather than a bactericidal effect, actual killing of *Streptococcus mutans* and *Vibrio cholerae* by lactoferrin has been reported (Arnold *et al*., 1977, 1981). The mechanism involved is not clear, but activity was demonstrated only against organisms in logarithmic growth phase, and at slightly acid pH, and appears to depend upon some factor other than the withholding of iron (Arnold *et al.*, 1982).

The extent to which the antimicrobial properties of the transferrins contribute to protection against infection is difficult to assess. Pathological conditions resulting in saturation of serum transferrin are invariably accompanied by other abnormalities which may also predispose to infection, and some siderophores may enter host cells and scavenge iron from sources other than transferrin. The studies of Miles and his colleagues (Miles and Khimji, 1975; Miles *et al*., 1976, 1979; Khimji and Miles, 1978) on the ability of iron and siderophores to enhance the virulence of various bacteria for guinea pigs suggest that the transferrin system may be important in reducing the pathogenicity of organisms with a limited capacity to produce siderophores. Virulent strains that produce siderophores readily and avirulent strains susceptible to other mechanisms such as complement-mediated bactericidal activity are probably little affected. It has also been proposed that the low serum pH of patients with diabetic ketoacidosis may labilise transferrin-bound iron and cause increased susceptibility to fungal infections (Artis *et al*., 1982).

Although lactoferrin may be bacteriostatic for *E. coli* (Bullen *et al*., 1972; Reiter *et al*., 1975; Bishop *et al*., 1976; Law and Reiter, 1977; Spik *et al*., 1978; Stephens *et al*., 1980) and *Neisseria* (Mickelesen *et al*., 1982), the activity may be abrogated in milk by the high citrate and low bicarbonate content relative to serum (Reiter *et al*., 1975; Griffiths and Humphreys, 1977), which probably combine to labilise the iron. This may be pertinent to the role of lactoferrin in protecting the mammary gland against infection: in cattle, the lactating gland can be infected by the introduction of as few as ten viable *Staphylococcus aureus* organisms (Brock *et al*., 1975), but the non-lactating gland, which contains high levels of lactoferrin (Welty *et al*., 1976) and an ionic composition closer to that of serum, is much more resistant to infection (Reiter, 1978). Lactoferrin may also play a part in the enhanced resistance of breast-fed infants to gastrointestinal infections. Experiments with suckling guinea pigs involving oral iron supplementation (Bullen *et al*., 1972) or addition of ovotransferrin to an artificial milk (Antonini *et al*., 1977) have supported this idea, as has a recent clinical study of ovotransferrin supplementation (Corda *et al*., 1983). Some of the evidence is, however, open to other interpretations, as discussed in an earlier review (Brock, 1980). Whether lactoferrin could remain biologically active in the intestine is also not certain (Brock, 1980), though the relative resistance of human apolactoferrin to tryptic digestion *in vitro* (Brines and Brock, 1983) may point to an evolutionary development designed to permit its survival.

Lactoferrin in the specific granules of neutrophils may contribute to the antimicrobial properties of these cells. Ingestion of ferritin-antiferritin immune complexes leads to a decrease in the bactericidal activity of neutrophils *in vitro* (Bullen and Wallis, 1977; Bullen and Armstrong, 1979) and *in vivo* (Bullen and Joyce, 1982), which can be related to saturation of the cellular lactoferrin with iron. Lactoferrin may also contribute to extracellular antimicrobial defences in an inflammatory lesion, as it has been shown that degranulation of neutrophils leads to the liberation of some 80 per cent of the lactoferrin into the extracellular fluid (Leffell and Spitznagel, 1975). This extracellular lactoferrin may enhance antimicrobial activity which would otherwise be impaired due to the lower pH of the lesion, which would lower the efficacy of transferrin.

5.6.9 Catalytic properties of the transferrins

Iron may act as a catalyst in the Haber–Weiss reaction, in which hydroxyl radicals are produced by the interaction of superoxide and hydrogen peroxide. The reaction may be represented thus:

$$O_2 \cdot^- + H_2O_2 + H^+ = \cdot OH + H_2O + O_2$$

Both transferrin- and lactoferrin-bound iron have been reported to enhance $\cdot$OH production (McCord and Day, 1978; Ambruso and Johnston, 1981; Bannister *et al*., 1982), and it has been suggested that this reaction may contribute to the bactericidal activity of neutrophils by promoting the formation of the toxic hydroxyl radical (Ambruso and Johnston, 1981). However, except in the study of McCord and Day (1978), in which no details were given, iron-saturated transferrins were used, and the possibility of intervention by small amounts of nonspecifically bound iron cannot be ruled out, given that only small amounts of the metal would be needed for catalytic activity. That 'free', rather than transferrin-bound, iron is in fact responsible is supported by the observations of Gutteridge *et al*. (1981) who showed that apotransferrin and 20 per cent saturated transferrin inhibited iron-mediated lipid peroxidation, whereas iron-saturated transferrin had no effect.

A different type of catalytic activity has been proposed by Anderson and Tomasi (1977), who reported that refolding of lysozyme was accelerated in the presence of transferrin or lactoferrin due to catalysis of thiol oxidation. However, copper rather than iron appeared to be the metal involved.

5.6.10 Modulation of immune responses by lactoferrin

Several recent studies have indicated that lactoferrin may play a role in the development and activity of cells involved in immune and inflammatory responses. In a series of papers, Broxmeyer and his associates have reported that human lactoferrin at concentrations as low as 10^{-17} M suppresses the production of mouse and human granulocytes and monocytes (Broxmeyer, 1978, 1979; Broxmeyer *et al*., 1978, 1980*a, b*, 1983; Pelus *et al*., 1979, 1981*a, b*; Zucali *et al*., 1979; Steinmann *et al*., 1982). Essentially similar effects on human bone-marrow cells were reported by Bagby *et al*. (1981), who also found that loss of activity at 'high' concentrations (i.e. $> 10^{-9}$ M) was associated with a tendency for lactoferrin to polymerise (Bennett *et al*., 1981; Bagby and Bennett, 1982). Although much has been published on the cell populations and effector mechanisms involved, some of the basic observations reported initially regarding the involvement of lactoferrin seem to require further investigation. Although Fe_2-lactoferrin was reported to be active at concentrations as low as 10^{-17} M (Broxmeyer *et al*., 1978; Bagby *et al*., 1981), the lowest concentrations reported in normal plasma do not fall below 10^{-8} to 10^{-10} M (Bennett and Mohla, 1976; Bennett and Kokocinski, 1978; Broxmeyer *et al*., 1983). Also, the minimum active concentration of Fe_2-lactoferrin exceeded that of apolactoferrin by ten orders of magnitude (Broxmeyer *et al*., 1978), implying a level of adventitious iron, which would bind to apolactoferrin, of $< 10^{-10}$ M in the culture system employed, which seems highly unlikely. The isoelectric point of the active component found in neutrophil extracts was reported to have a pI of 6.5 (Broxmeyer *et al*., 1978), which is substantially below the now generally accepted figure of 8.4-9.0 (section 5.4.2), and raises suspicions that lactoferrin might only be active when associated with other substances. It was also found that only lactoferrin from neutrophils bearing Fc receptors was active; FcR-negative neutrophils contained an inactive form of lactoferrin, though the biochemical basis for this difference was not established. Further investigations seem to be required before the role of lactoferrin as a regulator of granulopoiesis can be considered established, especially as Winton *et al*. (1981) failed to demonstrate any effect of lactoferrin on granulocyte production in the mouse. The relationship between this proposed role and lactoferrin's function as an iron-binding protein also poses an intriguing question.

Another, perhaps related, property of lactoferrin has been proposed by Oseas *et al*. (1981) who found that neutrophil chemotactic factors caused release of lactoferrin from the cells, which resulted in increased clumping and adherence of neutrophils to epithelial cells, and also induced binding of neutrophils to endothelium in hamster cheek pouches (Boxer *et al*., 1982). These observations might help to explain the mechanism by which neutrophils leave the circulation and accumulate at inflammatory sites, and perhaps provide a *raison d'être* for lactoferrin's propensity to bind to other substances. However, normal aggregation and margination of neutrophils was found in a patient whose neutrophils lacked lactoferrin (Gallin *et al*., 1982), so other factors may also be involved.

Finally, it has been reported that *in vitro* antibody production by murine spleen lymphocytes was inhibited by lactoferrin (Duncan and McArthur, 1981). The inhibition was mediated by an effect on the accessory macrophages rather than on the B-cells themselves, notwithstanding the fact that B-lymphocytes also bind lactoferrin (Bennett and Davis, 1981). Little difference was found in the activity of native (8 per cent Fe-saturated) and 100 per cent saturated lactoferrin, though the significance of this observation is doubtful in view of the low concentration of lactoferrin used (10^{-7} M) and the absence of any reported control of exogenous iron levels.

5.6.11 The role of lactoferrin in the anaemia of chronic disease

Masson and co-workers (Van Snick *et al*., 1974, 1975, 1977; Van Snick and Masson, 1976; Markowetz *et al*., 1979) have proposed that lactoferrin may be responsible, at least in part, for the increased iron content of cells of the MPS and the concomitant reduction in iron supply to the marrow which occurs in many chronic diseases (Roeser, 1980). Fe_2-lactoferrin is bound by mouse peritoneal macrophages by what appears to be a specific saturable receptor, there being 2×10^7 binding sites per cell with an affinity of $0.7–0.9 \times 10^6$ litres mol^{-1} (Van Snick and Masson, 1976). However, binding is unaffected by treating the cells with trypsin (Bennett and Davis, 1982), and no receptor molecule has yet been identified or isolated. Lactoferrin also binds to human blood monocytes (Birgens *et al*., 1983). Lactoferrin is released by degranulating neutrophils (Van Snick *et al*., 1974; Leffell and Spitznagel, 1975), and could remove iron from transferrin when both proteins were added to purulent pleural exudate. Lactoferrin bound by macrophages was ingested and digested, and the iron incorporated into ferritin (Van Snick *et al*., 1977). By this process, transferrin-bound iron would effectively be short-circuited to the MPS. *In vivo*, ^{125}I- or ^{59}Fe-labelled lactoferrin injected into humans was rapidly taken up by the liver and spleen, the iron being slowly returned to the bone marrow (Bennett and Kokocinski, 1979). However, studies of the rate of lactoferrin turnover in rabbits (Karle *et al*., 1979) suggest that this mechanism is unlikely to account for significant loading of macrophages in the spleen or liver with iron, and in any case most circulating lactoferrin appears to be eliminated by hepatocytes rather than macrophages (Prieels *et al*., 1978). On the other hand, the mechanism might be responsible for loading of macrophages in the localised environment of an inflammatory focus such as the inflamed joints in rheumatoid arthritis, in which levels of lactoferrin are many times higher than those in serum (Bennett *et al.*, 1973; Bennett and Skosey, 1977; Malmquist *et al.*, 1977).

5.6.12 Lactoferrin and iron absorption

Although transferrin acts as an acceptor of absorbed iron at the serosal surface (section 5.6.4), there is no clear evidence that transferrins are involved in uptake of iron by mucosal cells. On the face of it, lactoferrin, by virtue of its close association with the gut mucosa (Masson, 1970; Mason and Taylor, 1978), would seem a strong candidate for a mucosal iron donor, and indeed uptake of lactoferrin-bound iron by mucosal cells *in vitro* has been reported (Cox *et al.*, 1979). However, several other studies suggest that, on the contrary, lactoferrin may actually inhibit iron uptake. Apolactoferrin decreased iron uptake by everted guinea pig intestinal sacs, and Fe_2-lactoferrin had no effect (De Laey *et al.*, 1968), and essentially similar results have been obtained *in vivo* (De Vet and Van Gool, 1974; McMillan *et al.*, 1977; Fransson and Lönnerdal, 1982; Huebers *et al.*, 1982; Fransson *et al.*, 1983). It has been proposed that lactoferrin may act as an additional control of iron absorption when other controls are relaxed, for example in iron deficiency or during the neonatal period (Van Vugt *et al.*, 1975; Brock, 1980). This would be in accord with a number of reports indicating that passive diffusion is involved in iron uptake by everted gut sacs (Sheehan, 1976), isolated mucosal cells (Savin and Cook, 1978) and brush border membrane vesicles (Eastham *et al.*, 1977), since such a process is likely to be inhibited by apolactoferrin.

Huebers *et al.* (1976, 1982) have suggested that intracellular transferrin or transferrin-like proteins (section 5.7.1) may be involved in iron uptake at the mucosal surface, since transferrin-bound or inorganic iron were taken up more rapidly than iron bound to lactoferrin or ferritin by rat jejunal loops. Further studies will be required to reconcile these results with those indicating that passive diffusion is involved.

5.7 MISCELLANEOUS IRON BINDING PROTEINS

5.7.1 Mucosal iron-binding proteins

Intracellular iron-binding proteins in mucosal cells have been identified in the rat and guinea pig. In the former species the protein closely resembles serum transferrin, but differs slightly in amino acid composition and iron-binding properties (Huebers *et al.*, 1976). In the guinea pig, however, a quite distinct protein exists (Pollack and Lasky, 1976*a*, *b*), which, although similar to transferrin in its molecular weight (78 000) and ability to bind two iron atoms per molecule with an affinity of 10^{19} litres mol^{-1}, did not cross-react with transferrin immunologically. It consisted of two subunits, and no colour was produced as a result of iron binding. A similar protein has subsequently been identified in other guinea pig tissues (Pollack and Lasky, 1977). Although its function has not been established, it seems likely that it is involved in intracellular iron metabolism.

5.7.2 Iron-binding protein from melanoma cell membranes

Brown *et al.* (1982) have shown that the cell surface glycoprotein p97 found in many human melanoma cells binds iron and is structurally related to transferrin. It has a molecular weight of 97 000, and shows some N-terminal amino acid sequence homology with transferrin (figure 5.4). It did not cross-react antigenically with native transferrin or lactoferrin, but there was some cross-reaction between the denatured proteins. This may, however, be of limited significance since denatured transferrin also cross-reacts with denatured albumin and α-fetoprotein (Pekkala-Flagan and Ruoslahti, 1982). Its function is unknown, although clearly it might play a role in cellular iron uptake. The existence of this protein lends credence to the proposal of Williams *et al.* (1982*b*) that transferrin may have evolved from a cell membrane protein.

5.7.3 Uteroferrin

Uteroferrin is an iron-binding protein originally discovered in the porcine uterus (Murray *et al.*, 1972; Chen *et al.*, 1973) which, although not closely related to other members of the transferrin class of proteins, merits inclusion because of its role in iron transport to the fetus. It has a molecular weight of about 35 000 (Buhi *et al.*, 1982*a*), forms an intense purple-coloured complex with iron which has a λ_{max} at 545 nm, and possesses acid phosphatase activity (Schlosnagle *et al.*, 1974). Treatment with mercaptoethanol markedly enhances the enzymatic activity and causes the colour to change to pink with a λ_{max} at 508 nm, both changes being reversible by ferricyanide oxidation, whereas dithionite causes a loss of colour, iron, and enzymatic activity (Schlosnagle *et al.*, 1976; Keough *et al.*, 1980). Reactivation occurs by the addition of Fe^{3+}, and to some extent with Cu^{2+}, but not with Mg^{2+}, Mn^{2+}, Cr^{3+}, Ni^{2+} or Zn^{2+}. Campbell *et al.* (1978) and Keough *et al.* (1980) reported binding of two Fe atoms per molecule, these being either both in the ferrous form, or one ferrous and one ferric, but Schlosnagle *et al.* (1976) and Antanaitis *et al.* (1980) considered that only a single iron atom is bound, this being in the ferric state. However, Antanaitis and Aisen (1982) have shown by EPR spectroscopy that while native uteroferrin contains only a single ferric ion, a second atom may be bound, this time in the ferrous state, if care is not taken to eliminate adventitious iron. The change from purple to pink caused by mercaptoethanol appears to be due to a reorientation of the three tyrosyl residues thought to be coordinated to the iron atom, according to a study using resonance Raman and circular dichroism spectroscopy (Antanaitis *et al.*, 1982) and not to reduction of the bound Fe^{3+} to Fe^{2+}. Cleavage of a disulphide bridge and loss of about 40 per cent of the α-helical content of the protein also occur during the purple-to-pink conversion (Schlosnagle *et al.*, 1976; Antanaitis *et al.*, 1982). Further EPR studies by Antanaitis *et al.* (1983) now indicate that the pink diferric form contains one Fe^{2+} and one Fe^{3+} ion, the latter being responsible for the $g' = 1.74$ EPR signal, and that the purple form contains two Fe^{3+} ions in a binuclear cluster with opposed spins, resulting in a considerably lower EPR signal.

It is thought that uteroferrin serves as an intermediary in the transport of iron from mother to fetus in the pig. Under conditions comparable to those pertaining in allantoic fluid, i.e. pH ~ 6.8 and with about 50 μM ascorbic acid, iron is rapidly transferred to transferrin, and acquisition of uteroferrin-bound iron by the fetal pig has been demonstrated *in vivo* (Buhi *et al*., 1982*b*). Unlike transferrin, uteroferrin is degraded once it leaves the allantoic sac and enters the circulation. A similar protein has been isolated in cattle (Campbell and Zerner, 1973) and the horse (Zavy *et al*., 1979, 1982). In man, however, materno-fetal iron transport appears to be mediated entirely by transferrin without the intervention of any uteroferrin-like protein (Galbraith *et al*., 1980*d*).

5.8 CONCLUSIONS AND FUTURE PERSPECTIVES

The past decade has seen many advances in our knowledge of the structure and function of the transferrins. The elucidation of the primary structure of two transferrins, and advances in determining the three-dimensional structure by X-ray crystallography, mean that the day when the location of the iron-binding sites is established cannot be far away. This may provide further impetus to studies of the transferrin–receptor interaction at the molecular level, especially now that the structure of the transferrin receptor is becoming known. In particular, the structural features of the transferrin molecule which mediate the interaction with the receptor deserve more study than they have so far received. There has been an upsurge of interest in the interaction of transferrin with non-erythroid cells, which has provided some new insights into the mechanism of cellular iron uptake. Although much has been learnt about the way in which transferrin binds iron *in vitro*, and in particular the effects of experimental conditions, the manner in which transferrin acquires iron *in vivo* has still received little attention. The recently developed methods for loading cells of the mononuclear phagocyte system with radioiron both *in vivo* and *in vitro* may provide a new impetus to this type of study. Finally, several new functions for lactoferrin have been proposed in recent years, and its suggested roles in the modulation of immune and inflammatory responses merit further study.

ACKNOWLEDGEMENTS

The critical comments of my colleagues Tryfonia Mainou-Fowler and Xavier Alvarez are gratefully acknowledged.

REFERENCES

Aasa, R. (1972) *Biochem. Biophys. Res. Commun.* **49**, 806.
Aasa, R. and Aisen, P. (1968) *J. Biol. Chem.* **243**, 2399.

Aasa, R., Malmström, B. G., Saltman, P. and Vänngard, T. (1963) *Biochim. Biophys. Acta* **75**, 203.

Abola, J., Wood, M. K., Chweh, A., Abraham, D. and Pulsinelli, P. D. (1982) in *The Biochemistry and Physiology of Iron* (Saltman, P. and Hegenauer, J., eds). Elsevier Biomedical, New York, p. 27.

Ainscough, E. W., Brodie, A. M. and Plowman, J. E. (1979) *Inorg. Chim. Acta* **33**, 149.

Ainscough, E. W., Brodie, A. M., Plowman, J. E., Bloor, S. J., Loehr, J. S. and Loehr, T. M. (1980) *Biochemistry* **19**, 4072.

Aisen, P. (1977) in *Iron Metabolism* (Ciba Foundation Symp. 51). Elsevier, Amsterdam, p. 1.

Aisen, P. (1980) in *Iron in Biochemistry and Medicine* II (Jacobs, A. and Worwood, M., eds). Academic Press, London, p. 87.

Aisen, P. and Leibman, A. (1968) *Biochem. Biophys. Res. Commun.* **32**, 220.

Aisen, P. and Leibman, A. (1972) *Biochim. Biophys. Acta* **257**, 314.

Aisen, P. and Leibman, A. (1973) *Biochim. Biophys. Acta* **304**, 797.

Aisen, P. and Listowsky, I. (1980) *Ann. Rev. Biochem.* **49**, 357.

Aisen, P., Leibman, A. and Reich, H. A. (1966) *J. Biol. Chem.* **241**, 1666.

Aisen, P., Aasa, R., Malmström, B. G. and Vänngard, T. (1967) *J. Biol. Chem.* **242**, 2484.

Aisen, P., Aasa, R. and Redfield, A. G. (1969) *J. Biol. Chem.* **244**, 4628.

Aisen, P., Lang, G. and Woodworth, R. C. (1973*a*) *J. Biol. Chem.* **248**, 649.

Aisen, P., Leibman, A., Pinkowitz, R. A. and Pollack, S. (1973*b*) *Biochemistry* **12**, 3679.

Aisen, P., Leibman, A. and Zweier, J. (1978) *J. Biol. Chem.* **253**, 1930.

Alderton, G., Ward, W. H. and Fevold, H. L. (1946) *Arch. Biochem. Biophys.* **11**, 9.

Allerton, S. E., Renner, J., Colt, S. and Saltman, P. (1966) *J. Am. Chem. Soc.* **88**, 3147.

Ambruso, D. R. and Johnston, R. B. (1981) *J. Clin. Invest.* **67**, 352.

Anderson, N. L. and Anderson, N. G. (1979) *Biochem. Biophys. Res. Commun.* **88**, 258.

Anderson, W. L. and Tomasi, T. B. (1977) *Arch. Biochem. Biophys.* **182**, 705.

Antanaitis, B. C. and Aisen, P. (1982) *J. Biol. Chem.* **257**, 1855.

Antanaitis, B. C., Aisen, P., Lilienthal, H. R., Roberts, R. M. and Bazer, F. W. (1980) *J. Biol. Chem.* **255**, 11204.

Antanaitis, B. C., Strekas, T. and Aisen, P. (1982) *J. Biol. Chem.* **257**, 3766.

Antanaitis, B. C., Aisen, P. and Lilienthal, H. R. (1983) *J. Biol. Chem.* **258**, 3166.

Antonini, E., Orsi, N. and Valenti, P. (1977) *Giorn. Malatt. Infett. Parassit.* **29**, 481.

Archibald, F. S. (1983) *FEMS Microbiol. Lett.* **19**, 29.

Archibald, F. S. and DeVoe, I. W. (1979) *FEMS Microbiol. Lett.* **6**, 159.

Archibald, F. S. and DeVoe, I. W. (1980) *Infect. Immunity* **27**, 322.

Arnold, R. R., Cole, M. F. and McGhee, J. R. (1977) *Science* **197**, 263.

Arnold, R. R., Russell, J. E., Champion, W. J. and Gauthier, J. J. (1981) *Infect. Immunity* **32**, 655.

Arnold, R. R., Russell, J. E., Champion, W. J., Brewer, M. and Gauthier, J. J. (1982) *Infect. Immunity* **35**, 792.

Artis, W. M., Fountain, J. A., Delcher, H. K. and Jones, H. E. (1982) *Diabetes*, **31**, 1109.

Awai, M., Chipman, B. and Brown, E. B. (1975*a*) *J. Lab. Clin. Med.* **85**, 769.

Awai, M., Chipman, B. and Brown, E. B. (1975*b*) *J. Lab. Clin. Med.* **85**, 785.

Azari, P. R. and Feeney, R. G. (1958) *J. Biol. Chem.* **232**, 293.

Bagby, G. C. and Bennett, R. M. (1982) *Blood* **60**, 108.

Bagby, G. C., Rigas, V. D., Bennett, R. M., Vandenbark, A. A. and Garewal, H. S. (1981) *J. Clin. Invest.* **68**, 56.

Baggiolini, M., De Duve, C., Masson, P. L. and Heremans, J. F. (1970) *J. Exp. Med.* **131**, 559.

Bain, J. A. and Deutsch, H. F. (1948) *J. Biol. Chem.* **172**, 547.

Baker, E. and Morgan, E. H. (1969) *Biochemistry* **8**, 2954.

Baker, E., Shaw, D. C. and Morgan, E. H. (1968) *Biochemistry* **7**, 1371.

Baker, E., Vicary, F. R. and Heuhns, E. R. (1981) *Brit. J. Haematol.* **47**, 493.

Baker, E. N. and Rumball, S. V. (1977) *J. Mol. Biol.* **111**, 207.

Baldwin, D. A. (1980) *Biochim. Biophys. Acta* **623**, 183.

Baldwin, D. A. and De Sousa, D. M. R. (1981) *Biochem. Biophys. Res. Commun.* **99**, 1101.

Baldwin, D. A., De Sousa, D. M. R. and Ford, G. (1982) in *The Biochemistry and Physiology of Iron* (Saltman, P. and Hegenauer, J., eds). Elsevier Biomedical, New York, p. 57.

Bannister, J. V., Bannister, W. M., Hill, H. A. O. and Thornally, P. J. (1982) *Biochim. Biophys. Acta* **715**, 116.
Barnes, D. and Sato, G. (1979) *Nature, Lond.* **281**, 388.
Bartek, J., Vicklicky, V., Hradilek, A. and Neuwirt, J. (1983) in *Structure and Function of Iron Storage and Transport Proteins* (Urushizaki, I., Aisen, P., Listowski, I. and Drysdale, J. W., eds). Elsevier, Amsterdam, p. 291.
Bates, G. W. (1982) in *The Biochemistry and Physiology of Iron* (Saltman, P. and Hegenauer, J., eds). Elsevier Biomedical, New York, p. 3.
Bates, G. W. and Schlabach, M. R. (1973*a*) *J. Biol. Chem.* **248**, 3228.
Bates, G. W. and Schlabach, M. R. (1973*b*) *FEBS Lett.* **33**, 289.
Bates, G. W. and Schlabach, M. R. (1975) *J. Biol. Chem.* **250**, 2177.
Bates, G. W. and Wernicke, J. (1971) *J. Biol. Chem.* **246**, 3679.
Bates, G. W., Billups, C. and Saltman, P. (1967) *J. Biol. Chem.* **242**, 2810.
Beamish, M. R., Keay, L., Okigaki, T. and Brown, E. B. (1975) *Brit. J. Haematol.* **31**, 479.
Bearn, A. G. and Parker, W. C. (1966). in *The Glycoproteins* (Gottschalk, A., ed.). Elsevier, Amsterdam, p. 415.
Beaton, G. H., Selby, A. G. and Wright, A. M. (1961) *J. Biol. Chem.* **236**, 2001.
Bennett, R. M. and Davis, J. (1981) *J. Immunol.* **127**, 1211.
Bennett, R. M. and Davis, J. (1982) *J. Lab. Clin. Med.* **99**, 127.
Bennett, R. M. and Kokocinski, T. (1978) *Brit. J. Haematol.* **39**, 509.
Bennett, R. M. and Kokocinski, T. (1979) *Clin. Sci.* **57**, 453.
Bennett, R. M. and Mohla, C. (1976) *J. Lab. Clin. Med.* **88**, 156.
Bennett, R. M. and Skosey, J. L. (1977) *Arthritis Rheum.* **19**, 84.
Bennett, R. M., Eddie-Quartey, A. C. and Holt, P. L. J. (1973) *Arthritis Rheum.* **16**, 186.
Bennett, R. M., Bagby, G. C. and Davis, J. (1981) *Biochem. Biophys. Res. Commun.* **101**, 88.
Bezkorovainy, A. (1966) *Biochim. Biophys. Acta* **127**, 535.
Bezkorovainy, A. (1980) *Biochemistry of Nonheme Iron.* Plenum Press, New York.
Bezkorovainy, A. and Grohlich, D. (1967) *Biochim. Biophys. Acta* **147**, 497.
Birgens, H. S., Hansen, N. E., Karle, H. and Kristensen, L. O. (1983) *Brit. J. Haematol.* **54**, 383.
Bishop, J. G., Schanbacher, F. L., Ferguson, L. C. and Smith, K. L. (1976) *Infect. Immunity* **14**, 911.
Black, C., Glass, J., Núñez, M. T. and Robinson, S. H. (1979) *J. Lab. Clin. Med.* **93**, 645.
Bleil, J. D. and Bretscher, M. S. (1982) *EMBO J.* **1**, 351.
Block, T. and Bothwell, M. (1983) *Nature, Lond.* **301**, 342.
Bluard-Deconinck, J-M., Williams, J., Evans, R. W., Van Snick, J., Osinski, P. A. and Masson, P. L. (1978) *Biochem. J.* **171**, 321.
Booth, A. G. and Wilson, M. J. (1981) *Biochem. J.* **196**, 355.
Bottenstein, J. E. and Sato, G. H. (1979) *Proc. Natl Acad. Sci. USA* **76**, 514.
Bottenstein, J. E., Skaper, S. D., Varon, S. S. and Sato, G. H. (1980) *Exp. Cell Res.* **125**, 183.
Boxer, L. A., Björksten, B., Björk, J., Yang, H-H., Allen, J. M. and Baehner, R. L. (1982) *J. Lab. Clin. Med.* **99**, 866.
Boyett, J. D. and Sullivan, J. F. (1970) *Metabolism* **19**, 148.
Braun, V., Hancock, R. E. W., Hantke, K. and Hartmann, A. (1976) *J. Supramol. Struct.* **5**, 37.
Breitman, T. R., Collins, S. J. and Keene, B. R. (1980) *Exp. Cell Res.* **126**, 494.
Brines, R. D. and Brock, J. H. (1983) *Biochim. Biophys. Acta* **759**, 229.
Brock, J. H. (1980) *Arch. Dis. Childh.* **55**, 417.
Brock, J. H. (1981) *Immunology* **43**, 387.
Brock, J. H. and Arzabe, F. R. (1976) *FEBS Lett.* **69**, 63.
Brock, J. H. and Esparza, I. (1979) *Brit. J. Haematol.* **42**, 481.
Brock, J. H. and Mainou-Fowler, T. (1983) *Immunol. Today* **4**, 347.
Brock, J. H. and Rankin, M. C. (1981) *Immunology* **43**, 393.
Brock, J. H., Steel, E. D. and Reiter, B. (1975) *Res. Vet. Sci.* **19**, 152.
Brock, J. H., Arzabe, F. R., Lampreave, F. and Piñeiro, A. (1976) *Biochim. Biophys. Acta* **446**, 214.

Brock, J. H., Arzabe, F. R., Richardson, N. E. and Deverson, E. V. (1978*a*) *Biochem. J.* **171**, 73.
Brock, J. H., Lampreave, F. and Piñeiro, A. (1978*b*) *Ann. Rech. Vet.* **9**, 287.
Brock, J. H., Esparza, I., Oliver, R. A. and Spooner, R. L. (1980) *Biochem. Genet.* **18**, 851.
Brock, J. H., Pickering, M. G., McDowall, M. C. and Deacon, A. G. (1983) *Infect. Immunity* **40**, 453.
Brown, E. B., Okada, S., Awai, M. and Chipman, B. (1975) *J. Lab. Clin. Med.* **85**, 576.
Brown, J. P., Hewick, R. M., Hellström, I., Hellström, K. E., Doolittle, R. F. and Dreyer, W. J. (1982) *Nature, Lond.* **296**, 171.
Brown-Mason, A. and Woodworth, R. C. (1982) in *The Biochemistry and Physiology of Iron* (Saltman, P. and Hegenauer, J., eds). Elsevier Biomedical, New York, p. 301.
Brown-Mason, A. and Woodworth, R. C. (1984) *J. Biol. Chem.* **259**, 1866.
Broxmeyer, H. E. (1978) *Blood* **51**, 889.
Broxmeyer, H. E. (1979) *J. Clin. Invest.* **64**, 1717.
Broxmeyer, H. E., Smithyman, A., Eger, R. R., Meyers, P. A. and De Sousa, M. (1978) *J. Exp. Med.* **148**, 1052.
Broxmeyer, H. E., De Sousa, M., Smithyman, A., Ralph, P., Hamilton, J., Kurland, J. I. and Bognacki, J. (1980*a*) *Blood* **55**, 324.
Broxmeyer, H. E., Ralph, P., Bognacki, J., Kincade, P. W. and De Sousa, M. (1980*b*) *J. Immunol.* **125**, 903.
Broxmeyer, H. E., Gentile, P., Bognacki, J. and Ralph, P. (1983) *Blood Cells* **9**, 83.
Buhi, W. C., Gray, W. C., Mansfield, E. A., Chun, P. V., Ducsay, C. A., Bazer, F. W. and Roberts, R. M. (1982*a*) *Biochim. Biophys. Acta* **701**, 32.
Buhi, W. C., Ducsay, C. A., Bazer, F. W. and Roberts, R. M. (1982*b*) *J. Biol. Chem.* **257**, 1713.
Bullen, J. J. (1981) *Rev. Infect. Dis.* **3**, 1127.
Bullen, J. J. and Armstrong, J. A. (1979) *Immunology* **36**, 781.
Bullen, J. J. and Joyce, P. R. (1982) *Immunology* **46**, 497.
Bullen, J. J. and Wallis, S. N. (1977) *FEMS Microbiol.Lett.* **1**, 117.
Bullen, J. J., Cushnie, G. H. and Rogers, H. J. (1967) *Immunology* **12**, 303.
Bullen, J. J., Rogers, H. J. and Leigh, L. (1972) *Brit. Med. J.* **1**, 69.
Butler, J. E. (1973) *Biochim. Biophys. Acta* **295**, 341.
Butterworth, R. M., Gibson, J. F. and Williams, J. (1975) *Biochem. J.* **149**, 559.
Campbell, H. D. and Zerner, B. (1973) *Biochem. Biophys. Res. Commun.* **54**, 1498.
Campbell, H. D., Dionysius, H. D., Keough, D. T., Wilson, B. E., De Jersey, J. and Zerner, B. (1978) *Biochem. Biophys. Res. Commun.* **82**, 615.
Campbell, R. F. and Chasteen, N. D. (1977) *J. Biol. Chem.* **252**, 5996.
Cannon, J. C. and Chasteen, N. D. (1975) *Biochemistry* **11**, 4573.
Carey, P. R. and Young, N. M. (1974) *Can. J. Biochem.* **52**, 273.
Carmichael, A. and Vincent, J. S. (1979) *FEBS Lett.* **105**, 349.
Carrano, C. J. and Raymond, K. N. (1979) *J. Am. Chem. Soc.* **101**, 5401.
Carver, F. J. and Frieden, E. (1978) *Biochemistry* **17**, 167.
Charlwood, P. A. (1979) *Biochim. Biophys. Acta* **581**, 260.
Chasteen, N. D. (1983*a*) *Adv. Inorg. Biochem* **5**, 201.
Chasteen, N. D. (1983*b*) *Trends Biochem. Sci.* **8**, 272.
Chasteen, N. D. and Williams, J. (1981) *Biochem. J.* **193**, 717.
Chasteen, N. D., White, L. K. and Campbell, R. F. (1977) *Biochemistry* **16**, 363.
Chen, T-T., Bazer, F. W., Cetorelli, J. J., Pollard, W. E. and Roberts, R. M. (1973) *J. Biol. Chem.* **248**, 8560.
Chesters, J. K. and Will, M. (1981) *Brit. J. Nutr.* **46**, 111.
Christensen, A. C., Huebers, H. and Finch, C. A. (1978) *Am. J. Physiol.* **235**, R18.
Ciechanover, A., Schwartz, A. L., Dautry-Varsat, A. and Lodish, H. F. (1983) *J. Biol. Chem.* **258**, 9681.
Corda, R., Biddau, P., Corrias, A. and Puxeddu, E. (1983) *Int. J. Tissue Reac.* **5**, 117.
Cowart, R. E., Kojima, N. and Bates, G. W. (1982) *J. Biol. Chem.* **257**, 7560.
Cox, T. M., Mazurier, J., Spik, G., Montreuil, J. and Peters, T. J. (1979) *Biochim. Biophys. Acta* **588**, 120.
Dautry-Varsat, A., Ciechanover, A. and Lodish, H. F. (1983) *Proc. Natl Acad. Sci. USA* **80**, 2258.

De Abrew, S. (1981) *Int. J. Nucl. Med. Biol.* **8**, 217.
Debanne, M. T., Chindemi, P. A. and Regoeczi, E. (1981) *J. Biol. Chem.* **256**, 4929.
De Laey, P., Masson, P. L. and Heremans, J. F. (1968) *Protides Biol. Fluids* **16**, 627.
Delaney, T. A., Morgan, W. H. and Morgan, E. H. (1982) *Biochim. Biophys. Acta* **701**, 295.
DeLucas, L. J., Suddath, F. L., Gams, R. A. and Bugg, C. E. (1978) *J. Mol. Biol.* **123**, 285.
De Vet, B. C. J. M. and Van Gool, J. (1974) *Acta Med. Scand.* **196**, 393.
Dolby, J. M. and Honour, P. (1979) *J. Hyg. (Camb.)* **83**, 255.
Donovan, J. W. and Ross, K. D. (1975) *J. Biol. Chem.* **250**, 6026.
Donovan, J. W., Beardslee, R. A. and Ross, K. D. (1976) *Biochem. J.* **153**, 631.
Dorland, L., Haverkamp, J., Schut, B. L., Vliegenthart, J., Spik, G., Strecker, G., Fournet, B. and Montreuil, J. (1977) *FEBS Lett.* **77**, 15.
Dorland, L., Haverkamp, J., Vliegenthart, J. F. G., Spik, G., Fournet, B. and Montreuil, J. (1979) *Eur. J. Biochem.* **100**, 569.
Duncan, R. L. and McArthur, W. P. (1981) *Cell. Immunol.* **63**, 308.
Eastham, E. J., Bell, J. I. and Douglas, A. P. (1977) *Biochem. J.* **164**, 289.
Ecarot-Charrier, B., Grey, V. L., Wilczynska, A. and Schulman, H. M. (1980) *Can. J. Biochem.* **58**, 418.
Egyed, A. (1973) *Biochim. Biophys. Acta* **304**, 805.
Egyed, A. (1975) *Biochim. Biophys. Acta* **411**, 349.
Egyed, A. (1977) in *Proteins of Iron Metabolism* (Brown, E. B., Aisen, P., Fielding, J. and Crichton, R. R., eds). Grune & Stratton, New York, p. 237.
Egyed, A., May, A. and Jacobs, A. (1980) *Biochim. Biophys. Acta* **629**, 391.
Ekblom, P., Thesleff, I., Lehto, V-P. and Virtanen, I. (1983) *Int. J. Cancer* **31**, 111.
Elleman, T. C. and Williams, J. (1970) *Biochem. J.* **116**, 515.
El-Shobaki, F. A. and Rummel, W. (1979) *Res. Exp. Med. (Berl.)* **174**, 187.
Enns, C. A. and Sussman, H. H. (1981*a*) *J. Biol. Chem.* **256**, 9820.
Enns, C. A. and Sussman, H. H. (1981*b*) *J. Biol. Chem.* **256**, 12620.
Enns, C. A., Shindelman, J. E., Tonik, S. E. and Sussman, H. H. (1981) *Proc. Natl Acad. Sci. USA* **78**, 4222.
Enns, C. A., Suomalainen, H. A., Gebhart, J. E., Schroder, J. and Sussman, H. H. (1982) *Proc. Natl Acad. Sci. USA* **79**, 3241.
Enns, C. A., Larrick, J. W., Suomalainen, H., Schroder, J. and Sussman, H. H. (1983) *J. Cell Biol.* **97**, 579.
Esparza, I. (1979) Doctoral Thesis, University of Zaragoza.
Esparza, I. and Brock, J. H. (1980*a*) *Biochim. Biophys. Acta* **622**, 297.
Esparza, I. and Brock, J. H. (1980*b*) *Biochim. Biophys. Acta* **624**, 479.
Esparza, I. and Brock, J. H. (1981) *Brit. J. Haematol.* **49**, 603.
Evans, G. W. (1976) *Proc. Soc. Exp. Biol. Med.* **151**, 775.
Evans, G. W. and Winter, T. W. (1975) *Biochem. Biophys. Res. Commun.* **66**, 1218.
Evans, R. W. and Holbrook, J. J. (1975) *Biochem. J.* **145**, 201.
Evans, R. W. and Williams, J. (1978) *Biochem. J.* **173**, 543.
Evans, R. W., Donovan, J. W. and Williams, J. (1977) *FEBS Lett.* **83**, 19.
Evans, R. W., Williams, J. and Moreton, K. (1982) *Biochem. J.* **201**, 19.
Fedorko, M. E. (1974) *J. Cell Biol.* **62**, 802.
Feeney, R. E. and Komatsu, K. S. (1966) *Struct. Bonding* **1**, 149.
Feeney, R. E., Anderson, J. S., Azari, P. R., Bennett, N. and Rhodes, M. B. (1960) *J. Biol. Chem.* **235**, 2307.
Feeney, R. E., Osuga, D. T., Meares, C. F., Babin, D. R. and Penner, M. H. (1983) in *Structure and Function of Iron Storage and Transport Proteins* (Urushizaki, I., Aisen, P., Listowsky, I. and Drysdale, J. W., eds), Elsevier, Amsterdam, p. 231.
Fernandez-Pol, J. A. and Klos, D. J. (1980) *Biochemistry* **19**, 3904.
Fernandez-Pol, J. A., Dunn, A. M., Hamilton, P. D. and Klos, D. J. (1983) in *Structure and Function of Iron Storage and Transport Proteins* (Urushizaki, I., Aisen, P., Listowsky, I. and Drysdale, J. W., eds). Elsevier, Amsterdam, p. 371.
Fielding, J. and Speyer, B. E. (1974) *Biochim. Biophys. Acta* **363**, 387.
Figarella, C. and Sarles, H. (1975) *Scand. J. Gastroenterol.* **10**, 449.
Fillet, G., Cook, J. D. and Finch, C. A. (1974) *J. Clin. Invest.* **53**, 1527.
Finch, C. A., Huebers, H., Eng, M. and Miller, L. (1982) *Blood* **59**, 364.

Finkelstein, R. A., Sciortino, C. V. and McIntosh, M. A. (1983) *Rev. Infect. Dis.* **5** (Suppl. 4), S759.
Fitzgerald, S. P. and Rogers, H. J. (1980) *Infect. Immunity* **27**, 302.
Fletcher, J. and Huehns, E. R. (1967) *Nature, Lond.* **215**, 584.
Fletcher, J. and Huehns, E. R. (1968) *Nature, Lond.* **218**, 1211.
Folajtar, D. A. and Chasteen, N. D. (1982) in *The Biochemistry and Physiology of Iron* (Saltman, P. and Hegenauer, J., eds). Elsevier Biomedical, New York, p. 35.
Fontes, G. and Thivolle, L. (1925) *C.R. Seances Soc. Biol. Paris* **93**, 687.
Ford-Hutchinson, A. W. and Perkins, D. J. (1971) *Eur. J. Biochem.* **21**, 55.
Ford-Hutchinson, A. W. and Perkins, D. J. (1972) *Eur. J. Biochem.* **25**, 415.
Fransson, G-B. and Lönnerdal, B. (1980) *J. Pediat.* **96**, 380.
Fransson, G-B. and Lönnerdal, B. (1982) in *The Biochemistry and Physiology of Iron* (Saltman, P. and Hegenauer, J., eds). Elsevier Biomedical, New York, p. 305.
Fransson, G-B., Thoren-Tolling, K., Jones, B., Hambraeus, F. and Lönnerdal, B. (1983) *Nutr. Res.* **3**, 373.
Frazier, J. L., Caskey, J. H., Yoffe, M. and Seligman, P. A. (1982) *J. Clin. Invest.* **69**, 853.
Frieden, E. and Aisen, P. (1980) *Trends Biochem. Sci.* **5**, XI.
Froncisz, W. and Aisen, P. (1982) *Biochim. Biophys. Acta* **700**, 55.
Gaber, B. P., Miskowski, V. M. and Spiro, T. G. (1974) *J. Amer. Chem. Soc.* **96**, 6868.
Galbraith, G. M. P. and Galbraith, R. M. (1980) *Clin. Exp. Immunol.* **42**, 285.
Galbraith, G. M. P., Galbraith, R. M. and Faulk, W. P. (1980*a*) *Cell. Immunol.* **49**, 215.
Galbraith, G. M. P., Goust, J. M., Mercurio, S. M. and Galbraith, R. M. (1980*b*) *Clin. Immunol. Immunopathol.* **16**, 387.
Galbraith, R. M., Werner, P., Arnaud, P. and Galbraith, G. M. P. (1980*c*) *J. Clin. Invest.* **66**, 1135.
Galbraith, G. M. P., Galbraith, R. M. and Faulk, W. P. (1980*d*) *Placenta* **1**, 33.
Gallin, J. I., Fletcher, M. P., Seligmann, B. E., Hoffstein, S., Cehrs, K. and Mounessa, N. (1982) *Blood* **59**, 1317.
Gams, R. A., Webb, J. and Glickson, J. D. (1975) *Cancer Res.* **35**, 1422.
Garrett, N. E., Garrett, R. J. B. and Archdeacon, J. W. (1973) *Biochem. Biophys. Res. Commun.* **52**, 466.
Gelb, M. H. and Harris, D. C. (1980) *Arch. Biochem. Biophys.* **200**, 93.
Geoghegan, K. F., Dallas, J. L. and Feeney, R. E. (1980) *J. Biol. Chem.* **255**, 11429.
Gibbons, R. A., Dixon, S. N., Russell, A. M. and Sansom, B. F. (1976) *Biochim. Biophys. Acta* **437**, 301.
Giblett, E. R. (1969) *Genetic Markers in Human Blood*. F. A. Davis Co., Philadelphia, p. 135.
Glass, J., Núñez, M. T. and Robinson, S. H. (1977) *Biochem. Biophys. Res. Commun.* **75**, 226.
Glass, J., Núñez, M. T., Fischer, S. and Robinson, S. H. (1978) *Biochim. Biophys. Acta* **542**, 154.
Glass, J., Núñez, M. T. and Robinson, S. H. (1980) *Biochim. Biophys. Acta* **598**, 293.
Goding, J. W. and Burns, G. F. (1981) *J. Immunol.* **127**, 1256.
Goding, J. W. and Harris, A. W. (1981) *Proc. Natl Acad. Sci. USA* **78**, 4530.
Goodfellow, P. N., Banting, G., Sutherland, R., Greaves, M., Solomon, E. and Povey, S. (1982) *Somat. Cell Genet.* **8**, 197.
Gordon, A. H. and Louis, L. N. (1963) *Biochem. J.* **88**, 409.
Gorinsky, B., Horsburgh, C., Lindley, P. F., Moss, D. S., Parker, M. and Watson, J. L. (1979) *Nature, Lond.* **281**, 157.
Got, R., Font, J. and Goussault, Y. (1967) *Comp. Biochem. Physiol.* **23**, 317.
Graham, I. and Williams, J. (1975) *Biochem. J.* **145**, 263.
Griffiths, E. and Humphreys, J. (1977) *Infect. Immunity* **15**, 396.
Groen, R., Hendricksen, P., Young, S. P., Leibman, A. and Aisen, P. (1982) *Brit. J. Haematol.* **50**, 43.
Grohlich, D., Morley, C. G. D. and Bezkorovainy, A. (1979) *Int. J. Biochem.* **10**, 797.
Guérin, G., Vreeman, H. J. and Nguyen, T. C. (1976) *Eur. J. Biochem.* **76**, 433.
Gutteridge, J. M. C., Paterson, S. K., Segal, A. W. and Halliwell, B. (1981) *Biochem. J.* **199**, 259.

Hahn, D. and Ganzoni, A. M. (1975) *Acta Haematol.* **53**, 321.
Hahn, D., Baviera, B. and Ganzoni, A. M. (1975) *Acta Haematol.* **53**, 285.
Hamilton, T. A., Wada, H. G. and Sussman, H. H. (1979) *Proc. Natl Acad. Sci. USA* **76**, 6406.
Hancock, R. G. V., Evans, D. J. R. and Fritze, K. (1973) *Biochim. Biophys. Acta* **320**, 486.
Harding, C. and Stahl, P. (1983) *Biochem. Biophys. Res. Commun.* **113**, 650.
Harding, C., Heuser, J. and Stahl, P. (1983) *J. Cell Biol.* **97**, 329.
Harmon, R. J., Schanbacher, F. L., Ferguson, L. C. and Smith, K. L. (1975) *Am. J. Vet. Res.* **36**, 1001.
Harris, A. W. and Sephton, R. G. (1977) *Cancer Res.* **37**, 3634.
Harris, D. C. (1977*a*) *Biochemistry* **16**, 560.
Harris, D. C. (1977*b*) *Biochim. Biophys. Acta* **496**, 563.
Harris, D. C. and Aisen, P. (1975) *Biochemistry* **14**, 262.
Harris, D. C. and Gelb, M. H. (1980) *Biochim. Biophys. Acta* **623**, 1.
Harris, D. C., Gray, G. A. and Aisen, P. (1974) *J. Biol. Chem.* **249**, 5261.
Harris, D. C., Haroutunian, P. V. and Gutmann, S. M. (1977) *Brit. J. Haematol.* **37**, 302.
Harris, W. R., Carrano, C. J., Pecoraro, V. L. and Raymond, K. N. (1981) *J. Am. Chem. Soc.* **103**, 2231.
Harris, W. R. (1983) *Biochemistry* **22**, 3920.
Hatton, M. W. C., Regoeczi, E. and Wong, K-L. (1974) *Can. J. Biochem.* **52**, 845.
Hatton, M. W. C., Regoeczi, E., Wong, K-L. and Kraay, G. J. (1977) *Biochem. Genet.* **15**, 621.
Hatton, M. W. C., Regoeczi, E. and Kaur, H. (1978) *Can. J. Biochem.* **56**, 339.
Hatton, M. W. C., Marz, L., Berry, L. R., Debanne, M. T. and Regoeczi, E. (1979) *Biochem. J.* **181**, 633.
Hayashi, I. and Sato, G. (1976) *Nature, Lond.* **259**, 132.
Heaphy, S. and Williams, J. (1982*a*) *Biochem. J.* **205**, 611.
Heaphy, S. and Williams, J. (1982*b*) *Biochem. J.* **205**, 619.
Heide, K. and Haupt, H. (1964) *Behringwerk-Mitt.* **43**, 161.
Hekman, A. (1971) *Biochim. Biophys. Acta* **251**, 380.
Hemmaplardh, D. and Morgan, E. H. (1974) *Exp. Cell Res.* **87**, 207.
Hemmaplardh, D. and Morgan, E. H. (1976) *Biochim. Biophys. Acta* **426**, 385.
Hemmaplardh, D. and Morgan, E. H. (1977) *Brit. J. Haematol.* **36**, 85.
Hoffbrand, A. V., Ganeshaguru, K., Hooton, J. W. L. and Tattersall, M. H. N. (1976) *Brit. J. Haematol.* **33**, 517.
Hoffer, P. B., Huberty, J. and Khayam-Bashi, H. (1977) *J. Nucl. Med.* **18**, 713.
Holbein, B. E. (1981) *Curr. Microbiol.* **6**, 213.
Hopkins, C. R. (1983) *Cell* **35**, 321.
Hopkins, C. R. and Trowbridge, I. S. (1983) *J. Cell Biol.* **97**, 508.
Hopkins, L. L. and Schwartz, K. (1964) *Biochim. Biophys. Acta* **90**, 484.
Hosain, F., McIntyre, P. A., Poulose, K., Stern, H. S. and Wagner, H. N. (1969) *Clin. Chim. Acta* **24**, 69.
Hovanessian, A. G. and Awdeh, Z. L. (1976) *Eur. J. Biochm.* **68**, 333.
Howard, P. N., Wang, A-C. and Sutton, H. E. (1968) *Biochem. Genet.* **2**, 265.
Hu, H-Y. Y., Gardner, J., Aisen, P. and Skoultchi, A. I. (1977) *Science* **197**, 559.
Hudson, B. G., Ohno, M., Brockway, W. J. and Castellino, F. J. (1973) *Biochemistry*, **12**, 1047.
Huebers, H., Huebers, E., Rummel, W. and Crichton, R. R. (1976) *Eur. J. Biochem.* **66**, 447.
Huebers, H., Huebers, E., Linck, S. and Rummel, W. (1977) in *Proteins of Iron Metabolism* (Brown, E. B., Aisen, P., Fielding J. and Crichton, R. R., eds). Grune & Stratton, New York, p. 251.
Huebers, H., Huebers, E., Csiba, E. and Finch, C. A. (1978) *J. Clin. Invest.* **62**, 944.
Huebers, H., Bauer, W., Huebers, E., Csiba, E. and Finch, C. A. (1981*a*) *Blood* **57**, 218.
Huebers, H., Csiba, E., Johnson, B., Huebers, E. and Finch, C. A. (1981*b*) *Proc. Natl Acad. Sci. USA* 78, 621.
Huebers, H., Josephson, B., Huebers, E., Csiba, E. and Finch, C. A. (1981*c*) *Proc. Natl Acad. Sci. USA* 78, 2572.
Huebers, H., Huebers, E. and Finch, C. A. (1982) in *The Biochemistry and Physiology of*

Iron (Saltman, P. and Hegenauer, J., eds). Elsevier Biomedical, New York, p. 311.
Iacopetta, B. J., Morgan, E. H. and Yeoh, G. C. (1982) *Biochim. Biophys. Acta* **687**, 204.
Ichihara, S. and Mizushima, S. (1978) *J. Biochem. (Tokyo)* **83**, 137.
Inman, J. K. (1956) Doctoral Dissertation, Harvard University.
Iscove, N. N. and Melchers, F. (1978) *J. Exp. Med.* **147**, 923.
Iscove, N. N., Guilbert, L. J. and Weyman, C. (1980) *Exp. Cell Res.* **126**, 121.
Isobe, K., Isobe, Y. and Sakurami, T. (1981) *Acta Haematol.* **65**, 2.
Iwase, H. and Hotta, K. (1977) *J. Biol. Chem.* **252**, 5437.
Jacobs, A. (1977*a*) in *Iron Metabolism* (Ciba Foundation Symp. 51), Elsevier, Amsterdam, p. 91.
Jacobs, A. (1977*b*) in *Recent Advances in Haematology* 2 (Hoffbrand, A. V., Brain, M. C. and Hirsch, J., eds). Churchill-Livingstone, Edinburgh, p. 1.
Jandl, J. H. and Katz, J. H. (1963) *J. Clin. Invest.* **42**, 314.
Jandl, J. H., Inman, J. K., Simmons, R. L. and Allen, D. W. (1959) *J. Clin. Invest.* **38**, 161.
Jeltsch, J-M. and Chambon, P. (1982) *Eur. J. Biochem.* **122**, 291.
Jeppsson, J-O. (1967) *Acta Chem. Scand.* **21**, 1686.
Johansson, B. G. (1958) *Nature, Lond.* **181**, 996.
Jollés, J., Mazurier, J., Metz-Boutigue, H., Spik, J., Montreuil, J. and Jollés, P. (1976) *FEBS Lett.* **69**, 27.
Jones, H. D. C. and Perkins, D. J. (1965) *Biochim. Biophys. Acta* **100**, 122.
Kailis, S. G. and Morgan, E. H. (1977) *Biochim. Biophys. Acta* **464**, 389.
Kaminski, M., Didkowski, S. and Sykiotis, M. (1981) *Comp. Biochem. Physiol.* **68B**, 505.
Karin, M. and Mintz, B. (1981) *J. Biol. Chem.* **256**, 3245.
Karle, H., Hansen, N. E., Malmquist, J., Karle, A. H. and Larsson, I. (1979) *Scand. J. Haematol.* **23**, 303.
Keller, W. and Pennell, R. B. (1959) *J. Lab. Clin. Med.* **53**, 638.
Keough, D. T., Dionysius, D. A., De Jersey, J. and Zerner, B. (1980) *Biochem. Biophys. Res. Commun.* **94**, 600.
Kerckaert, J-P. and Bayard, B. (1982) *Biochem. Biophys. Res. Commun.* **105**, 1023.
Keung, W-M. and Azari, P. (1982) *J. Biol. Chem.* **257**, 1184.
Keung, W-M., Azari, P. and Phillips, J. L. (1982) *J. Biol. Chem.* **257**, 1177.
Khimji, P. L. and Miles, A. A. (1978) *Brit. J. Exp. Pathol.* **59**, 137.
Kinkade, J. M., Miller, W. W. K. and Segars, F. M. (1976) *Biochim. Biophys. Acta* **446**, 407.
Klausner, R. D., Ashwell, G., Van Renswoude, J., Harford, J. B. and Bridges, K. R. (1983*a*) *Proc. Natl Acad. Sci. USA*, **80**, 2263.
Klausner, R. D., Van Renswoude, J., Ashwell, G., Kempf, C., Schechter, A. N., Dean, A. and Bridges, K. R. (1983*b*) *J. Biol. Chem.* **258**, 4715.
Koenig, S. H. and Schillinger, W. (1969) *J. Biol. Chem.* **244**, 6520.
Kojima, N. and Bates, G. W. (1979) *J. Biol. Chem.* **254**, 8847.
Kojima, N. and Bates, G. W. (1981) *J. Biol. Chem.* **256**, 12034.
Koller, M. E., Prante, P. H., Ulvik, R. and Romslo, I. (1976) *Biochem. Biophys. Res. Commun.* **71**, 339.
Komatsu, S. K. and Feeney, R. E. (1967) *Biochemistry* **6**, 1136.
Konopka, K. (1978) *FEBS Lett.* **92**, 308.
Konopka, K. and Romslo, I. (1980) *Eur. J. Biochem.* **107**, 433.
Konopka, K. and Romslo, I. (1981) *Eur. J. Biochem.* **117**, 239.
Konopka, K. and Turska, E. (1979) *FEBS Lett.* **105**, 85.
Kornfeld, S. (1968) *Biochemistry* **7**, 945.
Kornfeld, S. (1969) *Biochim. Biophys. Acta* **194**, 25.
Kourilsky, F. M. and Burtin, P. (1968) *Nature, Lond.* **218**, 375.
Krusius, T. and Finne, J. (1981) *Carbohyd. Res.* **90**, 203.
Krysteva, M. A., Mazurier, J., Spik, G. and Montreuil, J. (1975) *FEBS Lett.* **56**, 337.
Krysteva, M. A., Mazurier, J. and Spik, G. (1976) *Biochim. Biophys. Acta* **453**, 484.
Kühnl, P. and Spielmann, W. (1978) *Hum. Genet.* **43**, 91.
Kvach, J. T., Wiles, T. I., Mellencamp, M. W. and Kochan, I. (1977) *Infect. Immunity* **18**, 439.
Lampreave, F. (1976) Master's Thesis, University of Zaragoza.
Landschulz, W., Thesleff, I. and Ekblom, P. (1984) *J. Cell Biol.* **98**, 596.

Lane, R. S. (1971) *Biochim. Biophys. Acta* **243**, 193.
Larrick, J. W. and Cresswell, P. (1979*a*) *J. Supramol. Struct.* **11**, 579.
Larrick, J. W. and Cresswell, P. (1979*b*) *Biochim. Biophys. Acta* **583**, 483.
Larrick, J. W. and Logue, G. (1980) *Lancet*, **ii**, 862.
Larson, S. M., Grunbaum, Z., Allen, D. R. and Rasey, J. S. (1978) *J. Nucl. Med.* **19**, 1245.
Larson, S. M., Rasey, J. S., Allen, D. R. and Nelson, N. J. (1979) *J. Nucl. Med.* **20**, 837.
Larson, S. M., Rasey, J. S., Allen, D. R., Nelson, N. J., Grunbaum, Z., Harp, G. D. and Williams, D. L. (1980) *J. Nat. Cancer Inst.* **64**, 41.
Law, B. A. and Reiter, B. (1977) *J. Dairy Res.* **44**, 595.
Lawrence, T. H., Biggers, C. J. and Simonton, P. R. (1977) *Ann. Hum. Biol.* **4**, 281.
Lebman, D., Trucco, M., Bottero, L., Lange, B., Pessano, S. and Rovera, G. (1982) *Blood* **59**, 671.
Lee, D. C., McKnight, G. S. and Palmiter, R. D. (1978) *J. Biol. Chem.* **253**, 3494.
Lee, M. Y., Huebers, H., Martin, A. W. and Finch, C. A. (1978) *J. Comp. Physiol.* **127**, 349.
Leffell, M. S. and Spitznagel, J. K. (1975) *Infect. Immunity* **12**, 813.
Leger, D., Tordera, V., Spik, G., Dorland, L., Haverkamp, J. and Vliegenthart, J. F. G. (1978) *FEBS Lett.* **93**, 255.
Lehrer, S. S. (1969) *J. Biol. Chem.* **244**, 3613.
Leibman, A. and Aisen, P. (1977) *Biochemistry* **16**, 1268.
Lestas, A. N. (1976) *Brit. J. Haematol.* **32**, 341.
Levine, P. H., Levine, A. J. and Weintraub, L. R. (1972) *J. Lab. Clin. Med.* **80**, 333.
Light, N. D. (1977) *Biochim. Biophys. Acta* **495**, 46.
Light, N. D. (1978) *Biochem. Biophys. Res. Commun.* **81**, 261.
Line, W. F., Grohlich, D. and Bezkorovainy, A. (1967) *Biochemistry* **6**, 3393.
Line, W. F., Sly, D. A. and Bezkorovainy, A. (1976) *Int. J. Biochem.* **7**, 203.
Lineback-Zins, J. and Brew, K. (1980) *J. Biol. Chem.* **255**, 708.
Logan, K. J., Ng, P. K., Turner, C. J., Schmidt, R. P., Terner, U. K., Scott, J. R., Lentle, B. C. and Noujaim, A. A. (1981) *Int. J. Nucl. Med. Biol.* **8**, 271.
Loh, T. T. (1982) *Clin. Exp. Pharmacol. Physiol.* **9**, 11.
Loh, T. T., Yeung, Y. G. and Yeung, D. (1977) *Biochim. Biophys. Acta* **471**, 118.
Loisillier, F., Burtin, P. and Grabar, P. (1968) *Ann. Inst. Pasteur* **115**, 829.
Luk, C. K. (1971) *Biochemistry* **10**, 2838.
Lush, I. E. (1966) *The Biochemical Genetics of Vertebrates except Man*, Elsevier/North-Holland, Amsterdam.
MacGillivray, R. T. A., Méndez, E. and Brew, K. (1977) in *Proteins of Iron Metabolism* (Brown, E. B., Aisen, P., Fielding, J. and Crichton, R. R., eds). Grune & Stratton, New York, p. 133.
MacGillivray, R. T. A., Méndez, E., Sinha, S. K., Sutton, M. R., Lineback-Zins, J. and Brew, K. (1982) *Proc. Natl Acad. Sci. USA* **79**, 2504.
MacGillivray, R. T. A., Méndez, E., Shewale, J. G., Sinha, S. K., Lineback-Zins, J. and Brew, K. (1983) *J. Biol. Chem.* **258**, 3543.
MacSween, R. N. M. and MacDonald, R. A. (1969) *Lab. Invest.* **21**, 230.
McCord, J. M. and Day, E. D. (1978) *FEBS Lett.* **86**, 139.
McMillan, J. A., Oski, F. A., Lourie, G., Tomarelli, R. M. and Landau, J. M. (1977) *Pediatrics* **60**, 896.
Maeda, K., McKenzie, H. A. and Shaw, D. C. (1980) *Anim. Blood Grps Biochem. Genet.* **11**, 63.
Makarechian, M. and Howell, W. E. (1966) *Can. J. Biochem. Physiol.* **44**, 1089.
Makey, D. G. and Seal, U. S. (1976) *Biochim. Biophys. Acta* **453**, 250.
Malmquist, J. and Johansson, B. G. (1971) *Biochim. Biophys. Acta* **236**, 38.
Malmquist, J., Thorell, J. I. and Wollheim, F. A. (1977) *Acta Med. Scand.* **202**, 313.
Manwell, C. and Baker, C. M. A. (1970) *Molecular Biology and the Origin of Species*, Sidgwick & Jackson, London, p. 100.
Markowetz, B., Van Snick, J. L. and Masson, P. L. (1979) *Thorax* **34**, 209.
Martel, P., Kim, S. M. and Powell, B. M. (1980) *Biophys. J.* **31**, 371.
Martin, A. W., Huebers, H. A., Huebers, E. and Finch, C. A. (1982) in *The Biochemistry and Physiology of Iron* (Saltman, P. and Hegenauer, J., eds). Elsevier Biomedical, New York, p. 79.

Martinez-Medellin, J. and Benavides, L. (1979) *Biochim. Biophys. Acta* **584**, 84.
Martinez-Medellin, J., Schulman, H. M., De Miguel, E. and Benavides, L. (1977) in *Proteins of Iron Metabolism* (Brown, E. B., Aisen, P., Fielding, J. and Crichton, R. R., eds). Grune & Stratton, New York, p. 305.
Marx, J. J. M., Klein-Gebbink, J. A. G., Nishisato, T. and Aisen, P. (1982) *Brit. J. Haematol.* **52**, 105.
März, L., Hatton, M. W. C., Berry, L. R. and Regoeczi, E. (1982) *Can. J. Biochem.* **60**, 624.
Mason, D. Y. and Taylor, C. R. (1978) *J. Clin. Pathol.* **32**, 316.
Masson, P. L. (1970) *La Lactoferrine.* Editions Arscia, Brussels.
Masson, P. L. and Heremans, J. F. (1968) *Eur. J. Biochem.* **6**, 579.
Masson, P. L., Heremans, J. F. and Dive, C. (1966) *Clin. Chim. Acta* **14**, 735.
Masson, P. L., Heremans, J. F. and Schonne, E. (1969) *J. Exp. Med.* **130**, 643.
Matsumoto, A., Yoshima, H., Takasaki, S. and Kobata, A. (1982) *J. Biochem., Tokyo* **91**, 143.
Mazurier, J. and Spik, G. (1980) *Biochim. Biophys. Acta* **629**, 399.
Mazurier, J., Aubert, J-P., Loucheux-Lefèvre, M-H. and Spik, G. (1976) *FEBS Lett.* **66**, 238.
Mazurier, J., Lhoste, J-M., Spik, G. and Montreuil, J. (1977) *FEBS Lett.* **81**, 371.
Mazurier, J., Legér, D., Tordera, V., Montreuil, J. and Spik, G. (1981) *Eur. J. Biochem.* **119**, 537.
Meares, C. F. and Ledbetter, J. E. (1977) *Biochemistry* **16**, 5178.
Mendelsohn, J., Trowbridge, I. and Castagnola, J. (1983) *Blood* **62**, 821.
Messer, A., Mazurkiewicz, J. E. and Maskin, P. (1981) *Cell. Mol. Neurobiol.* **1**, 99.
Messmer, T. O. (1973) *Biochim. Biophys. Acta* **320**, 663.
Metz-Boutigue, M-H., Mazurier, J., Jolles, J., Spik, G., Montreuil, J. and Jolles, P. (1981) *Biochim. Biophys. Acta* **670**, 243.
Mickelesen, P. A. and Sparling, P. F. (1981) *Infect. Immunity* **33**, 555.
Mickelesen, P. A., Blackman, E. and Sparling, P. F. (1982) *Infect. Immunity* **35**, 915.
Miles, A. A. and Khimji, P. L. (1975) *J. Med. Microbiol.* **8**, 477.
Miles, A. A., Pillow, J. and Khimji, P. L. (1976) *Brit. J. Exp. Pathol.* **57**, 217.
Miles, A. A., Khimji, P. L. and Maskell, J. (1979) *J. Med. Microbiol.* **12**, 17.
Montreuil, J. and Spik, G. (1975) in *Proteins of Iron Storage and Transport in Biochemistry and Medicine* (Crichton, R. R., ed.). Elsevier/North-Holland, Amsterdam, p. 27.
Moore, D. G., Yancey, R. J., Lankford, C. E. and Earhart, C. F. (1980) *Infect. Immunity* **27**, 418.
Morgan, E. H. (1964) *Brit. J. Haematol.* **10**, 442.
Morgan, E. H. (1977) *Biochim. Biophys. Acta* **499**, 169.
Morgan, E. H. (1979) *Biochim. Biophys. Acta* **580**, 312.
Morgan, E. H. (1981*a*) *Molec. Aspects Med.* **4**, 3.
Morgan, E. H. (1981*b*) *Biochim. Biophys. Acta* **642**, 119.
Morgan, E. H. and Appleton, T. C. (1969) *Nature, Lond.* **223**, 1371.
Morgan, E. H., Huebers, H. and Finch, C. A. (1978) *Blood* **52**, 1219.
Muller, C. and Shinitzky, M. (1979) *Brit. J. Haematol.* **42**, 355.
Multigner, L., Figarella, C. and Sarles, H. (1981) *Gut* **22**, 350.
Murray, F. A., Bazer, F. W., Wallace, H. D. and Warninck, A. C. (1972) *Biol. Reprod.* **7**, 314.
Najarian, R. C., Harris, D. C. and Aisen, P. (1978) *J. Biol. Chem.* **253**, 38.
Neilands, J. B. (1980) in *Iron in Biochemistry and Medicine* II (Jacobs, A. and Worwood, M., eds). Academic Press, London, p. 529.
Neilands, J. B. (1981) *Ann. Rev. Biochem.* **50**, 715.
Neuwirt, J., Borova, J. and Ponka, P. (1975) in *Proteins of Iron Storage and Transport in Biochemistry and Medicine* (Crichton, R. R., ed.). Elsevier/North-Holland, Amsterdam, p. 161.
Newman, R., Schneider, C., Sutherland, R., Vodinelich, L. and Greaves, M. (1982) *Trends Biochem. Sci.* 7, 397.
Nishisato, T. and Aisen, P. (1982) *Brit. J. Haematol.* **52**, 361.
Nishisato, T. and Aisen, P. (1983) in *Structure and Function of Iron Storage and Transport Proteins* (Urushizaki, I., Aisen, P., Listowsky, I. and Drysdale, J. W., eds). Elsevier, Amsterdam, p. 353.

Norrod, P. and Williams, R. P. (1978*a*) *Curr. Microbiol.* **1**, 281.
Norrod, P. and Williams, R. P. (1978*b*) *Infect. Immunity* **21**, 918.
Noujaim, A. A. and Weibe, L. I. (1981) *Int. J. Nucl. Med. Biol.* **8**, 215.
Noujaim, A. A., Lentle, B. C., Hill, J. R., Terner, U. K. and Wong, H. (1979) *Int. J. Nucl. Med. Biol.* **6**, 193.
Núñez, M. T. and Glass, J. (1983) *J. Biol. Chem.* **258**, 9676.
Núñez, M. T., Glass, J., Fischer, S., Lavidor, L. M., Lenk, E. M. and Robinson, S. H. (1977) *Brit. J. Haematol.* **36**, 519.
Núñez, M. T., Glass, J. and Cole, E. S. (1981) *Biochim. Biophys. Acta* **673**, 137.
O'Brien, E. G. and Gibson, F. (1970) *Biochim. Biophys. Acta* **215**, 393.
Octave, J-N., Schneider, Y-J., Hoffman, P., Trouet, A. and Crichton, R. R. (1979) *FEBS Lett.* **108**, 127.
Octave, J-N., Schneider, Y-J., Crichton, R. R. and Trouet, A. (1981) *Eur. J. Biochem.* **115**, 611.
Octave, J-N., Schneider, Y-J., Crichton, R. R. and Trouet, A. (1982*a*) *FEBS Lett.* **137**, 119.
Octave, J-N., Schneider, Y-J., Hoffman, P., Trouet, A. and Crichton, R. R. (1982*b*) *Eur. J. Biochem.* **123**, 235.
Octave, J-N., Schneider, Y-J., Trouet, A. and Crichton, R. R. (1983) *Trends Biochem. Sci.* **8**, 217.
O'Hara, P., Yeh, S. M., Meares, C. F. and Bersohn, R. (1981) *Biochemistry* **20**, 4704.
Ohkuma, S. and Poole, B. (1978) *Proc. Natl Acad. Sci. USA* **75**, 3327.
Okada, S. and Brown, E. B. (1977) in *Proteins of Iron Metabolism* (Brown, E. B., Aisen, P., Fielding, J. and Crichton, R. R., eds). Grune & Stratton, New York, p. 261.
Okada, S., Chipman, B. and Brown, E. B. (1977) *J. Lab. Clin. Med.* **89**, 51.
Okada, S., Rossman, M. D. and Brown, E. B. (1978) *Biochim. Biophys Acta* **543**, 72.
Okada, S., Jarvis, B. and Brown, E. B. (1979) *J. Lab. Clin. Med.* **93**, 189.
Omary, M. B. and Trowbridge, I. S. (1981*a*) *J. Biol. Chem.* **256**, 4715.
Omary, M. B. and Trowbridge, I. S. (1981*b*) *J. Biol. Chem.* **256**, 12888.
Oram, J. D. and Reiter, B. (1968) *Biochim. Biophys. Acta* **170**, 351.
Oseas, R., Yang, H-H., Baehner, R. L. and Boxer, L. A. (1981) *Blood* **57**, 939.
Palmiter, R. D. (1972) *J. Biol. Chem.* **247**, 6450.
Palmour, R. L. M. and Sutton, H. E. (1971) *Biochemistry* **10**, 4026.
Pan, B-T. and Johnstone, R. M. (1983) *Cell* **33**, 967.
Patch, M. G. and Carrano, C. J. (1982) *Biochim. Biophys. Acta* **700**, 217.
Paterson, S. and Morgan, E. H. (1980) *J. Cell. Physiol.* **105**, 489.
Pearse, B. M. F. (1982) *Proc. Natl Acad. Sci. USA* **79**, 451.
Pecoraro, V. L., Harris, W. R., Carrano, C. J. and Raymond, K. N. (1981) *Biochemistry* **20**, 7033.
Pekkala-Flagan, A. and Ruoslahti, E. (1982) *J. Immunol.* **128**, 1163.
Pelus, L. M., Broxmeyer, H. E., Kurland, J. I. and Moore, M. A. S. (1979) *J. Exp. Med.* **150**, 277.
Pelus, L. M., Broxmeyer, H. E., De Sousa, M. and Moore, M. A. S. (1981*a*) *J. Immunol.* **126**, 1016.
Pelus, L. M., Broxmeyer, H. E. and Moore, M. A. S. (1981*b*) *Cell Tissue Kinet.* **14**, 515.
Phelps, C. F. and Antonini, E. (1975) *Biochem. J.* **147**, 385.
Phillips, J. L. (1976) *Biochem. Biophys. Res. Commun.* **72**, 634.
Phillips, J. L. (1978) *Cell. Immunol.* **35**, 318.
Phillips, J. L. (1980) *Biol. Trace Element Res.* **2**, 291.
Phillips, J. L. and Azari, P. (1972) *Arch. Biochem. Biophys.* **151**, 445.
Phillips, J. L. and Azari, P. (1974) *Cell. Immunol.* **10**, 31.
Phillips, J. L. and Azari, P. (1975) *Cell. Immunol.* **15**, 94.
Pollack, J. R. and Neilands, J. B. (1970) *Biochem. Biophys. Res. Commun.* **38**, 989.
Pollack, S. and Lasky, F. D. (1976*a*) *Biochem. Biophys. Res. Commun.* **70**, 533.
Pollack, S. and Lasky, F. D. (1976*b*) *J. Lab. Clin. Med.* **87**, 670.
Pollack, S. and Lasky, F. D. (1977) in *Proteins of Iron Metabolism* (Brown, E. B., Aisen, P., Fielding, J. and Crichton, R. R., eds). Grune & Stratton, New York, p. 393.
Pollack, S., Vanderhoff, G. and Lasky, F. (1977) *Biochim. Biophys. Acta.* **497**, 481.
Pootrakul, P., Christensen, A., Josephson, B. and Finch, C. A. (1977) *Blood*, **49**, 957.
Price, E. M. and Gibson, J. F. (1972*a*) *J. Biol. Chem.* **247**, 8031.

Price, E.M. and Gibson, J.F. (1972*b*) *Biochem. Biophys. Res. Commun.* **46**, 646.
Prieels, J-P., Pizzu, S. V., Glasgow, L. R., Paulson, J. C. and Hill, R. C. (1978) *Proc. Natl Acad. Sci. USA* **75**, 2215.
Priest, N. D. and Haines, J. W. (1982) *Hlth Phys.* **42**, 415.
Princiotto, J. V. and Zapolski, E. J. (1975) *Nature, Lond.* **255**, 87.
Princiotto, J. V. and Zapolski, E. J. (1976) *Biochim. Biophys. Acta* **428**, 766.
Princiotto, J. V. and Zapolski, E. J. (1978) *Biochim. Biophys. Acta* **539**, 81.
Rasey, J. S., Nelson, N. J. and Larson, S. M. (1982) *Eur. J. Cancer Clin. Oncol.* **18**, 661.
Regoeczi, E. and Hatton, M. W. C. (1974) *Can. J. Biochem.* **52**, 645.
Regoeczi, E., Hatton, M.W.C. and Wong, K-L. (1974) *Can. J. Biochem.* **52**, 155.
Regoeczi, E., Hatton, M.W.C. and Wong, K-L. (1975) *J. Nucl. Biol. Med.* **19**, 149.
Regoeczi, E., Wong, K-L., Ali, M. and Hatton, M. W. C. (1977) *Int. J. Peptide Protein Res.* **10**, 17.
Regoeczi, E., Taylor, P., Debanne, M. T., März, L. and Hatton, M. W. C. (1979) *Biochem. J.* **184**, 399.
Reiter, B. (1978) *J. Dairy Res.* **45**, 131.
Reiter, B., Brock, J. H. and Steel, E. D. (1975) *Immunology* **28**, 83.
Richardson, N. E., Buttress, N., Feinstein, A., Stratil, A. and Spooner, R. L. (1973) *Biochem. J.* **135**, 87.
Roberts, T. K. and Boettcher, B. (1969) *J. Reprod. Fertil.* **18**, 347.
Roberts, T. K. and Boursnell, J. C. (1975) *J. Reprod. Fertil.* **42**, 579.
Roberts, T. K., Masson, P. L. and Heremans, J. F. (1973) in *Immunology of Reproduction* (Bratanov, K., ed.). Bulgarian Academy of Sciences Press, Sofia, p. 706.
Roeser, H. P. (1980) in *Iron in Biochemistry and Medicine* II (Jacobs, A., and Worwood, M., eds). Academic Press, London, p. 605.
Rogers, H. J. (1973) *Infect. Immunity* **7**, 445.
Rogers, H. J. (1976) *Immunology* **30**, 425.
Rogers, H. J. and Synge, C. (1978) *Immunology* **34**, 19.
Rogers, T. B., Feeney, R. E. and Meares, C. F. (1977*a*) *J. Biol. Chem.* **252**, 8108.
Rogers, T. B., Gold, R. A. and Feeney, R. E. (1977*b*) *Biochemistry* **16**, 2299.
Rogers, H. J., Synge, C., Kimber, B. and Bayley, P. M. (1977*c*) *Biochim. Biophys. Acta*, **497**, 548.
Rogers, T. B., Børresen, T. and Feeney, R. E. (1978) *Biochemistry* **17**, 1105.
Romslo, I. (1980) in *Iron in Biochemistry and Medicine* II (Jacobs, A. and Worwood, M., eds). Academic Press, London, p. 325.
Rosenmund, A., Gerber, S., Huebers, H. and Finch, C. (1980) *Blood* **56**, 30.
Rosseneu-Motref, M. Y., Soetewey, F., Lamote, R. and Peeters, H. (1971) *Biopolymers* **10**, 1039.
Rudland, P. S., Durbin, H., Clingan, D. and Jimènez de Asua, L. (1977) *Biochem. Biophys. Res. Commun.* **75**, 556.
Sabbioni, E. and Marafante, E. (1981) *J. Toxicol. Envir. Hlth* **8**, 419.
Samson, R. R., Mirtle, C. and McLelland, D. B. L. (1979) *Immunology* **38**, 367.
Savin, M. A. and Cook, J. D. (1978) *Gastroenterology* **75**, 688.
Schade, A. L. and Caroline, L. (1944) *Science* **100**, 14.
Schade, A. L. and Caroline, L. (1946) *Science* **104**, 340.
Schade, A. L., Reinhart, R. W. and Levy, H. (1949) *Arch. Biochem. Biophys.* **20**, 170.
Schade, S. G., Bernier, G. M. and Conrad, M. E. (1969) *Brit. J. Haematol.* **17**, 187.
Schlabach, M. R. and Bates, G. W. (1975) *J. Biol. Chem.* **250**, 2182.
Schlosnagle, D. C., Bazer, F. W., Tsibris, J. C. M. and Roberts, R. M. (1974) *J. Biol. Chem.* **249**, 7574.
Schlosnagle, D. C., Sander, E. G., Bazer, F. W. and Roberts, R. M. (1976) *J. Biol. Chem.* **251**, 4680.
Schneider, C., Sutherland, R., Newman, R. and Greaves, M. (1982) *J. Biol. Chem.* **257**, 8516.
Schreiber, G., Dryburgh, H., Millership, A., Matsuda, Y., Inglis, A., Phillips, J., Edwards, K. and Maggs, J. (1979) *J. Biol. Chem.* **254**, 12013.
Schreiber, G., Dryburgh, H., Weigand, K., Schreiber, M., Witt, I., Seydewitz, H. and Howlett, G. (1981) *Arch. Biochem. Biophys.* **212**, 319.

Schulman, H. M., Wilczynska, A. and Ponka, P. (1981) *Biochem. Biophys. Res. Commun.* **100**, 1523.
Seligman, P. A., Schleicher, R. B. and Allen, R. H. (1979) *J. Biol. Chem.* **254**, 9943.
Sephton, R. G. (1981) *Int. J. Nucl. Med. Biol.* **8**, 323.
Sephton, R. G. and Kraft, N. (1978) *Cancer Res.* **38**, 1213.
Sheehan, R. G. (1976) *Am. J. Physiol.* **231**, 1438.
Shewale, J. G. and Brew, K. (1982) *J. Biol. Chem.* **257**, 9406.
Shindelman, J. E., Ortmeyer, A. E. and Sussman, H. H. (1981) *Int. J. Cancer* **27**, 329.
Shisheva, A. C., Moutafchiev, D. A. and Sirakov, L. M. (1982) *Studia Biophys.* **88**, 27.
Skaper, S. D., Adler, R. and Varon, S. (1979) *Devel. Neurosci.* **2**, 233.
Skarberg, K., Eng, M., Huebers, H., Marsaglia, G. and Finch, C. (1978) *Proc Natl Acad. Sci. USA* **75**, 1559.
Smit, S., Leijnse, B. and Van der Kraan, A. M. (1981) *J. Inorg. Biochem.* **15**, 329.
Smith, K. J., Failla, M. L. and Cousins, R. J. (1978) *Biochem. J.* **184**, 627.
Speyer, B. E. and Fielding, J. (1974) *Biochim. Biophys. Acta* **332**, 192.
Spik, G., Vandersyppe, R., Tetart, D., Han, K. K. and Montreuil, J. (1974) *FEBS Lett.* **38**, 213.
Spik, G., Bayard, B., Fournet, B., Strecker, G., Bouquelet, S. and Montreuil, J. (1975) *FEBS Lett.* **50**, 296.
Spik, G., Cheron, A., Montreuil, J. and Dolby, J. M. (1978) *Immunology* **35**, 663.
Spik, G., Fournet, B. and Montreuil, J. (1979) *C.R. Acad. Sci. Ser. D* **288**, 967.
Spik, G., Strecker, G., Fournet, B., Bouquelet, S., Montreuil, J., Dorland, L., Van Halbeek, H. and Vliegenthart, J. F. G. (1982) *Eur. J. Biochem.* **121**, 413.
Spiro, T. G. (1977) in *Proteins of Iron Metabolism* (Brown, E. B., Aisen, P., Fielding, J. and Crichton, R. R., eds). Grune & Stratton, New York, p. xxiii.
Spiro, T. G. and Saltman, P. (1969) *Struct. Bonding* **6**, 116.
Spooner, R. L. and Baxter, G. (1969) *Biochem. Genet.* **2**, 371.
Spooner, R. L., Land, R. B., Oliver, R. A. and Stratil, A. (1970) *Anim. Blood Grps Biochem. Genet.* **1**, 241.
Spooner, R. L., Oliver, R. A. and Williams, G. (1977) *Anim. Blood Grps Biochem. Genet.* **8**, 21.
Steiner, M. (1980) *J. Lab. Clin. Med.* **96**, 1086.
Steinmann, G., Broxmeyer, H. E., De Harven, E. and Moore, M. A. S. (1982) *Brit. J. Haematol.* **50**, 75.
Stephens, S., Dolby, J. M., Montreuil, J. and Spik, G. (1980) *Immunology* **41**, 597.
Stevens, W., Bruenger, F. W. and Stover, B. J. (1968) *Radiat. Res.* **33**, 490.
Stibler, H., Allgulander, C., Borg, S. and Kjellin, K. G. (1978) *Acta Med. Scand.* **204**, 49.
Stibler, H., Borg, S. and Allgulander, C. (1979) *Acta Med. Scand.* **206**, 275.
Stjernholm, R., Warner, F. W., Robinson, J. W., Ezekiel, E. and Katayama, N. (1978) *Bioinorg. Chem.* **9**, 277.
Stover, B. J., Bruenger, F. W. and Stevens, W. (1968) *Radiat. Res.* **33**, 381.
Stratil, A. and Kubek, A. (1974) *Int. J. Biochem.* **5**, 895.
Stratil, A. and Spooner, R. L. (1971) *Biochem. Genet.* **5**, 347.
Strickland, D. K. and Hudson, B. G. (1978) *Biochemistry* **17**, 3411.
Strickland, D. K., Hamilton, J. W. and Hudson, B. G. (1979) *Biochemistry* **18**, 2549.
Sutherland, R., Delia, D., Schneider, C., Newman, R., Kemshead, J. and Greaves, M. (1981) *Proc. Natl Acad. Sci. USA* **78**, 4515.
Tabak, L., Mandel, I. D., Herrera, M. and Baurmash, H. (1978) *J. Oral. Pathol.* **7**, 91.
Takahashi, K. and Tavassoli, M. (1982) *Biochim. Biophys. Acta* **685**, 6.
Tan, A. T. and Woodworth, R. C. (1969) *Biochemistry* **8**, 3711.
Tei, I., Makino, Y., Sakagami, H., Kanamaru, I. and Konno, K. (1982) *Biochem. Biophys. Res. Commun.* **107**, 1419.
Tengerdy, C., Azari, P. and Tengerdy, R. P. (1966) *Nature, Lond.* **211**, 203.
Terner, U. K., Noujaim, A. A., Lentle, B. C. and Turner, C. J. (1981) *Int. J. Nucl. Med. Biol.* **8**, 357.
Tidmarsh, G. F. and Rosenberg, L. T. (1981) *Curr. Microbiol.* **6**, 217.
Tolleshaug, H., Chindemi, P. A. and Regoeczi, E. (1981) *J. Biol. Chem.* **256**, 6526.
Tomimatsu, Y. and Donovan, J. W. (1976) *FEBS Lett.* **71**, 299.

Tomimatsu, Y., Kint, S. and Scherer, J. R. (1976) *Biochemistry* **15**, 4918.
Tormey, D. C. and Mueller, G. C. (1972) *Exp. Cell Res.* **74**, 220.
Tormey, D. C., Imrie, R. C. and Mueller, G. C. (1972) *Exp. Cell Res.* **74**, 163.
Trowbridge, I. S. and Domingo, D. L. (1981) *Nature, Lond.* **294**, 171.
Trowbridge, I. S. and Lòpez, F. (1982) *Proc. Natl Acad. Sci. USA* **79**, 1175.
Trowbridge, I. S. and Omary, M. B. (1981) *Proc. Natl Acad. Sci. USA* **78**, 3039.
Trowbridge, I. S., Lesley, J. and Schulte, R. (1982) *J. Cell. Physiol.* **112**, 403.
Tsang, C. P., Boyle, A. J. F. and Morgan, E. H. (1975) *Biochim. Biophys. Acta* **386**, 32.
Tsao, D., Morris, D. H., Tengerdy, R. P. and Phillips, J. L. (1974*a*) *Biochemistry* **13**, 403.
Tsao, D., Azari, P. and Phillips, J. L. (1974*b*) *Biochemistry* **13**, 408.
Ulvik, R., Prante, P. H., Koller, M. E. and Romslo, I. (1976) *Scand. J. Clin. Lab. Invest.* **36**, 539.
Valenta, M., Stratil, A., Slechtova, V., Kalal, L. and Slechta, V. (1976) *Biochem. Genet.* **14**, 27.
Vallabhajosula, S. R., Harwig, J. F. and Wolf, W. (1981) *Int. J. Nucl. Med. Biol.* **8**, 363.
Van Baarlen, J., Brouwer, J. T., Leibman, A. and Aisen, P. (1980) *Brit. J. Haematol.* **46**, 417.
Van Bockxmeer, F. M. and Morgan, E. H. (1979) *Biochim. Biophys. Acta* **584**, 76.
Van Bockxmeer, F. M. and Morgan, E. H. (1982) *Comp. Biochem. Physiol.* **71A**, 211.
Van Bockxmeer, F. M., Yates, G. K. and Morgan, E. H. (1978) *Eur. J. Biochem.* **92**, 147.
Van der Heul, C., Kroos, M. J. and Van Eijk, H. G. (1978) *Biochim. Biophys. Acta* **511**, 430.
Van der Heul, C., Kroos, M. J., Van Noort, W. L. and Van Eijk, H. G. (1981) *Clin. Sci.* **60**, 185.
Van der Heul, C., Veldman, A., Kroos, M. J. and Van Eijk, H. G. (1984) *Int. J. Biochem.* **16**, 383.
Van Eijk, H. G., Van Noort, W. L., Kroos, M. J. and Van der Heul, C. (1978) *J. Clin. Chem. Clin. Biochem.* **16**, 557.
Van Eijk, H. G., Van Noort, W. L., Kroos, M. J. and Van der Heul, C. (1982) *Clin. Chim. Acta* **121**, 209.
Van Eijk, H. G., Van Noort, W. L., Dubelaar, M-L. and Van der Heul, C. (1983) *Clin. Chim. Acta* **132**, 167.
Van Halbeek, H., Dorland, L., Vliegenthart, J. F. G., Spik, G., Cheron, A. and Montreuil, J. (1981) *Biochim. Biophys. Acta* **675**, 293.
Van Renswoude, J., Bridges, K., Harford, J. and Klausner, R. D. (1982) *Proc. Natl. Acad. Sci. USA* **79**, 6186.
Van Snick, J. L. and Masson, P. L. (1976) *J. Exp. Med.* **144**, 1568.
Van Snick, J. L., Masson, P. L. and Heremans, J. F. (1974) *J. Exp. Med.* **140**, 1068.
Van Snick, J. L., Masson, P. L. and Heremans, J. F. (1975) in *Proteins of Iron Storage and Transport in Biochemistry and Medicine* (Crichton, R. R., ed.). North Holland, Amsterdam, p. 433.
Van Snick, J. L., Markowetz, B. and Masson, P. L. (1977) *J. Exp. Med.* **146**, 817.
Van Vugt, H., Van Gool, J., Ladiges, N. C. J. J. and Boers, W. (1975) *Q. J. Exp. Physiol.* **60**, 79.
Verhoef, N. J. and Noordeloos, P. J. (1977) *Clin. Sci. Mol. Med.* **52**, 87.
Verhoef, N. J., Kremers, J. H. W. and Leijnse, B. (1973) *Biochim. Biophys. Acta* **304**, 114.
Verhoef, N. J., Kottenhagen, M. J., Mulder, H. J. M., Noordeloos, P. J. and Leijnse, B. (1978) *Acta Haematol.* **60**, 210.
Verhoef, N. J., Kester, H. C. M., Noordeloos, P. J. and Leijnse, B. (1979) *Int. J. Biochem.* **10**, 595.
Wada, H. G., Hass, P. E. and Sussman, H. H. (1979) *J. Biol. Chem.* **254**, 12629.
Walsh, R. J., Thomas, E. D., Chow, S. K., Fluharty, R. G. and Finch, C. A. (1949) *Science* **110**, 396.
Wang, A-C. and Sutton, H. E. (1965) *Science* **149**, 435.
Wang, A-C., Sutton, H. E. and Biggs, A. (1966) *Am. J. Human Genet.* **18**, 454.
Wang, A-C., Sutton, H. E. and Howard, P. N. (1967) *Biochem. Genet.* **1**, 55.
Ward, J. H., Kushner, J. P. and Kaplan, J. (1982) *J. Biol. Chem.* **257**, 10317.
Warner, R. C. and Weber, I. (1953) *J. Am. Chem. Soc.* **75**, 5094.
Weinberg, E. D. (1978) *Microbiol. Rev.* **42**, 45.
Weiner, R. E. and Szuchet, S. (1975) *Biochim. Biophys. Acta.* **393**, 143.

Weiner, R. E., Schreiber, G. J., Hoffer, P. B. and Shannon, T. (1981) *Int. J. Nucl. Med. Biol.* 8, 371.
Welty, F. K., Smith, K. L. and Schanbacher, F. L. (1976) *J. Dairy Sci.* **59**, 224.
Wenn, R. V. and Williams, J. (1968) *Biochem. J.* **108**, 69.
Wilczynska, A. and Schulman, H. M. (1980) *Can. J. Biochem.* **58**, 935.
Wilkins, M., Williams, P. and Cavill, I. (1977) *Ann. Rheum. Dis.* **36**, 474.
Williams, J. (1962) *Biochem. J.* **83**, 355.
Williams, J. (1974) *Biochem. J.* **141**, 745.
Williams, J. (1975) *Biochem. J.* **149**, 237.
Williams, J. (1982*a*) *Trends Biochem. Sci.* 7, 394.
Williams, J. (1982*b*) *Biochem. J.* **201**, 647.
Williams, J. and Wenn, R. V. (1970) *Biochem. J.* **116**, 533.
Williams, J., Phelps, C. F. and Lowe, J. M. (1970) *Nature, Lond.* **226**, 858.
Williams, J., Evans, R. W. and Moreton, K. (1978) *Biochem. J.* **173**, 535.
Williams, J., Elleman, T. C., Kingston, I. B., Wilkins, A. G. and Kuhn, K. A. (1982*a*) *Eur. J. Biochem.* **122**, 297.
Williams, J., Grace, S. A. and Williams, J. M. (1982*b*) *Biochem. J.* **201**, 417.
Williams, J., Chasteen, N. D. and Moreton, K. (1982*c*) *Biochem. J.* **201**, 527.
Williams, S. C. and Woodworth, R. C. (1973) *J. Biol. Chem.* **248**, 5848.
Windle, J. J., Wiersma, A. K., Clark, J. R. and Feeney, R. E. (1963) *Biochemistry* **2**, 1341.
Winton, E. F., Kinkade, J. M., Vogler, W. R., Parker, M. B. and Barnes, K. C. (1981) *Blood* **57**, 574.
Wishnia, A., Weber, I. and Warner, R. C. (1961) *J. Am. Chem. Soc.* **83**, 2071.
Witt, D. P. and Woodworth, R. C. (1978) *Biochemistry* **17**, 3913.
Wolfson, D. R. and Robbins, J. B. (1971) *Pediat. Res.* **5**, 514.
Wollenberg, P., Huebers, H. and Rummel, W. (1982) in *The Biochemistry and Physiology of Iron* (Saltman, P. and Hegenauer, J., eds). Elsevier Biomedical, New York, p. 287.
Wong, H., Terner, U. K., English, D., Noujaim, A. A., Lentle, B. C. and Hill, J. R. (1980) *Int. J. Nucl. Med. Biol.* 7, 9.
Wong, K-L. and Regoeczi, E. (1977) *Int. J. Peptide Protein Res.* **9**, 241.
Wong, K-L., Debanne, M. T., Hatton, M. W. C. and Regoeczi, E. (1978) *Int. J. Peptide Protein Res.* **12**, 27.
Woodworth, R. C., Morallee, K. G. and Williams, R. J. P. (1970) *Biochemistry* **9**, 839.
Woodworth, R. C., Brown-Mason, A., Christensen, T. G., Witt, D. P. and Comeau, R. D. (1982) *Biochemistry* **21**, 4220.
Workman, E. F., Graham, G. and Bates, G. W. (1975) *Biochim. Biophys. Acta* **399**, 254.
Wyllie, J. C. (1977) *Brit. J. Haematol.* **37**, 17.
Yeh, C-J. G., Papamichael, M. and Faulk, W. P. (1982) *Exp. Cell Res.* **138**, 429.
Yeh, S. M. and Meares, C. F. (1980) *Biochemistry* **19**, 5057.
Yeh, Y., Iwai, S. and Feeney, R. E. (1979) *Biochemistry* **18**, 882.
Yeoh, G. C. T. and Morgan, E. H. (1979) *Cell Different.* **8**, 331.
Young, D. V., Cox, F. W., Chipman, S. and Hartman, S. C. (1979) *Exp. Cell Res.* **118**, 410.
Young, J. W. and Perkins, D. J. (1968) *Eur. J. Biochem.* **4**, 385.
Young, S. P. (1982) *Biochim. Biophys. Acta* **718**, 35.
Young, S. P. and Aisen, P. (1980) *Biochim. Biophys. Acta* **633**, 145.
Young, S. P. and Aisen, P. (1981) *Hepatology* **1**, 114.
Young, S. P., Bomford, A. and Williams, R. (1983) *J. Biol. Chem.* **258**, 4972.
Young, S. P., Bomford, A., Madden, A. D., Garratt, R. C., Williams, R. and Evans, R. W. (1984) *Brit. J. Haematol.* in press.
Zak, O., Leibman, A. and Aisen, P. (1983) *Biochem. Biophys. Acta* **742**, 490.
Zaman, Z., Heynen, M-J. and Verwilghen, R. L. (1980) *Biochim. Biophys. Acta* **632**, 553.
Zapolski, E. J. and Princiotto, J. V. (1976) *Biochim. Biophys. Acta* **421**, 80.
Zapolski, E. J. and Princiotto, J. V. (1977*a*) in *Proteins of Iron Metabolism* (Brown, E. B., Aisen, P., Fielding, J. and Crichton, R. R., eds). Grune & Stratton, New York, p. 205.
Zapolski, E. J. and Princiotto, J. V. (1977*b*) *Biochem. J.* **166**, 175.
Zapolski, E. J. and Princiotto, J. V. (1980*a*) *Biochemistry* **19**, 3599.
Zapolski, E. J. and Princiotto, J. V. (1980*b*) *Life Sci.* **27**, 739.
Zavy, M. T., Bazer, F. W., Sharp, D. C. and Wilcox, C. J. (1979) *Biol. Reprod.* **20**, 689.

Zavy, M. T., Sharp, D. C., Bazer, F. W., Fazleabas, A., Sessions, F. and Roberts, R. M. (1982) *J. Reprod. Fertil.* **64**, 199.
Zucali, J. R., Broxmeyer, H. E. and Ulatowski, J. A. (1979) *Blood* **54**, 951.
Zweier, J. L. (1978) *J. Biol. Chem.* **253**, 7616.
Zweier, J. L. and Aisen, P. (1977) *J. Biol. Chem.* **252**, 6090.
Zweier, J. L., Aisen, P., Peisach, J. and Mims, W. B. (1979) *J. Biol. Chem.* **254**, 3512.
Zweier, J. L., Wooton, J. B. and Cohen, J. S. (1981) *Biochemistry* **20**, 3505.

6
Oxygen carrier proteins

Maurizio Brunori, Massimiliano Coletta and Bruno Giardina

DEDICATION

The authors wish to dedicate this article to the memory of Professor Eraldo Antonini, who died of an incurable illness on 19 March 1983. It was Professor Antonini who introduced the authors to the study of oxygen-carrying proteins and inspired their scientific activity.

6.1 INTRODUCTION AND CLASSIFICATION

Studies on the structure-function relationships in respiratory proteins and on their physiological implications continue to represent a field of very active research. This is made clear by the large number of papers which have been published in the past few years, and by the rapid appearance of new books and review articles, which very often deal with one specific type of respiratory protein leaving aside all the others (Baldwin, 1975; Perutz, 1979). Among many of the available reviews, the paper by Parkhurst (1979) focused on the kinetics of ligand binding to haemoglobin, and the recent article by van Holde and Miller (1982), who reviewed extensively a great deal of the work on haemocyanins, should also be mentioned.

In this chapter we have focused our attention on molecular aspects of the structure and function of oxygen carriers, and have tried to give a unified view of the more recent advances in the area. The main difficulty of this approach is of course related to the great differences in the degree of our present understanding of the structure of the various oxygen carriers. However, it may be of interest to note that some general ideas first applied to haemoglobins may be successfully transferred to other more complex respiratory proteins. With this in mind we have directed our attention to the unifying aspects of the various problems. For this reason it is possible that we may have missed some important contributions and we offer in advance our apologies to those whose papers have not been quoted.

Within such a perspective, since we recognised the danger of an extensive overlap with the above-mentioned review by Parkhurst (1979), we have disregarded the otherwise very important aspect of the dynamics of ligand binding to haemoglobin.

During the course of evolution, living organisms, under the pressure of their metabolic needs, have developed a number of special systems for the transport and storage of oxygen. The so-called respiratory proteins, which are present throughout almost all the phyla in the animal kingdom, combine reversibly with oxygen, increasing by a factor of 50 to 100 the amount of gas delivered at the level of peripheral tissues.

On the basis of the chemical nature of the oxygen binding site, three different classes of these proteins have been identified: haemoglobins and myoglobins, haemerythrins and myohaemerythrins, and haemocyanins.

6.1.1 Haemoglobins and myoglobins

Haemoglobins are red pigments which are present either free in solution or within erythrocytes, and which contain as prosthetic group the haem, a porphyrin ring (protoporphyrin IX) with an iron atom (Fe(II)) at the centre. Molecular oxygen is bound reversibly at the level of the iron in a 1:1 ratio. The oxygen binding site, the haem, is the same in all of the respiratory haemoproteins, with the exception of a green pigment, chlorocruorin, present in the blood of some polychaete worms, in which the prosthetic group differs from that of haemoglobin in one of the side chains of the porphyrin ring (Antonini and Brunori, 1971). Intracellular haemoglobins are relatively small molecules, generally tetrameric, with a minimum (subunit) molecular weight of 15 000–18 000 Da. Extracellular haemoglobins (erythrocruorins) are composed of many subunits, forming noncovalent giant molecules with molecular weights ranging from 300 000 up to 4×10^6 Da. Vertebrate muscle myoglobin is monomeric.

6.1.2 Haemerythrins and myohaemerythrins

Haemerythrins and myohaemerythrins appear to have little in common with haemoglobins, and hence are thought to be a separate evolutionary development. Haemerythrins occur in erythrocytes of a few phyla, notably sipunculids, priapulids and brachiopods, and are non-haem proteins containing a pair of iron atoms at their active site. The metal–O_2 stoichiometric ratio is in this case 2 : 1. Haemerythrins generally occur as octamers (mol. wt ~ 108 000 Da) composed of identical sub-units (mol. wt @ 13 500 Da), each of which is able to react reversibly with molecular oxygen. However, not all haemerythrins are octomeric: a trimeric form has been found in the sipunculid *Phascolosoma lurco* and a monomeric one is present in muscles, known as myohaemerythrin (Wood, 1980).

6.1.3 Haemocyanins

Haemocyanins are quite widespread among arthropods and molluscs, which, however, may possess either haemocyanins or haemoglobins (van Holde and Miller, 1982). How this situation has arisen in the course of evolution is a matter of great interest. Haemocyanins, always found freely dissolved in the haemolymph, contain two copper atoms per oxygen binding site and are colourless in the deoxy state and blue when oxygenated. Although the chemical structure of the active site of haemocyanins from molluscs and arthropods seems quite similar, the molecular architecture is entirely different (see below).

Of the molluscan haemocyanins, those from gastropods are among the largest known protein molecules, with a molecular weight up to 9 million Da. That of cephalopod haemocyanin is about half this value. In the case of arthropods, the basic unit appears to be a 16S component with a molecular weight of about 450 000 Da, made up of six polypeptide chains each of approximately the same size (mol. wt 60 000–90 000 Da). Arthropod haemocyanins are formed of multiples up to 62S of this 16S component. No example of a low molecular weight oxygen storage pigment related to haemocyanin is known.

6.2 GENERAL FEATURES OF OXYGEN BINDING

In what follows we shall describe some general features of oxygen binding to oxygen carriers, with emphasis on the application of the two-state concerted model (Monod *et al.*, 1965). Although the two-state model is certainly an oversimplification, it is remarkable how much of the available data (even in the case of giant respiratory proteins such as erythrocruorins and haemocyanins) can be described within the conceptual framework of this simple model.

Methods in Enzymology (vol. LXXVI, 1981) provides a compact and complete description of the techniques most widely used in this field.

6.2.1 Equilibria of oxygen binding

Oxygen-carrying proteins may contain one or more (up to several hundred) oxygen binding sites per molecule. When several sites are present in the same molecule, the reversible reaction with oxygen may be characterised by interaction phenomena, either positive or negative, between the ligand binding sites.

A simple case is represented by a macromolecule P, which contains n identical and independent sites for the ligand X. Because of the absence of interaction phenomena and the equivalence of the sites, each of them is characterised by the same intrinsic association constant K_i, regardless of the state of occupancy of the other sites. The equilibria involved are:

$$
\begin{array}{ll}
P + X \rightleftharpoons PX_1 & (PX_1) = K_1\ (P)\ (X) \\
PX_1 + X \rightleftharpoons PX_2 & (PX_2) = K_2\ (PX_1)\ (X) = K_1 K_2\ (P)\ (X)^2 \\
\cdot & \cdot \\
\cdot & \cdot \\
\cdot & \cdot \\
PX_{n-1} + X \rightleftharpoons PX_n & (PX_n) = K_n\ (PX_{n-1})\ (X) = K_1 K_2 \cdots K_n (P)\ (X)^n
\end{array}
\quad (1)
$$

Equilibrium measurements of ligand binding typically yield the ratio of the bound ligand to the total number of ligand binding sites. This parameter, generally designated $\bar{Y}$ (fractional saturation), is given by:

$$
\bar{Y} = \frac{K_1\ (P)\ (X) + 2K_1 K_2\ (P)\ (X)^2 + \cdots + nK_1 K_2 \cdots K_n\ (P)\ (X)^n}{n\ [(P) + K_1\ (P)\ (X) + K_1 K_2\ (P)\ (X)^2 + \cdots + K_1 K_2 \cdots K_n\ (P)\ (X)^n]} \quad (2)
$$

and hence:

$$
\bar{Y} = \frac{K_1\ (X) + 2K_1 K_2\ (X)^2 + \cdots + nK_1 K_2 \cdots K_n\ (X)^n}{n\ [1 + K_1\ (X) + K_1 K_2\ (X)^2 + \cdots + K_1 K_2 \cdots K_n\ (X)^n]} \quad (3)
$$

For $n = 4$ this equation reduces to the Adair equation (Adair, 1925):

$$
\bar{Y} = \frac{K_1\ (X) + 2K_1 K_2\ (X)^2 + 3K_1 K_2 K_3\ (X)^3 + 4K_1 K_2 K_3 K_4\ (X)^4}{4\ [1 + K_1\ (X) + K_1 K_2\ (X)^2 + K_1 K_2 K_3\ (X)^3 + K_1 K_2 K_3 K_4\ (X)^4]} \quad (4)
$$

which has been applied to the description of the reaction of haemoglobin with several ligands. The macroscopic constants K_j (with $1 \leqslant j \leqslant n$) have to be distinguished from the single microscopic constant K_i (i stands for intrinsic), which is the same for all of the sites. Thus K_i refers to the equilibrium with respect to the particular microscopic species, whereas K_j involves the entire ensemble of species represented by P_j and P_{j-1}. In general it can be shown (by appropriate statistical considerations) that the experimental constant for the formation of the j complex (K_j) in a molecule containing n sites is related to the intrinsic constant (K_i) by the expression

$$
K_j = \frac{n - j + 1}{j} K_i \quad (5)
$$

Therefore if we consider the particular case of a macromolecule combining with four molecules of a ligand X, in the absence of interactions between the identical sites, the four macroscopic constants are, according to (5):

$$
K_1 = 4K_i \quad K_2 = \frac{3}{2} K_i \quad K_3 = \frac{2}{3} K_i \quad K_4 = \frac{1}{4} K_i
$$

It should be remarked that substitution of K_i with its appropriate statistical factor for the experimental constants K_j in equation (3) yields an expression identical to that which is obtainable for binding of ligand by a protein containing a single site per molecule with an equilibrium constant corresponding to the intrinsic constant, i.e.

$$\bar{Y} = \frac{K_i\,(X)}{1 + K_i\,(X)} \qquad (6)$$

However, when, in a multi-site protein, the ratios of the equilibrium constants do not correspond to their statistical values, the behaviour of the system is complex and the ligand equilibrium cannot be described by the simple equation reported above. In this case: (a) the sites may not be intrinsically equivalent, (b) positive or negative interactions may exist, and (c) non-equivalence of the sites may be associated with functional interactions. In a cooperative system with four oxygen binding sites, the apparent association constant for one or more of the successive steps should increase as saturation progresses. Assuming that all four steps involve progressively stronger binding, this implies:

$$\frac{1}{4}K_1 > \frac{2}{3}K_2 > \frac{3}{2}K_3 > 4K_4$$

Application of Adair's scheme to haemoglobin has revealed that cooperativity in oxygen binding results from a value of $K_4 \sim 200\,K_1$. It should be remarked that, although Adair's formulation contained no hypothesis about the molecular mechanism underlying site-site interactions, it represented a very important step in haemoglobin studies since it postulated the presence of functional interactions between the haems and correlated the behaviour of the protein to its tetrameric structure.

A useful way of analysing the experimental data is based on Hill's empirical equation (Hill, 1910):

$$\frac{\bar{Y}}{1 - \bar{Y}} = KX^n \qquad (7)$$

where K and n are empirical constants. This equation has no direct physical meaning, but it is able to fit empirically the experimental data with only two parameters: log $p_{1/2}$ (i.e. the value of pressure p required to yield 50 per cent saturation), which indicates the overall affinity for the ligand, and n, which is related to the shape of the ligand binding curve.

The use of the Hill equation makes the comparison between different binding curves easy and provides an immediate description of the character of the process. Thus a plot of log $(\bar{Y}/1 - \bar{Y})$ versus log pO_2 in the case of most haemproteins gives a straight line for values of $\bar{Y}$ which are easily accessible to measurements $(0.1 \leqslant \bar{Y} \leqslant 0.9)$. Furthermore, the slope of the curve, n, gives an indication of the equivalence and interactions among the sites. Thus $n = 1$ corresponds to a hyperbolic binding curve, indicating absence of functional interactions; $n > 1$ corresponds to a sigmoidal curve and is related to the presence of positive site-site interactions. Values of $n < 1$ may indicate either independent sites with different ligand affinities (i.e. intramolecular heterogeneity) or negative interactions.

It should be noted that in the limits of very low ($\bar{Y} < 0.1$) and very high ($\bar{Y} > 0.9$) saturations, the Hill plot would also be a straight line with unitary slope in the case of cooperative systems. Thus, at low p we have:

$$\log(\bar{Y}/1 - \bar{Y}) \longrightarrow \log K_1 + \log p$$

and at high p:

$$\log(\bar{Y}/1 - \bar{Y}) \longrightarrow \log K_4 + \log p$$

Therefore extrapolation of the asymptotes of the oxygen binding curve allows the estimate of the first and the last Adair constants. The difference between the free energies of binding the first and the last ligand, i.e. the so-called free energy of interaction (ΔF_I), is therefore obtainable by:

$$\Delta F_I = -RT \ln K_1 + RT \ln K_4 = RT \ln (K_4/K_1) \quad (8)$$

This parameter, introduced by Wyman (1964), has been widely used in studies on haemoglobins and haemocyanins.

6.2.1.1 The two-state concerted model of Monod–Wyman and Changeux

The basis of this model (often referred to as the MWC model) is the new idea that the intrinsic ligand affinity is determined by the quaternary state of the macromolecule rather than by the number of ligand molecules already bound. The model assumes that only two quaternary states of the molecule, referred to as the low-affinity T state and high-affinity R state, are populated all along the binding curve. According to these assumptions, only two microscopic constants are required to specify the binding of oxygen to the proteins. These are designated as K_R and K_T, being usually $K_R \gg K_T$.

Since all the sites are taken as identical, and only two states are populated, the corresponding equilibria are:

$$T_0 \rightleftharpoons R_0$$

$$\begin{array}{ll} T_0 + O_2 \rightleftharpoons T_1 & R_0 + O_2 \rightleftharpoons R_1 \\ T_1 + O_2 \rightleftharpoons T_2 & R_1 + O_2 \rightleftharpoons R_2 \\ T_2 + O_2 \rightleftharpoons T_3 & R_2 + O_2 \rightleftharpoons R_3 \\ T_3 + O_2 \rightleftharpoons T_4 & R_3 + O_2 \rightleftharpoons R_4 \end{array}$$

Within this framework, five states of ligation can be defined for each quaternary state, as illustrated in figure 6.1. In the energy level diagram these states are equally spaced in free energy, K_T and K_R specifying the separation in energy between successive states respectively in the T and R quaternary structure. One

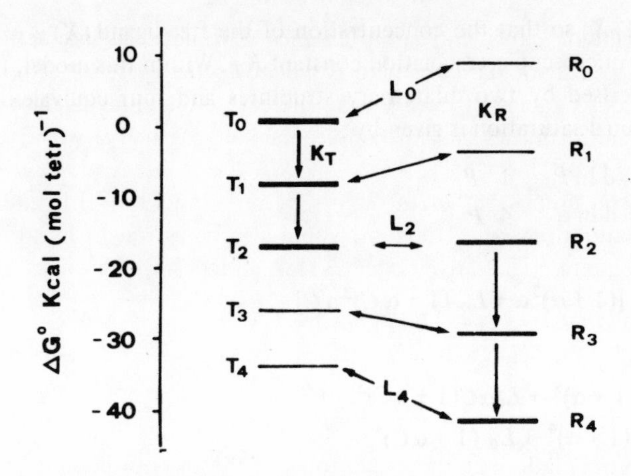

Figure 6.1 Energy levels diagram for the 'two-state model' applied to haemoglobin. K_T and K_R represent the oxygen binding constants to the low-affinity T and the high-affinity R quaternary states. L_i is the equilibrium constant for the interconversion of the quaternary state for a molecule having i oxygen molecules bound.

more parameter is necessary to specify the relative energy of R with respect to T. This parameter is defined as the equilibrium ratio between T_0 and R_0, i.e.

$$L_0 = \frac{[R_0]}{[T_0]} \qquad (9)$$

A value of $L_0 < 1$ therefore implies that the T quaternary state is more populated in the absence of a ligand.

Each of the other constants, L_i, is related to L_0 by the simple equation:

$$L_i = L_0\, C^i$$

where

$$C = \frac{K_R}{K_T} \qquad (10)$$

In the case of haemoglobin, the binding polynomial for the two-state model is given by:

$$P = \nu_T\,(1 + K_T X)^4 + \nu_R\,(1 + K_R X)^4 \qquad (11)$$

or:

$$P = (1 + K_T X)^4 + L_0\,(1 + K_R X)^4$$

where $L_0 = \nu_R/\nu_T$; or, in the equivalent form:

$$P = (1 + \alpha)^4 + L_0\,(1 + C\,\alpha)^4$$

where $\alpha = K_T X$; so that the concentration of the free ligand (X) is expressed in units of the microscopic association constant K_T. Within this model, for a molecule characterised by two quaternary structures and four equivalent sites, the ligand fractional saturation is given by:

$$\bar{Y} = \frac{1}{4}\frac{d \ln P}{d \ln \alpha} = \frac{1}{4}\frac{P'}{P} \tag{12}$$

where

$$P' = 4\,[(1+\alpha)^3 \alpha + L_0 (1 + \alpha C)^3 \alpha C] \tag{13}$$

hence:

$$\bar{Y} = \frac{\alpha(1+\alpha)^3 + L_0 \alpha C(1+\alpha C)^3}{(1+\alpha)^4 + L_0 (1+\alpha C)^4} \tag{14}$$

In general, for a protein with n subunits, the fractional saturation with oxygen will be given by

$$\bar{Y} = \frac{\alpha(1+\alpha)^{n-1} + L_0\, C\alpha(1+\alpha C)^{n-1}}{(1+\alpha)^n + L_0 (1+\alpha C)^n} \tag{15}$$

The switchover point, i.e. the state of ligation at which R and T are equally populated, can be simply calculated imposing $L_i = 1$:

$$i_s = -\frac{\log L_0}{\log C} \tag{16}$$

It should be noted that if $L_0 = 0$, then $\bar{Y} = \alpha/(1+\alpha)$, which is the usual expression for a hyperbolic saturation curve, i.e. for the oxygen binding to identical and independent sites on a macromolecule.

Another point of interest is represented by the expression of the Adair constants in terms of the allosteric parameters. Thus, they are given by:

$$K_1 = \frac{(1+L_0 C)}{(1+L_0)} K_T \qquad K_2 = \frac{(1+L_0 C^2)}{(1+L_0 C)} K_T$$

$$K_3 = \frac{(1+L_0 C^3)}{(1+L_0 C^2)} K_T \qquad K_4 = \frac{(1+L_0 C^4)}{(1+L_0 C^3)} K_T$$

It may be seen that if $C > 1$, the successive Adair constants increase, yielding cooperative binding. It is apparent that as C approaches unity, the sigmoidicity of the saturation curve decreases. At the limit of $C = 1$, the two conformational forms are indistinguishable as far as that specific ligand is concerned.

A graphical interpretation of the allosteric parameters is shown in figure 6.2.

6.2.1.2 Oxygen affinity modifiers

According to the model, effectors bind preferentially to either the R or the T state at sites distinct from the oxygen binding sites, giving rise to the so-called

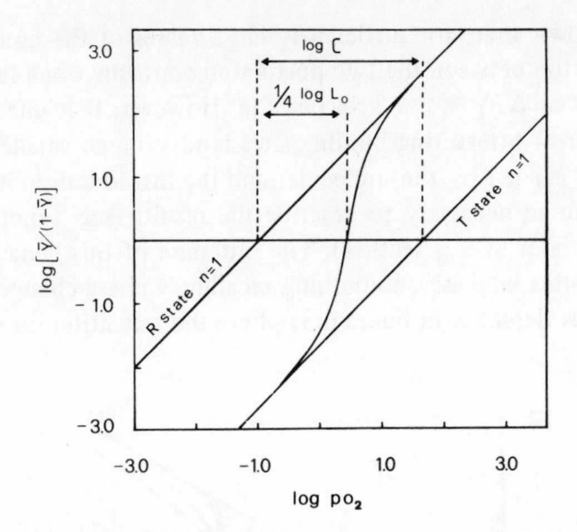

Figure 6.2 Hill plot of the oxygen dissociation curve of haemoglobin. Careful measurements at the extreme saturations allow the experimental determination of the asymptotes and thus of the binding constants of R and T conformational states. The graphical interpretation of $C = K_R/K_T$ and of L_0 is indicated.

heterotropic interactions. Within the strict boundaries of the MWC model, these effectors act only on the allosteric equilibrium constant (Rubin and Changeux, 1966).

If an affinity modifier, M, is characterised by a binding constant β to the T structure and a binding constant γ to the R structure, the total concentration of molecules in the two states is given by:

$$[T_0]_{tot} = [T_0]\,(1 + \beta x)$$
$$[R_0]_{tot} = [R_0]\,(1 + \gamma x) \qquad (18)$$

where $[T_0]$ and $[R_0]$ represent the concentration of molecules with no bound M, and x is the concentration of free M. The new allosteric constant L' is given by

$$L' = \frac{[R_0]_{tot}}{[T_0]_{tot}} = L_0 \frac{(1 + \gamma x)}{(1 + \beta x)} \qquad (19)$$

6.2.1.3 Allosteric description of giant respiratory proteins

A case of interest is provided by the extremely large respiratory proteins, which contain even more than 100 oxygen binding sites. In this case a very large value of n (the slope of the Hill plot) is often coupled with a relatively small total interaction free energy. In other words the Hill plot shows a rather sharp upward bend, often in the middle range of saturation, associated with rather closely spaced asymptotes. The sharpness of the transition in relation to the spacing of the asymptotes, i.e. n in relation to the total interaction free energy (ΔF_I), has been analysed in the framework of the MWC model (Colosimo *et al.*, 1974). This analysis has shown that, for sufficiently large values of the number of binding sites, the transition between the two postulated conformations can be extremely sharp for values of $\Delta F_I \geqslant \sim 1$ kcal per site. However, it is unequivocally clear that the number of interacting binding sites is very much smaller than the total number of sites carried by the molecule, and the introduction of a new concept has been considered necessary to describe the results (see 'functional constellations' discussed later in this section). The outcome of this behaviour is that, as the number of sites increases, something recalling a phase change is approached. This behaviour is depicted in figure 6.3, where the transition for different values

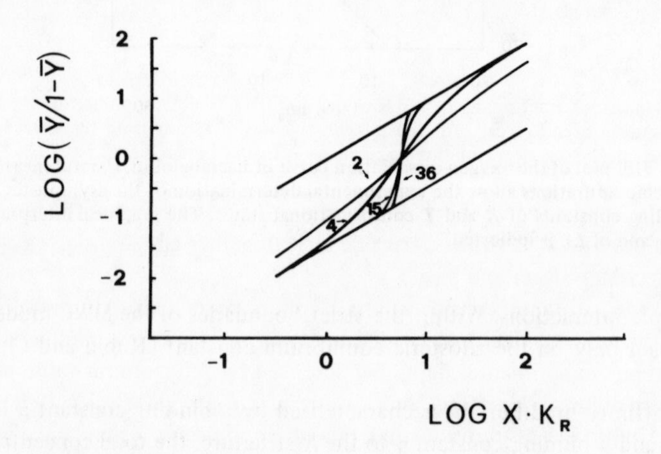

Figure 6.3 Hill plots for the symmetrical case calculated as a function of the number of binding sites (r) as indicated in the figure. The curves all correspond to $C = K_R/K_T = 30$. The interaction energy ΔF_I is 2000 cal, except for $r = 2$ where $\Delta F_I = 1235$ cal. (Reprinted with permission from Colosimo *et al.*, 1974.)

of the number of binding sites (r) is reported, imposing the condition of symmetry of the binding curve, i.e.

$$L_0 = C^{-r/2} \tag{20}$$

It might be noticed that the binding curve for a molecule containing only two sites is always symmetrical. However, for every value of r greater than 2, this is not demanded by the model, and hence the symmetry of the binding curve must be verified experimentally.

The condition of symmetry always corresponds to the highest degree of cooperativity observable with a given number of interacting sites. Thus, when the constraint of symmetry is relaxed and the switchover point is no longer at $\bar{Y} = 0.5$, the value of n decreases as asymmetry increases.

Figure 6.4 shows another aspect of the situation, indicating how the value of n increases towards its limiting value, given by the total number of interacting

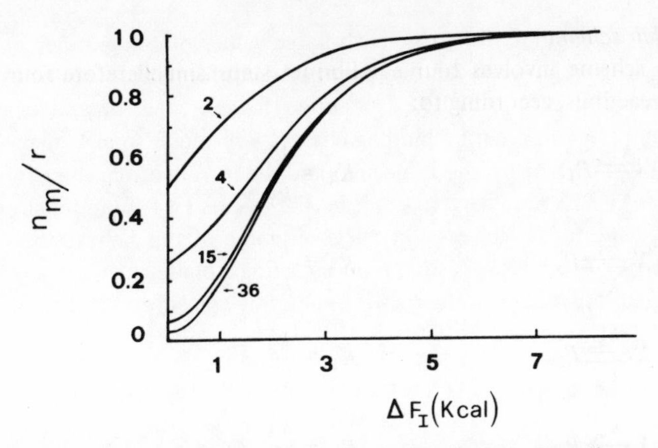

Figure 6.4 Values of n normalised to its upper limit r (i.e. corresponding to the binding sites as a function of ΔF_I(kcal), for the different values of r shown on the curves. This is for the symmetrical case. (Reprinted with permission from Colosimo *et al.*, 1974.)

sites r, as the apparent free energy of interaction increases. It should be remarked that for values of $r \geqslant 4$, all the curves lie close together and approach a common limit as r becomes bigger and bigger.

Application of these considerations to the case of chlorocruorin from the worm *Spirographis spallanzanii*, which contains ~ 72 oxygen binding sites, has shown that the experimental data, characterised by a maximum value of $n = 5$, a switchover point at $\bar{Y} \sim 0.64$ and a value of $\Delta F_I \sim 1800$ cal per site, are incompatible with the assumption that the whole molecule behaves according to the simple MWC model. If this were the case, the value of ΔF_I would demand a value of $n_m = 32$ as compared with the observed value of $n = 5$. This fact has led to the hypothesis that the sites interact in independent 'functional constellations', each containing r sites. The value of r, calculated assuming that the MWC model is applicable to each 'functional constellation', is ~ 10. Thus, the applicability of the MWC model to the giant respiratory proteins (apart from other considerations, see van Holde and Miller, 1982), demanded the introduction of an additional concept which implies that a two-state transition involves groups of strongly interacting sites called functional constellations (see also section 6.8).

6.2.2 Kinetics of ligand binding

The kinetics of the reactions of respiratory proteins with oxygen or other ligands (such as carbon monoxide) is a necessary complement to equilibrium studies in so far as they provide a different type of information on the structure–function relationships and on the mechanism of reaction. The kinetic approach will be described here in terms of the Adair scheme and within the two-state MWC model. This description is limited to the case of a four-site protein for simplicity and for historical reasons, since most of the experimental work in the field has been carried out on haemoglobin.

6.2.2.1 Adair scheme

The Adair scheme involves four equilibrium steps and therefore four 'on' and four 'off' reactions, according to:

$$
\begin{aligned}
P + X &\underset{k_1}{\overset{k'_1}{\rightleftharpoons}} P_1 & K_1 &= \frac{k'_1}{k_1} \\
P_1 + X &\underset{k_2}{\overset{k'_2}{\rightleftharpoons}} P_2 & K_2 &= \frac{k'_2}{k_2} \\
P_2 + X &\underset{k_3}{\overset{k'_3}{\rightleftharpoons}} P_3 & K_3 &= \frac{k'_3}{k_3} \\
P_3 + X &\underset{k_4}{\overset{k'_4}{\rightleftharpoons}} P_4 & K_4 &= \frac{k'_4}{k_4}
\end{aligned}
\tag{21}
$$

The kinetics of ligand binding following Adair may be formulated along the following lines (Roughton, 1949). The saturation of the molecule with oxygen is given by

$$\bar{\theta} = \frac{P_1 + 2P_2 + 3P_3 + 4P_4}{P_0} \tag{22}$$

where $\bar{\theta}$ corresponds to 4 $\bar{Y}$, and P_0 represents the concentration of the protein at time zero.

On this basis the velocity of the jth reaction is given by:

$$\frac{\mathrm{d}P_j}{\mathrm{d}t} = K'_j P_{j-1} [X] - k_j P_j \tag{23}$$

which is intractable in the absence of simplifying conditions such as those that may be made to apply in the dissociation and combination experiments, respectively:

(a) when the oxygen concentration is immediately brought to zero by dithionite, so that the combination processes can be neglected;

(b) at a ratio of ligand to protein binding sites high enough to consider the ligand concentration constant throughout the combination process and to allow the contribution of the back reaction to the velocity to be neglected.

In fact, in the latter case, the differential equations describing the time course become:

$$\frac{\mathrm{d}P}{\mathrm{d}t} = - k'_1 P [X]$$

$$\frac{\mathrm{d}P_1}{\mathrm{d}t} = k'_1 P [X] - k'_2 P_1 [X]$$

$$\frac{dP_2}{dt} = k_2' P_1 [X] - k_3' P_2 [X] \tag{24}$$

$$\frac{dP_3}{dt} = k_3' P_2 [X] - k_4' P_3 [X]$$

$$\frac{dP_4}{dt} = k_4' P_3 [X]$$

which represents a system of first-order sequential differential equations that are analytically soluble, as shown by Bateman (1910).

In the absence of functional interactions for intrinsically equivalent binding sites, the ratios of the combination and dissociation velocity constants will correspond to their statistical values. On the other hand, for a cooperative system, the shape of the progress curve for the ligand combination does not correspond, in general, to a second-order reaction. Therefore, in the case of co-operative interaction, the value of the apparent second-order rate constant increases with the progress of the reaction, the change being evident especially in the initial stages of the time course. This phenomenon, which gives rise to the so-called autocatalytic shape of the time course, is clear kinetic evidence of site-site interactions, and reflects the fact that one or more of the combination velocity constants in the reaction increases over and above the corresponding statistical values as a result of the binding of the ligand to other sites.

It should be pointed out that the absence of an autocatalytic time course (in either the combination or the dissociation) does not imply unequivocally that cooperative phenomena are absent. To illustrate this point, let us consider a protein with only two ligand binding sites, which is described by the following Adair scheme:

$$P + X \underset{k_1}{\overset{k_1'}{\rightleftharpoons}} PX_1 \qquad K_1 = k_1'/k_1$$

$$PX_1 + X \underset{k_2}{\overset{k_2'}{\rightleftharpoons}} PX_2 \qquad K_2 = k_2'/k_2 \tag{25}$$

At sufficiently high ligand concentrations, the kinetics of ligand binding is essentially dominated by k_1' and k_2' since the contribution of the dissociation processes to the overall velocity may be negligible (see also above). The ratio determines the shape of the progress curve (whether autocatalytic or not), which within an Adair scheme is independent of the ligand concentration (i.e. assuming that any conformational transition, if present, is very fast and never rate limiting). A series of computed curves showing values of the apparent second-order rate constant against the saturation with the ligand for different ratios of k_1'/k_2' is shown in figure 6.5.

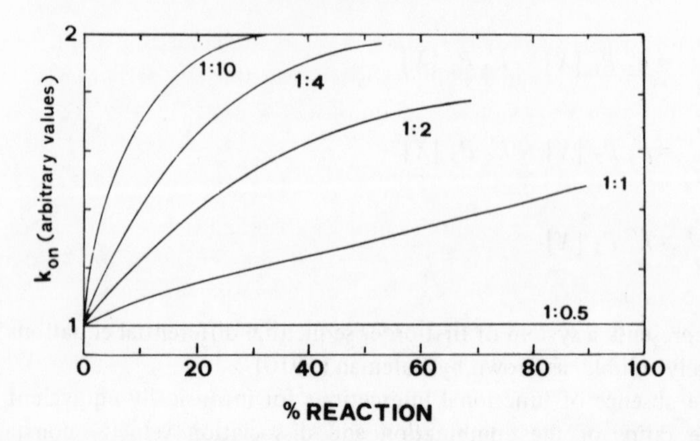

Figure 6.5 A series of computed curves, showing values of the apparent second-order rate constant versus progress of reaction for different ratios of k'_1/k'_2 in the case of a simple dimer scheme (see text). It is evident that k'_1/k'_2 determines the shape of the progress curve.

Two limiting cases are immediately obvious: (a) when the ratios $k'_2/k'_1 = 0.5$, the two sites are intrinsically identical and independent in their kinetic behaviour, and therefore the combination time course is (pseudo) first order with no sign of autocatalytic behaviour; (b) when the ratio $k'_2 : k'_1 \geqslant 10$, the time course is again (pseudo) first order, but with an apparent value equal to twice that of k'_1 (since the first combination rate constant is effectively rate limiting). In-between these two cases, an autocatalytic behaviour may be observed, and the time course assumes different shapes depending on the exact ratio of the two rate constants.

It should not be forgotten, however, that when the ratio between the two equilibrium constants (K_2/K_1) exceeds the statistical value (giving rise to a cooperative ligand-binding isotherm), this may depend on the contribution of combination or dissociation rate constants (or both). Thus, cooperative phenomena, observed at equilibrium, may result from (a) an increase of the combination velocity constants as the reaction proceeds; (b) an increase (higher than that expected on statistical considerations) of the dissociation velocity constants; or (c) a combination of (a) and (b).

In conclusion, in the absence of a complete equilibrium and kinetic characterisation, it may be very difficult to discriminate between a cooperative and a non-cooperative system. In fact it cannot be completely excluded that a non-cooperative equilibrium curve may result from complete compensation effects involving 'on' and 'off' kinetic constants in opposite directions.

6.2.2.2 Two-state model

Within the MWC model, the molecule has ten significant states which are the i-fold ligated T and R forms (where $0 \leqslant i \leqslant 4$) (see figure 6.1). The four subunits of haemoglobin have no interactions in either the T or the R state, and hence the kinetics of ligand binding to each subunit is assumed to be independent of the ligation of the other subunits within each one of the two conformational states. In other words, the ligand-binding kinetics depends only on the quaternary state of the molecule, and each subunit may be characterised by the following intrinsic rate constants:

$$
\begin{aligned}
&R + X \underset{R_k}{\overset{R_{k'}}{\rightleftharpoons}} RX \\
&T + X \underset{T_k}{\overset{T_{k'}}{\rightleftharpoons}} TX
\end{aligned}
\qquad (26)
$$

In addition to the above parameters, rate constants for the transitions

$$T_n \underset{RT_{k_n}}{\overset{TR_{k_n}}{\rightleftharpoons}} R_n \quad (\text{with } 0 \leqslant n \leqslant 4) \qquad (27)$$

should be included to describe completely the kinetic system.

These ten additional rate constants may be disregarded if the rates of ligand binding are always slow compared with the rate of the establishment of $R_i \rightleftharpoons T_i$ equilibrium. Thus the ratio of the populations defined by

$$f_i = \frac{T_i}{R_i + T_i} \qquad (28)$$

will always have its equilibrium value, and the two-state kinetic model can be described in terms of an Adair model, which, as described, does not consider the existence of rate-limiting steps due to intramolecular rearrangements. Thus:

$$\text{Hb } X_{i-1} + X \underset{k_i}{\overset{k_i'}{\rightleftharpoons}} \text{Hb } X_i \ (\text{with } 1 \leqslant i \leqslant 4) \qquad (29)$$

The Adair kinetic constants are then related to those of the MWC model through the following equations (Hopfield *et al.*, 1971).

$$\begin{aligned}
k'_1 &= 4\,[f_0\,{}^{T}k' + (1-f_0)\,{}^{R}k'] \\
k'_2 &= 3\,[f_1\,{}^{T}k' + (1-f_1)\,{}^{R}k'] \\
k'_3 &= 2\,[f_2\,{}^{T}k' + (1-f_2)\,{}^{R}k'] \\
k'_4 &= [f_3\,{}^{T}k' + (1-f_3)\,{}^{R}k'] \\
k_1 &= [f_1\,{}^{T}k + (1-f_1)\,{}^{R}k] \\
k_2 &= 2\,[f_2\,{}^{T}k + (1-f_2)\,{}^{R}k] \\
k_3 &= 3\,[f_3\,{}^{T}k + (1-f_3)\,{}^{R}k] \\
k_4 &= 4\,[f_4\,{}^{T}k + (1-f_4)\,{}^{R}k]
\end{aligned} \tag{30}$$

The conditions for an autocatalytic progress curve of ligand binding will be given by: $k'_1/4 < k'_2/3 < k'_3/2 < k'_4$. Such a situation will occur when: (a) ${}^{R}k'$ and ${}^{T}k'$ are sufficiently different (at least by a factor of 5), and when (b) f_i is dependent on the extent of ligation. Therefore, an autocatalytic behaviour cannot be observed for $i_s < 1$, because the switch occurs very early in the binding, and for $i_s \geqslant 3$ since $f_i \sim 1$ for $i < 4$. In the latter case, ligand binding takes place entirely within the T state with a quaternary switch only after the binding of the fourth ligand.

Only over the range $1 < i_s < 3$ is a clear autocatalytic behaviour seen, the value of $\bar{Y}$ at which acceleration becomes appreciable depending on the value of i_s.

Exactly identical considerations prevail for ligand dissociation experiments. The autocatalytic character of the ligand release is described by the two-state kinetic model similarly to that given above for the forward kinetics. In the case of the reaction of oxygen with human haemoglobin A, the first ligand molecule is released slowly from the fully liganded state R_4, and the switch of the quaternary conformation to the T state is associated with a large increase in the dissociation rate constant, giving rise to the autocatalytic shape of the progress curve.

6.3 THE ACTIVE SITE

6.3.1 The haem: some general considerations

The structure of the haem (protoporphyrin IX), the prosthetic group of haemoglobins, is constituted by four pyrrole rings, connected by methine bridges (α, β, γ, δ) and substituted by different groups in the external positions. In metalloporphyrins the central position of the porphyrin ring is occupied by a metal atom, which in the case of haemproteins is, in fact, an iron atom, coordinated to the four pyrrole nitrogens of the porphyrin. One of the axial coordination sites of the metal makes a covalent bond with the N_ϵ of the 'proximal' histidine residue in position F8 (note helical regions are referred to by letters A to H), the opposite axial coordination site (the 'distal' site) is the binding site of oxygen which comes to lie between the iron atom and His E7 ('distal' histidine) and valine E11. It should be recalled that haem iron can assume the ferrous (+2) or the ferric (+3) oxidation state and that only ferrous haemoglobins reversibly bind dioxygen. Moreover, the oxygen bound to the iron is within hydrogen bonding distance of the N_ϵ of the distal histidine (Shaanan, 1982).

X-ray studies of iron porphyrin compounds (Perutz, 1979) have shown that the length of the bond from the iron to the nitrogen atoms of the porphyrin ring varies between 2.061 Å and 1.99 Å, respectively, in high-spin and low-spin ferric compounds. Moreover, in five-coordinated high-spin ferric compounds, the iron atom is forced to lie out of the plane of the four nitrogen atoms by distances of between 0.38 and 0.47 Å, whereas in six-coordinated low-spin ferric complexes the iron atom tends to lie within 0.05 Å of the same plane. In the case of the high-spin ferrous ion, the iron atom seems further displaced from the plane of the four nitrogens. This case is representative of deoxyhaemoglobin, in which the high-spin ferrous ion is displaced from the haem plane by 0.55–0.63 Å. On the other hand, in oxyhaemoglobin the iron is both low-spin and six-coordinated so that its displacement is greatly decreased. However, it should be remarked that this is not necessarily a general rule. In fact structural studies (Steigemann and Weber, 1979) on erythrocruorin from *Chironomus thummi thummi* have shown that in this protein the iron atom in oxyerythrocruorin is displaced slightly more (~ 0.3 Å) from the plane of the haem than in deoxy (~ 0.2 Å). This has been related to the particular features of the helix E and the haem pocket, which allow a water molecule to be hydrogen bonded to the molecular oxygen bound to the iron atom.

The haem is embedded in a cleft provided by the polypeptide chain near the surface of the molecule. The molecular interactions are very complex and involve about 90 van der Waals contacts with non-polar groups of the polypeptide chain. It is to be noted that several aromatic rings are very near and parallel to the haem, and probably interact with it through π interactions.

The affinity constant of the haem for the protein is very high and it has been estimated to be of the order of 10^{12}–10^{15} M^{-1} at neutral pH (see Antonini and Brunori, 1971). Thus, the reversible dissociation of the prosthetic group from the protein moiety can be demonstrated to be finite only in haem transfer experiments of the type haemprotein* + globin → globin* + haemprotein.

6.3.2 Reactivity and its determinants

The haem and the amino acid residues to which the iron atom is linked are the same in almost all haemoglobins, which also display a very similar overall molecular architecture. In spite of this, haemoglobins show large differences in their affinities and rates of reaction with molecular oxygen. For example, the oxygen affinity is about $p_{1/2}$ = 0.04 mmHg in soybean leghaemoglobin and about $p_{1/2}$ = 1000 mmHg in some fish haemoglobins possessing a Root effect. The great variability observed in nature has been thought to arise from several factors, involving stereochemical and electronic effects related to the interaction of the metalloporphyrin with the globin. Furthermore, cooperativity in tetrameric haemoglobins occurs through the existence and interconversion of two alternative quaternary structures of the tetramer: a relaxed structure (*R*), with high oxygen affinity, and a tense structure (*T*), with a much lower affinity (see the discussion of the MWC model, section 6.2.1.1). The difference in ligand affinity between these two structures has been attributed, according to Perutz (1970), to various degrees of out-of-plane displacement of the iron atom with respect to the porphyrin plane.

In order to discriminate between the role of the protein moiety and the electronic effects by the metalloporphyrin on the ligand reactivity, suitable haem compounds have been synthesised. Unfortunately, a limited amount of data on oxygen binding to model systems is available, mainly because of the great tendency of their iron atoms to oxidise in the presence of air. Most of the data on determinants of haem reactivity refer to carbon monoxide, which can be used as a ligand of Fe(II) also in the presence of reductants. Among several reported synthetic haem compounds, we shall focus our attention mainly on two models which turn out to mimic more closely than others the ligand binding reactions occurring in ferrous haemproteins: 'picket fence' porphyrin (Collman *et al.*, 1975) and 'chelated' protohaem (Chang and Traylor, 1973).

'Picket fence' porphyrin has been shown to bind reversibly oxygen (Collman *et al.*, 1975) and carbon monoxide (Collman *et al.*, 1979). This haem model has been reported to display an affinity change for oxygen similar to that between *T*- and *R*-state Hb by simply substituting a residue on the proximal side (Collman *et al.*, 1978). Thus, if the imidazole C-atom between N_ϵ and N_δ has the hydrogen atom replaced by a methyl group, the oxygen affinity drops by a factor of $\sim$ 70. Moreover, since the crystal structure of such substituted derivatives has been solved for the unliganded and oxygenated form, this affinity decrease has been observed to correspond to a residual out-of-plane displacement of the Fe even when it is bound to oxygen (Jameson *et al.*, 1978). Thus, on the basis of the geometry of the complex, it has been possible to show that a sterically active substituent on the proximal imidazole decreases the oxygen affinity of the haem compound almost uniquely by weakening the Fe–O–O bond and leaving intact the Fe–Im bond (Jameson *et al.*, 1980).

On the other hand, in 'chelated' protohaem, which is more similar than the 'picket fence' to the active site of haemproteins, the dioxygen complex is quickly oxidised. In this case it has been possible to study the oxygen reactivity kinetically, exploiting the photodissociation of the carbon monoxide adduct of this haem compound in the presence of oxygen (Traylor, 1981).

From an overview of the experimental results obtained with such synthetic compounds, the following remarks may be made:

(a) the protein is of crucial importance in keeping the haem iron reduced;
(b) the coordination of a proximal imidazole to the metal as a fifth ligand increases the carbon monoxide affinity with respect to the four-coordinated protohaem; this depends on the fact that five-coordination

yields a decrease of both association and dissociation rate constants, by a factor of 30 times and 10^4 times respectively;

(c) a strain at the proximal imidazole lowers the ligand affinity of oxygen and carbon monoxide, by both decreasing the association rate and raising the dissociation rate;

(d) a decrease in solvent polarity enhances the oxygen dissociation rate, whereas carbon monoxide binding is not affected by this parameter;

(e) substitutions of the haem vinyls by ethyl or acetyl residues have negligible effects on the carbon monoxide binding but increase progressively the oxygen dissociation rates;

(f) the deprotonation of N_δ of the proximal imidazole decreases the carbon monoxide association rate, which, however, is not affected by an increase of π-electron density on the porphyrin (Stanford *et al.*, 1980).

Points (d) and (e), along with the rise in oxygen dissociation rate upon substitution of proximal imidazole by pyridine, strongly suggest the dipolar character of the Fe-O-O bond. Moreover, together with point (f), they indicate a substantial difference in the control of the reactivity between carbon monoxide and oxygen.

Distal steric effects, present in haemproteins, have been simulated in these synthetic compounds by hydrophobic groups which reduce the association rate for both oxygen and carbon monoxide, but do not seem to discriminate between the two ligands.

In conclusion, from functional studies carried out on haem models, the existence of common as well as ligand-specific reactivity determinants is well established. Thus, a decrease of association rate constant brought about by distal steric effects is shared by every ligand, along with a marked effect on combination and dissociation rate caused by a strain at the proximal side. Such behaviour has also been reported for sperm-whale myoglobin, which displays an enhancement of the carbon monoxide association rate as pH is lowered (Giacometti *et al.*, 1977), possibly arising from the breaking of the Fe—imidazole bond as a consequence of the protonation of N_ϵ of the proximal imidazole. More recently (Traylor *et al.*, 1983), a role of the Fe-haem plane distance in the deoxy form has been proposed for the carbon monoxide reactivity in haemproteins. The imidazole itself would then represent a strain for the iron, pulling it out of the porphyrin plane and altering the planarity of the four-coordinated protohaem, and the release of this strain may be responsible for the increased carbon monoxide reactivity of sperm-whale myoglobin at low pH (Giacometti *et al.*, 1977). Such a model may also account for the very high carbon monoxide association rate constant of *Chironomus thummi thummi* erythrocruorin (Amiconi *et al.*, 1972) for which the Fe-haem plane distance is much reduced with respect to other monomeric haemproteins (Takano, 1977; Steigemann and Weber, 1979).

On the other hand, the existence of ligand-specific determinants of reactivity is also demonstrated, since only the carbon monoxide association rates are decreased by the deprotonation of proximal imidazole N_δ, whereas only oxygen dissociation rates are affected by the solvent polarity and substitution at the side chains of the haem.

The overall view leads to the suggestion that, although steric effects play a role in determining the reactivity of haemproteins mainly with bulky and branched ligands such as isocyanides, electronic effects at the haem are equally important for discriminating between different ligands.

6.3.3 Role of the metal

One of the most obvious and effective methods of probing the functional role of the haem iron is to study the properties of artificial haemproteins with other metals substituted for iron. In this perspective the iron atom of haemoglobin has been replaced by manganese, copper, nickel and cobalt (O'Hagen, 1961; Fabry *et al.*, 1968; Yonetani *et al.*, 1970). However, only the substitution of cobalt yields a molecule capable of reacting reversibly with molecular oxygen.

In the deoxy state the cobalt ion (Co II) differs from iron in that it has a low spin configuration. In this configuration the ionic radius of cobalt (Co II) is sufficiently small to permit the ion to fit into the central hole of the porphyrin, and hence it is probably not located significantly out of the porphyrin plane (Fermi *et al.*, 1982). Thus the characteristic features of cobalt-porphyrin recombined with globin are likely to be the nearly in-plane position of the cobalt nucleus and a rather long Co—N_ϵ *axial* bond distance. The low spin configuration is maintained in oxy-cobalt-Hb and the unpaired electron is largely delocalised to dioxygen, which can be considered as almost a superoxide ion.

The oxygen affinity of cobalt-myoglobin is 50 to 100 times lower than that of the corresponding iron-containing protein. No other relevant differences are observed. The oxygen affinity of haemoglobin is reduced 10 to 25 times, whereas the Bohr effect is essentially the same as that of the native iron protein. Furthermore, cobalt-substituted haemoglobin binds oxygen cooperatively, with a Hill coefficient of $n = 2.3$. In conclusion, cobalt-haemoglobin displays functional properties very similar to those of the native protein, although the oxygen affinity is very much reduced (Yonetani *et al.*, 1974). In this connection it should be noted that the reduction of oxygen affinity upon cobalt substitution is unique among modified haemoglobins. Most of the other modifications, at the level of both the haem and the protein moiety, result in a large increase of oxygen affinity.

The cooperative behaviour of oxygen binding in cobalt-haemoglobin seems to be in contrast with the proposal that the trigger for the quaternary transition resides in the 'out-of-plane'–'in plane' transition of the iron. However, since the unpaired electron of the d_{z^2} orbital of Co II is almost completely delocalised to oxygen, the Co-N_ϵ axial bond distance in oxy-cobalt-Hb is probably shorter (by about 0.4 Å) than it is in deoxy-cobalt-Hb (Dickinson and Chien, 1973). It is therefore the contraction of this bond that is probably responsible for the triggering of the homotropic allosteric interaction observed in cobalt-substituted haemoglobin.

In the past few years, hybrid tetramers obtained by recombining a cobalt-containing subunit (either α or β) with an iron-containing partner have been systematically investigated from the functional and spectroscopic points of view. The importance of this type of investigation is connected to the problem of determining independently the individual microscopic constants of a complete Adair scheme which differentiates among the α- and β-chains (Yamamoto *et al.*, 1974; Imai *et al.*, 1980). This is made possible by the differences in the intrinsic reactivity and spectral properties of the cobalt-substituted subunits. Moreover, whereas the iron-containing subunits combine reversibly with both oxygen and carbon monoxide, the cobalt subunits bind exclusively with oxygen and with an affinity very much lower than that of their partners.

6.3.4 The protein envelope

The comparative systematic work on sequence analysis of haemproteins from different species has represented, and continues to represent, a fruitful approach for elucidating the structure–function relationships of the molecule. Thus, comparison of the primary structure of haemoglobins and myoglobins from different sources has unequivocally established that there is a certain number of residues, generally described as 'invariant', which are present in an identical position along the sequence. Most of these are in the haem pocket, where they may play a 'key' role in locking the haem into a specific spatial arrangement and in ensuring the stability of the oxygenated derivative.

On the other hand, in spite of extensive amino acid substitutions in other regions of the molecule, the overall folding of the polypeptide chain is very similar to that reported originally for sperm-whale myoglobin, suggesting that the tertiary structure of the molecule is the result of an extreme degree of specialisation acquired during the course of evolution. In this respect, although evolutionary aspects are outside the scope of this article, we wish to outline the new perspectives opened up by the recent observation that many eukaryotic genes possess a mosaic structure, in which expressed sequences (or exons) are separated by non-coding intervening sequences (or introns).

One important question raised by this discovery concerns the relationship, in terms of structure and function, between the protein segment that would be coded for by an exon and the corresponding segment in the intact protein. Because the genes coding for haemoglobin have been extensively investigated and the protein structure is well known, this protein is an ideal case for attacking this important question. Globin genes consist of five segments, and precisely three exons separated by two introns (Konkel *et al.*, 1978; van den Berg, 1978). In the case of the α-chain, the three exons code for residues 1–31, 32–99 and 100–141; in β-globin, the corresponding positions are 1–30, 31–104 and 105–146 (figure 6.6).

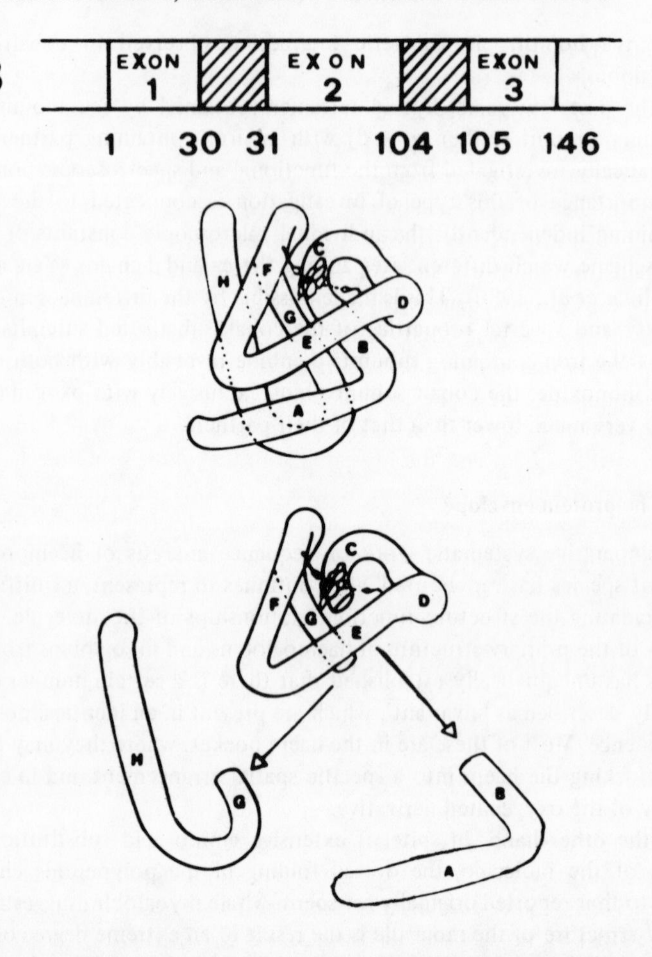

Figure 6.6 Top: diagrammatic representation of the human β-globin gene. The exons and the corresponding amino acid residues are indicated. The intervening segments (introns) are shown by the hatched areas. Bottom: schematic representation of a β-subunit and its structural division into the three exonic transcripts. The α-helical regions are indicated alphabetically. (Modified from Blake, 1983.)

A fascinating hypothesis envisages that the protein segments coded for by the three exons are functional domains which play different roles (Gilbert, 1979; Eaton, 1980; Gò, 1981). Starting from the basic property of the reversible binding of oxygen, related to the central coding sequence (which contains essentially all the residues in contact with the haem), replacement of the third and first coding sequences by mutation and recombination processes may have originated cooperative binding and heterotropic effects. It has recently been demonstrated that the central fragment of β-globin (residues 31–104) binds haem tightly and with spectral features comparable to those of the haemoglobin chains (Craik *et al.*, 1980). However, it appears that the complex recovers more precisely the structural features of the haem pocket whenever the two extreme fragments are added, and bind through noncovalent forces. In other words, the central fragment has the structural potential of providing a tight and specific binding site for the haem, but the fit is sharpened by the addition of the N- and C-terminal fragments. Reversible oxygen binding, however, was not obtained under these conditions because of the irreversible binding of the distal histidine to the iron. To obtain formation of a stable dioxygen complex, two requisites are essential in the case of globins, namely (a) the presence of the side fragments, and (b) the addition of the complementary subunit (i.e. the α-chain) (Craik *et al.*, 1981). It seems possible that the partner subunit serves to overcome the tendency of the distal histidine to bind at the haem with formation of an irreversible hemichromogen. These results, in relation to many others already known, outline the complexity of the system and the importance of the whole protein envelope for the exact modulation of the function. However, it is beyond doubt that further studies in this direction will provide new and relevant information on the structure-function relationships in haemproteins and on the process of evolution of proteins in general.

6.4 HAEMOGLOBIN: HOMOTROPIC INTERACTIONS

6.4.1 The iron-oxygen bond

The electronic structure of deoxyHb is represented as: $(d_{xz})^2$ $(d_{yz})^1$ $(d_{xy})^1$ $(d_{z^2})^1$ $(d_{x^2-y^2})^1$, with S = 2, whereas in oxyHb the ground state has S = 0, which would suggest pairing of the electrons in the three lowest orbitals.

The nature of the iron-oxygen bond is still the subject of a controversial debate after the early finding that the oxyHb ground state is diamagnetic, with a bent Fe-O-O bond (135°) (Pauling and Coryell, 1936). According to this model the iron-oxygen bond would be made by the hybridisation between the π^* orbitals of the oxygen and the d_{xz} and d_{yz} orbitals of the iron, with some net transfer of charge from the oxygen to the iron.

Later, a different model was formulated (Weiss, 1964), according to which the Fe-O-O bond would be made between a ferric ion and a superoxide anion with a full transfer charge from the iron to the oxygen ($Fe^{3+}O_2^-$). This seems supported by observations of the infrared stretching frequency of HbO_2, which is at 1107 cm^{-1} (Barlow *et al.*, 1973), lower than that of the oxygen molecule (1556 cm^{-1}) and within the range observed for superoxide (1150–1100 cm^{-1}). The possibility of a metal-dioxygen complex with a doubly coordinated structure made by equivalent oxygen atoms (Griffith, 1956), and giving a stretching frequency at 850 cm^{-1} can be ruled out. Moreover, the liberation of superoxide ion during the autoxidation of oxyhaemoglobin (Misra and Fridovich, 1972) and single subunits (Brunori *et al.*, 1975) as well as the presence of an unpaired electron density on the iron atom by X-ray fluorescence (Koster, 1975) seem in accord with the Weiss model. Further support to this hypothesis comes from observations of K-edge absorption (Michalowicz *et al.*, 1983) along with optical absorption at low temperatures (Wittenberg *et al.*, 1970), which report a strong similarity between oxyHb and alkaline metHb.

The diamagnetism of the HbO_2 ground state in the Pauling model has been challenged by measurements of the magnetic susceptibility of oxyhaemoglobin, which indicate (Cerdonio *et al.*, 1978) the existence of a low-lying triplet state with a consequent weak paramagnetism. A spin equilibrium between a quasi-degenerate singlet and triplet ground state of oxyhaemoglobin has been suggested also by semiempirical calculations (Herman and Loew, 1980), possibly accounting (Bacci *et al.*, 1979) for the quadrupole splitting of oxyHb Mössbauer spectra, which display a marked temperature dependence similar to that of ferric compounds (Lang and Marshall, 1966). The paramagnetic character of oxyHb has not been confirmed by Philo *et al.* (1983), who again report a full diamagnetism of HbO_2 ground state when the sample is entirely free of metHb. The diamagnetic character of both oxy and carbonmonoxy Hb is finally agreed upon, thanks to Sawicki and Lang (1984), who have shown that the results reported by Cerdonio *et al.* (1978) were vitiated by the presence of deoxyHb in their sample (see also Cerdonio *et al.*, 1984).

On the other hand, the Pauling model seems to be supported by *ab initio* quantum mechanical calculations on model systems (Olafson and Goddard, 1977; Case *et al.*, 1979), which predict a bent geometry for the Fe–O–O bond and a weak transfer of charge from iron to oxygen. Furthermore, this geometry has been demonstrated in oxymyoglobin by X-ray analysis (Phillips, 1980).

The controversy about the mode of the Fe–O–O bond in oxyhaemoglobin may be, at least partially, reconciled by proposing a mixture of $[Fe(III) \cdot (O_2^-)]$ and $[Fe(II) \cdot (O_2)]$ species (Reed and Cheung, 1977), which also seems to be supported by recent Mössbauer data (Tsai *et al.*, 1981).

6.4.2 Quaternary structures

The existence of *at least* two different conformations is related to the characteristic sigmoidal shape of the oxygen binding equilibrium isotherm displayed by haemoglobin. According to the allosteric MWC model (see section 6.2.1.1), cooperativity may be explained by the statistical predominance of the low-affinity (*T*) quaternary structure in the absence of oxygen. Every addition of the ligand increases the oxygen binding potential of the protein, progressively decreasing the energetic difference between the *T* and the *R* conformations, and in fully liganded haemoglobin the *R* state is favoured. In HbA the difference in oxygen affinity corresponds to a free energy of binding of about 4 kcal per mole (haem) higher in the *R* than in the *T* state.

Studies on the structure–function relationships in haemoglobin have provided an explanation of the structural mechanism by which (a) the binding of oxygen to the iron can first decrease and then reverse the energetic relation between the *T* and *R* conformations (homotropic interactions) and (b) the binding of non-haem ligands can affect this process (heterotropic interactions).

6.4.2.1 Structural information

The basis of such an approach is represented by the available three-dimensional structure of liganded haemoglobin and deoxyHb. According to the Perutz (1970) model, the parts of the molecule that are particularly important in regulating the functional behaviour of haemoglobin are (a) the haem complex and the geometry of its surroundings, and (b) the intersubunit interfaces and the carboxy-terminals.

In the human carbon monoxy haemoglobin (HbCO) (Baldwin, 1980), the Fe atom is six-coordinated, low spin and is 0.04 Å out of the α-haem plane towards the proximal histidine, whereas in β-subunits this distance is larger, namely 0.21-0.22 Å from the mean porphyrin plane. The ligand (CO) lies off the normal to the haem plane because of steric hindrance represented by two residues of E helix on the distal side, His E7 (distal histidine) and Val E11. According to a different interpretation, based on structural data on $Fe{-}CN^-$ complex (Deatherage *et al.*, 1976), the carbon monoxide would remain perpendicular and the porphyrin plane would be tilted, allowing the ligand to keep a linear geometry without forcing the nearby residues off their minimum energy position (Moffat *et al.*, 1979). This interpretation is supported also by theoretical studies (Hoffman *et al.*, 1977).

The structure of human oxyhaemoglobin has been resolved by Shaanan (1982) at −2°C; from the structural analysis of the haem pocket region, reported in a preliminary note (Shaanan, 1982), it is clear that there is a different geometry of the iron–oxygen complex with respect to other haemproteins, such as sperm-whale myoglobin (Phillips, 1980) and oxyerythrocruorin (Steigemann and Weber, 1979). Thus, the Fe-O-O bond angle in oxyHb is 156° as compared with 115° in oxymyoglobin and 180° in oxyerythrocruorin. Moreover, the position of N_ϵ of His E7 in the α-chains, as in oxymyoglobin, is at an H-bond distance from the distal oxygen, whereas in the β-subunits it is further from oxygen, making unlikely the occurrence of an H-bond.

In deoxyhaemoglobin (Fermi, 1975) the Fe atom is high-spin and five-coordinated, and in both subunits its position is out of the centre of the porphyrin ring by 0.60–0.63 Å. On the distal side, the pocket for the ligand is significantly reduced in the β-deoxy chains, but not in the α.

As a consequence of ligand binding and the ensuing change of quaternary structure, the α-haem moves 0.6 Å and the β-haem 1.5 Å into the respective pockets, rotating slightly (Baldwin and Chothia, 1979). Furthermore, in both subunits, the F helix follows the haem in a parallel fashion, transmitting the motion also to the initial portion of the G helix (figure 6.7). This movement brings about a twofold structural change: (a) the imidazole ring of the proximal histidine, which in deoxyHb is in an asymmetric position with respect to the haem, tilts by 7°; this, together with the haem rotation, removes the steric

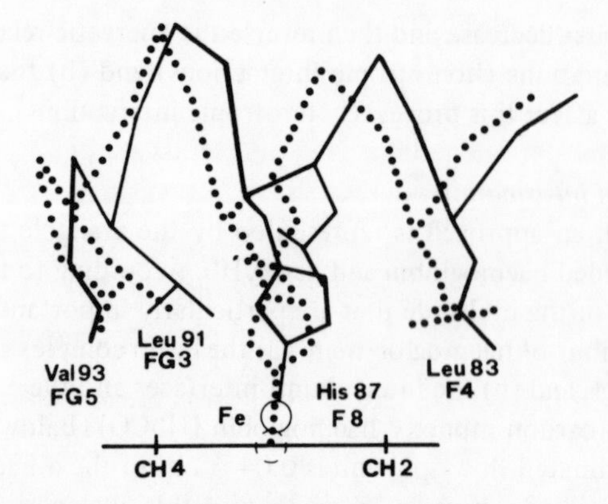

Figure 6.7 The positions of the F helix relative to the haem in the subunit of deoxyHb (solid lines) and HbCO (broken lines). The positions are shown relative to superimposed haems. The C_α atoms of residues 81–93 are shown and also the side chains of those in contact with the haem; Leu 83, Leu 86, His 87, Leu 91 and Val 93. The haem direction CH1–CH3 is perpendicular to the figure. The figure illustrates how, on ligand binding, the F helix moves in the CH2–CH4 direction, and tilts towards the haem plane. The tilt is dissipated in residues 89 and 90 so that Leu 91 and Val 93 remain at the same distance from the haem plane. A figure drawn for the β-subunits of deoxyHb and HbCO would be very similar. (Modified from Baldwin and Chothia, 1979.)

hindrance represented by the imidazole carbons and the porphyrin pyrrole nitrogens, and possibly allows the iron to move freely into the haem plane without stretching the Fe-N_ϵ bond; (b) the intersubunit contacts at the $\alpha_1\beta_2$ (and $\alpha_2\beta_1$) interface are altered because most of the residues involved in these contacts belong to F and G helices. Besides minor rearrangements around some other residues, the main change occurring at the $\alpha_1\beta_2$ interface upon the quaternary transition is represented by the disappearance, in the *R* structure, of the H-bond between Asp G1 (99)β and Tyr C7 (42)α, present in the *T* conformation, and the appearance of a new H-bond between Asn G4 (102)β and Asp G1 (94)α.

As a consequence of these structural rearrangements, the cavity between the two β-subunits is very much reduced in the R structure. This region of the molecule forms the binding site for the polyphosphates (Arnone, 1972) which bind stronger in the *T* conformation, accounting for their action as allosteric effectors (Benesch and Benesch, 1967). Moreover, the structural change occurring at the $\alpha_1\beta_2$ interface leads to a different stabilisation energy of this intersubunit contact. As a matter of fact, the tetramer–dimer dissociation constant is different in human deoxyHb as compared with the liganded form, the latter being more easily dissociated into dimers (Guidotti, 1967; Andersen *et al.*, 1971; Kellett, 1971). On the other hand, the $\alpha_1\beta_1$ contacts are much more stable and not significantly affected by the conformational change (Baldwin and Chothia, 1979).

An additional structural rearrangement involves the penultimate tyrosines of all subunits, which in deoxyHb are firmly anchored into a pocket between F and

H helices by van der Waals contacts and by an H-bond between the —OH group of the tyrosine and the carbonyl group of Val FG5, whereas they are free in the liganded form. Moreover, the *R* structure differs from the *T* structure in the absence of several salt bridges involving the carboxyterminals; these are depicted in figure 6.8, from which intra- and intersubunit contacts are evident.

$$\text{Arg HC3 (141)}\alpha_1\text{ (or }\alpha_2\text{)}\begin{cases}\text{COO}^- \cdots {}^+\text{NH}_3\text{ Val NA1 (1)}\alpha_2\text{ (or }\alpha_1\text{)}\\ \text{Gua}^+ \cdots {}^-\text{OOC Asp H9 (126)}\alpha_2\text{ (or }\alpha_1\text{)}\end{cases}$$

$$\text{His HC3 (146)}\beta_1\text{ (or }\beta_2\text{)}\begin{cases}\text{COO}^- \cdots {}^+\text{NH}_3\text{ Lys C6 (40)}\alpha_2\text{ (or }\alpha_1\text{)}\\ \text{Im}^+ \cdots {}^-\text{OOC Asp FG1 (94)}\beta_1\text{ (or }\beta_2\text{)}\end{cases}$$

Figure 6.8 Salt bridges made by carboxyterminal residues of both types of chain of human HbA in the deoxy (*T*) conformation.

Another structural difference between deoxy- and liganded Hb is represented by the accessibility of the reactive sulphydryl group of Cys F9 (93)β, which in deoxyHb is shielded from the solvent by the salt bridge between His HC3β and Asp FG1β. In the liganded state the cysteine occupies two different positions (McConnell *et al.*, 1969), either fully exposed to the solvent or inserted into the pocket between F and H helices, from which the penultimate Tyr HC2 (145)β has been pushed off (Moffat, 1971).

6.4.2.2 Spectroscopic probes

A number of spectroscopic methods have been used in the study of the ligand-linked quaternary change in haemoglobin, largely with the aim of providing a measurement of the state of the molecule in solution. The vast amount of work available cannot be summarised here; we shall limit ourselves to a consideration of some recent results and direct the reader to more extensive reviews (see *Methods in Enzymology* (1981), vol. LXXVI).

The two H-bonds at the $\alpha_1\beta_2$ interface, characteristic of two conformational states, give rise to two distinct NMR proton resonances, Asn G4β–Asp G1α at –5.8 ppm from water and Tyr C7α–Asp G1β at –10.0 ppm (Fung and Ho, 1975).

Circular dichroism spectra of deoxyHb in the *T* conformation have shown a negative ellipticity with a single peak at 287 nm, whereas in the *R* structure two dips near 285 and 290 nm were observed (Perutz *et al.*, 1974*a*). The negative band has been attributed to a quaternary-linked change of environment of Trp C3 (37)β at the $\alpha_1\beta_2$ contact. The same event has been considered in connection with the blue shift observed in the aromatic region (at 294 and 302 nm) (Perutz *et al.*, 1974*a*) by difference spectroscopy of deoxyHb in T_0 (normal human HbA) and R_0 (occurring in mutant or modified haemoglobins which do not undergo the ligand-linked conformational transition, like Hb Kempsey (Bunn *et al.*, 1974)). Analogue peaks in the difference spectra, at 278 and 287 nm, have been correlated with a similar variation involving Tyr C7α (M. F. Perutz and S. R. Simon, unpublished).

Blue shifts of the difference spectra between two quaternary states of unliganded Hb have been observed also in the Soret absorption region (400–450 nm) and in the visible range (500–700 nm) (Gibson, 1959; Brunori *et al.*, 1968). These have been referred (Perutz *et al.*, 1974*b*) to a possible lengthening of the Fe-N_ϵ bond distance in the *T* structure. Such an interpretation is consistent with low-frequency resonance Raman spectra (Nagai and Kitagawa, 1980), which indicate a definite correlation between the quaternary transition in the absence of ligand and the strength of the iron-proximal imidazole bond. However, in view of the small frequency shift observed, which accounts for approximately 0.3 kcal per site energy difference in-between the two quaternary states, these results limit quantitatively the role of this bond in the energetics of the *T-R* conformational change.

The assignment of this Raman frequency at 223 cm^{-1} to the iron-proximal imidazole bond has been strongly criticised by Stein *et al.* (1980), who suggest that this frequency shift is explained by a partial donation of the imidazole N_δ proton to an H-bond acceptor, e.g. a carbonyl oxygen of the peptide backbone. This process would be expected to increase the electron-donating properties of the N_ϵ and to strengthen the Fe-N_ϵ bond. Although this might be a model that works for other non-O_2-carrying haem proteins, such as horseradish peroxidase (La Mar and de Ropp, 1982), the role of this proton donation has been minimised in haemoglobin by observations on the NMR N_δ proton shifts (Nagai *et al.*, 1982), which do not display any definite correlation with the quaternary change.

A significant contribution to the understanding of such a frequency change comes from transient Raman spectra of quickly photolysed liganded haemoglobin (~ 10 ns) (Friedman *et al.*, 1982). On the assumption that 10 ns after photodissociation the Fe-N_ϵ bond is representative of a deoxyHb still in the *R* quaternary conformation (R_0), the observed difference in frequency between transient and static spectra has been tentatively assigned to change of the geometrical relationship between the proximal imidazole and the haem plane. If this interpretation is correct, it would lead to the conclusion that both quaternary switch and ligand binding contribute to the release of steric repulsion between the porphyrin ring and proximal histidine.

6.4.3 Energetics of ligand-linked conformational changes

The problem of quantitatively describing the homotropic interactions, viewed in the previous section from the structural and spectroscopic standpoints, may be summarised by two basic questions: (a) How does the binding of oxygen to the haem iron change the quaternary structure of the globin from *T* to *R*, and why is the ligand affinity of the T state lower than that of the R state? (b) Why is the T_0 state more stable than the R_0 state?

6.4.3.1 Stereochemical and electronic aspects

It was initially proposed (Perutz, 1970) that the high-spin ion in deoxyHb must be outside the porphyrin plane because of its larger radius, as compared with the low-spin state (Zerner *et al*., 1966). Subsequent quantum mechanical calculations (Olafson and Goddard, 1977) and structural studies of a ferrous high-spin haem model compound have shown that the high-spin metal can be in the porphyrin plane (Reed *et al*., 1980). Moreover, deoxyHb from *Chironomus thummi thummi* has been reported to have an Fe–haem plane distance of only 0.17 Å (Steigemann and Weber, 1979), four times smaller than that of human deoxyHb (Fermi, 1975). Calculations of the distribution of potential energy of the haem complex (Warshel, 1977) have indicated that it is energetically more expensive to pull a low-spin iron out of plane than to push a high-spin iron into the plane.

EXAFS measurements also showed a change of the Fe–N_{pyrr} bond length upon binding of oxygen within a given quaternary conformation (Eisenberger *et al*., 1978). On the other hand, these bond distances are unaltered in the T_0–R_0 structural transition and this excludes the possibility that Fe–N_{pyrr} interactions could play a key role in triggering the quaternary switch.

This information and a complete analysis of the structural data (Baldwin and Chothia, 1979) suggest that a more plausible explanation for the different geometry between unliganded and liganded haemoglobin may reside in the steric repulsions between the proximal histidine imidazole and the pyrrole nitrogen N1. The importance of the position of the proximal imidazole with respect to the porphyrin plane was also emphasised by Gelin and Karplus (1977), who pointed out that a change in the tilt of the haem was necessary in order to allow the imidazole to come closer to the porphyrin following the iron.

All available data indicate very clearly that this steric disturbance does not exert any tension in deoxyHb, i.e. in the T structure, but steric hindrance comes into play when oxygen binds in the T conformation, because the motion of the iron towards the porphyrin plane is constricted by the clash of the imidazole with the pyrrole nitrogens. These steric effects have two consequences: (a) the Fe–O–O bond in the T state is much weaker than in the R state, which accounts for the lower oxygen affinity of the T state, and (b) the tendency of the six-coordinated low-spin iron to move into the plane, although constricted by the steric repulsions mentioned earlier, puts some strain on the proximal side.

The strain induced by ligation in the T state is associated with a movement of the F helix (Baldwin and Chothia, 1979). Upon quaternary switch, large structural rearrangements at the $\alpha_1\beta_2$ interface (see above) and modifications of the tertiary structure of the liganded subunit are brought about, leading to a stronger Fe–O–O bond and a consequent higher oxygen affinity. The role of the intersubunit contacts in the energetics of the T-R transition has been well established, mainly by comparative studies on mutant haemoglobins which have amino acid substitutions at the interfaces. Accurate thermodynamic measurements of oxygen binding to several of these mutants (Pettigrew *et al.*, 1982) support the view that on ligand binding every single subunit contributes to cooperativity by transmitting the strain to the $\alpha_1\beta_2$ interface.

According to a different view, based on NMR measurements (Viggiano and Ho, 1979), binding of oxygen to one subunit is indeed transmitted to the interface, which, however, acts essentially as a propagator of the perturbation to the other unliganded subunits whose tertiary structure is inevitably altered. Thus ligand binding produces a sequential modification of the tertiary structures in the unliganded subunits, even within the same quaternary conformation, in line with the KNF model (Koshland *et al.*, 1966). Unless a similarity in oxygen affinity of the different NMR-detectable tertiary structures is assumed, this view is difficult to reconcile with observations on ligand binding of the valency hybrids (i.e. molecules which have one type of chain in the ferrous state and the other in the ferric state). In the absence of oxygen these valency hybrids were shown to assume either the T or the R conformation, depending on the presence of allosteric effectors, such as inositol hexaphosphate (IHP) (Ogawa and Shulman, 1971; Shulman *et al.*, 1975; Perutz, 1982). Ligand binding to the ferrous haem therefore simulates the process occurring in a doubly liganded tetramer, and the possibility of a quaternary switch, at a constant ligation stage, allows discrimination between tertiary and quaternary determinants of the ligand affinity. The presence of the allosteric effector brings about a decrease in the combination rate constant for carbon monoxide similar to that observed in going from the R to the T state, showing that quaternary conformation is more effective than the number of liganded sites in controlling the ligand reactivity (Cassoly *et al.*, 1971). Such a hypothesis is further strengthened by the observation that haemproteins which keep the T quarternary state even in the liganded form (T_4), like Hb Kansas or Hb M Iwate in the presence of IHP, display the same kinetic constant independently of the number of liganded subunits (Salhany *et al.*, 1976).

In addition to the release of the steric strain present in the T-liganded structure, a significant contribution to the increased oxygen affinity of the R_0 conformation and the consequent strengthening of the Fe–O–O bond has been proposed to arise from an increase in the porphyrin electron density (Shelnutt *et al.*, 1979). This is suggested by the Raman high-frequency shifts at the oxidation state marker lines (1300–1700 cm^{-1}) observed in going from T_0 to R_0. The increased electron density on the porphyrin π^* orbitals may originate from a closer approach of the Phe CD1 to the haem in the R conformation. Such a hypothesis finds support in the theoretical calculations of Warshel and Weiss (1981) according to which the closer approach of the main-chain dipoles of the G helix could stabilise a positive charge distribution on the vinyl side chains of the porphyrin. Such a role is clearly supported by experimental evidence, for example: (a) an enhanced oxygen affinity of ferrous haems upon increase of the solvent polarity (Chang and Traylor, 1975); (b) a decreased cooperativity associated with a higher affinity of oxygen binding if the vinyls are substituted by either a formyl (Asakura and Sono, 1974) or a methyl group (Rossi-Fanelli and Antonini, 1959; Brunori *et al.*, 1969). A different, but not necessarily alternative, explanation for the increased electron density of the porphyrin may be related to a difference in the electron-donation propensity of the proximal histidine (Nagai and Kitagawa, 1980; Perutz, 1980).

The overall picture that emerges from these observations indicates that the *R* quaternary structure, regardless of the presence of the ligand, is characterised by a higher electron distribution on the porphyrin ring (originating via an electron donation from the protein either through the Fe–N_ϵ bond and/or the van der Waals contacts between the protein and the porphyrin), which results in an increased affinity for oxygen, this being a good electron acceptor.

6.4.3.2 Thermodynamics of oxygen binding

The energetics of oxygen binding and ligand-linked conformational changes demand a careful consideration of the thermodynamic aspect. A very accurate study on human HbA has been carried out by Imai (1979) using the Adair model (Adair, 1925), which, unfortunately, does not readily connect the thermodynamic parameters with physical events, contrary to the MWC model.

From analysis of these results within the framework of the MWC model (Imai, 1979) it is possible to conclude that the *T* structure appears to be stabilised by enthalpy terms, because the T_0–R_0 transition is an entropy-driven endothermic process; conversely, entropy contributions seem to make the *R* conformation more favourable. Therefore, during oxygenation the *T*–*R* transition occurs at that step at which the entropy contribution exceeds the enthalpic one. Because the equilibrium binding constants for the first three steps are similar, an enthalpy–entropy compensation must be implied, whereas binding of the fourth oxygen molecule, which occurs with much higher affinity, is not associated with such a compensation. It should be pointed out very clearly that the observed amount of heat exchanged at each step, once corrected for the contribution of proton and anion binding, becomes uniform, thus indicating that homotropic interactions are largely entropy driven. This finding (Imai, 1979) indicates the very important role played by non-haem ligands in the expression of cooperativity in oxygen binding.

6.4.4 Subunit interactions

Since the largest conformational change occurs at the same intersubunit contact ($\alpha_1\beta_2$) involved in the dissociation of the tetramer into dimers, the relationships between the energetics of oxygen binding and the extent of subunit dissociation have been investigated. Furthermore, at least in the unliganded state, the affinity of the α- and β-chains, as well as the dimers, approaches that of the *R* conformation (Brunori *et al.*, 1966; Mills and Ackers, 1979), and thus is much higher than that of the *T* state of the tetramer. This suggested a strict structural similarity between single chains and dimers, even in the deoxygenated form, and fully liganded tetrameric haemoglobin, as substantiated by kinetic studies (Brunori and Schuster, 1969; Noble *et al.*, 1969). On these premisses it follows that the assembly of the unliganded subunits into the tetramer involves structural rearrangements similar to those underlying the R_0–T_0 transition, which dramatically lowers the affinity for oxygen. Thus, the higher dimer–tetramer association constant for the T state implies that deoxyHb, in the concentration range usually employed, is virtually all tetrameric, whereas the progressive increase of the R state while the molecule is saturated is associated, at sufficiently low protein concentrations, with the appearance of non-cooperative dimers (Hewitt *et al.*, 1972).

Ackers and colleagues (Mills and Ackers, 1979) have noticed that the oxygen affinity of the fourth site (in the R conformation) is higher than that of the $\alpha\beta$ dimer. This effect, called 'quaternary enhancement', is also consistent with the higher tendency to dimerise of the triliganded species with respect to the fully liganded form. This fivefold increase in affinity of the R tetramer over the dimer has been incorporated into a self-consistent mechanism in which the quaternary enhancement is operative at each stage of ligation. However, during the first binding steps it would be masked by the constraints present in the T state, constraints which are progressively released during saturation with oxygen, leading to a full expression of the quaternary enhancement in the last binding step.

Within this model (Mills and Ackers, 1979), quaternary constraints have been proposed to be located largely at the $\alpha_1\beta_2$ intersubunit contacts, whereas quaternary enhancement would be mainly expressed at $\alpha_1\beta_1$ interfaces.

Such a hypothesis is formulated on the assumption that the energetic difference between unliganded and liganded haemoglobin resides completely in the differential arrangement of the subunit interface, expressed by variations of tetramer–dimer dissociation equilibrium. This hypothesis is by no means proven, although it is consistent with the difference in the association equilibria between unliganded and fully liganded mutant haemoglobins (Pettigrew *et al.*, 1982). As a matter of fact, these authors claim that only mutants with replacements at the $\alpha_1\beta_2$ interface display a meaningful decrease of cooperativity associated with a smaller difference in the tetramer–dimer equilibrium between deoxy- and oxyHb.

Since tetramer–dimer equilibrium in deoxyHb is strictly related to the $R_0 \rightleftharpoons T_0$ conformational transition, quantitative studies of assembly contribute an answer to the question of the greater stability of the unliganded T state (T_0) as compared with the R state (R_0). On the basis of the temperature, pH and salt dependence (Ip and Ackers, 1977; Chu and Ackers, 1981) of the tetramer–dimer dissociation constant of deoxyHb, the role of hydrogen bonding in stabilising T_0 has been emphasised (Ackers, 1980), based on the enthalpic and entropic contributions to the free energy of the assembly being more negative in deoxy- than in fully oxyHb. This conclusion would tend to reduce significantly the role of contributions from either hydrophobic interactions (Chothia *et al.*, 1976) or salt bridges (Perutz, 1978) to the stabilisation energy of tetrameric deoxyHb.

The latter hypothesis, originally formulated by Perutz (1970), has also been questioned on the basis of potentiometric titrations (Flanagan *et al.*, 1981), which, on the other hand, suggest a significant role of salt bridges in the stabilisation of the tetramer form in oxyHb.

An interesting contribution to the explanation of the different tetramer-dimer dissociation constants in the two conformations comes from recent simulations of dimers associating into tetramers in which the stability of tetramers is estimated from nonbonded interactions and from the surface area buried in $\alpha_1\beta_2$ contacts (Janin and Wodak, 1984). A thorough analysis of two-fold symmetrical tetramers reveals that when monomers (and dimers) have *R* tertiary structure, only one tetrameric form with an *R*-like quaternary conformation is stable. On the other hand, dimers with the *T* tertiary structure can assemble into several, different, equally stable, quaternary structures. An important point is represented by the close packing of the $\alpha_1\beta_2$ interface, which is maintained in these different *T* structures especially at the contact region between the C helix of α-chains and the FG corner of β-chains. It allows one to suggest the possibility of the existence of a manifold *T* quaternary structure, in which the assembly to tetramers is entropically favoured with respect to the *R* conformation.

It should be realised, however, that the role of the salt bridges as the sole source of free energy difference between the two quaternary states has already been questioned on the basis of results obtained with trout Hb I (Brunori, 1975). This haemoglobin, which lacks most of the specific groups involved in these links and does not show any appreciable dimerisation in either of two conformations, none the less keeps a normal cooperative behaviour (Giardina *et al.*, 1973). Finally, even the essential and unique role of different subunits (α and β) in the expression of a cooperative haemoglobin has to be revised, since homotropic interactions in a dimer formed by two identical subunits (homodimer) have been unequivocally established (Ohnoki *et al.*, 1973; Chiancone *et al.*, 1981*a*).

6.4.5 General overview of the oxygen binding mechanism

Oxygen binding to human haemoglobin can be described satisfactorily with the help of the experimental results reported above. In the absence of the ligand, T_0 is preferred to R_0 because of additional H-bonds, salt bridges and/or non-polar interactions which can be made at the $\alpha_1\beta_2$ interface in the *T* conformation, with a contribution also from interactions between Tyr HC2 and Val FG5 of all subunits (quaternary constraints). Moreover, the multiple symmetrical quaternary conformations which can be made by the assembly of deoxy dimers into tetramers represent an additional favourable entropic factor.

As a ligand binds in the T_0 structure, steric hindrance at the proximal side of the haem brings about (when the iron tends to approach the mean haem plane) a strain on the F helix, transmitted through the Fe–N_ϵ bond. On the distal side, at least in the β-subunits, an additional steric disturbance may be represented by His E7 (63) and Val E11 (67). All these effects lead to a relatively weak Fe-O-O bond and therefore to the low affinity of the *T* state.

However, ligand binding seems also to affect the salt bridges present in the *T* conformation as well as other electrostatic interactions, not necessarily only those involving the carboxyterminals, some of which are weakened. It leads to a reduced tightness of the tertiary structure of the liganded subunit and affects also the $\alpha_1\beta_2$ contacts, making less unfavourable the quaternary switch. It is worth noting that the reduced strength of these bridges in the very early binding steps can be responsible for the increased exposure to the solvent of Cys F9β, as observed after ligand binding even in the *T* conformation (Antonini and Brunori, 1969).

The additional binding of ligand molecules to the other subunits in the tetramer further reduces the energy gap between the two conformations, the quaternary enhancement becoming additionally more evident as more subunits become liganded, and the quaternary constraints are progressively released. As soon as the entropic contribution from the enhancement overcomes the enthalpic contribution from constraints, the quaternary switch occurs. Such a conformational change then brings about a release of the steric hindrance on the proximal side of the haem and a consequent increased electron donation from the proximal histidine to the iron atom. Together with the dipole interaction from the G helix in the *R* conformation, it raises the electron density on the π^* orbitals of the porphyrin and then on the iron, increasing its tendency to make the Fe-O-O bond and thus raising the oxygen affinity.

Besides the rearrangement at the $\alpha_1\beta_2$ interface, the conformational change leads to the breaking of the H-bond between Tyr HC2 and Val FG5, also because Cys F9β tends to replace the tyrosine in the pocket between F and H helices. The role of these residues in the quaternary switch of human haemoglobin seems substantiated by the loss of cooperativity after enzymatic removal of the penultimate tyrosine (Bonaventura, J. *et al*., 1974; Kilmartin *et al*., 1975). It is also indirectly confirmed by the observation that in fish haemoglobins that maintain, under suitable conditions, the *T* conformation also in the liganded form, Cys F9β is consistently substituted by a serine that is less bulky and hydrophobic. This substitution brings about a reduced competition with Tyr HC2β for the site between helices F and H, and is accompanied by a more favourable energy for the *T* structure. The stabilisation of the *T* structure is also favoured by the possibility of Ser F9β donating an H-bond to the free oxygen atom of His HC3β and accepting one from the peptide NH of the same residue (Perutz and Brunori, 1982). This different mechanism for the modulation of the quaternary structural equilibrium in fish haemoglobins parallels the previously reported observations on the tetramer–dimer dissociation differences between human haemoglobin and other species (see above).

6.5 HAEMOGLOBIN: HETEROTROPIC INTERACTIONS

According to the MWC model the effect of non-haem ligands on the binding of oxygen to human haemoglobin (the heterotropic effect) has to be correlated to the different affinity of the heterotropic ligand towards a haemoglobin in the two different quaternary conformations (see also above).

6.5.1 Bohr effect

A traditional example of a heterotropic ligand is the proton, which is known to affect (by the Bohr effect) the overall oxygen affinity of human haemoglobin. Early observations showed a linear proton release upon carbon monoxide binding (Antonini *et al.*, 1963); this result matched the reported invariance of the shape of oxygen equilibrium curves (Antonini *et al.*, 1962), and also allowed the equivalence of the Bohr and Haldane coefficients to remain (Wyman, 1964):

$$\underbrace{\left(\frac{\partial\, H^+}{\partial\, Y_{O_2}}\right)_{pH}}_{\text{Haldane coefficient}} = - \underbrace{\left(\frac{\partial \log pO_2}{\partial\, pH}\right)_{\bar{Y}_{O_2}}}_{\text{Bohr coefficient}}$$

This shape invariance would have implied that the proton affinity ratio between the two quaternary structures of haemoglobin was unaffected by ligand binding. However, precise oxygen equilibrium isotherms (extending the observations to the very low and very high degrees of saturation) showed a nonuniform contribution to the Bohr effect of different binding steps, indicating that the binding constant for the fourth oxygen molecules is much less pH dependent than the others (Roughton and Lyster, 1965; Imai and Yonetani, 1975). Such a nonuniformity was shown to have almost undetectable effects on the proton release linearity (Imai and Yonetani, 1975), possibly also because the different statistical distribution of partially saturated populations during ligand binding strongly reduces the effects of their unequal contribution (Tyuma and Ueda, 1975).

The alkaline Bohr effect (i.e the decrease of oxygen affinity as the pH drops from 9.0 to 7.0) has been related in structural terms to the role of the carboxyterminal region of the molecule (Perutz, 1970).

A predominant role for β-chain carboxyterminal residue His HC3β was strongly supported by NMR measurements of the pK change displayed by this residue in going from CO-Hb (pK = 7.1) to deoxyHb (pK = 8.0) (Kilmartin *et al.*, 1973*a*), and by the dramatic reduction of the Bohr effect when this group is either replaced (Perutz *et al.*, 1971*b*) or enzymatically cleaved (Bonaventura, J. *et al.*, 1974; Kilmartin *et al.*, 1975). From accurate equilibrium measurements (Imai and Yonetani, 1975) it was suggested that the Bohr effect is a mixture of tertiary (i.e. affecting the intrinsic ligand affinity of a single conformation) and quaternary (i.e. influencing the equilibrium between the two conformations) contributions. Such a twofold effect demonstrates the strict interrelationship between these two events and also accounts for the Bohr effect observed in the *R* state (Kwiatkowski and Noble, 1982) attributable to His HC3β. Finally, a tertiary contribution to the Bohr effect is unequivocally demonstrated by a pH dependence of oxygen binding to dimers (Rollema *et al.*, 1980; Chu and Ackers, 1981; Kurtz *et al.*, 1981), even though the effect is greatly reduced.

On the other hand, a definite correlation between the Bohr effect and the quaternary conformational change comes from crystallographic work on the carbon monoxy derivative of Hb Kansas ($\alpha_2\beta_2{}^{\mathrm{Asn\ G4-Thr}}$), which, at pH below 7 and in the presence of IHP, is in the quaternary *T* state also when fully liganded (T_4). Thus, it has been shown that in the T_4 state, salt bridges appear to be intact (Anderson, 1975) and subsequently proton NMR measurements of the nitrosyl derivative displayed no p*K* change of His HC3β between T_0 and T_4 (Kilmartin *et al.*, 1978).

This structural interpretation of the Bohr effect, which tends to focus attention on a few critical residues, has been challenged by the lack of a p*K* difference for His HC3β between deoxy- and CO-Hb when experiments are carried out at low ionic strength (0.01 M Cl^-) (Russu *et al.*, 1980). This result suggests that the previously observed change (Kilmartin *et al.*, 1973*a*) has to be attributed to a p*K* difference mediated by Cl^-, and that the role of salt bridges involving His HC3β may have been overestimated. This criticism of the original Perutz model seems to be supported also by potentiometric titration curves of human oxy- and deoxyHb (Matthew *et al.*, 1979) analysed, according to the modified Tanford-Kirkwood theory, using the X-ray coordinates of the respective crystal structures. According to these authors the concentration of Cl^- affects electrostatic interactions, and at high ionic strength (0.1 M Cl^-) 10 groups per tetramer would contribute to the observed Bohr effect, whereas at lower ionic strength the number of groups rises to about 28 (Matthew *et al.*, 1979). This interpretation, however, has overlooked the fact that the Bohr effect is reduced by 60 per cent at low ionic strength after the removal of His HC3β, reassessing an important role of this residue even under the experimental conditions tested by the NMR and potentiometric measurements (Kilmartin *et al.*, 1980). Further convincing support for Perutz's model comes from the absence of heterotropic interactions in the component I of the trout Hb system. Such a peculiar feature has been successfully correlated with the substitution of residues (Barra *et al.*, 1981), which had previously been proposed to play an important role in modulating the influence of non-haem ligands on the functional properties of human Hb (Perutz, 1970).

6.5.2 Anion binding

Such a debate illustrates that the main difficulty in assessing the role of different groups involved in the Bohr effect is related to the linkage in the binding of protons and other ions, mediated by electrostatic effects. The effect of Cl^-, and of anions generally, on the proton release in going from deoxy- to oxyHb has been extensively studied, in parallel with the changes in ligand affinity, by a number of authors (De Bruin *et al.*, 1974; Rollema *et al.*, 1975; Amiconi *et al.*, 1981). The effect of Cl^- has been ascribed to a modulation of affinity related to two or possibly three specific binding sites. As a matter of fact, the magnitude of the alkaline Bohr effect is significantly affected by the concentration of chloride ions, being maximal at 0.1 M and depressed as the concentration is either increased or decreased (Rossi-Fanelli *et al.*, 1961; De Bruin *et al.*, 1974), in apparent agreement with the different p*K* changes observed at varying concentrations of Cl^- (Kilmartin *et al.*, 1973; Russu *et al.*, 1980).

The influence of anions on the Bohr effect can be understood as an increase in the p*K* of the groups involved because of the negative charges of the anion. The reduction of the alkaline Bohr effect at ionic strength higher than 0.1 M may depend on a relative increase of proton affinity for the sites in the *R* state as compared to the *T* state, possibly because in this conformation they are already saturated.

Conversely, a higher proton affinity in the oxy- than in deoxyHb is responsible for the acid Bohr effect (i.e. for the increase of oxygen affinity observed at pH below 6.5). This effect has been demonstrated to be associated almost completely with a Cl^--induced p*K* increase more pronounced in the *R* than in the *T* conformation. This difference would then be dramatically reduced as the chloride concentration is decreased, making the acid Bohr effect vanishingly small, at least below 1 mM Cl^- (Amiconi *et al.*, 1981).

Specific anion binding sites have been tested by X-ray diffraction (O'Donnell *et al.*, 1979), ^{35}Cl NMR measurements (Norne *et al.*, 1976), and studies on abnormal haemoglobins. It has been shown that the presence of Val NA1 (1)α and Arg HC3 (141)α is important for the expression of the oxygen-linked effect of Cl^- (O'Donnell *et al.*, 1979; Poyart *et al.*, 1980), and Lys EF6 (82)β is likewise very important (Nigen and Manning, 1975; Bonaventura *et al.*, 1976). Nigen *et al.* (1980), in a study carried out on a hybrid made by carbamylated α-chains (see below) and β-chains from Hb Providence ($\alpha_2\beta_2{}^{\text{Lys EF6-Asp}}$), have demonstrated the important role of residues Val NA1α (specifically inactivated by the carbamylation) and Lys EF6β (replaced by a negative charge in this mutant Hb) in Cl^- binding. However, from these data a third oxygen-linked binding site has been identified, which may involve Arg Hc3α or Ser H13 (131)α, in accordance with results on human Hb specifically carbamylated at the α-amino group (O'Donnell *et al.*, 1979). These various binding sites seem to have different affinities, at least at pH 7.0 (Poyart *et al.*, 1980), and that contributed by Val NA1α and Arg HC3α displays a higher binding constant. Careful oxygen equilibrium studies also suggest that this site plays most of its oxygen-linked role in the binding to the *T*-quaternary conformation. It might imply that the second site, i.e. that located in the diad axis between the two β-subunits (which comes into play at higher anion concentrations), accounts to a large extent for the role of Cl^- as allosteric effector.

It is certainly consistent with this view that this anion low-affinity site, i.e. Lys EF6β, is also one of the positive charges that coat the polyphosphate's binding site. Therefore, competition between anion and polyphosphate binding at this residue, at least in the T state, is shown. As a matter of fact, it is only in the T state that the stereochemistry of the site (Arnone, 1972) allows a perfect fit of the bulky polyphosphate molecule, as demonstrated also by calorimetric measurements of IHP binding (Gill *et al.*, 1980), and this accounts for the strong effect of 2,3-DPG and IHP on the allosteric equilibrium constant. Such a twofold linkage between protons and anions, as well as that between polyphosphates and anions, explains the relationships between polyphosphate binding and the Bohr effect (De Bruin and Janssen, 1973; Kilmartin, 1974). This linkage is expressed in a manner very similar to the influence of anions on proton affinity, 2,3-DPG increasing both the alkaline and the acid Bohr effect. The alkaline enhancement has been attributed to His H21 (143)β, Val NA1 (1)β and His NA2 (2)β, whereas for the acid part the second protonation of the phosphate is thought to be responsible (De Bruin and Janssen, 1973; Bucci *et al.*, 1978).

6.5.3 Carbon dioxide

The second anion binding site, i.e. that in-between Val NA1α and Arg HC3α, is also involved in the binding of a different and physiologically more important, heterotropic effector, carbon dioxide. The predominant binding site, at least at physiological pH (7.2), has been demonstrated to be the α-amino group of all four chains (Kilmartin *et al.*, 1973*b*), which forms a carbamino compound according to the equation:

$$Hb{-}NH_2 + CO_2 \rightleftharpoons Hb{-}NHCOO^- + H^+$$

However, at more alkaline pH the ϵ-amino groups also show an appreciable affinity for carbon dioxide (Gros *et al.*, 1981), as shown very convincingly by ^{13}C NMR measurements (Gurd *et al.*, 1980).

Binding of carbon dioxide at the amino terminals interferes with the heterotropic effects displayed by Cl^- and polyphosphates because of the overlapping of the binding sites (Perrella *et al.*, 1975); such competitive effect has recently been demonstrated by experiments in which the influence of one effector on the oxygen binding isotherms was measured, leaving constant the activity of all other effectors (Imaizumi *et al.*, 1982).

Moreover, it has been shown that, in the absence of other heterotropic effectors, the four Adair constants are not affected uniformly by carbon dioxide, K_4 remaining unchanged and all others decreasing. Since carbon dioxide lowers the alkaline Bohr effect by an extent which is progressively reduced as the pH is lowered (and disappears below pH 6.5), bicarbonate is also thought to act as an anion, roughly similar to Cl^- (Imaizumi *et al.*, 1982).

Bicarbonate has indeed been observed to play an important role in the allosteric properties of crocodilian haemoglobins (Bauer *et al.*, 1981). Such an effect was suggested to be related to specific amino acid substitutions occurring in crocodile Hb, which weaken the binding of polyphosphates at the site in the central cavity of deoxyHb. However, binding of bicarbonate to the crevice in the *T* state is not impaired by these specific amino acid substitutions, making this anion a major allosteric effector of crocodile haemoglobin. This specific effect of bicarbonate in crocodile haemoglobin has been very beautifully correlated with the physiology of respiration in this reptile (Perutz *et al.*, 1981).

An effort to dissect the role of different residues has been made by Perutz *et al.* (1981), who also attempted an identification of the groups responsible for the Bohr effect in the presence and absence of some of these heterotropic effectors. Under all experimental conditions a significant role has been proposed for His H5 (122)α (Nishikura, 1978) and His HC3β. However, in the presence of 0.1 M Cl^-, a contribution to the acid Bohr effect from His H21β is suggested, possibly due to an abnormally low p*K* of this residue in the *T* conformation because of the close vicinity to other positive charges (such as Lys EF6β and Lys HC1 (144)β). Moreover, a significant part of the alkaline Bohr effect seems to be played by Lys EF6β, although its contribution, like that of Val NA1α, is definitely linked to anion binding. Under physiological conditions (i.e. in the presence of 2,3-DPG), the role of Lys EF6β in the Bohr effect becomes much less significant, except for the modulation in connection with the binding of polyphosphate.

6.5.4 Comparative aspects

Heterotropic interactions indeed represent additional and important degrees of freedom in the control of the functional behaviour of haemproteins and this explains why they represent the major means of connecting functional properties to evolutionary adaptation. As a matter of fact, most of the fishes and all of the mammals, birds and amphibians, although living in completely different environments, use haemoglobins as oxygen carriers, controlling the delivery of oxygen to the tissues largely by exploiting the heterotropic interactions.

In this perspective, a type case of molecular adaptation is represented by the haemoglobins from trout (Brunori, 1975), in which the two major components of the haemolysate seem to play different roles in the physiology of oxygen delivery by displaying dramatic differences in their heterotropic features. Thus, trout Hb I shows no heterotropic effects, whereas trout Hb IV is strongly affected by a decrease in pH which lowers its oxygen affinity to such an extent that at pH < 7.0 the protein is partially saturated with oxygen even at atmospheric pressure (Binotti *et al.*, 1971). This marked Bohr effect (called the Root effect (Root, 1931)), which is common to most of the fish haemoglobins, has been related, from the physiological standpoint, both to the presence and function of the swimbladder (Scholander and van Dam, 1954) and to oxygen secretion in the eye (Wittenberg and Wittenberg, 1962). Thus in fish haemoglobins the Root effect facilitates oxygen release against very high back pressures because of the dramatic oxygen affinity decrease as the pH is lowered (Scholander and van Dam, 1954; Brunori *et al.*, 1978).

On the other hand, trout Hb I, whose oxygen affinity is unaffected by pH and organic phosphates, acts as a pH-independent oxygen-supply system in case of heavy oxygen demand such as may occur in very active fishes.

The trout haemoglobin system represents only one example of a number of cases of molecular adaptation to specific physiological requirements, but in this context serves the purpose of illustrating how heterotropic effects are influential in modulating oxygen supply to tissues under different conditions (Perutz, 1984).

6.6 GIANT OXYGEN CARRIERS

6.6.1 Haemocyanins: structural aspects

In this section we will limit ourselves to an overview of the essential structural and functional features of haemocyanins, largely because a recent and excellent paper by van Holde and Miller (1982) and the review by Lontie and Witters (1981) deal with this subject in great detail. In haemocyanins the active site contains two copper ions capable of binding one molecule of oxygen. As indicated by resonance Raman spectroscopy, molecular oxygen appears to be bound in the form of peroxide (O_2^{-2}) with a nonplanar μ-dioxygen structure (Loehr *et al.*, 1974; Freedman *et al.*, 1976). Hence, in agreement with other evidence previously reported, upon oxygen binding the formal oxidation state of copper changes from Cu(I) to Cu(II), as indicated in the following scheme:

$$\mathrm{Cu(I)} \quad \mathrm{Cu(I)} + O_2 \rightleftharpoons \mathrm{Cu(II)} \quad O_2^{-2} \quad \mathrm{Cu(II)}$$

This valence-state change was proposed many years ago on the basis of the fact that the oxy form of the pigment is blue (λ_{max} = 570 nm) and the deoxy form colourless. Oxygenated haemocyanin displays two main absorption bands, at about 340 and 570 n. The use of labelled oxygen (O_2^{18-16}) has indicated that both oxygen atoms are bound to copper in a rather symmetrical fashion (Thamann *et al.*, 1977). In addition there is evidence that each copper ion at the active site is coordinated to histidine imidazole groups. Resonance Raman studies, as well as other spectroscopic data, have led to an assignment of the two absorption bands of oxyhaemocyanin, the transition at ~ 340 nm being due to charge transfer from the imidazole nitrogen of the histidine residues to Cu(II) [Im → Cu(II)], and that at 570 nm to O_2^{-2} → Cu(II) charge transfer.

Apart from being bridged by dioxygen, the two copper ions are magnetically coupled by an internal ligand, as indicated by experimental evidence from several sources, and particularly by the possibility of obtaining a met-form of haemocyanin which is EPR silent.

A potential candidate for this endogenous bridging ligand would be a phenolate ion on the basis of the absorption spectra of oxy- and met-haemocyanins, which display a shoulder around 400 nm that may reasonably be attributed to a copper–tyrosine energy transfer (Eickman *et al.*, 1979). Very recent evidence from EXAFS studies (Brown *et al.*, 1980) confirms that the oxidation state of the copper is Cu(I) in deoxy- and Cu(II) in oxyhaemocyanin. Moreover, whereas the Cu-Cu distance in oxyhaemocyanin is different from that observed for deoxyhaemocyanin (3.67 vs 3.38 Å), the average distance between the copper atoms and their immediate neighbours is the same (1.95 Å) in both ligation states. However, the coordination number of copper is 4 or 5 for oxy and 2 or 3 for deoxy. This last result suggests that the two bridging ligands of the copper site, i.e. dioxygen and the endogenous ligand, are lost on deoxygenation.

A proposed structural model for the active site of haemocyanin is shown in figure 6.9, where the oxygen-linked transition of the copper from a square-pyramidal coordination group to a trigonal planar arrangement is indicated.

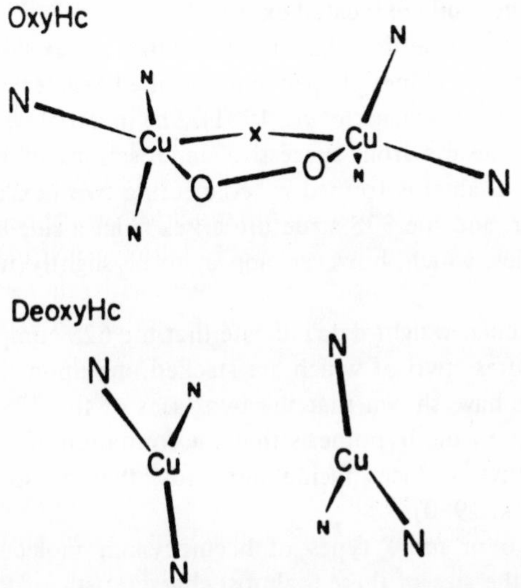

OxyHc

DeoxyHc

Figure 6.9 A model for the binding site of haemocyanin. In oxyhaemocyanin both copper ions Cu(II) have a square-pyramidal coordination group whose basal plane is defined by the bridging dioxygen molecule, the internal oxygen atom (X), and two of the nitrogen atoms of imidazole. In the deoxy state, both Cu(I) have a trigonal planar arrangement, being bound only to the three imidazole residues. (Reprinted with permission from Brown *et al.*, 1982. Copyright 1982 American Chemical Society.)

It should be noted that the binding sites of arthropod and mollusc haemocyanins, although similar, are not identical since they display strong differences in the kinetics of copper removal (by cyanide) and significant quantitative differences in the absorption spectra (Nickerson and van Holde, 1971). However, the main features appear to be the same and those distinctions that are observed may be interpreted on the basis of slight geometric alterations rather than of major structural changes in the sites.

6.6.1.1 Arthropod haemocyanins

Arthropod haemocyanins are formed of multiples of 16S structures (i.e ~ 25S, ~ 37S, ~ 62S) with squared, rectangular or hexagonal views in the electron microscope and a maximum dimension of 100–120 Å (van Holde and van Bruggen, 1971). The basic unit, the 16S component (mol. wt ~ 450 000 Da), is made up of six polypeptide chains whose molecular weights are in the range 70 000–90 000 Da. Subunit diversity, which has been clearly revealed by immunoelectrophoretic studies, seems to depend on the species. Thus five to eight subunits have been detected for cheliceratan haemocyanins, two to seven for crustacean haemocyanins, seven for *Tachypleus tridentatus* and eight for both *Limulus polyphemus* and *Androctonus australis* (Hoylaert *et al.*, 1979; Lamy *et al.*, 1979*a, b*). In the latter case it was demonstrated that half of the eight types of subunit, that constitute the 37S native molecule (tetramer of 16S units) are present in two copies and occupy internal positions, and the remaining four are present in four copies and are located externally.

X-ray studies (5 Å resolution) have indicated that the six subunits constituting the 16S component are kidney-shaped and that the hexamer is in a trigonal anti-prism arrangement (van Schaick *et al.*, 1981). The material with higher sedimentation coefficients results from successive dimerisations of the hexamers: the 25S dimer (i.e. 12 chains) is formed by connecting two hexamers with trigonal axes perpendicular, and the 37S structure arises from a side-by-side association of two 25S particles, which, however, appear to be slightly tipped with respect to one another.

Moreover, molecular weight data indicate that the 62S components are dimers of the 37S structures, two of which are stacked one upon the other. Electron microscope studies have shown that the two faces of the 37S molecule are not identical; this leads to the hypothesis that the formation of the 62S component leaves less interactive surfaces facing outwards, thereby stopping aggregation (van Heel and Frank, 1980).

Frequently, two or more types of haemocyanin molecule coexist in the haemolymph and the size of these is almost characteristic of the specific class of animal. Thus, crustaceans and spiders usually have 16S and 25S components, shrimps 16S and 37S, scorpions 37S and 60S. It therefore seems that different haemocyanins may yield a wide variety of aggregation states, although the range of accessible states is clearly dependent on the specific haemocyanin. In most cases, arthropod haemocyanins dissociate into individual polypeptide chains by simply raising the pH to 9 (or above) with concomitant removal of divalent cations (Ca^{2+} and Mg^{2+}). Dissociation may occur even in the presence of Ca^{2+} (or Mg^{2+}), but in this case a much higher pH is required, a fact which outlines the stabilising effect of divalent ions.

In most cases dissociation into subunits can be reversed by either decreasing the pH and/or restoring the concentration level of Ca^{2+} (or Mg^{2+}) (van Holde and Miller, 1982). However, reassociation very often leads to a heterogeneous mixture of hexamers, even if the native molecules represented an homogeneous set of molecular species. This finding is not surprising if one considers the high degree of subunit heterogeneity displayed by these molecules. Thus each of the six polypeptide chains that constitute the native hexamer probably occupies a well defined and specific location, a condition which is not always achieved upon reassociation. This is further demonstrated by the common observation that whereas native hexamers are fully capable of dimerisation to form dodecamers, the hexamers reconstituted from monomer chains fail totally or partially to dimerise. Thus, when monomers are brought back to reassociating conditions, it is often found that hexamers are quantitatively re-formed but that only partial (~ 15 per cent) reconstitution of the dodecamer is achieved. Of course the percentage of dodecamer that will be reconstituted will depend on the number of different chains involved and the specificity of their location for the higher order assembly. All these considerations strongly suggest the existence *in vivo* of a specific mechanism of control of quaternary assembly whose molecular basis is still obscure.

6.6.1.2 Molluscan haemocyanin

Ultracentrifuge studies show that gastropod haemocyanins have sedimentation coefficients of ~ 100S and those from cephalopods ~ 60S. The molecular weight is nearly 9×10^6 in the former case and about half this value in the latter. The native 100S molecules are stable only within a certain pH range ($5 \leqslant pH \leqslant 8$). Outside this range, the molecules dissociate into submultiples of one-half (60S), one-tenth (~ 19S) and one-twentieth (~ 11S) the original size. In this case as well the pH-stability region is generally larger when divalent cations are present. When examined under the electron microscope, the 60S particle is a hollow right cylinder of diameter 350 Å and height about 170 Å; the 100S component has the same diameter but is twice as high. The 60S cylinder of gastropod haemocyanins is composed of 10 subunits (~ 11S) arranged helically and is characterised by a marked asymmetry. Thus at one end of the cylinder it is evident as a structural feature, with tenfold symmetry, usually referred to as a 'collar' (van Bruggen, 1978). The asymmetry of the cylinder is believed to prevent further association beyond the 100S dimer. The whole molecule of gastropod haemocyanin is therefore made up of 20 subunits arranged in two sets of 10 to form two half-molecules which are joined, the collars pointing outwards (Mellema and Klug, 1972; Siezen and van Bruggen, 1974). This model is supported by electron microscopic and low-angle X-ray scattering studies.

Sedimentation equilibrium studies indicate that the one-twentieth molecule has a molecular weight between 400 000 and 450 000 Da. Since one oxygen binding site per ~ 50 000 Da has been determined, the 11S component should contain eight oxygen binding units. The basic units may be isolated only after proteolytic digestion and therefore have been considered as covalently bound to form a collar-type structure. In conclusion the 11S particle is a very large polypeptide chain that seems to consist of a string of (non-identical) domains which carry one oxygen binding site; each string folds up and combines with nine others to form the 60S particle. (Brouwer and Kuiper, 1973; Gielens *et al.*, 1973, 1975; Brouwer *et al.*, 1979). It is noteworthy that tryptic cleavage of the collar domains (which are believed to be those towards the C-terminal end) is followed by polymerisation into long tubes; this substantiates the function of the collar in preventing infinite polymerisation under physiological conditions (van Breemen *et al.*, 1975).

In conclusion, molluscan haemocyanin can be visualised as being made up of multi-domain subunits (11S), which can dimerise to yield 19S particles. These are in turn associated into pentamers in the 60S molecule. In gastropods the 60S structures can dimerise to yield the 100S structures and sometimes even higher polymers have been observed.

Particularly interesting from a comparative and evolutionary point of view is the demonstration of a multi-domain structure in one of the giant invertebrate haemoglobins (Terwilliger *et al.*, 1977). This finding outlines once more the strong similarities that exist in terms of structure–function relationship between these two classes of oxygen binding proteins (see section 6.8).

6.6.2 Haemocyanins: functional features

6.6.2.1 Equilibrium features

Under physiological conditions, which include the presence of Ca^{2+} and/or Mg^{2+} ions, haemocyanins generally display cooperative oxygen binding. Hill plots of the binding curves are characterised by a steep transition from a low (T) to a high (R) oxygen affinity state of the protein, with maximum values of n as high as 7 to 9. The ligand-linked transition occurs over a narrow range of ligand concentrations and it is characterised by relatively small values of the overall free energy of interaction per site ($\Delta F = 0.9$–2.5 kcal per site) (Colosimo *et al.*, 1974; Klarman and Daniel, 1980; Brunori *et al.*, 1982; van Holde *et al.*, 1982). As outlined in a previous section, this behaviour is characteristic of allosteric systems possessing a large number of oxygen binding sites, and it is indicative of the 'explosive' character of the quaternary conformational change.

Quaternary structural changes have been demonstrated by optical diffraction studies of the electron microscopic images of tubular polymers of β-haemocyanin from *Helix pomatia*. Thus it has been observed that upon deoxygenation the diameter of the tubes decreases with a simultaneous increase of their length (van Breemen *et al.*, 1979).

The shape and position of the oxygen binding curves of haemocyanin from both arthropods and molluscs may be generally affected by the presence of various ionic components and typically by protons. Many haemocyanins display a 'normal' Bohr effect, that is, a decrease in oxygen affinity as the pH is brought towards acid values. However, in some cases (and usually among gastropods), a 'reverse' Bohr effect has been observed at physiological pH values. An interesting example is that of *Helix pomatia*, which possesses two different types of haemocyanins, α and β, which exhibit respectively a positive (or reverse) and a negative (or normal) Bohr effect. Moreover, whereas increasing Ca^{2+} increases the affinity of α-haemocyanin, it decreases the affinity of the β-component (Zolla *et al.*, 1978).

An extreme example of linkage between proton and oxygen binding sites is represented by the dependence of the binding behaviour of *Panulirus interruptus* haemocyanin. This protein is characterised by the presence of the so-called Root effect, i.e. an extreme decrease in oxygen affinity upon lowering the pH, causing it progressively to lose oxygen. The presence of this particular type of Bohr effect has also been reported for the haemocyanin of *Callianassa californiensis, Octopus vulgaris* and *Buccinum undatum* (Brunori, 1971; Arisaka and van Holde, 1979; Brix *et al.*, 1979; Kuiper *et al.*, 1980*a*).

The effect of allosteric effectors like protons and divalent ions is mostly exerted by shifting the allosteric equilibrium constant L_0. However, in some cases a specific effect on the properties of either the T or the R state of the protein has also been observed. For example, high pH, besides stabilising the low oxygen-affinity T state of the β-haemocyanin from *Helix pomatia*, shifts the position of the T-state asymptote itself. Moreover, in the case of *Panulirus interruptus*, calcium ions shift the allosteric equilibrium constant L_0, simultaneously affecting both asymptotic values. Since the MWC model does not assume a variation of the binding properties of the extreme states (T and R) (see section 6.2), a modified version of the model has been applied to describe the binding data. In the case of *Callianassa californiensis* and *Penaeus setiferus*, in which the allosteric unit contains six sites, the data were described assuming three different states, R, T and a hybrid state, R_3T_3. This seems a reasonable assumption on the basis of the two-tiered trigonal anti-prism structure of arthropod haemocyanin (Miller and van Holde, 1974; Brouwer *et al.*, 1978; Arisaka and van Holde, 1979). It should be noted once more that cooperative oxygen binding in the larger multi-subunit haemocyanins is generally confined to subgroups of interacting sites which have been named 'functional constellations'. No unequivocal correlation has been established between the structural domains of the molecule and the functional constellations which appear from the analysis of the data (see section 6.8).

The remarkable similarity of the overall behaviour of arthropod and mollusc haemocyanins breaks down if one compares the functional properties of the subunits obtained by dissociation and/or digestion. The more striking difference is that subunits from arthropods generally display an oxygen affinity significantly lower than that of the whole molecule (and similar to their T state), whereas in the case of molluscs just the opposite is observed. In the case of arthropod haemocyanins the quaternary enhancement proposed for human haemoglobin A (Mills and Ackers, 1979) is completely applicable, as documented for *Panulirus interruptus*.

It should be noted that except for the case of haemocyanin from *Panulirus interruptus*, which has recently been characterised in great detail (Antonini *et al.*, 1983), information about the thermodynamic parameters of oxygen binding is still scarce. However, it has been clearly established that the driving force for the quaternary structural change may be different for different proteins. For example, in the case of β-haemocyanin from *Helix pomatia*, the allosteric transition seems to be entropy-driven since the heats of oxygenation of the *T* and *R* states are very similar to each other (Zolla *et al.*, 1981). On the contrary, the *T* and *R* states of haemocyanin from *Levantina hierosolima* exhibit enthalpy values for oxygenation of different sign ($\Delta H_T \sim +3$ kcal mol^{-1} and $\Delta H_R \sim -7.5$ kcal mol^{-1}), a fact which outlines the enthalpic nature of the quaternary structural change (Er-El *et al.*, 1972).

In the case of *Panulirus interruptus* haemocyanin, analysis of the temperature dependence of ligand-binding equilibria and kinetics within the context of the two-state model (Antonini *et al.*, 1983) has provided an estimate of the thermodynamic parameters for oxygen binding by the two states (*R* and *T*) and for the ligand-linked conformational change. The values for ΔG_T, ΔH_T, ΔG_R and ΔH_R (expressed in kilocalories per mole of binding site) are −6.3, −11.0, −7.6 and −6.0, respectively. Those for $\Delta G(T_0 - R_0)$ and $\Delta H(T_0 - R_0)$ are respectively +3.7 and −18 (expressed in kilocalories per mole of hexamer). The activation parameters of the oxygenation and deoxygenation reactions of both states (table 6.1) correspond satisfactorily to the thermodynamic parameters obtained by equilibrium analysis.

Table 6.1 Activation parameters for the oxygenation and deoxygenation of *Panulirus interruptus* haemocyanin in the *T* and *R* states.

	Oxygenation			Deoxygenation		
	$\Delta G\ddagger$	$\Delta H\ddagger$	$\Delta S\ddagger$	$\Delta G\ddagger$	$\Delta H\ddagger$	$\Delta S\ddagger$
T state	7.1	7.4	0.84	13.2	18.1	16.6
R state	7.1	3.0	−14.0	14.8	14.0	−3.1

Notes: Temperature 20°C. $\Delta G\ddagger$ and $\Delta H\ddagger$ are expressed in kcal per mole of binding site. $\Delta S\ddagger$ is given in cal deg^{-1} mol^{-1}

The most significant result that emerges from this analysis is the strong exothermic character of the allosteric transition which displays a ΔH_{L_0} of − 18 kcal per mole of hexamer. This is at variance to what is usually observed in vertebrate haemoglobins and namely in human haemoglobin A, whose allosteric transition is characterised by a positive enthalpy change ($\Delta H_{L_0} \sim +17$ kcal per mole of tetramer; Imai, 1979). As a consequence the entropy change associated with the allosteric transition is favourable in the case of haemoglobins and unfavourable in the case of *Panulirus interruptus* haemocyanin. This finding seems fully consistent with the fact that (a) the isolated subunits of *Panulirus interruptus* haemocyanin are characterised by low oxygen affinity contrary to the isolated α- and β-chains of HbA, and (b) the deoxygenated derivative of *Panulirus interrup-*

tus dissociates into subunits more easily than the oxygenated one, a behaviour that is also opposite to that of human haemoglobin.

Finally, it must be recalled that, contrary to haemoglobins, binding of carbon monoxide to haemocyanin shows no significant cooperativity ($n \leqslant 1$) (Bonaventura, J. *et al.*, 1974; Kuiper *et al.*, 1976; Brunori *et al.*, 1981*b*). The marked difference between oxygen and carbon monoxide may find a structural basis in the different modes of binding at the active site. Thus, infrared studies performed on both arthropod and mollusc haemocyanin using $^{13}C^{18}O$ have shown that carbon monoxide is bound to only one of the copper atoms, in all probability through the oxygen, and without interactions with the other copper atom in a site (Fager and Alben, 1972; van der Deen and Hoving, 1979). The cooperative effects have therefore been correlated with the distortion of the Cu–Cu distance which may be induced only by a bridging ligand such as oxygen (Brunori *et al.*, 1982*b*). Moreover, spectroscopic evidence (Bonaventura, J. *et al.*, 1974; Kuiper *et al.*, 1980*b*) suggests that the copper remains in the Cu(I) state on binding carbon monoxide.

6.6.2.2 Kinetics of ligand binding

Analysis of the kinetics of oxygen binding has provided estimates of the rate constants for the allosteric states and in some cases of the rates of the allosteric transition (van Driel *et al.*, 1974, 1978; Kuiper *et al.*, 1977; Wood *et al.*, 1977). Kinetic data obtained for a number of haemocyanins under conditions where either the *R* or the *T* state prevails indicate that the 'on' rate constants are very similar for both states, whereas the 'off' rate constants differ by a factor of 25-140 depending upon the species (table 6.2). Therefore cooperativity in the binding of oxygen seems to be kinetically controlled by the dissociation rates. The kinetic analysis of the binding reaction for *Helix pomatia* (αHc) in the low-affinity *T* state has indicated the presence of a slow monomolecular component, in addition to the expected bimolecular process. The corresponding rates (20-150 s^{-1}) may be related to the rate of allosteric transition ($T \rightarrow R$). Similarly, the rate of uptake of Bohr protons associated with the binding of oxygen to the β-haemocyanin of *Helix pomatia* was shown to be too slow compared with oxygen binding, and therefore has been interpreted along the same lines (Kuiper *et al.*, 1978).

Such slow kinetic components have not been detected in the case of the haemocyanins from arthropods, which are much smaller than those from gastropods. This finding has suggested the possibility of obtaining a better insight into the fundamental mechanisms responsible for the transfer of allosteric information by studying molecules characterised by functional constellations of different sizes.

6.6.3 Giant haemoglobins: erythrocruorins

Giant extracellular haemoglobins, often referred to as erythrocruorins, are found among annelids, molluscs and arthropods. Annelid erythrocruorins show molecular weights of the order of millions (2-4 × 10^6), and therefore structural studies have taken advantage of information obtained by electron microscopy (Terwilliger, 1974; Garlick and Terwilliger, 1977; Wood, 1980). In the micrographs erythrocruorins generally appear as dodecamers with 12 submultiples arranged at the vertices of two hexagons one on top of the other (Roche *et al.*, 1960). At least in one case, the central cavity, generally observed in the electron microscope, appears to be filled by an extra subunit. The hexagon is about 270 Å across and 180 Å thick, and the submultiples have an apparent molecular weight of 250 000-320 000 Da. The minimum molecular weight, based on the haem content, shows some variability, ranging from 20 000 to 29 000 Da for erythrocruorins, and from 25 000 to 35 000 Da for chlorocruorins (Antonini and Chiancone, 1977; Vinogradov *et al.*, 1977). These differences obviously imply variations in the total number of haems carried by the whole molecule; in this respect it should be mentioned that there are indications that the polypeptide chains are not always associated with a haem group. In spite of this structural complexity, sequence data are now emerging (Garlick and Riggs, 1982) and in the future will certainly allow a structural comparison between vertebrate and invertebrate haemoglobins to be made. Various models for the organisation of the subunits in the molecule have been proposed on the basis of electron micrographs and physico-chemical studies (Antonini and Chiancone, 1977; Vinogradov *et al.*, 1977). Although not definitive, they may be thought to represent a good starting point for the interpretation of the crystallographic data now emerging. There may well be differences between different erythrocruorins, but the overall similarity of the aggregates from a wide variety of species points to a basically similar mode of assembly.

An interesting case is represented by the extracellular erythrocruorin from the planorbid snail *Helisoma trivolvis*, whose multi-domain arrangement seems to display structural aspects very similar to those of molluscan haemocyanins (Terwilliger *et al.*, 1977). This molecule has a molecular weight of 1.75 × 10^6 Da and a quaternary structure represented by a decameric ring with tenfold symmetry. The constituent subunits are very large, with a molecular weight of ~ 175 000 Da, which is about ten times that of most haemoglobins or myoglobins (15 000-18 000 Da). Limited proteolysis with subtilisin cleaves the protein into haem-containing fragments whose molecular weight is in the range 15 000-17 000 Da and integral multiples of this value. Isolated fragments consisting of one haem group bind oxygen reversibly with high oxygen affinity. Furthermore, heterotropic and homotropic interactions, present in the intact protein, are lost at the level of the subunits, indicating that the intact molecule is necessary to express completely its functional behaviour. Although more detailed studies are needed to elucidate the structural characteristics of the various fragments, it is very significant that the minimum functional units have molecular weights and amino acid compositions which are comparable to those of the myoglobin of the radular muscle of the same animal.

As in the case of haemocyanins, all possible situations occur with respect to oxygen affinity, cooperativity and Bohr effect of erythrocruorins (Ching Ming Chung and Ellerton, 1979). Thus in some cases the oxygen binding curves are characterised by strong haem–haem interactions (*Arenicola, Lumbricus*), (Waxman, 1971; Giardina *et al.*, 1975), and in others they exhibit very little, if any, homotropic interactions (*Cirraformia, Neanthes*) (Swaney and Klotz, 1971; Economides and Wells, 1975).

The actual interpretation of the functional data follows the lines previously outlined in relation to haemocyanins. Thus we may consider (a) the ligand binding sites of the molecule as if they were segregated into functional constellations, which display little interactions among them, and (b) the overall apparent cooperativity as a result of the very strong interactions within each constellation of sites and the opposing effect of the heterogeneity of constellations (if present).

As far as the heterotropic interactions are concerned, the allosteric effectors of vertebrate haemoglobins (such as 2,3-DPG, IHP and ATP) were found not to have significant effects on oxygen binding properties of erythrocruorin. However, although insensitive to the organic phosphates, the functional behaviour of erythrocruorin is regulated by inorganic salts (Weber, 1981). This effect, which has been studied in detail in the case of erythrocruorin from *Arenicola marina* (Chiancone *et al.*, 1981*b*), may be of great physiological importance in so far as annelids lack a significant capacity for osmotic regulation and therefore experience large fluctuations in blood electrolyte levels. It has been shown that inorganic cations modulate the oxygen affinity of erythrocruorin (*Arenicola marina*) by preferentially modifying the association constant of the high-affinity state of the molecule (K_R). This finding may be interpreted as a mechanism of adaptation to hypoxic conditions that animals are forced to experience in the natural habitat.

6.7 HAEMERYTHRIN

6.7.1 Active site

The two iron atoms contained in each of the eight subunits of the octameric molecule provide the site for the reversible binding of molecular oxygen. In deoxy-haemerythrin the oxidation state of iron has been unequivocally established, by chemical means, as Fe(II), and chemical and spectroscopic studies (Okamura and Klotz, 1973) have indicated a close similarity between oxy- and methaemerythrin. This similarity has led to the suggestion that the oxidation state of the iron in oxyhaemerythrin should be Fe(III). Therefore we may describe the uptake of oxygen by the following scheme:

$$[\mathrm{Fe(II)\,Fe(II)}] + O_2 \rightleftharpoons [\mathrm{Fe(III)\,Fe(III)}]\,O_2^-$$

which indicates some analogy with haemocyanins. Resonance Raman spectra support this description, providing direct evidence of the peroxide-type electronic state of the bound oxygen (Loehr and Loehr, 1979).

The bound peroxide can be displaced irreversibly by small anions like N_3^-, CN^- and F^-; thus we have:

$$[Fe(III)\ Fe(III)]\ O_2^- \xrightarrow{X^-} [Fe(III)\ Fe(III)]\ X^- + O_2^-$$

that is, the irreversible conversion of oxyhaemerythrin into the physiologically inactive methaemerythrin.

The iron atoms are held at the active site by direct coordination to side chains from amino acid residues that have been identified on the basis of crystallographic studies on methydroxo- and metazidohaemerythrin (Stenkamp *et al.*, 1976, 1981; Hendrickson, 1978). The complex consists of two octahedrally coordinated iron atoms bridged by two carboxylate groups (from Asp 106 and Glu 58) and an oxygen atom. However, the two iron atoms are not equivalent, since one of them is coordinated to three histidine residues (73, 77, 101) and the other to two histidine (25, 54 in the sequence).

Mössbauer spectroscopic data have shown that the two iron atoms are in the +3 state in both met- and oxyhaemerythrin (Garbett *et al.*, 1969; York and Bearden, 1970). However, oxyhaemerythrin is unique in so far as the Mössbauer spectrum shows two doublets (Okamura *et al.*, 1969), which indicates that in this derivative the two metals are chemically unequivalent (although both in the Fe(III) state). Moreover, in both derivatives a strong antiferromagnetic coupling between the iron ions is supported by magnetic susceptibility measurements (Moss *et al.*, 1971; Dawson *et al.*, 1972), Mössbauer data (Okamura and Klotz, 1973), and the presence of an intense ligand field band in the visible absorption spectrum (Garbett *et al.*, 1969; Dawson *et al.*, 1972).

Comparison with model iron complexes indicates that the magnetic and spectroscopic properties of methaemerythrin can be explained by the presence of a bridging ligand between the two metal ions, which from the X-ray structure of the metazido form is confirmed to be an oxygen atom (Stenkamp *et al.*, 1981).

Based on what is known from model compounds, the most likely candidate for this ligand would be a μ-oxo group derived from the aqueous environment. That this could be the case is supported by the observation of an isotope-sensitive Fe–O vibration which shifts when the solvent is changed from H_2O^{16} to H_2O^{18} (Kurtz *et al.*, 1977). However, discrimination between a μ-oxo and a carboxylate bridge is made difficult by the fact that a carboxyl group may also exchange with solvent oxygen.

As far as the mode of binding of the exogenous ligand is concerned, resonance Raman experiments have indicated that azide (N_3^-) binds to the iron complex in methaemeryrthrin in an end-on configuration (Loehr and Loehr, 1979). This is based on the two distinct N═N stretching vibrations which are observable when $(N^{15}N^{14}N^{14})^-$ is used as exogenous ligand. In oxyhaemerythrin the same type of experiments using $O^{16}O^{18}$ yield two O—O stretching frequencies, indicating that the peroxide coordinates with the iron complex through a single oxygen atom (Kurtz *et al.*, 1977). Moreover, a structural study at 2.2 Å resolution (Stenkamp *et al.*, 1981) has clearly shown that azide ion binds to only one iron atom rather than bridging the two, as previously suggested (Stenkamp *et al.*, 1978).

The end-on binding of the exogenous ligand (both O_2 and N_3^-) is further supported by the electronic spectra of oriented crystals. In fact in oxyhaemerythrin the Fe(III)-O_2^- charge transfer band located at 525 nm by the resonance Raman excitation profile (Dunn *et al.*, 1977) is seen to be polarised perpendicularly to the Fe–Fe axis (Gay and Solomon, 1978). The same applies in metazidohaemerythrin for the absorption band at 503 nm, which has been attributed to an azide-to-iron charge transfer (Dunn *et al.*, 1977).

In conclusion, the active site is characterised by the presence of a dimeric iron centre, in which one iron atom is octahedrally coordinated to protein imidazole, carboxylate groups and an oxygen atom, and the other is penta-coordinated and can accept an exogenous ligand (such as azide) or possibly be unoccupied in methydroxohaemerythrin (Stenkamp *et al.*, 1981). As far as the bridging oxygen atom is concerned, it is impossible, at the present resolution (2.2 Å), to differentiate between a single oxygen atom, a hydroxide ion or a water molecule (Stenkamp *et al.*, 1981). Dioxygen attaches to one or both of the Fe(II) atoms of deoxyhaemerythrin through a single oxygen atom, being reduced to peroxide and situated perpendicular to the iron–iron axis.

As is indicated by the Mössbauer spectroscopic parameters (Garbett *et al.*, 1969; York and Bearden, 1970), upon deoxygenation the iron atoms become divalent high-spin Fe(II). Moreover, the large value of magnetic susceptibility at room temperature (Okamura *et al.*, 1969; York and Bearden, 1970) is indicative of a large decrease in antiferromagnetic coupling between the two iron atoms. This fact could be explained by the loss of a bridging group which acted as an exchange mediator, or it could be a consequence of the lower oxidation state of iron and the potentially smaller exchange interaction of divalent iron atoms. However, the mechanism underlying the mode of ligand binding, described above, does not find unequivocal support in the crystallographic data so far obtained. It would be helpful in solving this problem to test model compounds in terms of mechanisms both for mediating magnetic exchange and for the reversible binding of oxygen.

6.7.2 The protein

Amino acid sequences have been determined for several haemerythrins and a myohaemerythrin (Loehr *et al.*, 1978*b*). Comparison of the data shows that haemerythrin sequences have a much greater degree of homology than each of them does with myohaemerythrin. The degree of sequence identity between haemerythrins and myohaemerythrin appears to be just over 40 per cent with many differences presumably in surface residues, which are responsible for the tendency of the molecule to form octamers or trimers in one case and to remain monomeric in the other. In spite of these large differences, the tertiary structures by X-ray crystallography seem to be very similar and the situation is therefore reminiscent of that already found in vertebrate myoglobins and haemoglobins.

The greater degree of homology in haemerythrin sequences is found between residues 19 and 66, i.e. in one of the regions which have been supposed to play an important role in protein assembly. An interesting feature is represented by the presence at position 50 of a cysteine residue which is absent both in myohaemerythrin and in those haemerythrins which are known to possess a trimeric quaternary structure (as in the case of haemerythrin from *Phascolosoma lurco*).

Another point of great interest from an evolutionary point of view is represented by the fact that molecular variants may be found in erythrocytes of a single species (up to five in *Phasocolopsis gouldii*). As far as the tertiary structure is concerned, the major structural feature of both haemerythrin and myohaemerythrin chains is the possession of four approximately parallel sections of α-helix, which together account for roughly 70 per cent of the total sequence. The arrangement of these helices can be seen in figure 6.10. It seems likely that the five extra amino acid residues characterising the myohaemerythrin molecule may be accommodated by forming a small loop between those positions which in haemerythrin are designated by 90 and 91. In the model this would represent an insertion into the CD corner. A region of particular importance for the maintenance of the integrity of the tertiary structure appears to be the NH_2-terminal region, which shows the most extensive sequence homologies and is characterised by many aromatic residues which have their nonpolar portions buried inside the molecule (Loehr *et al.*, 1978*a*). X-ray studies have shown that the octameric molecule is composed of two layers, each being a square of four subunits arranged in an end-to-side fashion. This arrangement leads to a large central cavity approximately 20 Å in diameter. The two helices between residues 19 and 66 provide major points of subunit contact for octamer formation.

In myohaemerythrin these positions are sufficiently altered to eliminate most of the noncovalent interactions, explaining its occurrence as a monomeric species. In this respect it may be noted that the sulphydryl group of cysteine 50 is very near to the residue involved in the above-mentioned contacts and in fact its chemical modification by specific SH reagents gives rise to subunit dissociation.

6.7.3 Functional properties

The functional properties of oxygen binding to haemerythrin have not been extensively studied and investigations have been limited mostly to the octameric form of *Sipunculus nudus* Hmr (Bates *et al.*, 1968) and *Golfingia gouldii* Hmr (de Phillips, 1971; de Waal and Wilkins, 1976). In spite of the multimeric structure of the protein, the binding of oxygen to haemerythrin shows very weak, if any, homotropic interactions since $n \leqslant 1.4$ has always been reported (Manwell, 1960; Bates *et al.*, 1968; de Phillips, 1971).

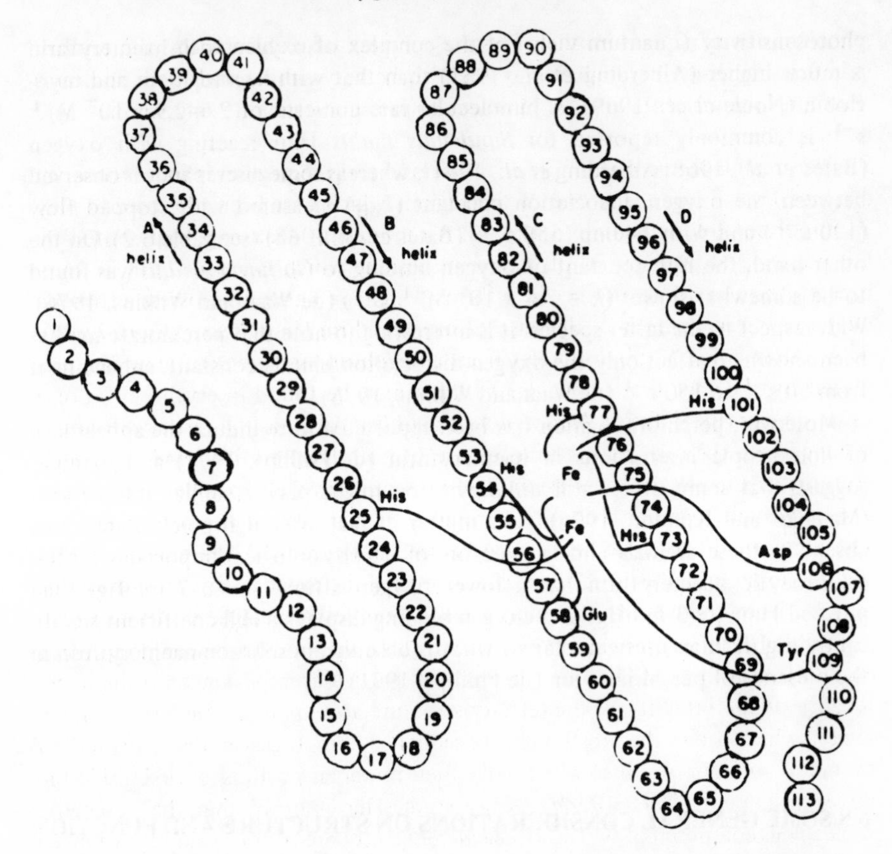

Figure 6.10 Schematic drawing of haemerythrin subunit, indicating helical regions and Fe ligands. (Reprinted with permission from Stenkamp *et al.*, 1978. Copyright 1978 American Chemical Society.)

Furthermore, no Bohr effect has been noticed, indicating a lack of relationship in this protein between oxygen affinity and eventual p*K* differences in the proton binding constants between the unliganded and liganded form.

With respect to the quaternary structure, it must be pointed out that the monomer–octamer association process has been studied only for the metHmr, since the octameric form is very stable in oxy- and possibly also deoxyHmr (Tan *et al.*, 1975). Calorimetric studies (Langerman and Sturtevant, 1971) have shown that the enthalpy of the monomer–octamer association of the ferric form of *Golfingia gouldii* haemerythrin is positive and pH-dependent, the octamer being stabilised by a group with a p*K* of 8.0. Such a residue has been tentatively identified as the side chain of cysteine in position 50, which is in the region involved in the intersubunit contacts (see section 6.7.2).

The kinetics of oxygen binding have been studied for both these haemerythrins, using stopped flow and T-jump as well as laser photolysis, since the photosensitivity (quantum yield) of the complex of oxygen with haemerythrin is much higher (Alberding *et al.*, 1981) than that with haemoglobin and myoglobin (Noble *et al.*, 1969). A bimolecular rate constant of $2.6\text{–}2.9 \times 10^7\ M^{-1}\ s^{-1}$ is commonly reported for *Sipulculus nudus* Hmr reacting with oxygen (Bates *et al.*, 1968; Alberding *et al.*, 1981), whereas some discrepancy is observed between the oxygen dissociation constant (k_{off}) measured with stopped flow ($120\ s^{-1}$) and with T-jump ($650\ s^{-1}$) (Bates *et al.*, 1968) (see table 6.2). On the other hand, the rate constant of oxygen binding to *Golfingia gouldii* was found to be somewhat slower ($k = 7.4 \times 10^6\ M^{-1}\ s^{-1}$) (de Waal and Wilkins, 1976). With respect to the latter species, it is interesting to note that perchlorate ion has been shown to affect only the oxygen dissociation kinetic constant, enhancing it from $50\ s^{-1}$ to $150\ s^{-1}$ (de Waal and Wilkins, 1976).

Moreover, perchlorate anion has been reported also to induce the appearance of homotropic interactions in haemerythrin (de Phillips, 1971) and it might suggest that some other, still unknown, ion might play a similar role *in vivo* (Mangum and Kondon, 1975). As a matter of fact, although perchlorate ion is absent in these animals and is therefore of no physiological importance, intraerythrocytic haemerythrin has a lower oxygen affinity (~ 6-7 mmHg) than purified Hmr (~ 3 mmHg) and oxygen binding displays a Hill coefficient significantly higher than unity similar to what is observed in solution haemerythrin in the presence of perchlorate ion (de Phillips, 1971).

6.8 SOME GENERAL CONSIDERATIONS ON STRUCTURE AND FUNCTION

It is customary to describe the properties of haemoglobins, haemocyanins, haemerythrins and erythrocruorins as if the only element in common between these various metalloproteins were the fact that they are all oxygen carriers. In fact very often they are the subjects of different books. In our view this is largely a result of the fact that the structural information available on these various metalloproteins is at a very different degree of resolution, being very detailed for haemoglobins and still rather limited for the other respiratory proteins. However, in spite of the very large chemical differences in the nature of the prosthetic group and of the protein moiety, similarities in functional properties and stereochemical effects can be foreseen.

In this section we shall try to point out some of these similarities in function and, whenever possible, in structure, in an attempt to provide a unified view of the basic features controlling the reactivity and the overall behaviour of all oxygen carriers. Indeed it may be proposed, as a result of these considerations, that this group of metalloproteins provides a good example of different chemical strategies which have been adopted during evolution to achieve a consistent (and possibly optimal) solution to the same vital problem, i.e. the supply of oxygen to tissues for the metabolic demands.

Table 6.2 Association (k') and dissociation (k) kinetic constants for the reaction of several oxygen-carrier proteins with oxygen at 20°C and pH 7.0

Protein	k'_T ($\mu M^{-1} s^{-1}$)	k_T (s^{-1})	k'_R ($\mu M^{-1} s^{-1}$)	k_R (s^{-1})	Method
Monomers					
Human Hb α-chain	–	–	50[a]	20[a]	Flash–flow
	–	–	48[b]	28[b]	*T*-jump
Human Hb β-chain	–	–	71[a]	17[a]	Flash–flow
	–	–	65[b]	16[b]	T-jump
Sperm-whale Mb	–	–	19[b]	11[b]	T-jump
Tetramers					
Human Hb α-chain	2.9[c]	18[c]	59[c]	12[c]	Laser photolysis
Human Hb β-chain	12[c]	2500[c]	59[c]	21[c]	
Giant O_2 carriers					
Mollusc haemocyanins:					
Helix pomatia β-Hcy	5.0[d]	700[d]	5.0[d]	5.0[d]	T-jump
Arthropod haemocyanins:					
Panulirus interruptus Hcy hexamer	–	–	31[e]	60[e]	T-jump
P. interruptus Hcy monomer	46[e]	1500[e]	–	–	
Erythrocruorins:					
Lumbricus terrestris Ery	–	–	30[f]	60[f]	T-jump
Haemerythrins					
Sipunculus nudus Hmr	–	–	26[g]	650[g]	T-jump
				120[g]	Stopped flow

[a]Noble *et al.*, 1969; [b]Brunori and Schuster, 1969; [c]Sawicki and Gibson, 1977; [d]Brunori *et al.*, 1981*a*; [e]Kuiper *et al.*, 1977 (*T*-state rates were obtained from the dissociated molecule at pH 9.6); [f]Chiancone *et al.*, 1973; [g]Bates *et al.*, 1968.

6.8.1 Stereochemical effects in ligand binding

Comparison of molecular events occurring on ligand binding in the various respiratory proteins is not easy because of the different level of structural information available. On the basis of the extensive knowledge obtained on haemoglobins, however, some correlations may be proposed.

The description of the ligand-linked structural changes occurring in haemoglobin have been reported earlier in this chapter, making use of the crystallographic data on deoxy- and CO-haemoglobins. This is because, in the case of oxygen, information on the orientation of the ligand and its interaction with the protein side chains has become available only recently (Shaanan, 1982). The model of Baldwin and Chothia summarises the relative movements of the atoms in the porphyrin–metal system and in the nearby residues on the protein (figure 6.7). The upshot of the results is that a structural perturbation of the active site (involving especially the interactions of the proximal histidine with the porphyrin) is necessary for an optimal fit of the ligand. The difference in free energy of binding among the two allosteric states (T and R) has been associated by Perutz (1972) with a tension developed when the ligand binds to the low-affinity state of haemoglobins. Although the mechanism of energy flux from one haem to the others is still a matter for investigation, in the case of haemoglobin, ligand binding to the iron and the ensuing structural perturbation extend to the boundaries between chains and alter the subunit contacts (see Bolton and Perutz, 1970; Ackers, 1980).

Haemocyanins also display local changes in structure upon binding of oxygen, as shown by spectroscopic observations (especially EXAFS, Raman and IR). A model for oxy- and deoxyhaemocyanin has been proposed largely on the basis of EXAFS carried out on native and modified haemocyanin. These studies indicate that oxygen is bound as a bridging ligand to both copper atoms in a site, and the structure of the metal–dioxygen–metal complex is that of a coplanar Cu_2(II) peroxo. Charge transfer from the metals is very well supported in oxyhaemocyanins. In deoxyhaemocyanins the copper–copper distance is different from that of the oxygenated complex (3.4 Å in the latter and 3.7 Å in the former, according to Brown *et al.* (1980) and at the same time the number of coordinated internal ligands decreases with breakage of the oxo-bridge provided by the protein. Thus extensive structural changes in going from deoxy- to oxyhaemocyanin are well documented, and they seem to be related to the necessity to optimise the copper-to-copper distance to allow a best fit of dioxygen.

At the level of the quaternary structure, extensive changes between oxy- and deoxyhaemocyanins are also documented, although the lack of crystallographic information at the necessary level of resolution makes the identification of the structural changes impossible. Ligand-linked subunit dissociation, proteolytic attack, optical probes and electron microscopy all indicate a quaternary conformational change in the case of cooperative haemocyanins.

Therefore the local perturbation required to fit dioxygen in-between the two metals in haemocyanin is well substantiated and indeed may represent an initial event in the onset of the quaternary changes that underlie the allosteric phenomena. On the basis of the knowledge that carbon monoxide in haemocyanin binds to only one of the two copper atoms in a site, we have proposed a correlation between the bridging nature of oxygen and the onset of structural changes which start at the active site and extend to the whole haemocyanin molecule. This proposal was substantiated by the experiments showing that carbon monoxide binding is not associated with a quaternary transition (Brunori *et al.*, 1982*b*). It is interesting in this respect that in haemerythrin, which is also binuclear carrier, dioxygen is bound in an end-on fashion at the active site, and homotropic interaction phenomena are not observed, in spite of its polymeric nature (Loehr and Loehr, 1979).

6.8.2 Functional properties: homotropic interactions

A superficial comparison of the oxygen binding properties of haemoglobins and haemocyanins shows differences in behaviour; however, analysis of the available equilibrium and kinetic data carried out within the framework of a two-state MWC allosteric model indicates their fundamental similarity.

The differences in the ligand binding curves between the two classes of co-operative oxygen carriers are quantitative rather than qualitative; they have been correlated to (a) the values of the free energy of interaction per site, and (b) the values of the maximum Hill coefficient. These parameters have been interpreted on the basis of a model implying that the number of interacting sites in a co-operative unit is not the same as the total number of sites per macromolecule.

A non-trivial problem in comparing the homotropic interactions of these different oxygen carriers and in the quantitative analysis of the binding curves is related to the definition of the 'functional constellation', i.e. the minimum number of binding sites which interact preferentially, within the polymer, contributing the largest part to the free energy of interaction. This set of strongly interacting sites has been referred to as a 'functional constellation' of (presumably) contiguous sites, within which most of the cooperative homotropic interactions are operative. The term 'function consellation', so defined, was introduced to describe some properties of erythrocruorin (Colosimo *et al.*, 1974) and has also been applied, quite extensively, to the case of molluscan haemocyanins (Colosimo *et al.*, 1977). The model was developed by Colosimo *et al.* (1974), who proposed that in the case of the giant respiratory proteins only a small fraction of the total number of sites on the macromolecule is part of a 'functional constellation'.

Analysis of the oxygen binding data excludes unequivocally the possibilities that homotropic interactions involve the whole molecule (e.g. 160 oxygen binding sites), given a simple relationship between the free energy of interaction per site and the maximum Hill coefficient (n) measured at equilibrium (see figure 6.4 for this relationship). As the next possible assumption, therefore, it has been proposed that the sites interact in independent constellations, each one containing a number of sites much smaller than the total and behaving according to a simple MWC model.

Application of this model to (for example) *Helix pomatia* haemocyanins provides the objective estimate that a 'functional constellation' is made up of 12 binding sites, and the whole molecule carries a total of 160 sites. The size of a functional constellation is often close to the dimension of a structural domain, which contains a fixed number of polypeptides identified by hydrodynamic measurements and electron microscopy. In the case of molluscan haemocyanins this number is 16 (i.e. one-tenth of the total number of binding sites in a macromolecule of molecular weight 9×10^6 Da). Although the correspondence between 'functional constellation' and 'structural domain' is by no means certain, it is valid to propose that the physical basis of the preferentially interacting units is represented by sets of nearby groups of polypeptides.

This model is approximate in two respects. (a) It takes no account of any possible (and likely) functional heterogeneity between sites; thus the sites contained in the collar of the Hcy molecule (i.e. those at the two extremes of the cylinder) may form a group with distinct functional properties, although not necessarily so in the polymer. (b) It takes no account of interactions at the boundaries between 'functional constellations'; these interactions may well be energetically small but they are certainly not zero, since the dissociated structural domains (i.e. 1/10 molecule) do not display cooperative oxygen binding (van Driel, 1973).

In spite of these approximations, this model provides a consistent description of the equilibrium data and allows a uniform treatment of results obtained on erythrocruorins, chlorocruorins and haemocyanins from different species, with different molecular weights and prosthetic groups.

In the case of tetrameric Hbs the structural basis of homotropic interactions was summarised earlier. However, we wish to point out that conceptually there are analogies with the model developed for the giant oxygen carriers, which in fact was proposed as an attempt to apply the two-state MWC model to very large functioning macromolecules (Colosimo *et al.*, 1974).

It is nowadays accepted that the free $\alpha_1\beta_1$ dimer does not display cooperative effects, and that the tetramer is necessary to account for the observed Hill coefficient of $n = 3$. However, it has been debated for a long time as to whether a dimer within the tetramer may still be the major source of cooperative interactions, the model going back to the so-called rectangular scheme proposed by Wyman (1948), and thereafter taken up by the Rome group. This may be depicted as an assembled tetramer with much stronger functional interactions within the $\alpha_1\beta_2$ dimer and much weaker between $\alpha_1\beta_1$ dimers, the latter extending over the boundaries between the two strongly interacting 'functional constellations' (which in this case would be the $\alpha_1\beta_2$ or $\alpha_2\beta_1$ dimers). In the last few years, a substantial amount of work has been invested in an attempt to establish the distribution of the free energy of interaction (12 kcal per tetramer) within the haemoglobin molecule. Spectroscopic and thermodynamic measurements have been correlated to crystallographic data, and some of the more relevant conclusions have been given earlier (section 6.4). Although it seems clear that the amount of energy 'stored' at the level of the haem is quite small (possibly only 0.3 kcal per site), and although the role of the subunit interfaces is unequivocally established, the partition of the bulk of the free energy of interaction among the possible subunit contacts is not quantitatively known. Ackers' approach, based on accurate dissociation data of oxy- and deoxyHb, cannot prove that in the assembled tetramer the free energy of interaction is stored exclusively at the $\alpha_1\beta_2$ (or $\alpha_2\beta_1$) boundaries, although it indicates an asymmetry in the distribution of energy among the two possible dimers (Ackers, 1980). However, recently, Ackers and co-workers (G. K. Ackers, personal communication) have observed that in an asymmetric hybrid tetramer from HbA and Hb Kempsey ($\alpha_1{}^H\alpha_2{}^K\beta_1{}^H\beta_2{}^K$) (i.e. where the interfaces $\alpha_1\beta_2$ and $\alpha_2\beta_1$ have in the deoxy form an *R*-like and a *T*-like structure, respectively), the energetic difference between the unliganded and liganded forms is approximately half that of normal human HbA. This would suggest that the cooperativity is mostly transmitted through these interfaces, and that the information for this structural change is transferred only to a negligible extent through other inter-subunit contacts, such as $\alpha_1\beta_1$, $\alpha_2\beta_2$ and $\alpha_1\alpha_2$.

In fact Perutz's analysis of the tridimensional structures of deoxy and liganded haemoglobin (Perutz and Ten Eyck, 1971) has shown that very relevant and diagnostic structural changes occur at the $\alpha_1\beta_2$ interface, where a few amino acid side chains seem to acquire critical positions in the two allosteric states of haemoglobin.

Thus a rectangular model, with strong interactions within the $\alpha_1\beta_2$ dimer (the 'functional constellation') and weaker interactions between dimers, is still an intriguing hypothesis and asymmetry of the binding curve is not inconsistent with such a view.

6.8.3 Allosteric states and heterotropic effects

The description of the cooperative properties of haemoglobins and haemocyanins with the MWC model has been extended to the effect of heterotropic ligands. In the case of human HbA, the effect of pH on the ligand binding curve has been investigated in detail (see section 6.5). Quite apart from the identification of the amino acid residues contributing to the Bohr effect, the shape of the oxygen binding curve (notably the maximum Hill coefficient) changes *very little* with pH over the range 6–9. Some of the conclusions reported earlier in this chapter demanded a careful correlation of very accurate oxygen binding, spectroscopic and structural data on normal and chemically modified haemoglobins.

In the case of some fish haemoglobins, which display a very large effect of pH on the shape of the ligand binding curve (the Root effect, section 6.5), the relative stability of the two allosteric states was shown to be dependent on heterotropic effectors (Noble *et al.*, 1970; Wyman *et al.*, 1978). We shall not discuss here the role of chain functional heterogeneity in the manifestation of the Root effect (see Brunori *et al*., 1978), but point out that a stereochemical interpretation of the stabilisation of the *T*-state at low pH has been proposed on the basis of considerations of the tridimensional model of HbA and sequence information on fish haemoglobins.

In a recent publication, Perutz and Brunori (1982) have indicated that the presence of Ser at position F9 (93)β leads to a stabilisation of the *T*-quaternary state at low pH values (say ~ 6.5) because of the formation of two additional weak interactions per β-chain when His HC3 (146)β is protonated (figure 6.11).

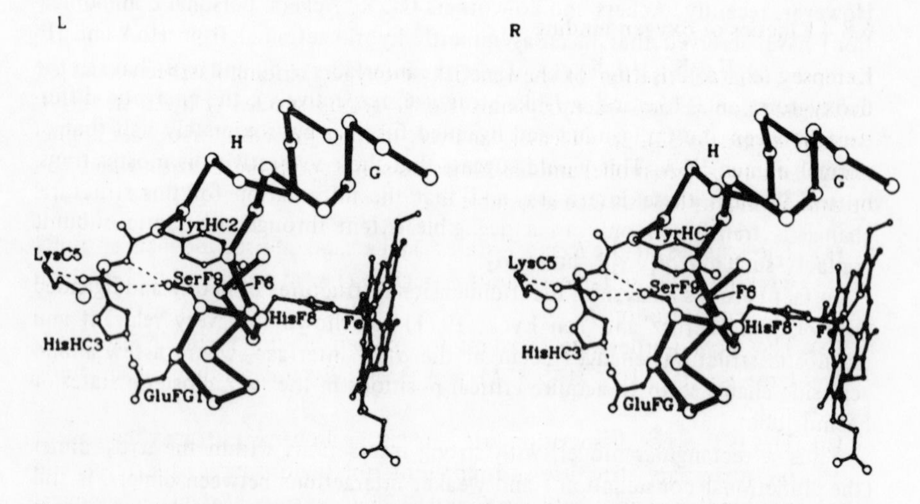

Figure 6.11 Stereodiagram showing the bonds at the C-terminus of the β-chain in the *T* structure of fish haemoglobin possessing a Root effect. The amino group of Lys C5α donates a hydrogen bond to one of the carboxylate oxygens of His HC3β, and the OH of Ser F9β donates a hydrogen bond to the other oxygen. The imidazole donates a hydrogen bond to the carboxylate of Glu FG1. The main chain NH of His HC3 donates a hydrogen bond to the lone pair electrons of the OH of Ser F9. Note that the haem iron is directly linked via His F8 and Ser F9 to the C-terminus.

The formation of these bonds (and possibly other unidentified contributions) with the protonated form of His HC3β leads to the pH-dependence of the allosteric equilibrium constant, with a predominance of the *T*-state at low pH in both the unliganded and liganded forms. At high pH values many fish haemoglobins often show a reverse effect, displaying high affinity for oxygen and again a Hill coefficient approaching unity (Tan *et al*., 1973); this behaviour has been interpreted in terms of a stabilisation of the *R* state at high pH.

A similar type of phenomenon is often observed with haemocyanins, where it has been shown that not only protons but anions and cations as well may dramatically affect the shape of the oxygen binding curve (Brouwer *et al*., 1978; Zolla *et al*., 1978; Kuiper *et al*., 1979).

This general behaviour is shown to be consistent with the basic features of a two-state MWC model, and emphasises the overall similarities between different oxygen carriers, independent of detailed chemical features. Thus heterotropic effectors display in both classes of respiratory proteins a similar pattern of phenomena, which have been interpreted, to a first approximation, within the framework of the MWC model. This type of behaviour is more clearly observed in haemocyanins because the larger size of the functional constellation (including the possession of many more binding sites) increases the steepness of the phenomenon (Colosimo *et al.*, 1974).

6.8.4 Kinetics of oxygen binding

Extensive characterisation of the kinetics of oxygen binding has been reported for haemoglobins and haemocyanins from various species (Parkhurst, 1979; Brunori *et al.*, 1982). In all cases, analysis has confirmed a first-order applicability of a two-state model and has indicated some common features, which can be summarised as follows (see also table 6.2):

(a) The kinetics of the R and T states can be characterised independently by working in a suitable saturation regime (i.e. at very low or very high oxygen saturations).

(b) The second-order rate constant for oxygen binding is, in all cases, very large ($\sim 10^7$–10^8 M^{-1} s^{-1}) and essentially independent of allosteric state.

(c) The first-order dissociation rate constant is, however, strongly dependent on quaternary structure and represents the major source of difference in the binding energy between R and T. Thus, in spite of the large chemical differences, the process of control is exerted in both classes of protein on the 'off' rates.

(d) The quaternary switch is generally very fast compared with binding or dissociation of oxygen, at least for those proteins which are characterised by a smaller 'functional constellation' (i.e. $\leqslant$ six sites). In haemoglobin and in hexameric haemocyanins the relaxation time for this process is $\geqslant$ $10^4 s^{-1}$ (Sawicki and Gibson, 1977). However, when the size of the allosteric unit becomes larger, the quaternary conformational change becomes slower, and in some haemocyanins it seems to be rate limiting (van Driel *et al.*, 1974; Kuiper *et al.*, 1978). Thus a relationship between the overall size of the functional constellation and the speed of the quaternary conformational change has been detected for the first time (van Driel *et al.*, 1978).

Some of the representative rate constants for the reaction with oxygen of haemoglobin and haemocyanin in both quaternary states, as well as the corresponding non-cooperative subunits, are reported in table 6.2.

The set of data reported above indicates the overall similarity in the kinetic behaviour of these various oxygen carriers, and seems to confirm the general rule that, from the kinetic viewpoint, an essential control is effective on the dissociation rate constants, whereas the combination rate constants tend to be diffusion controlled.

6.8.5 Evolutionary considerations

In this section we have attempted to provide some evidence for the fact that, in spite of the differences in the protein moiety and the metal site, all oxygen carriers share a number of common basic features. The capability to form a reversible oxygen complex is obviously basic to all of them, being the unique requirement of *any* oxygen carrier. However, it is clear that homotropic and heterotropic effects, and the kinetic basis of cooperative oxygen binding, are common and have been described taking into account ligand-linked conformational changes which may have a similar stereochemical basis. The capability of the two-state MWC model, in providing – with suitable modifications – a first-order description of functional behaviour of these different oxygen carriers, speaks for the generality of this model.

Some considerations about the evolutionary significance of what has been reported here are in order. It is well known that haemoglobins (myoglobins) and haemocyanins are both widespread in nature. The distribution of haemerythrins is considerably less extensive, and we cannot offer an explanation for this. Moreover, it is clear that, among the protostomes, arthropods and molluscs may utilise either haemocyanins or haemoglobins as oxygen carriers. Among the latter for example, the bivalves have haemoglobins, sometimes cooperative in their ligand binding properties, and the cephalopods have haemocyanins. Finally, several species are known to possess haemocyanin in the haemolymph and myoglobin in their muscles; thus the same individual has the capability of producing both types of oxygen carrier, as exemplified by the case of *Aplysia limicina.*

All of these arguments present some questions in relation to the appearance of oxygen carriers in the course of evolution. It is known that haemoglobins from plants (e.g. *Lupinus*) and animals (many) must have originated from a common ancestor on the basis of similarity in their structures.

An identical and equally stringent consideration cannot be made for haemocyanins; however, it seems that the oxygen binding site of *Neurospora* tyrosinase is very similar to that of haemocyanins (Schoot-Uiterkamp and Mason, 1973; Lerch, 1981), which would lead to a similar conclusion as to the possibility for the existence of a common ancestor for the binuclear oxygen-binding copper proteins.

Given all of this, we are confronted with the interpretation of the mechanism by which two chemically very distinct oxygen carriers appear today in the same phylum and/or in the same species. It is not possible to provide an unequivocal answer to this question, but two possible solutions may be proposed at this stage. A first one is the development, during evolution, of *two independent chemical species* capable of acting as oxygen transporters, which have both been used to cope with a very important vital problem and which only thereafter have been 'selected' for, so that nowadays some species use haemocyanins and others use haemoglobins. If this were the case, it might be possible that in several species the information for synthesis of both types of oxygen carrier is still present, but selectively repressed.

A second possibility is that both haemoglobins and haemocyanins have the *same ancestor*, i.e. that an 'original' oxygen carrier has thereafter developed to use either copper or iron as a metallic prosthetic group, leading to different classes of respiratory protein. It may be possible to probe this second hypothesis by objectively comparing similarities and differences in the structure of the active site of haemocyanins with that of haemoglobins, when the former is sufficiently well understood. Although the latter hypothesis is unlikely, it should not be ignored, and the isolation of functional domains by limited proteolysis may add interesting information to test this possibility.

REFERENCES

Ackers, G. K. (1980) *Biophys. J.* **32**, 331.

Adair, G. S. (1925) *J. Biol. Chem.* **63**, 529.

Alberding, N., Lavalette, D. and Austin, R. H. (1981) *Proc. Natl Acad. Sci. USA* **78**, 2307.

Amiconi, G., Antonini, E., Brunori, M., Formaneck, H. and Huber, R. (1972) *Eur. J. Biochem.* **31**, 52.

Amiconi, G., Antonini, E., Brunori, M., Wyman, J. and Zolla, L. (1981) *J. Mol. Biol.* **152**, 111.

Andersen, M. E., Moffat, J. K. and Gibson, Q. H. (1971) *J. Biol. Chem.* **246**, 2796.

Anderson, L. (1975) *J. Mol. Biol.* **94**, 33.

Antonini, E. and Brunori, M. (1969) *J. Biol. Chem.* **244**, 3909.

Antonini, E. and Brunori, M. (1971) *Hemoglobin and Myoglobin in their Reaction with Ligands.* Elsevier/North-Holland, Amsterdam.

Antonini, E. and Chiancone, E. (1977) *Ann. Rev. Biophys. Bioengng.* **6**, 239.

Antonini, E., Wyman, J., Rossi-Fanelli, A. and Caputo, A. (1962) *J. Biol. Chem.* **237**, 2773.

Antonini, E., Wyman, J., Brunori, M., Bucci, E., Fronticelli, C. and Rossi-Fanelli, A. (1963) *J. Biol. Chem.* **238**, 2950.

Antonini, E., Brunori, M., Colosimo, A., Kuiper, H. A. and Zolla, L. (1983) *Biophys. Chem.* **18**, 117.

Arisaka, F. and van Holde, K. E. (1979) *J. Mol. Biol.* **134**, 41.

Arnone, A. (1972) *Nature, Lond.* **237**, 146.

Asakura, T. and Sono, M. (1974) *J. Biol. Chem.* **249**, 7087.

Bacci, M., Cerdonio, M. and Vitale, S. (1979) *Biophys. Chem.* **10**, 113.

Baldwin, J. M. (1975) *Prog. Biophys. Molec. Biol.* **29**, 225.

Baldwin, J. M. (1980) *J. Mol. Biol.* **136**, 103.

Baldwin, J. M. and Chothia, C. (1979) *J. Mol. Biol.* **129**, 175.

Barlow, C. H., Maxwell, J. C., Wallace, W. J. and Caughey, W. S. (1973) *Biochim. Biophys. Res. Commun.* **55**, 91.

Barra, D., Bossa, F. and Brunori, M. (1981) *Nature, Lond.* **293**, 587.

Bateman, H. (1910) *Proc. Cambridge Phil. Soc.* **15**, 423.

Bates, G., Brunori, M., Amiconi, G., Antonini, E. and Wyman, J. (1968) *Biochemistry* 7, 3016.

Bauer, C., Forster, M., Gros, G., Mosca, A., Perrella, M., Rollema, H. S. and Vogel, D. (1981) *J. Biol. Chem.* **256**, 8429.

Benesch, R. and Benesch, R. E. (1967) *Biochim. Biophys. Res. Commun.* **26**, 162.

Binotti, I., Giovenco, S., Giardina, B., Antonini, E., Brunori, M. and Wyman, J. (1971) *Arch. Biochem. Biophys.* **142**, 274.
Blake, C. (1983) *Trends Biochem. Sci.* **8**, 11.
Bolton, W. and Perutz, M. F. (1970) *Nature, Lond.* **228**, 551.
Bonaventura, C., Sullivan, B., Bonaventura, J. and Bourne, S. (1974) *Biochemistry* **13**, 4784.
Bonaventura, J., Bonaventura, C., Brunori, M., Giardina, B., Antonini, E., Bossa, F. and Wyman, J. (1974) *J. Mol. Biol.* **82**, 499.
Bonaventura, J., Bonaventura, C., Sullivan, B., Ferruzzi, G., McCurdy, P. R., Fox, J. and Moo-Penn, W. F. (1976) *J. Biol. Chem.* **251**, 7563.
Brix, O., Lykkeboe, G. and Johansen, K. (1979) *J. Comp. Physiol.* **129**, 97.
Brouwer, M. and Kuiper, H. A. (1973) *Eur. J. Biochem.* **35**, 428.
Brouwer, M., Bonaventura, C. and Bonaventura, J. (1978) *Biochemistry* **11**, 2148.
Brouwer, M., Wolters, M. and van Bruggen, E. F. J. (1979) *Arch. Biochem. Biophys.* **193**, 487.
Brown, J. M., Powers, L., Kincaid, B., Larrabee, J. A. and Spiro, T. G. (1980) *J. Am. Chem. Soc.* **102**, 4210.
Brunori, M. (1971) *J. Mol. Biol.* **55**, 39.
Brunori, M. (1975) *Curr. Topics Cell. Reg.* **9**, 1.
Brunori, M. and Schuster, T. M. (1969) *J. Biol. Chem.* **214**, 4046.
Brunori, M., Noble, R. W., Antonini, E. and Wyman, J. (1966) *J. Biol. Chem.* **241**, 5238.
Brunori, M., Antonini, E., Wyman, J. and Anderson, S. R. (1968) *J. Mol. Biol.* **34**, 357.
Brunori, M., Antonini, E., Phelps, C. and Amiconi, G. (1969) *J. Mol. Biol.* **44**, 563.
Brunori, M., Falcioni, G., Fioretti, E., Giardina, B. and Rotilio, G. (1975) *Eur. J. Biochem.* **53**, 99.
Brunori, M., Coletta, M., Giardina, B. and Wyman, J. (1978) *Proc. Natl Acad. Sci. USA* **75**, 4310.
Brunori, M., Kuiper, H. A., Antonini, E., Bonaventura, C. and Bonaventura, J. (1981*a*) in *Invertebrate Oxygen-binding Proteins: Structure, Active Site and Function* (Lamy, J. and Lamy, J., eds). Marcel Dekker, New York, p. 693.
Brunori, M., Zolla, L., Kuiper, H. A. and Finazzi-Agrò, A. (1981*b*) *J. Mol. Biol.* **153**, 1111.
Brunori, M., Giardina, B. and Kuiper, H. A. (1982*a*) in *Inorganic Biochemistry*, Vol. III. (H. A. O. Hill, ed.). Royal Society of Chemistry, London, p. 126.
Brunori, M., Kuiper, H. A. and Zolla, L. (1982*b*) *EMBO J.* **1**, 329.
Bucci, E., Salahuddin, A., Bonaventura, J. and Bonaventura, C. (1978) *J. Biol. Chem.* **253**, 821.
Bunn, H. F., Wohl, R. C., Bradley, T. B., Cooley, M. and Gibson, Q. H. (1974) *J. Biol. Chem.* **249**, 7402.
Case, D. A., Huynh, B. H. and Karplus, M. (1979) *J. Am. Chem. Soc.* **101**, 4433.
Cassoly, R., Gibson, Q. H., Ogawa, S. and Shulman, R. G. (1971) *Biochim. Biophys. Res. Commun.* **44**, 1015.
Cerdonio, M., Congiu-Castellano, A., Calabrese, L., Morante, S., Pispisa, B. and Vitale, S. (1978) *Proc. Natl Acad. Sci. USA* **75**, 4916.
Cerdonio, M., Morante, S., Torrosini, D., Vitale, S., De Young, A. and Noble, R. W. (1984) *Proc. Natl Acad. Sci. USA* in press.
Chang, C. K. and Traylor, T. G. (1973) *J. Am. Chem. Soc.* **95**, 5810.
Chang, C. K. and Traylor, T. G. (1975) *Proc. Natl Acad. Sci. USA* **72**, 1166.
Chiancone, E., Giardina, B., Brunori, M. and Antonini, E. (1973) in *Comparative Physiology* (Bolis, L., Schmidt-Nielsen, K. and Maddrell, S. H. P., eds). Elsevier/North-Holland, Amsterdam.
Chiancone, E., Vecchini, P., Verzili, D., Ascoli, F. and Antonini, E. (1981*a*). *J. Mol. Biol.* **152**, 577.
Chiancone, E., Ferruzzi, G., Bonaventura, C. and Bonaventura, J. (1981*b*) *Biochim. Biophys. Acta* **670**, 84.
Ching Ming Chung, M. and Ellerton, H. D. (1979) *Prog. Biophys. Molec. Biol.* **35**, 53.
Chothia, C., Wodak, S. and Janin, J. (1976) *Proc. Natl Acad. Sci. USA* **73**, 3793.
Chu, A. H. and Ackers, G. K. (1981) *J. Biol. Chem.* **256**, 1199.
Collman, J. P., Gagné, R. R., Reed, C. A., Halbert, T. R., Lang, G. and Robinson, W. T. (1975) *J. Am. Chem. Soc.* **97**, 1427.
Collman, J. P., Brauman, J. I., Doxsee, K. M., Halbert, T. R. and Suskick, K. S. (1978) *Proc. Natl Acad. Sci. USA* **75**, 564.

Collman, J. P., Brauman, J. I. and Doxsee, K. M. (1979) *Proc. Natl Acad. Sci. USA* **76**, 6035.
Colosimo, A., Brunori, M. and Wyman, J. (1974) *Biophys. Chem.* **2**, 338.
Colosimo, A., Brunori, M. and Wyman, J. (1977) in *Structure and Function of Haemocyanins* (Bannister, J. V., ed.). Springer-Verlag, Berlin, p. 189.
Craik, C. S., Buchanan, S. R. and Beychok, S. (1980) *Proc. Natl Acad. Sci. USA* **77**, 1384.
Craik, C. S., Buchanan, S. R. and Beychok, S. (1981) *Nature, Lond.* **291**, 87.
Dawson, J. W., Gray, H. B., Hoenig, H. E., Rossman, G. R., Schredder, J. M. and Wang, R. H. (1972) *Biochemistry* **11**, 461.
Deatherage, J. F., Loe, R. S., Anderson, C. M. and Moffat, K. (1976) *J. Mol. Biol.* **104**, 687.
De Bruin, S. H. and Janssen, L. H. M. (1973) *J. Biol. Chem.* **248**, 2774.
De Bruin, S. H., Rollema, H. S., Janssen, L. H. M. and van Os, G. A. J. (1974) *Biochim. Biophys. Res. Commun.* **58**, 210.
de Phillips, H. A. (1971) *Arch. Biochem. Biophys.* **144**, 122.
de Waal, D. J. A. and Wilkins, R. G. (1976) *J. Biol. Chem.* **251**, 2339.
Dickinson, J. C. and Chien, J. C. W. (1973) *J. Biol. Chem.* **248**, 5005.
Dunn, J. B. R., Addison, A. W., Bruce, R. E., Loehr, J. S. and Loehr, T. M. (1977) *Biochemistry* **16**, 1743.
Eaton, W. A. (1980) *Nature, Lond.* **284**, 183.
Economides, A. P. and Wells, R. M. G. (1975) *Comp. Biochem. Physiol.* **A51**, 219.
Eickman, N. C., Himmelwright, R. S. and Solomon, E. I. (1979) *Proc. Natl Acad. Sci. USA* **76**, 2094.
Eisenberger, P., Shulman, R. G., Kincaid, B. M., Brown, G. S. and Ogawa, S. (1978) *Nature, Lond.* **274**, 30.
Er-el, Z., Shaklai, N. and Daniel, E. (1972) *J. Mol. Biol.* **64**, 341.
Fabry, T. L., Simo, C. and Javaherian, K. (1968) *Biochim. Biophys. Acta* **160**, 118.
Fager, L. Y. and Alben, J. O. (1972) *Biochemistry* **11**, 4786.
Fermi, G. (1975) *J. Mol. Biol.* **97**, 237.
Fermi, G., Perutz, M. F., Dickinson, L. C. and Chien, J. C. W. (1982) *J. Mol. Biol.* **155**, 495.
Flanagan, M. A., Ackers, G. K., Matthew, J. B., Hanania, G. I. H. and Gurd, F. R. N. (1981) *Biochemistry* **20**, 7439.
Freedman, T. B., Loehr, J. S. and Loehr, T. M. (1976) *J. Am. Chem. Soc.* **98**, 2809.
Friedman, J. M., Rousseau, D. L., Ondrias, M. R. and Stepnoski, R. A. (1982) *Science* **218**, 1244.
Fung, L. W.-M. and Ho, C. (1975) *Biochemistry* **14**, 2526.
Garbett, K., Darnall, D. W., Klotz, I. M. and Williams, R. J. P. (1969) *Arch. Biochem. Biophys.* **135**, 419.
Garlick, R. L. and Riggs, A. F. (1982) *J. Biol. Chem.* **257**, 9005.
Garlick, R. L. and Terwilliger, R. C. (1977) *Comp. Biochem. Physiol.* **57B**, 177.
Gay, R. R. and Solomon, E. I. (1978) *J. Am. Chem. Soc.* **100**, 1973.
Gelin, B. R. and Karplus, M. (1977) *Proc. Natl Acad. Sci. USA* **74**, 801.
Giacometti, G. M., Traylor, T. G., Ascenzi, P., Brunori, M. and Antonini, E. (1977) *J. Biol. Chem.* **252**, 7447.
Giardina, B., Brunori, M., Binotti, I., Giovenco, S. and Antonini, E. (1973) *Eur. J. Biochem.* **39**, 571.
Giardina, B., Chiancone, E. and Antonini, E. (1975) *J. Mol. Biol.* **93**, 1.
Gibson, Q. H. (1959) *Biochem. J.* **71**, 293.
Gielens, C., Préaux, G. and Lontie, R. (1973) *Arch. Int. Physiol. Biochim.* **81**, 182.
Gielens, C., Préaux, G. and Lontie, R. (1975) *Eur. J. Biochem.* **60**, 271.
Gilbert, W. (1979) in *Eukaryotic Gene Regulation* ICN-UCLA *Symp. Molecular and Cell Biology* **14** (1). Academic Press, New York.
Gill, S. J., Gaud, H. T. and Barisas, B. G. (1980) *J. Biol. Chem.* **255**, 7855.
Gò, M. (1981) *Nature, Lond.* **291**, 90.
Griffith, J. S. (1956) *Proc. R. Soc. Lond. Ser. A* **235**, 23.
Gros, G., Rollema, H. S. and Forster, R. E. (1981) *J. Biol. Chem.* **256**, 5471.
Guidotti, G. (1967) *J. Biol. Chem.* **242**, 3685.
Gurd, F. R. N., Matthew, J. B., Wittebort, R. J., Morrow, J. S. and Friend, S. H. (1980) in *Biophysics and Physiology of* CO_2 (Bauer, C., Gros, G. and Bartels, H., eds). Springer-Verlag, Berlin, p. 89.
Hendrickson, W. A. (1978) *Naval Res. Reviews* **31**, 1.

Herman, Z. S. and Loew, G. H. (1980) *J. Am. Chem. Soc.* **102**, 1815.
Hewitt, J. A., Kilmartin, J. V., Ten Eyck, L. F. and Perutz, M. F. (1972) *Proc. Natl Acad. Sci. USA* **69**, 203.
Hill, A. V. (1910) *J. Physiol.* **40**, iv–vii.
Hoffman, R., Chen, M. M.-L. and Thorn, D. L. (1977) *Inorg. Chem.* **16**, 503.
Hopfield, J. J., Shulman, R. G. and Ogawa, S. (1971) *J. Mol. Biol.* **61**, 425.
Hoylaert, M., Préaux, G., Witters, R. and Lontie, R. (1979) *Arch. Int. Physiol. Biochim.* **87**, 417.
Imai, K. (1979) *J. Mol. Biol.* **133**, 233.
Imai, K. and Yonetani, T. (1975) *J. Biol. Chem.* **250**, 2227.
Imai, K., Ikeda-Saito, M., Yamamoto, H. and Yonetani, T. (1980) *J. Mol. Biol.* **138**, 635.
Imaizumi, K., Imai, K., and Tyuma, I. (1982) *J. Mol. Biol.* **159**, 703.
Ip, S. H. C. and Ackers, G. K. (1977) *J. Biol. Chem.* **252**, 82.
Jameson, G. B., Molinaro, F. S., Ibers, J. A., Collman, J. P., Brauman, J. I., Rose, E. and Suslick, K. S. (1978) *J. Am. Chem. Soc.* **100**, 6769.
Jameson, G. B., Molinaro, F. S., Ibers, J. A., Collman, J. P., Brauman, J. I., Rose, E. and Suslick, K. S. (1980) *J. Am. Chem. Soc.* **102**, 3224.
Janin, J. and Wodak, S. J. (1984) *EMBO J.* in press.
Kellett, G. L. (1971) *J. Mol. Biol.* **59**, 401.
Kilmartin, J. V. (1974) *FEBS Lett.* **38**, 147.
Kilmartin, J. V., Breen, J. J., Roberts, G. C. K. and Ho, C. (1973*a*) *Proc. Natl Acad. Sci. USA* **70**, 1246.
Kilmartin, J. V., Fogg, J., Luzzana, M. and Rossi-Bernardi, L. (1973*b*) *J. Biol. Chem.* **248**, 7039.
Kilmartin, J. V., Hewitt, J. A. and Wootton, J. F. (1975) *J. Mol. Biol.* **93**, 203.
Kilmartin, J. V., Anderson, N. L. and Ogawa, S. (1978) *J. Mol. Biol.* **123**, 71.
Kilmartin, J. V., Fogg, J. H. and Perutz, M. F. (1980) *Biochemistry* **19**, 3189.
Klarman, A. and Daniel, E. (1980) *Biochemistry* **19**, 5176.
Konkel, D. A., Tilghmann, S. M. and Leder, P. (1978) *Cell* **15**, 1125.
Koshland, D. E., Nemethy, G. and Filmer, D. (1966) *Biochemistry* **5**, 365.
Koster, A. S. (1975) *J. Chem. Phys.* **63**, 3284.
Kuiper, H. A., Torensma, R. and van Bruggen, E. F. J. (1976) *Eur. J. Biochem.* **68**, 425.
Kuiper, H. A., Antonini, E. and Brunori, M. (1977) *J. Mol. Biol.* **116**, 569.
Kuiper, H. A., Brunori, M. and Antonini, E. (1978) *Biochim. Biophys. Res. Commun.* **82**, 1062.
Kuiper, H. A., Forlani, L., Chiancone, E., Antonini, E., Brunori, M. and Wyman, J. (1979) *Biochemistry* **18**, 5849.
Kuiper, H. A., Coletta, M., Zolla, L., Chiancone, E. and Brunori, M. (1980*a*) *Biochim. Biophys. Acta* **626**, 412.
Kuiper, H. A., Finazzi-Agrò, A., Antonini, E. and Brunori, M. (1980*b*) *Proc. Natl Acad. Sci. USA* **77**, 2387.
Kurtz, A., Rollema, H. S. and Bauer, C. (1981) *Arch. Biochem. Biophys.* **210**, 200.
Kurtz, D. M., Shriver, D. F. and Klotz, I. M. (1977) *Coord. Chem. Revs* **24**, 145.
Kwiatkowski, L. D. and Noble, R. W. (1982) *J. Biol. Chem.* **257**, 8891.
La Mar, G. N. and de Ropp, J. S. (1982) *J. Am. Chem. Soc.* **104**, 5203.
Lamy, J., Lamy, J. and Weill, J. (1979*a*) *Arch. Biochem. Biophys.* **193**, 140.
Lamy, J., Lamy, J. and Weill, J. (1979*b*) *Arch. Biochem. Biophys.* **196**, 324.
Lang, G. and Marshall, W. (1966) *Proc. Phys. Soc. London* **87**, 3.
Langerman, N. and Sturtevant, J. M. (1971) *Biochemistry* **10**, 2809.
Lerch, K. (1981) in *Metal Ions in Biological Systems* (Sigel, H., ed.). Marcel Dekker, New York, p. 143.
Loehr, J. S. and Loehr, T. M. (1979) in *Advances in Inorganic Biochemistry* Vol. I (Eichhorn, G. L. and Marzilli, L. O. eds). Elsevier/North Holland, Amsterdam, pp. 235–52.
Loehr, J. S., Freedman, T. B. and Loehr, T. M. (1974) *Biochim. Biophys. Res. Commun.* **56**, 510.
Loehr, J. S., Lammers, P. J., Brimhall, B. and Hermandson, M. A. (1978*a*) *Biochemistry* **7**, 3868.
Loehr, J. S., Lammers, P. J., Brimhall, B. and Hermandson, M. A. (1978*b*) *J. Biol. Chem.*

253, 5726.
Lontie, R. and Witters, R. (1981) in *Metal Ions in Biological Systems* (Sigel, H., ed.). Marcel Dekker, New York, p. 229.
McConnell, H., Deal, W. J. and Ogata, R. G. (1969) *Biochemistry* **8**, 2580.
Mangum, C. P. and Kondon, M. (1975) *Comp. Biochem. Physiol.* **50A**, 777.
Manwell, C. (1960) *Science* **132**, 550.
Matthew, J. B., Hanania, G. I. H. and Gurd, F. R. N. (1979) *Biochemistry* **18**, 1928.
Mellema, J. E. and Klug, A. (1972) *Nature, Lond.* **239**, 145.
Michalowicz, A., Pin, S. and Alpert, B. (1983) *Biophys. J.* **41**, 413a.
Miller, K. and van Holde, K. E. (1974) *Biochemistry* **13**, 1668.
Mills, F. C. and Ackers, G. K. (1979) *Proc. Natl Acad. Sci. USA* **76**, 273.
Misra, H. P. and Fridovich, I. (1972) *J. Biol. Chem.* **247**, 6960.
Moffat, J. K. (1971) *J. Mol. Biol.* **55**, 135.
Moffat, K., Deatherage, J. F. and Seybert, D. W. (1979) *Science* **206**, 1035.
Monod, J., Wyman, J. and Changeux, J.-P. (1965) *J. Mol. Biol.* **12**, 88.
Moss, T. H., Maleski, C. and York, J. L. (1971) *Biochemistry* **10**, 840.
Nagai, K. and Kitagawa, T. (1980) *Proc. Natl Acad. Sci. USA* **77**, 2033.
Nagai, K., La Mar, G. N., Jue, T. and Bunn, H. F. (1982) *Biochemistry* **21**, 842.
Nickerson, K. W. and van Holde, K. E. (1971) *Comp. Biochem. Physiol.* **39**, 855.
Nigen, A. M. and Manning, J. M. (1975) *J. Biol. Chem.* **250**, 8248.
Nigen, A. M., Manning, J. M. and Alben, J. O. (1980) *J. Biol. Chem.* **255**, 5525.
Nishikura, K. (1978) *Biochem. J.* **173**, 671.
Noble, R. W., Gibson, Q. H., Brunori, M., Antonini, E. and Wyman, J. (1969) *J. Biol. Chem.* **241**, 3905.
Noble, R. W., Parkhurst, L. J. and Gibson, Q. H. (1970) *J. Biol. Chem.* **245**, 6628.
Norne, J.-E., Chiancone, E., Forsén, S., Antonini, E. and Wyman, J. (1976) *FEBS Lett.* **94**, 410.
O'Donnell, S., Mandaro, R., Schuster, T. M. and Arnone, A. (1979) *J. Biol. Chem.* **254**, 12204.
Ogawa, S. and Shulman, R. G. (1971) *Biochim. Biophys. Res. Commun.* **42**, 9.
O'Hagen, J. E. (1961) in *Haematin Enzymes*. Pergamon Press, London, p. 173.
Ohnoki, S., Mitomi, T., Hata, R. and Satake, K. (1973) *J. Biochem.* **73**, 717.
Okamura, M. Y., Klotz, I. M., Johnson, C. E., Winter, M. R. C. and Williams, R. J. P. (1969) *Biochemistry* **8**, 1951.
Okamura, M. Y. and Klotz, I. M. (1973) in *Inorganic Biochemistry* (Eichhorn, G. L., ed.). Elsevier, New York.
Olafson, B. D. and Goddard III, W. A. (1977) *Proc. Natl Acad. Sci. USA* **74**, 1315.
Parkhurst, L. J. (1979) *Ann. Rev. Phys. Chem.* **30**, 503.
Pauling, L. and Coryell, C. D. (1936) *Proc. Natl Acad. Sci. USA* **22**, 210.
Perrella, M., Kilmartin, J. V., Fogg, J. and Rossi-Bernardi, L. (1975) *Nature, Lond.* **256**, 759.
Perutz, M. F. (1970) *Nature, Lond.* **228**, 726.
Perutz, M. F. (1972) *Nature, Lond.* **237**, 495.
Perutz, M. F. (1978) *Science* **201**, 1187.
Perutz, M. F. (1979) *Ann. Rev. Biochem.* **48**, 327.
Perutz, M. F. (1980) *Proc. R. Soc. Lond. Series B* **208**, 135.
Perutz, M. F. (1984) *Adv. Prot. Chem.* in press.
Perutz, M. F. and Brunori, M. (1982) *Nature, Lond.* **299**, 421.
Perutz, M. F. and Ten Eyck, L. F. (1971) *Symp. Quant. Biol.* **XXXVI**, 295.
Perutz, M. F., del Pulsinelli, P., Ten Eyck, L., Kilmartin, J. V., Shibata, S., Iucki, I., Miyaji, T. and Hamilton, H. B. (1971) *Nature New Biol.* **232**, 147.
Perutz, M. F., Ladner, J. E., Simon, S. R. and Ho, C. (1974*a*) *Biochemistry* **13**, 2163.
Perutz, M. F., Heidner, E. J., Ladner, J. E., Beetlestone, J. G., Ho, C. and Slade, E. F. (1974*b*) *Biochemistry* **13**, 2187.
Perutz, M. F., Kilmartin, J. V., Nishikura, K., Fogg, J. H., Butler, P. J. G. and Rollema, H. S. (1980) *J. Mol. Biol.* **138**, 649.
Perutz, M. F., Bauer, C., Gros, G., Leclercq, F., Vandecasserie, C., Schnek, A. G., Braunitzer, G., Friday, A. E. and Joysey, K. A. (1981) *Nature, Lond.* **291**, 682.
Pettigrew, D. W., Romeo, P. H., Tsapis, A., Thillet, J., Smith, M. L., Turner, B. W. and

Ackers, G. K. (1982) *Proc. Natl Acad. Sci. USA* **79**, 1849.
Phillips, S. E. V. (1980) *J. Mol. Biol.* **142**, 531.
Philo, J. S., Dreyer, U. and Schuster, T. M. (1984) *Biochemistry* **23**, 865.
Poyart, C., Bursaux, E., Bohn, B. and Guesnon, P. (1980) *Biochim. Biophys. Acta* **626**, 417.
Reed, C. A. and Cheung, S. K. (1977) *Proc. Natl Acad. Sci. USA* **74**, 1780.
Reed, C. A., Mashiko, T., Scheidt, W. R., Spartalian, K. and Lang, G. (1980) *J. Am. Chem. Soc.* **102**, 2302.
Roche, J., Bessis, M. and Thiery, J. P. (1960) *Biochim. Biophys. Acta* **41**, 182.
Rollema, H. S., De Bruin, S. H., Janssen, L. H. M. and van Os, G. A. J. (1975) *J. Biol. Chem.* **250**, 1333.
Rollema, H. S., Gros, G. and Bauer, C. (1980) *J. Biol. Chem.* **255**, 2756.
Root, R. W. (1931) *Biol. Bull. (Woods Hole, Mass.)* **61**, 427.
Rossi-Fanelli, A. and Antonini, E. (1959) *Arch. Biochem. Biophys.* **80**, 269.
Rossi-Fanelli, A., Antonini, E. and Caputo, A. (1961) *J. Biol. Chem.* **236**, 391.
Roughton, F. J. W. (1949) in *Hemoglobin*. Butterworth, London, p. 83.
Roughton, F. J. W. and Lyster, R. L. J. (1965) *Hvalradets Skr.* **48**, 185.
Rubin, M. M. and Changeux, J.-P. (1966) *J. Mol. Biol.* **21**, 265.
Russu, I. M., Tseng Ho, N. and Ho, C. (1980) *Biochemistry* **19**, 1043.
Salhany, J. M., Castillo, G. L. and Ogawa, S. (1976) *Biochemistry* **15**, 5344.
Sawicki, C. A. and Gibson, Q. H. (1977) *J. Biol. Chem.* **252**, 5783.
Sawicki, J. and Lang, G. (1984) *Proc. Natl Acad. Sci. USA* in press.
Scholander, P. F. and van Dam, L. (1954) *Biol. Bull. (Woods Hole, Mass.)* **107**, 247.
Schoot-Uiterkamp, A. J. M. and Mason, H. S. (1973) *Proc. Natl Acad. Sci. USA* **70**, 993.
Shaanan, B. (1982) *Nature, Lond.* **296**, 683.
Shelnutt, J. A., Rousseau, D. L., Friedman, J. M. and Simon, S. R. (1979) *Proc. Natl Acad. Sci. USA* **76**, 4409.
Shulman, R. G., Ogawa, S. and Hopfield, J. J. (1975) *Quart. Rev. Biophys.* **8**, 325.
Siezen, R. J. and van Bruggen, E. F. J. (1974) *J. Mol. Biol.* **90**, 91.
Stanford, M. A., Swartz, J. C., Phillips, T. E. and Hoffman, B. M. (1980) *J. Am. Chem. Soc.* **102**, 4492.
Steigemann, W. and Weber, E. (1979) *J. Mol. Biol.* **127**, 309.
Stein, P., Mitchell, M. and Spiro, T. G. (1980) *J. Am. Chem. Soc.* **102**, 7795.
Stenkamp, R. E., Siecker, L. C. and Jensen, L. H. (1976) *Proc. Natl Acad. Sci. USA* **73**, 349.
Stenkamp, R. E., Siecker, L. C., Jensen, L. H. and McQueen, Jr, J. E. (1978) *Biochemistry* **17**, 2499.
Stenkamp, R. E., Siecker, L. C., Jensen, L. H. and Sanders-Loehr, J. (1981) *Nature, Lond.* **291**, 263.
Swaney, J. B. and Klotz, J. M. (1971) *Arch. Biochem. Biophys.* **147**, 475.
Takano, T. (1977) *J. Mol. Biol.* **110**, 569.
Tan, A. L., Noble, R. W. and Gibson, Q. H. (1973) *J. Biol. Chem.* **248**, 2880.
Tan, K. H., Keresztes-Nagy, S. and Frankfater, A. (1975) *Biochemistry* **14**, 4280.
Terwilliger, R. C. (1974) *Comp. Biochem. Physiol.* **48A**, 745.
Terwilliger, R. C., Terwilliger, N. B., Bonaventura, C. and Bonaventura, J. (1977) *Biochim. Biophys. Acta* **494**, 416.
Thamann, T. J., Loehr, J. S. and Loehr, T. M. (1977) *J. Am. Chem. Soc.* **99**, 4187.
Traylor, T. G. (1981) *Acc. Chem. Res.* **14**, 102.
Traylor, T. G., Deardurff, L. A., Coletta, M., Ascenzi, P., Antonini, E. and Brunori, M. (1983) *J. Biol. Chem.* **258**, 12147.
Tsai, T. E., Groves, J. L. and Wu, C. S. (1981) *J. Chem. Phys.* **74**, 4306.
Tyuma, I. and Ueda, Y. (1975) *Biochim. Biophys. Res. Commun.* **65**, 1278.
van Breemen, J. F. L., Wichterjes, T., Muller, M. F. J., van Driel, R. and van Bruggen, E. F. J. (1975) *Eur. J. Biochem.* **60**, 129.
van Breemen, J. F. L., Ploegman, J. H. and van Bruggen, E. F. J. (1979) *Eur. J. Biochem.* **100**, 61.
van Bruggen, E. F. J. (1978) *Ninth Intern. Congr. Electron Microscopy*, Toronto, **13**, p. 450.
van den Berg (1978) *Nature, Lond.* **276**, 37.
van der Deen, H. and Hoving, H. (1979) *Biophys. Chem.* **9**, 169.
van Driel, R. (1973) *Biochemistry* **12**, 2696.
van Driel, R., Brunori, M. and Antonini, E. (1974) *J. Mol. Biol.* **89**, 103.

van Driel, R., Kuiper, H. A., Antonini, E. and Brunori, M. (1978) *J. Mol. Biol.* **121**, 431.
van Heel, M. and Frank, J. (1980) in *Pattern Recognition in Pratica* (Gelsema, E. S. and Kanal, L. N., eds). Elsevier/North-Holland, Amsterdam.
van Holde, K. E. and Miller, K. I. (1982) *Quart. Rev. Biophys.* **15**, 1.
van Holde, K. E. and van Bruggen, E. F. J. (1971) in *Biological Macromolecules* vol. V (Timasheff, S. N. and Fasman, G. A., eds). Marcel Dekker, New York, p. 1.
van Schaick, E. J. M., Schutter, W. G., Gaykema, W. P. J., van Bruggen, E. F. J. and Hal, W. G. J. (1981) in *Invertebrate Oxygen-binding Proteins: Structure, Active Site and Function* (Lamy, J. and Lamy, J., eds). Marcel Dekker, New York, p. 353.
Viggiano, G. and Ho, C. (1979) *Proc. Natl Acad. Sci. USA* **76**, 3673.
Vinogradov, S. N., Shlom, J. M., Hall, B. C., Kapp, O. S. and Mizukami, H. (1977) *Biochim. Biophys. Acta* **492**, 136.
Warshel, A. (1977) *Proc. Natl Acad. Sci. USA* **74**, 1789.
Warshel, A. and Weiss, R. M. (1981) *J. Am. Chem. Soc.* **103**, 446.
Waxman, L. (1971) *J. Biol. Chem.* **246**, 7318.
Weiss, J. J. (1964) *Nature, Lond.* **202**, 83.
Wittenberg, J. B. and Wittenberg, B. A. (1962) *Nature, Lond.* **194**, 106.
Wittenberg, J. B., Wittenberg, B. A., Peisach, J. and Blumberg, W. E. (1970) *Proc. Natl Acad. Sci. USA* **67**, 1846.
Wood, E. J. (1980) *Essays in Biochem.* **16**, 1.
Wood, E. J., Cayley, G. R. and Pearson, J. S. (1977) *J. Mol. Biol.* **109**, 1.
Wyman, J. (1948) *Adv. Prot. Chem.* **4**, 407.
Wyman, J. (1964) *Adv. Prot. Chem.* **19**, 224.
Wyman, J. (1969) *Nobel Symposia* **11**, 266.
Wyman, J., Gill, S. T., Gaud, H. T., Colosimo, A., Giardina, B., Kuiper, H. A. and Brunori, M. (1978) *J. Mol. Biol.* **124**, 161.
Yamamoto, H., Kayne, F. J. and Yonetani, T. (1974) *J. Biol. Chem.* **249**, 691.
Yonetani, T., Drott, H. R., Leigh, J. S., Reed, G. H., Waterman, M. R. and Asakura, T. (1970) *J. Biol. Chem.* **245**, 2998.
Yonetani, T., Yamamoto, H. and Woodrow III, G. V. (1974) *J. Biol. Chem.* **249**, 682.
York, J. L. and Bearden, A. J. (1970) *Biochemistry* **9**, 4549.
Zerner, M., Gouterman, M. and Kobayashi, H. (1966) *Theoret. Chim. Acta* **6**, 363.
Zolla, L., Kuiper, H. A., Vecchini, P., Antonini, E. and Brunori, M. (1978) *Eur. J. Biochem.* **87**, 467.
Zolla, L., Kuiper, H. A., Brunori, M. and Antonini, E. (1981) in *Invertebrate Oxygen-binding Proteins: Structure, Active Site and Function* (Lamy, J. and Lamy, J., eds). Marcel Dekker, New York, p. 719.

Index